Maximizing Autodesk®
Mechanical Desktop®

Maximizing Autodesk® Mechanical Desktop®

Daniel T. Banach
Ron K. C. Cheng

autodesk®
press

THOMSON LEARNING™

Australia • Canada • Mexico • Singapore • Spain • United Kingdom • United States

THOMSON LEARNING

autodesk press

Maximizing Autodesk® Mechanical Desktop®
Daniel T. Banach / Ron K. C. Cheng

Autodesk Press Staff

Business Unit Director:
Alar Elken

Executive Editor:
Sandy Clark

Acquisitions Editor:
James DeVoe

Development Editor:
John Fisher

Editorial Assistant:
Jasmine Hartman

Executive Marketing Manager:
Maura Theriault

Channel Manager:
Mary Johnson

Marketing Coordinator:
Karen Smith

Executive Production Manager:
Mary Ellen Black

Production Manager:
Larry Main

Production Editor:
Stacy Masucci

Art and Design Coordinator:
Mary Beth Vought

COPYRIGHT © 2002 Thomson Learning™.

Printed in Canada
1 2 3 4 5 XXX 04 03 02 01

For more information, contact Autodesk Press, 3 Columbia Circle, PO Box 15015, Albany, New York, 12212-15015.

Or find us on the World Wide Web at www.autodeskpress.com

All rights reserved. No part of this work covered by the copyright hereon may be reproduced or used in any form or by any means—graphic, electronic, or mechanical, including photocopying, recording, taping, Web distribution or information storage and retrieval systems—without written permission of the publisher.

To request permission to use material from this text contact us by
Tel: 1-800-730-2214
Fax: 1-800-730-2215
www.thomsonrights.com

Library of Congress Cataloging-in-Publication Data
Cheng, Ron
  Maximizing Autodesk mechanical desktop / Ron K.C. Cheng.
    p. cm.
  ISBN 0-7668-3307-0
  1. Computer graphics. 2. Engineering design. 3. Mechanical Desktop (Computer file) I. Title

T385 .C477 2002
620'.0042'02855369—dc21
                                    2001043626

Notice To The Reader

The publisher and author do not warrant or guarantee any of the products described herein or perform any independent analysis in connection with any of the product information contained herein. The publisher and author do not assume and expressly disclaim any obligation to obtain and include information other than that provided to it by the manufacturer.

The reader is expressly warned to consider and adopt all safety precautions that might be indicated by the activities described herein and to avoid all potential hazards. By following the instructions contained herein, the reader willingly assumes all risks in connection with such instructions.

The publisher and author make no representations or warranties of any kind, including but not limited to, the warranties of fitness for particular purpose or merchantability, nor are any such representations implied with respect to the material set forth herein, and the publisher and author take no responsibility with respect to such material. The publisher and author shall not be liable for any special, consequential, or exemplary damages resulting, in whole or in part, from the readers' use of, or reliance upon, this material.

Trademarks

Autodesk, the Autodesk logo, and AutoCAD are registered trademarks of Autodesk, Inc., in the USA and other countries. Thomson Learning is a trademark used under license. Online Companion is a trademark and Autodesk Press is an imprint of Thomson Learning. Thomson Learning uses "Autodesk Press" with permission from Autodesk, Inc., for certain purposes. All other trademarks, and/or product names are used solely for identification and belong to their respective holders.

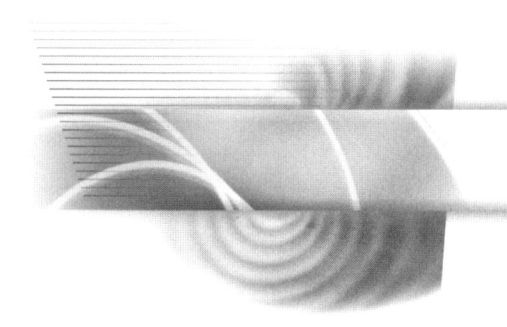

CONTENTS

Preface .... **XV**
FEATURES OF THIS EDITION XV
STYLE CONVENTIONS XVII
HOW TO USE THIS BOOK XVII
ONLINE COMPANION XVII
ABOUT THE AUTHOR XVII
ACKNOWLEDGMENTS XVIII

Chapter 1 Computer Modeling **1**
OVERVIEW 1
COMPUTER MODELING CONCEPTS 2
    Representation 2
    Traditional 2D Approach 2
    3D Model 3
    Choice of Model 7
    Modeling Method 7
    Assembly Modeling 8
    Engineering Drafting 9
USING MECHANICAL DESKTOP 9
    File Type 9
    User Interface 10
    Command Entry 14
    Layers 14
    AutoCAD 2002 tools 15
    Parametric Solid Modeling Tool 15
    Assembly Modeling Tool 17
    Surface Modeling Tool 18

Associative Engineering Drafting Tool 22
Compatibility and Interoperation 23
Options 24
SUMMARY 25
REVIEW QUESTIONS 26

Chapter 2 Parametric Sketching 27
OVERVIEW 27
PARAMETRIC SKETCHING CONCEPTS 27
Sketching 28
Manipulating Geometric Constraints 30
Adding Parametric Dimensions 30
OPTIONS 31
Distance and Angular Tolerances 31
Apply Constraint Rules and Assume Rough Sketch 33
Other Part Modeling Options 33
CONSTRUCTING A PARAMETRIC SKETCH 34
Sketching 34
Resolving a Sketch 35
Manipulating Geometric Constraints 36
Manipulating Parametric Dimensions 38
DESKTOP BROWSERS 43
Types of Browsers 43
SKETCHING TECHNIQUES 44
Construction Geometry 45
Append and Re-solve 47
Multiple Sketches 48
Text Sketch 49
OTHER SKETCHING TECHNIQUES 50
Copy Sketch 50
Copy Edge 50
SUMMARY 51
PROJECTS 51
Steering Shaft 52
Clip 52
Axle 54
Screw Cap 54
Wheel Cap 55
Nut 56
Wheel 57
Frontal Frame 59
REVIEW QUESTIONS 60

Chapter 3 Part Modeling I 61
OVERVIEW 61
VISUALIZATION 62
Orbit 62
Display Mode 63
Lights 63
SKETCHED FEATURES: EXTRUDE, REVOLVE, AND BEND 64
Extruded Solid Feature 64
Revolved Solid Feature 66
Bend Feature 68
BOOLEAN OPERATIONS 70
Join 70
Cut 77
Intersect 78
Split 81
EDITING 83
Edit Sketch of a Feature 83
Update 84
Edit Feature 85
Reorder Features 86
Feature Suppression 89
Delete Feature 90
Replay 91
Base Part 91
MASS PROPERTIES 91
ATTRIBUTES 94
SUMMARY 95
PROJECTS 95
Clip 95
Axle 96
Screw Cap 96
Wheel Cap 97
Wheel 98
Nut 99
Bolt 100
Frontal Frame 102
Parts for the Infant Scooter 105
REVIEW QUESTIONS 107

Chapter 4 Part Modeling II 109
OVERVIEW 109
WORK FEATURES 110
Work Planes 110
Work Axis 121
Work Points 124
Visibility of Work Features 125
SKETCHED FEATURES: RIB, SWEEP, AND LOFT 126
Rib Feature 126
Sweep Features 129
Copy Edge, 3D Edge Path, and 3D Pipe Path 150
Loft Feature 154
PLACED FEATURES: HOLE, THREAD, FILLET, CHAMFER, SHELL, AND PATTERN 157
Hole Feature 157
Thread Feature 162
Fillet Feature 164
Chamfer Feature 166
Shell Feature 168
Pattern Feature 170
SUMMARY 178
PROJECTS 178
Wheel 178
Central Frame 179
Frontal Frame 180
Rear Frame 183
Steering Shaft 185
Tire 186
Seat 187
Threaded Bolt 191
Handle 195
Steering Wheel 199
Component Parts for the Infant Scooter 204
REVIEW QUESTIONS 206

Chapter 5 Assembly Modeling 207
OVERVIEW 207
ASSEMBLY MODELING CONCEPTS 207
Linking to Solid Part Files 208
Assembly Catalog 209
Instance 210
Assembling the Component Parts 211

DESIGN APPROACHES	219
Bottom-Up Approach	219
Top-Down Approach	220
Hybrid Approach	228
Choice of Approach	232
MANIPULATING ASSEMBLY AND COMPONENT DEFINITIONS	232
External and Local Definitions	232
Copy In and Copy Out	233
In-place Editing	234
Basic 3D Work Planes	237
Replace	237
Where Used	238
Query	238
Audit	239
Refreshing External Component Parts	239
ANALYSIS	240
Mass Properties	240
Interference	241
Minimum Distance	241
ASSEMBLY OPTIONS	242
SUMMARY	244
PROJECTS	245
Constraint Manipulation	245
Front End of the Infant Scooter	250
Central Frame of the Infant Scooter	251
Rear End of the Infant Scooter	251
Seat Assembly of the Infant Scooter	253
Infant Scooter	256
REVIEW QUESTIONS	258

Chapter 6 Assembly Scene 259

OVERVIEW	259
ASSEMBLY SCENE CONCEPTS	260
Scene	260
Scene Construction	260
Activate Scene	262
Scene Visibility	262
EXPLODED VIEWS	263
Explosion Factor	263
Tweak	264
TRAIL LINES	267

SCENE MANIPULATION 268
    Copy Scene 269
    Rename Scene 269
    Lock Scene 270
    Update Scene 270
    Suppress Sections 270
SCENE OPTIONS 271
SUMMARY 272
PROJECTS 273
    Front End of the Infant Scooter 273
    Rear End of the Infant Scooter 274
    Infant Scooter 275
REVIEW QUESTIONS 276

Chapter 7 Surface Modeling I 277

OVERVIEW 277
SURFACE MODELING CONCEPTS 278
    Polygon Mesh 278
    NURBS Surface 279
    Modeling Approach 279
    Significance of Surface Model 280
    Volume Representation 282
NURBS SURFACE CONSTRUCTION 283
    Primitive Surfaces 283
    Free-Form Surfaces 285
    Derived Surfaces 288
    Conversion 290
NURBS SURFACE EDITING 290
    Boundary Editing 290
    Profile Editing 296
    Normal Direction 299
    Solid Operation 300
    Conversion 301
3D WIRES CONSTRUCTION 301
    Augmented Lines 302
    Joining Wires 302
    Spline Fitting 303
    Tangent Spline 303
    Offsetting 3D Wires 304
    Filleting 3D Wires 304
    Intersection of Two Surfaces 304
    Projecting a Wire 305

Copying Surface Edges	305
Extract Trim Boundaries	305
Cutting a Series of Section Lines	306
Generating Flow Lines	306
Generating a Parting Line	307

3D WIRES EDITING 307
Spline Edit	307
Unsplining	311
Refining a Wire	311
Changing the Direction of a Wire	312
Editing Augmented Lines	313

UTILITIES 313
Surface Analysis	313
Surface Display	315
Show Edge Nodes	315
Surface Mass Properties	315
Check Fit	316

SURFACE MODELING OPTIONS 316
VISIBILITY 319
SUMMARY 320
PROJECTS 321
Oil Cooler	321
Helical Spring	322
Remote Control Surface Model	326
Remote Control Solid Model 1	345
Remote Control Solid Model 2	347
Mouse	348

REVIEW QUESTIONS 356

Chapter 8 Surface Modeling II 357

OVERVIEW 357
SURFACE MODELING TOOLS 357
Surface Construction	357
Surface Manipulation	358
3D Wire Construction	359
Wire Manipulation	359
Solid Operation	360

SUMMARY 360
PROJECTS 360
Model Car A	360

  Model Car B 379
  Model Car C 416
REVIEW QUESTIONS 418

Chapter 9 Part Modeling III 419

OVERVIEW 419
SURFACE CUT 420
COMBINATION OF SOLID PARTS 423
PART SPLITTING 429
FACE DRAFT 436
FACE SPLITTING 437
DESIGN VARIABLES 441
  Design Variable File 441
  Table-Driven Parts 447
SCALE 451
MIRROR 452
MAKE BASE PART 452
CONVERT AUTOCAD SOLID 453
BASE FEATURE EDITING 462
SOLID PART EXPORT 466
  ACIS 466
  VRML 467
  STL 467
  IGES 467
  STEP 468
FEATURE RECOGNITION 469
COLLABORATION 474
SUMMARY 475
PROJECTS 475
  Camera Casing 475
  Camera Lens Housing 488
REVIEW QUESTIONS 490

Chapter 10 Engineering Drafting I 491

OVERVIEW 491
ENGINEERING DRAFTING CONCEPTS 492
  Engineering Drawing Construction 492
  Layout 492
  Engineering Standards 496
  Drawing Views 499
  Manipulating Drawing Views 506

ENGINEERING DRAWING VIEWS 509
    Multiple Views for a Surface Model 509
    Auxiliary View 513
    Detail View 515
    Broken View 518
    Section View 521
ANNOTATIONS 539
    Centerlines 539
    Text 541
    Leader 542
    Hole Note 543
    Hole Chart 545
    Dimensions 548
    Tolerance 553
    Surface Finish Symbol 566
EXPORTS 568
SUMMARY 569
PROJECTS 569
    Infant Scooter 569
    Toy Car 569
    Model Car Bodies 569
    Camera Casing 569
REVIEW QUESTIONS 570

Chapter 11 Engineering Drafting II 571

OVERVIEW 571
ASSEMBLY DRAWING CONSTRUCTION 571
    Layout and Engineering Standard 572
    Drawing Views 572
    Hatch Pattern of Components in an Assembly 576
ANNOTATION 581
    Bill of Material 582
    Balloon 585
SUMMARY 586
PROJECTS 586
    Infant Scooter 586
    Toy Car 586
REVIEW QUESTIONS 587

Index 589

PREFACE

Mechanical Desktop 6 runs inside AutoCAD 2002. It is a three-dimensional design application that consists of four major modules: part modeling, assembly modeling, surface modeling, and engineering drafting. The part modeling module is a feature-based parametric solid modeler. The assembly modeling module manages and constrains component parts that are assembled together. The surface modeling module is a Non-Uniform Rational B-Spline (NURBS) surface modeler. The engineering drafting module is an associative engineering drafting tool that enables you to construct two-dimensional engineering drawings from three-dimensional objects (parts, assemblies, and surfaces).

FEATURES OF THIS EDITION

This book illustrates the computer modeling concepts found in Mechanical Desktop 6 and gives you an opportunity to practice applying those concepts by constructing 3D parametric solid parts; assembling solid parts and 3D NURBS surface models; outputting 2D engineering drawings; and becoming familiar with the utilities provided. It is assumed that you have already installed the Mechanical Desktop application properly in your computer and know basic AutoCAD commands such as lines, arcs, circles, polylines, and so on.

The first chapter provides an overview of computer modeling concepts, the different ways of representing 3D objects in the computer, and how the four Mechanical Desktop modules are applied in computer modeling. It also explains the use of native AutoCAD file formats and introduces the Mechanical Desktop user interface, including the use of the Desktop Browser. (Throughout the book, the use of this Browser is emphasized because it has been greatly enhanced and provides all the necessary command shortcuts.)

Chapters 2 through 4 concern parametric solid part modeling. Chapter 2 explains the parametric sketching concepts, delineates the procedures behind constructing a parametric sketch, and outlines various sketching techniques. Chapter 3 depicts the basic ways to use parametric sketches in making sketched solid features (extrude, revolve, and bend) and explains how a complex solid part is built from a set of solid features combined together. It also teaches you how to edit a solid part, evaluate mass properties, and add attributes. Chapter 4 is a continuation of Chapter 3 in greater detail. It explains the use of work features (work axis, work point, and work plane) as construction objects, and illustrates the use of work features in making three kinds of sketched features (rib, sweep, and loft). In addition, it introduces the use of placed (pre-constructed) solid features (Hole, Thread, Fillet, Chamfer, Shell, and Pattern). Among them, the hole feature in particular has been enhanced, and the thread feature is new to Mechanical Desktop. The independent pattern instance is also discussed.

After discussing how to construct parametric solid parts, Chapter 5 explains assembly modeling concepts and design approaches, and equips you with the know-how to construct virtual assemblies in the computer. In particular, it delineates the use of the enhanced 3D manipulator. This chapter also shows you how to manipulate an assembly and components within an assembly, including the use of a template when externalizing a local component; and how to evaluate mass properties of an assembly. Chapter 6 illustrates assembly scene concepts, and shows you how to construct assembly scenes, exploded views, and trail lines.

Chapter 7 introduces the NURBS surface modeling concepts and shows you how to construct and edit NURBS surfaces. Because surfaces are constructed on a framework of wires, 3D wire construction and editing are also discussed. Chapter 8 also concerns surface modeling. It guides you a through a set of more in-depth tutorials to familiarize you with using the NURBS surface modeling tool in making free-form surfaces and solids.

After covering assembly modeling in Chapters 5 and 6 and surface modeling in chapters 7 and 8, Chapter 9 addresses more advanced solid modeling techniques. These include the use of a surface to cut a solid, combining two solid parts to form a single solid part, splitting a solid part into two, making a face draft, splitting a face, using design variables to control the dimensions of a single solid part and a set of solid parts, modifying the size of a solid part by scaling, making a mirror part, making a parametric solid part static, converting AutoCAD solids to Mechanical Desktop solids, editing static base solid features by leveraging AutoCAD solid editing tools, and using the feature exchange add-on application.

Chapter 10 explains associative engineering drafting concepts and shows you how to construct engineering drawing views of solid parts and surface models. It also discusses adding annotations to a drawing and outputting a purely 2D drawing. The book concludes with Chapter 11, which also addresses associative engineering drafting and details the way to construct an engineering drawing of an assembly.

STYLE CONVENTIONS

There are several ways to run a Mechanical Desktop command: selecting an item from the pull-down menu, selecting a button from a toolbar, typing a command at the command line interface, and selecting an item from a right-click menu.

In this book, commands selected from a pull-down menu are illustrated by specifying the pull-down menu item and an item from the pull-down menu or the cascading menu of a pull-down menu. For example, selecting the Power Manipulator item from the Assembly pull-down menu is depicted by:

> Select Assembly>Power Manipulator

When a command is required to be typed, the command name will be specified as follows:

> Command: AMNEW

HOW TO USE THIS BOOK

There are 11 chapters in this book. Each chapter introduces a set of topics, takes you through basic, step-by-step examples, and builds on the materials learned in the previous chapters. At the end of each chapter, you will find a summary, projects for you to work on to enhance your knowledge, and review questions. Accompanying this book is a CD-ROM in which all the drawing files constructed in the tutorials and projects are available for your reference.

ONLINE COMPANION

In the future, updated materials for this book will be posted on the Internet as an Online Companion. From time to time, please visit the following Web site for updates:

- http://www.autodeskpress.com

You may also contact the author, Ron K. C. Cheng, via his email address: icrcheng@polyu.edu.hk.

ABOUT THE AUTHOR

Ron K. C. Cheng leads the Product Design Unit of the Industrial Center of The Hong Kong Polytechnic University. His main areas of interest are computer-aided design and computer-based teaching and learning. He is actively involved in developing computer-based learning materials and online self-learning packages. He also leads his team in setting up and maintaining an instructional Web site (http://pdu.ic.polyu.edu.hk) in which learning materials on product data management, product design, computer modeling, and engineering drafting are made available to his students as well as to the general public. He is the author of eight official training books from Autodesk Press (USA). These titles include *AutoCAD*, *Mechanical Desktop*, and *Autodesk Inventor*.

ACKNOWLEDGMENTS

This book never would have been realized without the contributions of many individuals.

I am grateful to Jeff Hill from Sinclair Community College in Middleton, Ohio, for reviewing the manuscript and providing thoughtful suggestions. I would especially like to thank Lee Seroka for performing a technical edit on the material and providing numerous helpful suggestions, comments, and corrections.

Several people at Delmar Publishing Company also deserve special mention: particularly Alar Elken, publisher; Jim DeVoe, acquisitions editor; John Fisher, developmental editor, who worked closely with me on this book; Stacy Masucci, production editor; Mary Beth Vought, art and design coordinator; and Jasmine Hartman, editorial assistant.

I would also like to thank John Shanley of Phoenix Creative Graphics for carrying out the composition, and Rachel Pearce Anderson for performing the copyedit.

Ron K. C. Cheng

CHAPTER 1

Computer Modeling

OBJECTIVES

The aims of this chapter are to give an overview of computer modeling; to outline the key functions of Mechanical Desktop; to introduce feature-based parametric solid modeling, assembly modeling, surface modeling, and associative engineering drafting; and to familiarize you with the user interface of Mechanical Desktop. After studying this chapter, you should be able to do the following:

- Explain the three kinds of computer models
- State the key functions of Mechanical Desktop
- Outline the basic steps to construct parametric solid parts, assemble solid parts, and produce associative engineering drawing
- State the use of NURBS (Non-Uniform Rational B-Spline) surfaces in free-form modeling

OVERVIEW

In industrial and engineering design, you use models to express your design idea. To represent your design precisely and to facilitate downstream computerized manufacturing processes, you use three-dimensional computer models. To represent products or systems, you use virtual assembly of models. From 3D models and assemblies of 3D models, you generate traditional two-dimensional engineering drawings.

Mechanical Desktop 6 sits on top of AutoCAD 2002 and uses the same DWG file format. In addition to the basic computer-aided design functions of AutoCAD, Mechanical Desktop has four sets of design tools for construction of 3D solid models, 3D surface models, assemblies and subassemblies of 3D solid models, and engineering documents of 3D solids, assemblies, and surfaces. The user interface of Mechanical Desktop is very similar to that of AutoCAD. It has, in addition to the basic window areas of AutoCAD, a browser interface displaying the objects and hierarchy of objects in the solid, assembly, and engineering drawing.

COMPUTER MODELING CONCEPTS

In various design stages before you actually make a product or system, you use models as a substitute of the real thing to express your design idea. You use models to explore, visualize, and learn more about the products or systems. Sometimes, at a later design stage when the design is near completion, you may use a physical mock-up model to further evaluate your design. However, using a physical model is more expensive and inflexible. With the advent of computer and computer-aided design applications, more and more use is made of the computer in modeling.

REPRESENTATION

Before starting to construct a computer model, you need to think about what you will represent in the computer model: geometry (shape, profile, and silhouette), appearance (color and texture), and properties (mass, center of gravity, and other physical properties). Here we will use the computer to construct a virtual model of the 3D object and focus mainly on using computer models to represent the geometry of products and systems.

TRADITIONAL 2D APPROACH

The traditional approach of using 2D orthographic views to represent a 3D object requires a thorough understanding of the principles of projection and an interpretation of the 3D object by associating the 2D views. Furthermore, it is usually not possible to use 2D drawing views to represent a complex 3D free-form object precisely. Figure 1–1 shows the 2D engineering drawing of a car body. You can see that the silhouettes and profiles are not clearly represented. Naturally, such a drawing cannot be used in the factory to make the product precisely.

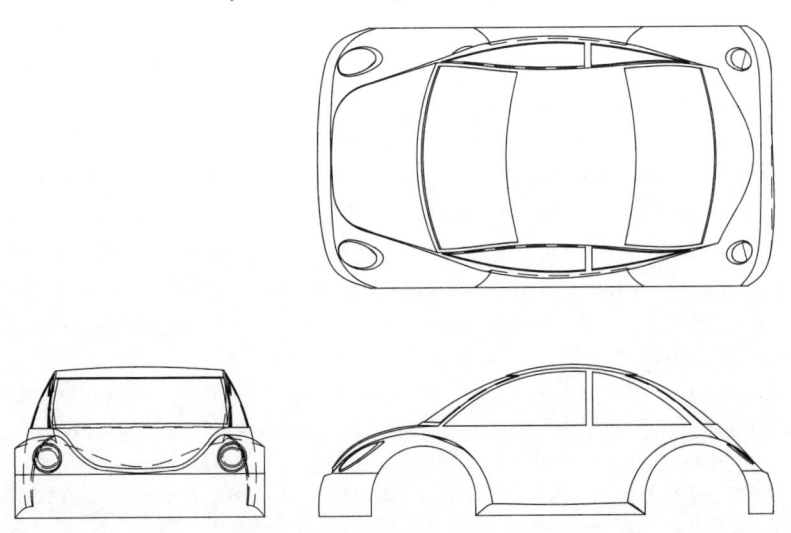

Figure 1–1 *Engineering drawing of a free-form object*

3D MODEL

To represent the geometry of a 3D object precisely in the computer, you should use a 3D computer model. Basically, there are three kinds of 3D computer models: wireframe, surface, and solid models. Take the example of the model of a rectangular box. A wireframe model of the box has 12 line segments. A surface model of the box has six surfaces. A solid model of the box has one integrated solid data.

Wireframe Model

Wireframe modeling is the most primitive kind of 3D model in computer-aided design. It represents a 3D object by using a set of unassociated curves depicting the edges of the 3D object. The model only gives the pattern of a 3D object. There is no relationship among the curves. Hence, the wireframe model has no surface and volume information, apart from the edges and vertices. You have to associate the curves with one another to implicitly perceive the surface and volume of the object. Figure 1–2 shows a wireframe model of a free-form object. It shows only the edges where the faces intersect, but the faces are not represented in the model. Because there is no surface and volume data and the wires do not have any thickness, you cannot perform rendering and hidden line removal operations on the model.

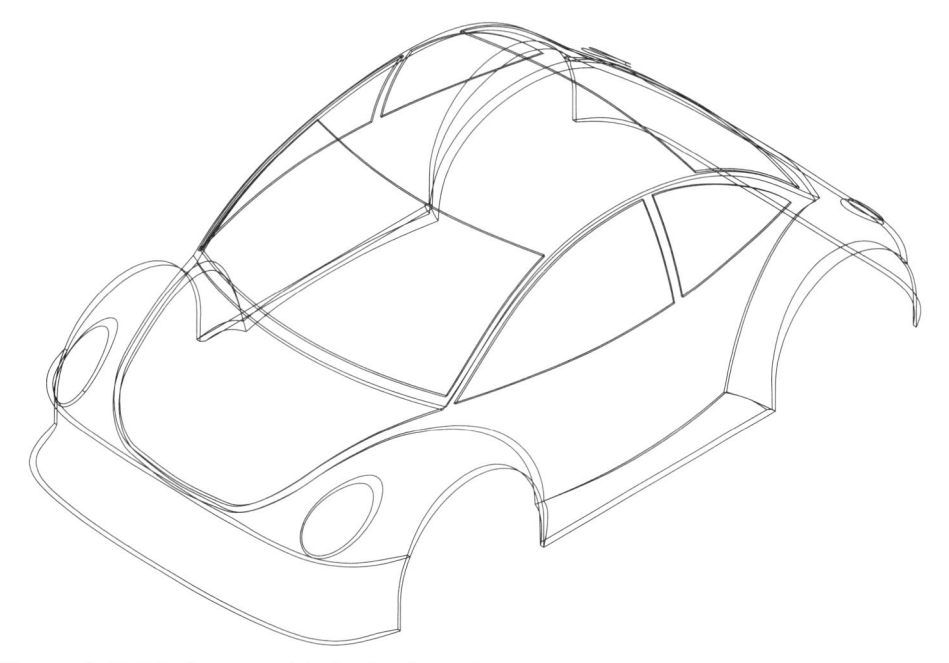

Figure 1–2 *Wireframe model of a free-form object*

Sometimes a wireframe model can be very ambiguous. Figure 1–3 shows a simple cubical box represented as a wireframe model (left) and the possible objects (right).

WIREFRAME MODEL

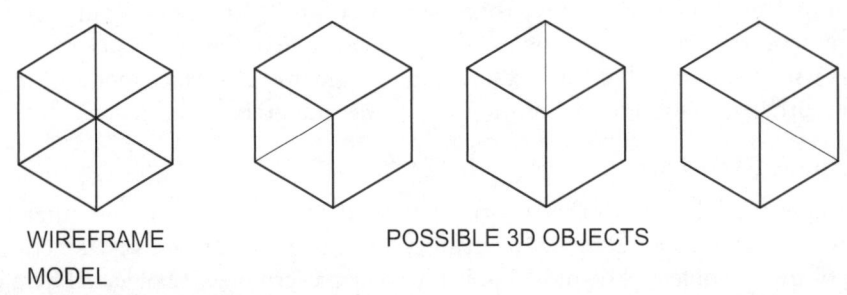

POSSIBLE 3D OBJECTS

Figure 1–3 *Ambiguous representation*

For downstream manufacturing operations, a wireframe model has limited usage. You may only use it in 2-1/2D CNC machines and 2D profiling. With regard to analysis, you cannot use a wireframe model to evaluate mass properties, perform collision detection among component parts, or other important analyses.

Surface Model

The second type of 3D model, the 3D surface model, is a set of 3D surfaces that are put together in a 3D space to create the figure of a 3D object. When compared to a 3D wireframe model, a 3D surface model has information on the contour and silhouette of the surfaces, in addition to edge data. You can use 3D surface models in a computerized manufacturing system or to generate photo-realistic rendering or animation.

Because a surface model is a set of surfaces, making a surface model concerns constructing the surfaces one by one. The fundamental way to construct a surface is to construct a set of curves and let the computer construct the surface on the curves. Figure 1–4 shows a set of curves and a surface constructed from the curves.

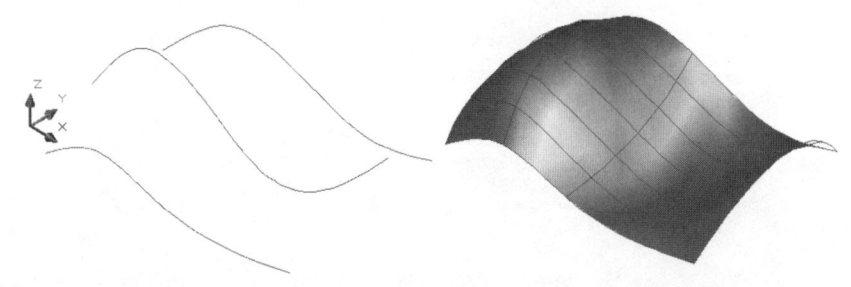

Figure 1–4 *Curves and surfaces constructed from the curves*

With a set of surfaces, you compose a surface model. Figure 1–5 shows a surface model and the surfaces exploded apart and Figure 1–6 shows the rendered image of the surface model.

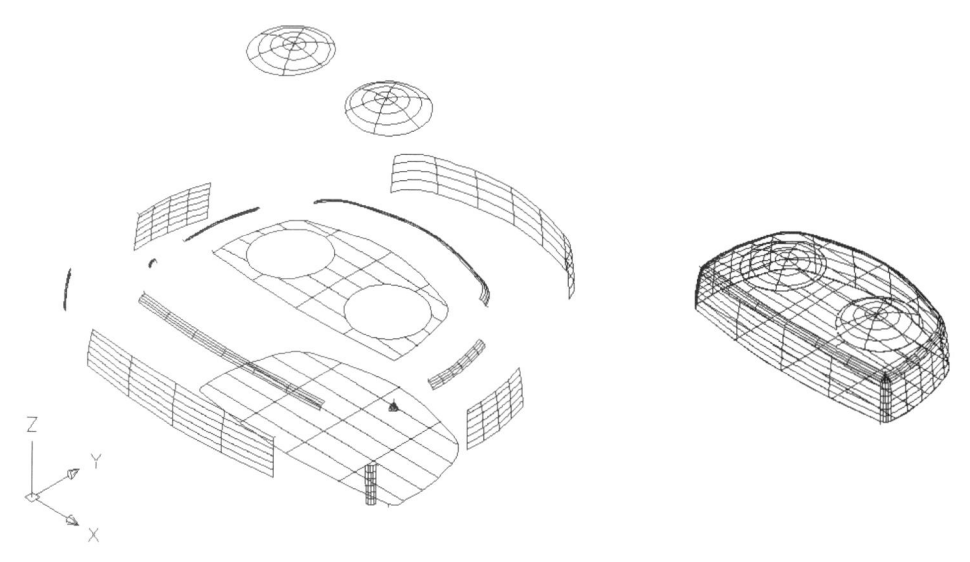

Figure 1-5 *Surface model and surfaces exploded apart*

Figure 1-6 *Rendered image of the surface model*

Solid Model

With regard to comprehensiveness of information, a 3D solid model is the superior model to use because a solid model in a computer creates integrated mathematical data about the surfaces, edges, and volume of the object that the model describes. In addition to visualization and manufacturing, you can use the data of a solid model for design calculation. Figure 1–7 shows the solid model of a spur gear.

Figure 1–7 *Solid model of a spur gear*

Because each individual 3D object is unique in shape, the integrated data needed to represent a solid in the computer is more complicated as compared to the surface and wireframe models. Among the many ways to construct solids with a computer, the most common method is to construct sets of curves and treat the curves in four fundamental ways: extrude, revolve, loft, and sweep (see Figures 1–8 and 1–9).

EXTRUDE   REVOLVE   LOFT   SWEEP

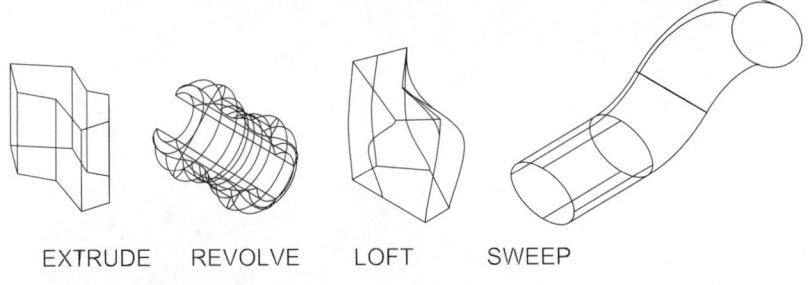

Figure 1–8 *Four basic kinds of solid*

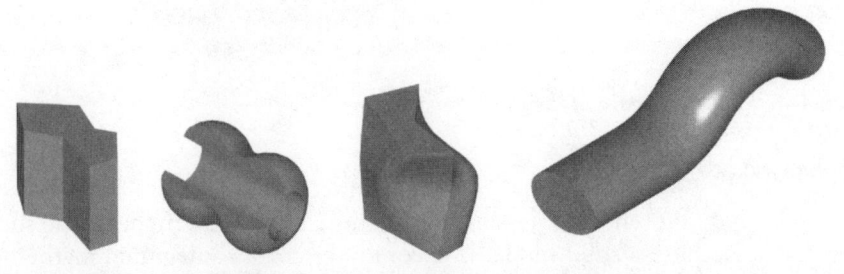

Figure 1–9 *Rendered image of extruded, revolved, loft, and sweep solids*

Basically, construction of complex solids concerns making solids of basic shapes and combining the solids by using Boolean operations. For example, the spur gear shown in Figure 1–7 is a combination of extruded and revolved solid features; the main body of the spur gear is a revolved solid feature; and each gear tooth is an extruded solid feature (see Figure 1–10).

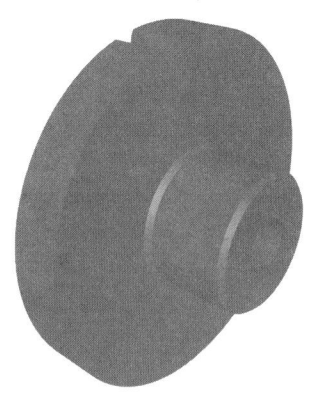

Figure 1–10 *An extruded solid (gear tooth) cut on a revolved solid (gear blank)*

CHOICE OF MODEL

A wireframe model only represents the edges and vertices of a 3D object. In addition to edges and vertices, a surface model includes the surfaces and silhouettes of a 3D object. A 3D solid model is the superior modeling choice, since it provides volume data as well as surface, edge, and vertex data.

Nowadays, a wireframe model alone has very little use. Comparing the surface and solid models, solid models are easier to construct. However, it has limited capability in representing complex free-form shapes. To make 3D objects of complicated profiles and silhouettes, you must use a 3D surface model.

In general, 3D solids are more appropriate for constructing 3D objects of regular shape and 3D surfaces are more suitable for constructing 3D free-form objects. (If you have a surface model that consists of a set of surfaces enclosing an "air-tight" volume, you can convert the surface model into a solid model.) Although you do not use 3D curves to construct 3D wireframe models, you use them as frameworks for making 3D surfaces and 3D solids.

MODELING METHOD

To construct a 3D computer model, consider the 3D object holistically and think about how to decompose it into elements that can be represented in the computer. Then construct the elements and combine the elements to form the model. Depending on the kind of 3D model that you are going to construct, decompose the object into different kinds of elements.

To construct a wireframe model, think about the edges that two faces of a 3D object meet. Sometimes, you also have to consider the implicit edges at the tangent of surfaces. The work to construct a wireframe is tedious. You have to input 3D coordinates to define the wires. To construct a simple cube, for example, you need to construct 12 lines in the 3D space. For curved edges, construction work is even more difficult and exhausting.

To construct a surface model, think about the surfaces and decompose the 3D object into surfaces. To make the surfaces, construct sets of curves. With the surfaces, compose a surface model. Making the surface is almost automatic but construction of the curves for the surfaces is difficult.

To make a solid model, decompose the 3D object into solid elements, construct the solid elements, and combine the elements. By using the parametric approach (explained in more detail later in this chapter), solid modeling is much easier than surface and wire modeling.

ASSEMBLY MODELING

A product or a system usually has more than one component. To represent a product or system, you use assembly models in the computer by putting together a number of components in an assembly file. Because the definitions of the 3D models are already depicted in the individual 3D component files, the assembly file only delineates the way that the components are put together and the geometric relationship among the components. Figure 1–11 shows the assembly of an infant scooter.

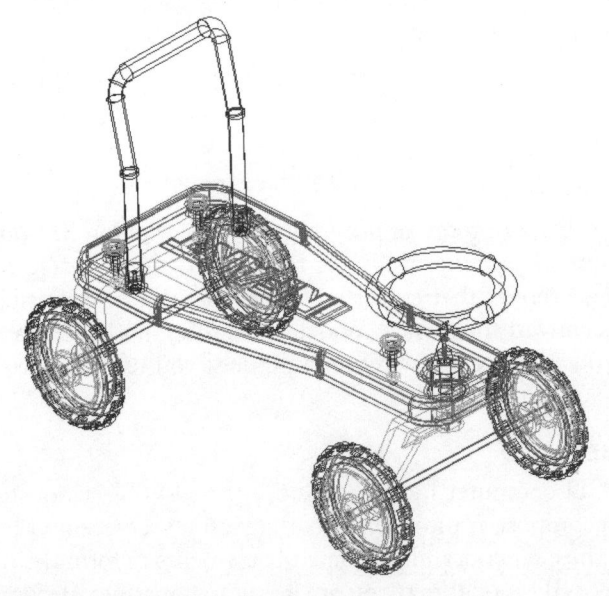

Figure 1–11 *Assembly of an infant scooter*

ENGINEERING DRAFTING

Although computer models are widely used in most contemporary factories, it is sometimes necessary to output 2D engineering drawings to meet individual needs. With 3D computer models, you can construct 2D orthographic drawings in a partially automated way. You specify the kinds of view and the locations, the system generates the orthographic views, and you add appropriate annotations to complete the engineering drawing. Figure 1–12 shows an engineering drawing constructed from the assembly of the infant scooter.

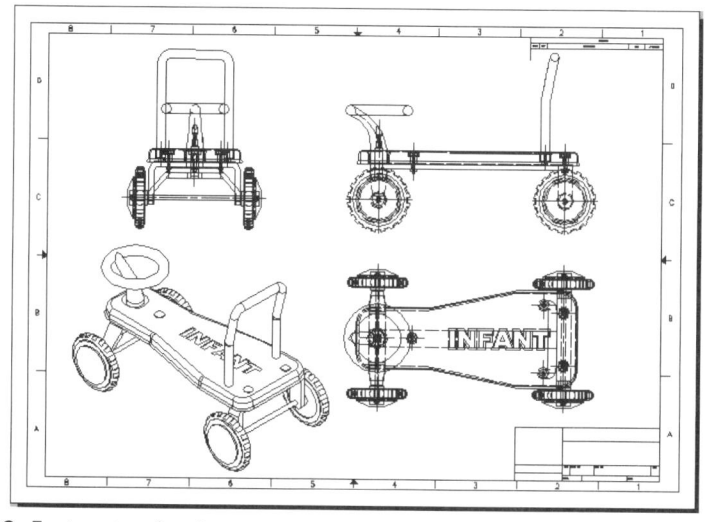

Figure 1–12 *Engineering drawing*

USING MECHANICAL DESKTOP

Mechanical Desktop 6 is a 3D industrial and engineering design tool that is integrated in AutoCAD 2002. It uses the same AutoCAD native DWG file format and extends the capability of AutoCAD by providing four major functions:

1. Parametric solid modeling
2. Assembly modeling
3. Surface modeling
4. Associative engineering drafting

FILE TYPE

Although Mechanical Desktop builds on AutoCAD 2002, it has two distinct kinds of files: part files and assembly files. You use a part file for the construction of solid parts and an assembly file for the construction of assemblies. For the sake of consistency, you should use part files for making surfaces, although you can use either kind of file.

Part File

Similar to AutoCAD 2002, a part file has two working environments: Model and Drawing. You construct solid parts in the Model environment and construct engineering drawing of the solid parts in the Drawing environment. To start a new part file, select New Part File from the File pull-down menu.

> **File>New Part File**

Assembly File

An assembly file has three working environments: Model, Scene, and Drawing. You construct solid parts and assembly and sub-assemblies of solid parts in the Model environment, construct scenes of assembly in the Scene environment, and construct engineering drawings in the Drawing environment. To start an assembly file, select New from the File pull-down menu.

> **New>File**

USER INTERFACE

The Mechanical Desktop user interface is very similar to the AutoCAD user interface, with the exception of a Desktop browser dialog box shown at the right (or the left). (See Figure 1–13.)

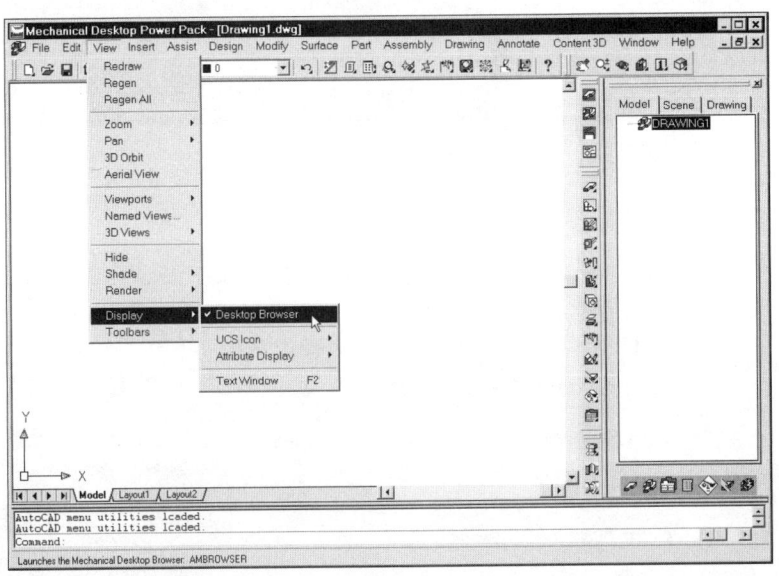

Figure 1-13 *Mechanical Desktop user interface*

Desktop Browser

The Desktop browser is a unique dialog box that displays the objects that you construct in various working environments (model, scene, and drawing) and provides command shortcuts in right-click menus. Corresponding to the two kinds of files, there are two kinds of desktop browsers: the part file browser and the assembly file browser. The part file desktop browser has two tabs (model and drawing), and the assembly file desktop browser has three tabs (model, scene, and drawing). The model tab displays the objects of the solid part or component parts of the assembly, the scene tab displays the objects of the assembly scenes, and the drawing tab displays the drawing objects. The desktop browser shown in Figure 1–13 has three tabs. It is an assembly file browser. Figure 1–14 shows the part file user interface. It has two tabs.

To manipulate the objects in your file, you select one from the browser, right-click, and select the appropriate command button (see Figure 1–15).

If the browser is not shown in your screen, select Desktop Browser from the Display cascading menu of the View pull-down menu.

View>Display>Desktop Browser

Pull-Down Menu

At the top of your screen, you will find 12 pull-down menu items. In particular, the Surface, Part, Assembly, and Drawing pull-down menus concern the four major design tools of Mechanical Desktop. If you do not find these pull-down menus, type **MENULOAD** at the command line interface to display the Menu Customization dialog box (see Figure 1–16). In the Menu Customization dialog box, select the pull-down menu items to be displayed and select the Insert button.

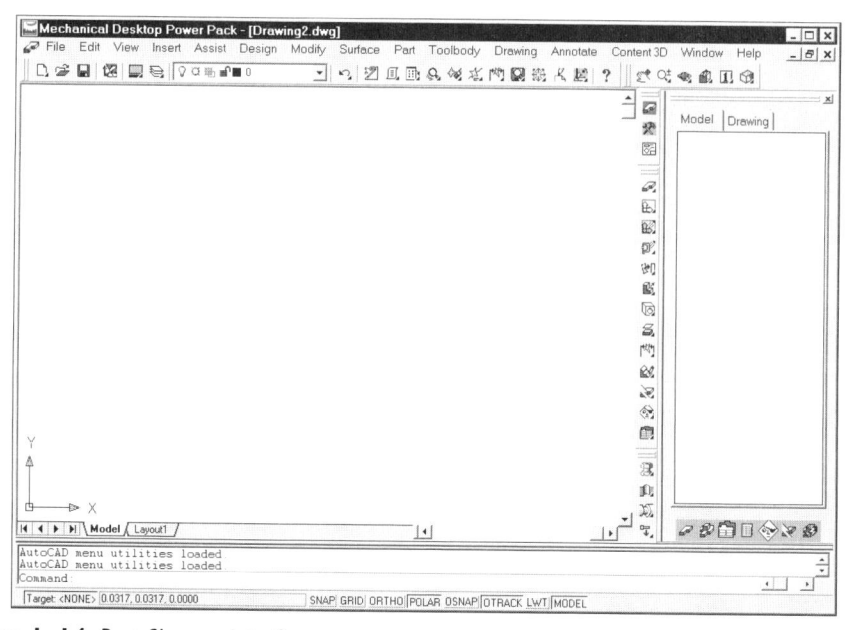

Figure 1–14 *Part file user interface*

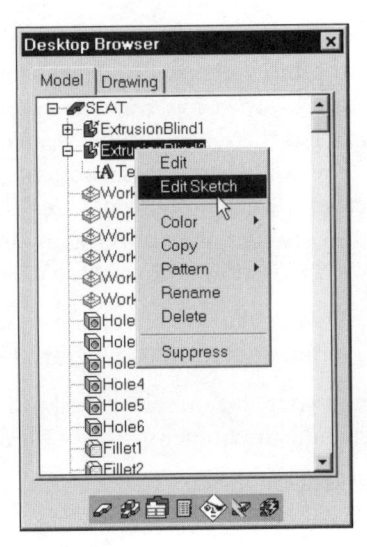

Figure 1–15 *Right-click menu in the browser*

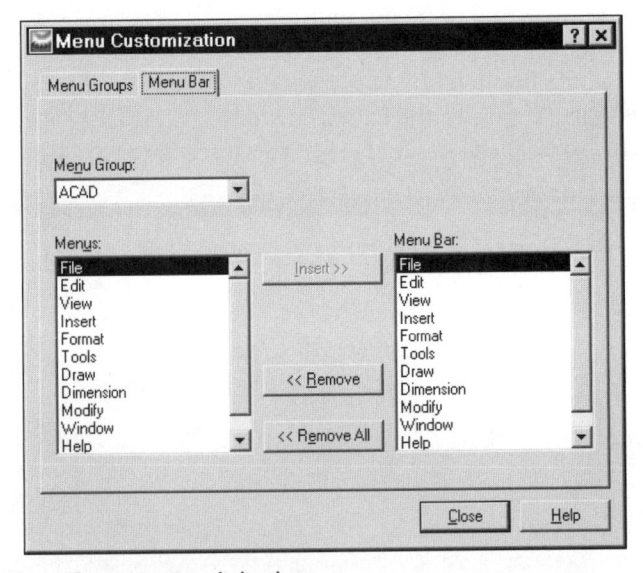

Figure 1–16 *Menu Customization dialog box*

Toolbars

Below the pull-down menus is a Mechanical Main toolbar (see Figure 1–17). Because Mechanical Desktop and AutoCAD commands coexist, both Mechanical Desktop and AutoCAD commands are available. Owing to the limited space in the pull-down menu and the window area, only some fundamental AutoCAD command menus are displayed in conjunction with the Mechanical Desktop menus and toolbars.

Toolbars are displayed corresponding to the respective kinds of tasks to be carried out (part modeling, assembly model, scene construction, and engineering drafting). (See Figures 1–18 through 1–21.) To display the surface modeling toolbar (see Figure 1–22), select Launch Toolbar from the Surface pull-down menu. (Here MDT6 is running with the Power Pack; if you are using MDT6 without the Power Pack, the toolbars will look different.)

Figure 1–17 *Mechanical Main toolbar*

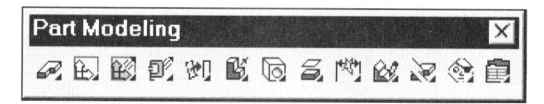

Figure 1–18 *Part Modeling toolbar*

Figure 1–19 *Assembly Modeling toolbar*

Figure 1–20 *Scenes toolbar*

Figure 1–21 *Drawing Layout toolbar*

Figure 1–22 *Surface Modeling toolbar*

To display other toolbars not shown on the screen, select Customize Toolbars from the Toolbars cascading menu of the View pull-down menu (see Figure 1–23).

View>Toolbars>Customize Toolbars

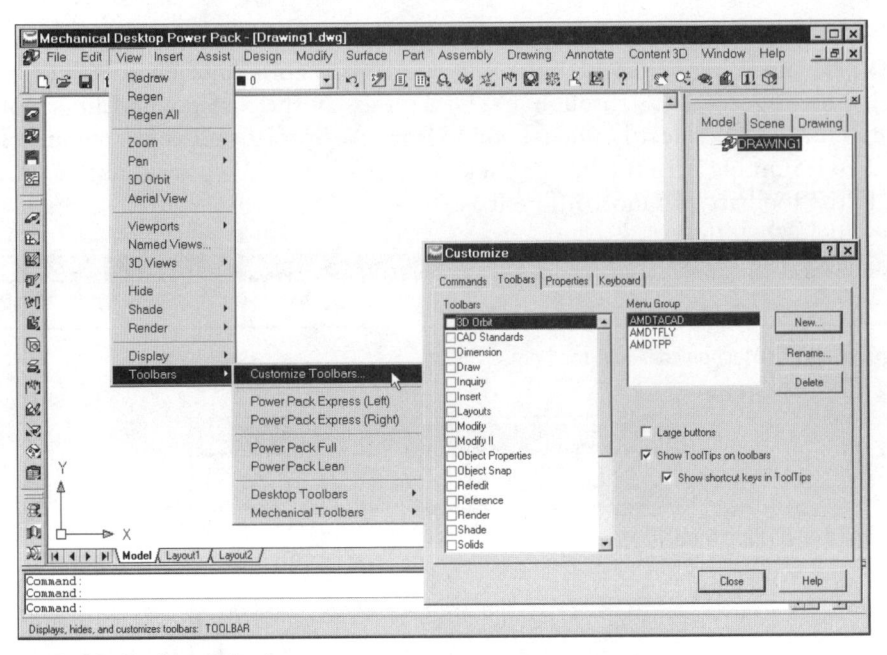

Figure 1–23 *Toolbar dialog box*

The alternative to displaying a toolbar is to right-click any toolbar and select from the pop-up display of all the toolbars available.

COMMAND ENTRY

With the pull-down menu and toolbars in their proper positions, you can use Mechanical Desktop commands in several ways:

- Select an item from the pull-down menu or the cascading menu from the pull-down menu
- Select a button from the toolbars
- Type the command name at the command line interface
- Right-click in various areas of the user interface to use the right-click menu
- Double-click an object from the browser

LAYERS

As you work along with Mechanical Desktop, you will find a number of layers prefixed by the letters AM. Mechanical Desktop uses these layers for specific reasons. These layers are automatically created to hold various objects, such as visible lines, hidden lines, and viewport borders of an engineering drawing constructed from a solid part or an assembly of solid parts. Do not change these layer names.

AUTOCAD 2002 TOOLS

You use basic AutoCAD tools for two purposes. You use the basic commands of AutoCAD to construct curves in order to make surfaces and solids, and you construct engineering title blocks and related annotations in order to make engineering drawings.

PARAMETRIC SOLID MODELING TOOL

The first generation of PC-based solid models are generally static, in the sense that the solid model cannot be changed after it is constructed from the curves (by extruding, revolving, lofting, and sweeping). With the advent of more powerful computers and computer applications, the parameters of solids can now be modified anytime during the design process. The former kind of solids, which are static in nature, are normally referred to as non-parametric solids (AutoCAD native solid), while the latter kind of solids, which are modifiable, are referred to as parametric solids. To modify the size of a parametric solid, change the dimension value. Hence, the modeling system is also known as a dimension-driven parametric solid model system.

The Mechanical Desktop part modeling module is a parametric feature-based solid modeling system. To construct the solid model of a 3D object, analyze and decompose the object into solid features and construct the features one by one. Because a solid is decomposed into features and the parameters of the features can be modified, the solids are called parametric feature-based solids.

To construct a solid feature, construct rough sketches by using AutoCAD tools. Using Mechanical Desktop, resolve the sketch to a parametric sketch and construct solid features from the parametric sketches. To make a complex solid feature, construct more features and combine them by using Boolean operations. Figure 1–24 shows a solid object with four kinds of features (extrude, revolve, sweep, and loft) joined together. Solid features constructed from sketches are called sketched solid features.

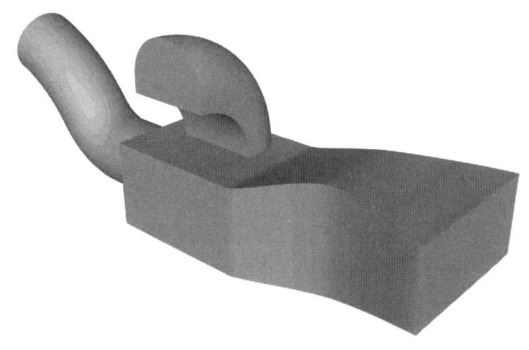

Figure 1–24 *Solid with four sketched features: extruded (center), revolved (top), sweep (left), and loft (right)*

In addition to sketched solid features, you can use pre-constructed solid features. You construct them by selecting a feature from the menu and specifying parameters and location. Because you construct these features by placing them in your model, they are called placed solid features. Figure 1–25 shows a solid model with four kinds of placed solid features (chamfer, fillet, hole, and shell).

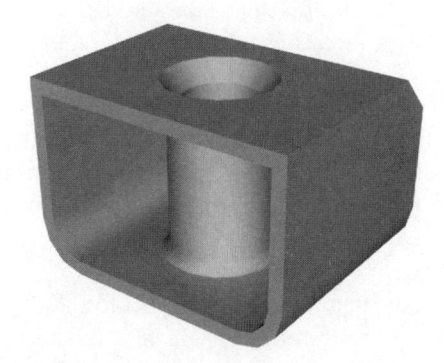

Figure 1–25 *Solid with four kinds of placed solid features: chamfer, fillet, hole, and shell*

To help construct solid features, you use work features as geometric references. Therefore, a solid part can have three kinds of features: sketched solid features, placed solid features, and work features.

Figure 1–26 shows the Mechanical Desktop Part pull-down menu and toolbar. Using the menu and toolbar, you can construct parametric feature-based solid parts.

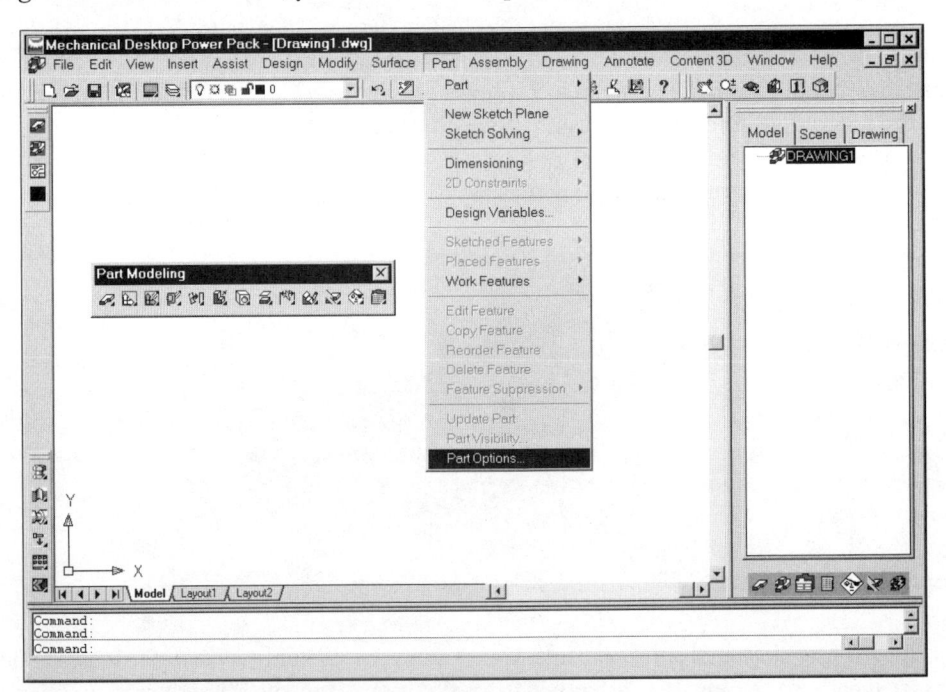

Figure 1–26 *Mechanical Desktop Part pull-down menu*

If you already know how to use AutoCAD, you may know that AutoCAD also has a set of solid modeling tools. (See Figure 1–27, the Solids cascading menu of the Design pull-down menu.) Unlike the parametric tools, this set of tools only enables you to construct non-parametric solids (AutoCAD native solids). You construct parametric solids by using Mechanical Desktop and you construct non-parametric solids by using AutoCAD.

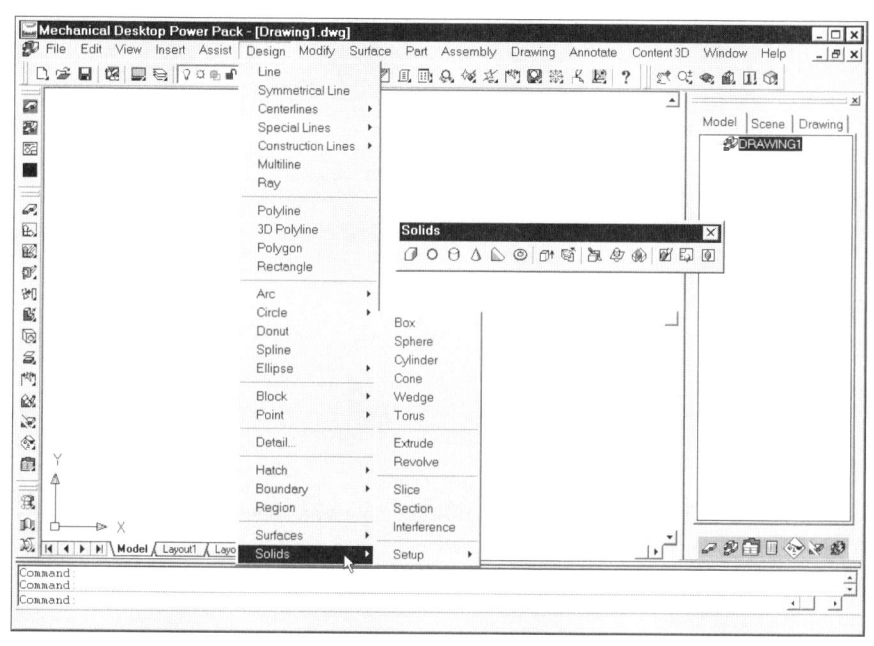

Figure 1–27 *Solids cascading menu of the Design pull-down menu*

You will learn how to construct parametric sketches in Chapter 2 and parametric solids in Chapters 3, 4, and 9.

ASSEMBLY MODELING TOOL

The assembly modeling module is a unique set of design tools for putting together a set of components to form an assembly of a product or system. Basically, the components of an assembly are solid parts of individual components. However, if you are dealing with a product or system that has a large number of parts, it is more convenient and practical to put parts into smaller subassemblies and gather together the subassemblies into the final assembly. Therefore, the meaning of components in the context of assembly can be a solid part or a subassembly.

Construction of an assembly in the computer involves linking a collection of components (solid parts or subassemblies) that makes up the assembly, translating the components to their appropriate locations, and assigning relationships among the components by applying assembly constraints to selected pairs of features of the components.

With an assembly, you model the entire product or system, evaluate the validity of the assembly, check interference among the components, and display an exploded view of the product or system. Figure 1–28 shows the Mechanical Desktop Assembly pull-down menu and toolbar.

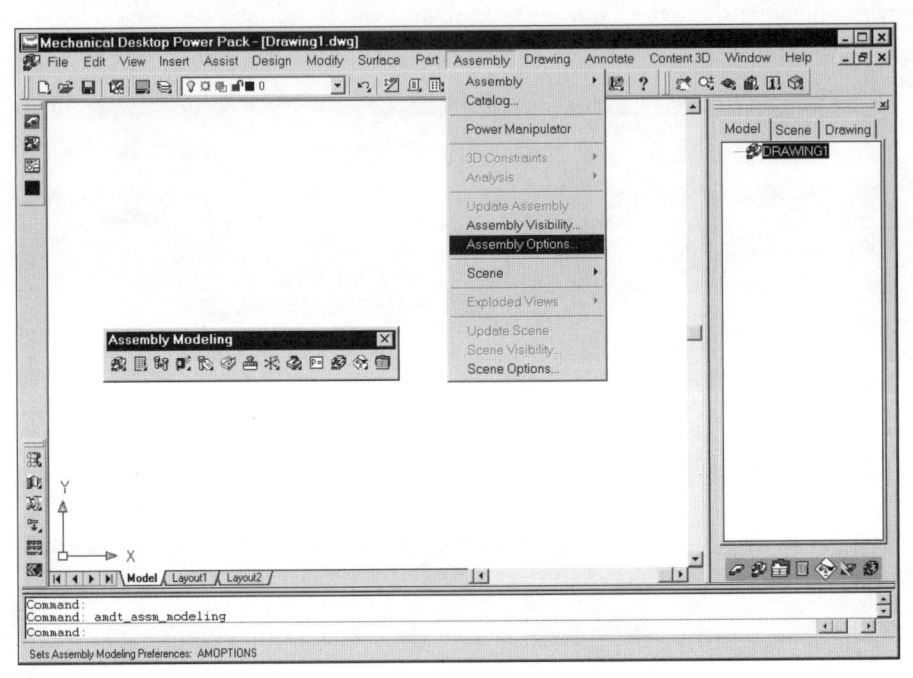

Figure 1–28 *Mechanical Desktop Assembly pull-down menu and toolbar*

You will learn how to construct assemblies of components in Chapter 5 and scenes of assemblies in Chapter 6.

SURFACE MODELING TOOL

Surface models are best used to represent 3D free-form objects. To construct surface models in the computer, you construct a framework of 3D wires, specify the kind of surfaces to be produced, and let the computer construct the surfaces.

The surface modeling module is a Non-Uniform Rational B-Spline (NURBS) surface modeling system for design and manufacturing. You use it to construct free-form surfaces in a computer (see Figure 1–29). It also incorporates a set of 3D wire construction and editing tools for you to construct complex frameworks of 3D wires for creating surfaces.

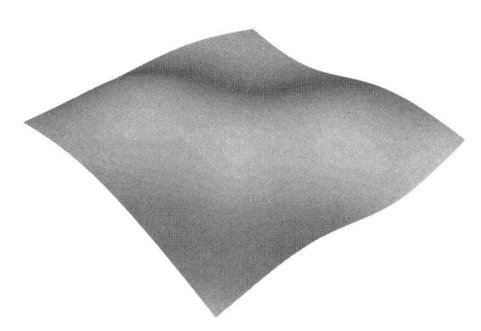

Figure 1–29 *Free-form surface*

With regard to integration of a surface to a solid, the surface modeling module allows you to apply a thickness to a surface to convert it to a solid with a 3D free-form profile, to stitch together a set of "air-tight" surfaces into a 3D free-form solid, and to use a surface as a cutting tool to shave a solid. Working in the opposite direction, you can convert a stitched surface, an AutoCAD native solid, or a Mechanical Desktop solid to a set of NURBS surfaces (see Figures 1–30 through 1–33).

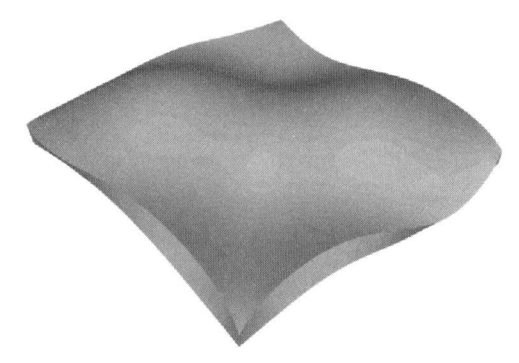

Figure 1–30 *Solid constructed by assigning a thickness to a surface*

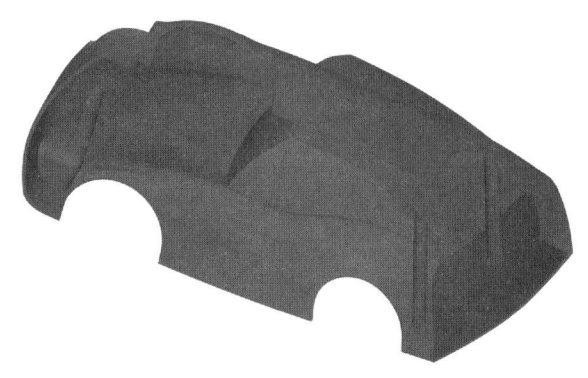

Figure 1–31 *Bottom view of the solid model of a car body constructed from a set of stitched surfaces*

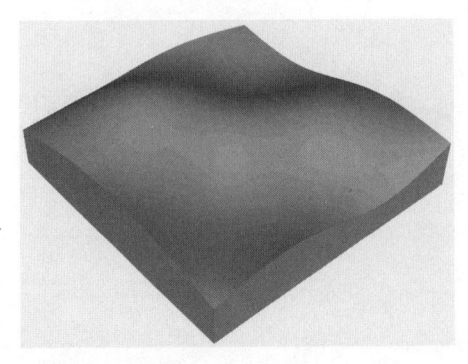

Figure 1-32 *A solid shaved by a surface*

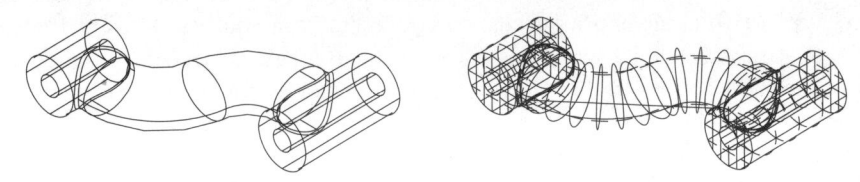

Figure 1-33 *A solid converted into a set of surfaces*

Figure 1–34 shows the Mechanical Desktop Surface pull-down menu and toolbar. You should not confuse this set of tools with the AutoCAD surface tools. Figure 1–35 shows the Surface cascading menu of the Design pull-down menu and the Surfaces toolbar.

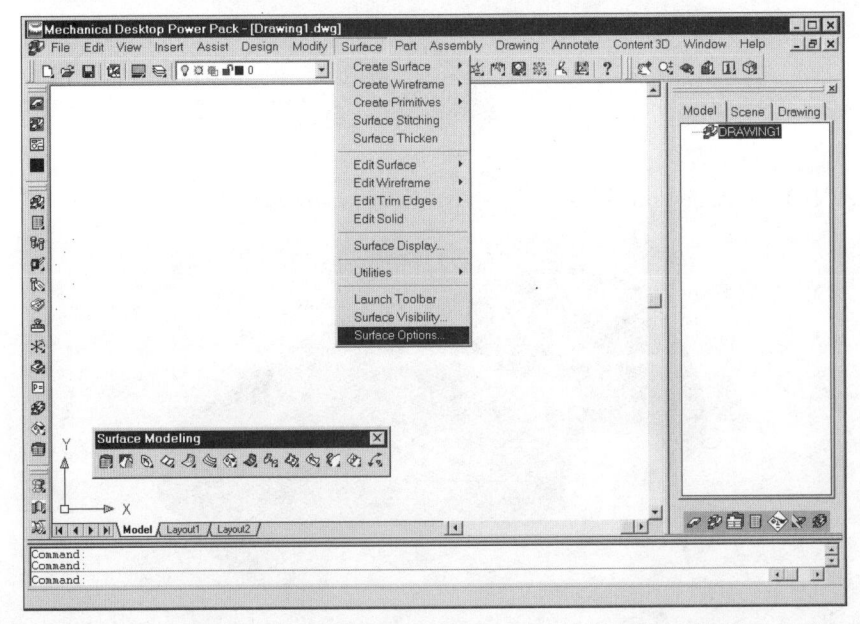

Figure 1-34 *Mechanical Desktop Surface pull-down menu and Surface Modeling toolbar*

In essence, the AutoCAD surface tools are earlier, simpler tools that define a surface by using polygon meshes. Any complex smooth surface is decomposed into a mesh of polygonal faces and curvilinear edges into segments of straight lines. Figure 1–36 shows a set of surface meshes. They are approximations of the original surfaces. On the contrary, the surfaces that you construct by using Mechanical Desktop surface tools are continuously smooth surfaces. You can retrieve accurate coordinates from any point on the surface. You will learn how to construct surface models in Chapters 7 and 8.

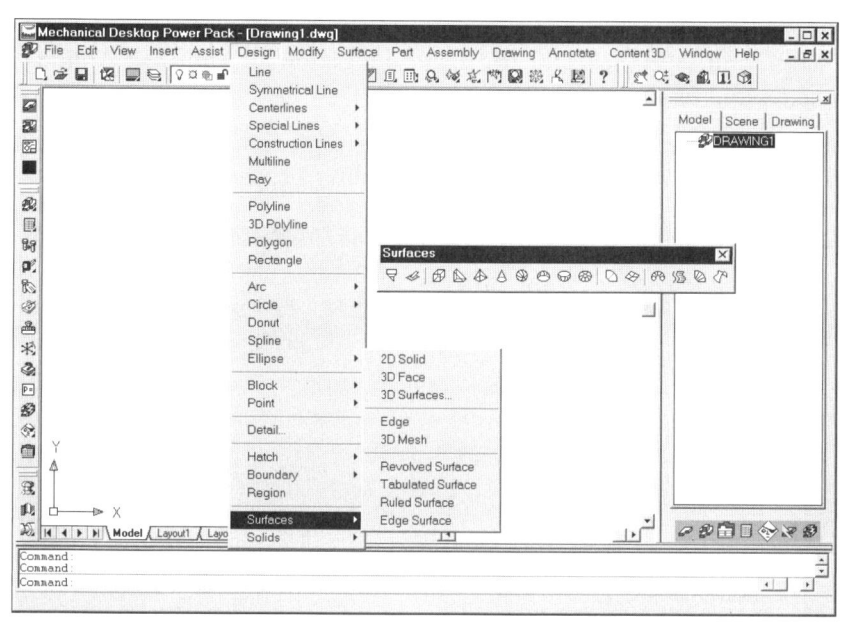

Figure 1–35 *Surfaces cascading menu of the Design pull-down menu and the Surfaces toolbar*

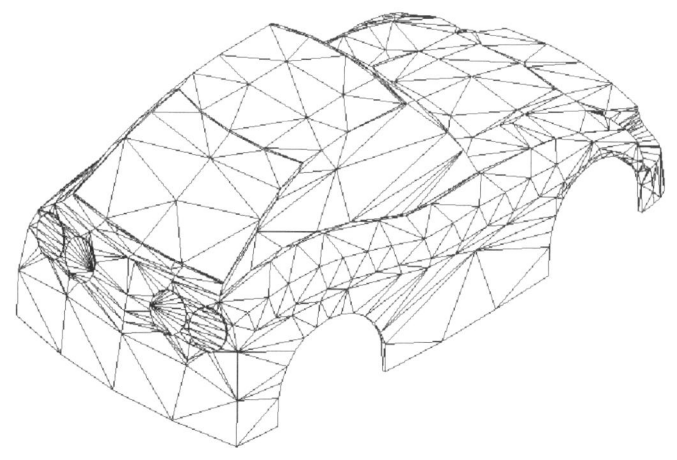

Figure 1–36 *Polygon meshed representation of surfaces by the AutoCAD surface tools*

ASSOCIATIVE ENGINEERING DRAFTING TOOL

The Mechanical Desktop drawing and annotation module generates 2D engineering drawing views from surfaces, solids, and assemblies, and enables you to include annotations to the engineering drawing.

After you have constructed 3D models in the computer, production of engineering drawing becomes semi-automated. Use the drawing mode and specify the kind of drawing view and display scale. Then the computer generates the drawings automatically for you. After generating the drawing views, you add appropriate annotations. Figures 1–37 and 1–38 shows the Mechanical Desktop Drawing and Annotate pull-down menus and toolbars.

You will learn how to construct engineering drawings for solid parts and surfaces in Chapter 10 and drawings for assemblies in Chapter 11.

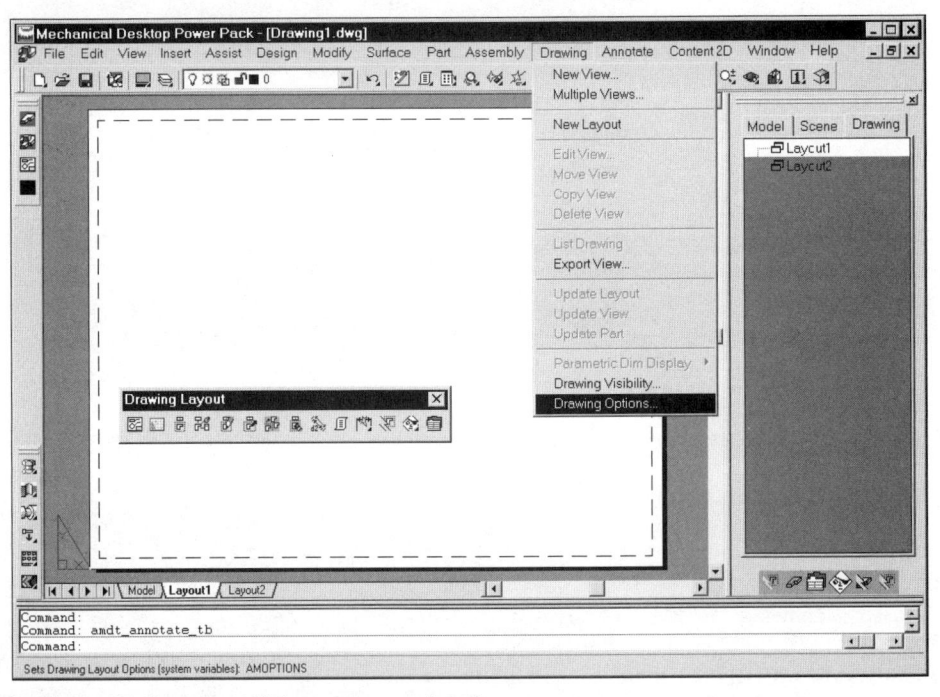

Figure 1–37 *Drawing pull-down menu and toolbar*

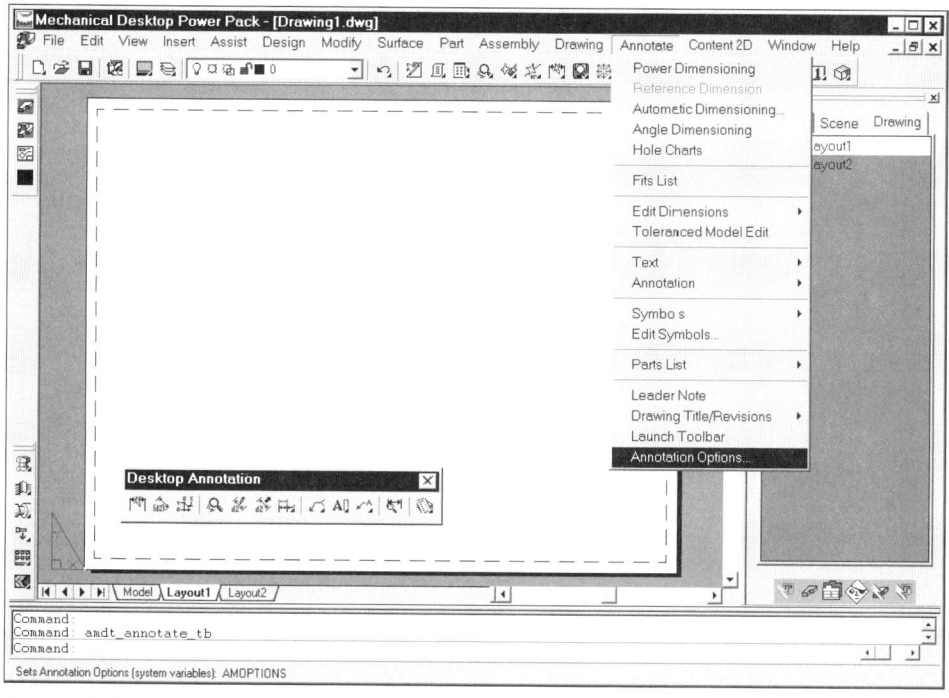

Figure 1-38 *Annotation pull-down menu and toolbar*

COMPATIBILITY AND INTEROPERATION

To reiterate, Mechanical Desktop 6 builds on AutoCAD 2002 and uses native DWG file format. Hence, Mechanical Desktop solids and surfaces and AutoCAD native solids are fully compatible. You can interoperate among them.

AutoCAD Native Solids

You can convert an AutoCAD native solid into a Mechanical Desktop base solid feature on which you add more parametric solid features. Also, you can derive a set of NURBS surfaces from the faces of a native solid.

Mechanical Desktop Surfaces

You stitch a set of water-tight surfaces to form a solid and use a NURBS surface as a surface feature to cut a parametric solid or a native solid.

Mechanical Desktop Parametric Solids

You change a parametric solid to a native solid and convert a parametric solid to a set of NURBS surfaces.

OPTIONS

Settings in the Options dialog box affect how the system operates and what you see on screen (see Figure 1–39).

Select Assist>Mechanical Options

Command: AMOPTIONS

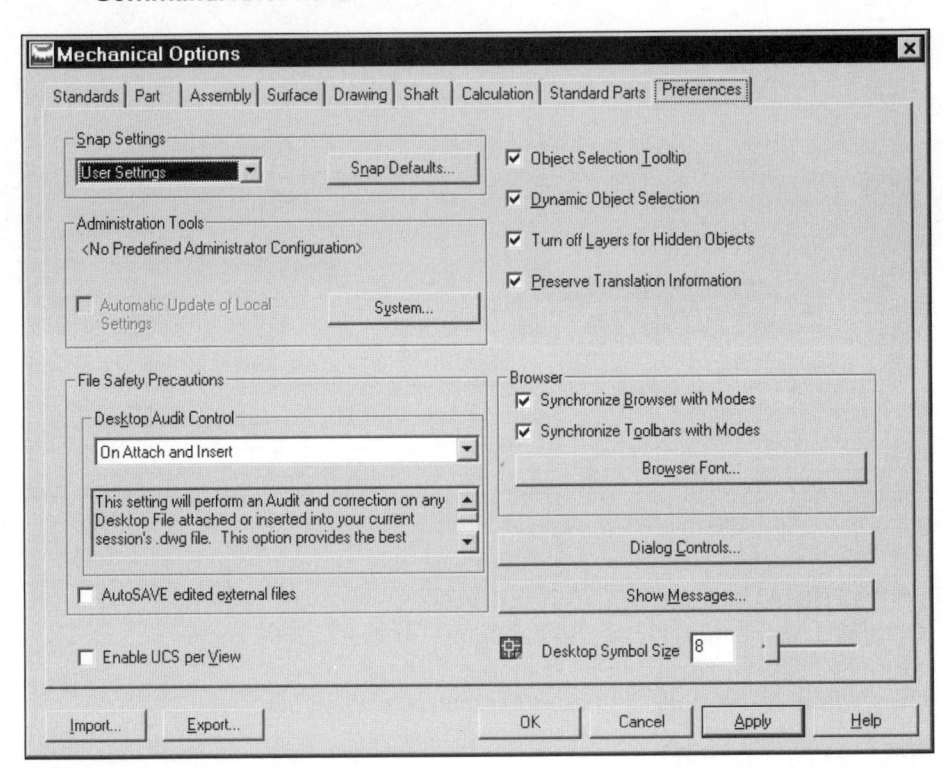

Figure 1–39 *Mechanical Options dialog box*

The Mechanical Options dialog box has a number of tabs that are relevant to various tasks. To begin with, select the Preferences tab. Among various settings here, you should ensure that the Synchronize Browser with Modes and Synchronize Toolbars with Modes checkboxes are checked. Checking these boxes causes the browser and the toolbars synchronize with the working mode (model, scene, and drawing). Select the OK button to close the dialog box.

SUMMARY

The prime objective in using a computer model is to represent a 3D object precisely in the computer. There are three kinds of computer models. With regard to comprehensiveness and integrity in data, solid modeling is superior and is best suitable for most downstream manufacturing operations. However, surface models can better represent complex 3D free-form shapes than solid models. Therefore, use solid modeling methods for 3D objects of general shapes (extrude, revolve, loft, and sweep), use surface modeling methods for complex free-form objects, and convert a surface model with surfaces enclosing a volume without any opening or gap to a solid model if a solid of complex shape is required. Wireframe modelng alone has very little use in computer modeling. However, you sometimes need to construct a framework of wireframes for making surfaces and solids.

Mechanical Desktop 6 is an engineering design tool that runs on top of AutoCAD 2002. It has four design modules: parametric solid modeling, assembly modeling, surface modeling, and associative engineering drafting.

The parametric solid modeling module enables you to construct parametric feature-based solid parts. Solids are constructed from rough sketches and can be modified. Surfaces can be included in the solid. The assembly modeling module enables you to construct assemblies of components which can be solid parts or subassemblies of solid parts. It also enables you to construct exploded views of the assemblies. The surface modeling module enables you to construct free-form surfaces and use the free-form surfaces in making solid parts. The associative engineering drafting module enables you to construct engineering drawings from solids, assemblies, and surfaces.

Basically, the Mechanical Desktop user interface is much the same as that of AutoCAD, with a few differences. It has additional pull-down menu items, a Desktop Browser that lists objects in the part or assembly file and provides command shortcuts, and the pull-down menu items are rearranged.

REVIEW QUESTIONS

1. What are the three kinds of 3D models? Briefly describe their characteristics.

2. What are the four design tools of Mechanical Desktop? State their prime functions.

3. How many ways can you run a Mechanical Desktop command? List them.

4. How would you model a product or system with a number of component parts?

5. How would you construct an engineering drawing from 3D models?

CHAPTER 2

Parametric Sketching

OBJECTIVES

The aims of this chapter are to introduce the key concepts of parametric sketching, to delineate the basic steps to constructing a parametric sketch, and to let you master various sketching techniques. After studying this chapter, you should be able to do the following:

- Explain the key concepts of parametric sketching
- Construct a parametric sketch
- Explain the meaning of constraints
- Manipulate geometric constraints and parametric dimensions
- Use various sketching techniques

OVERVIEW

The starting point for making a parametric feature-based solid model is to make a rough sketch, which does not have to be accurate. Using the sketch, construct a solid. While making the sketch, concentrate on shape and form and don't pay any attention to exact geometry and size. By default, the sketch you construct is non-parametric because you are working in the AutoCAD environment and Mechanical Desktop is an add-on application to AutoCAD. To convert a non-parametric AutoCAD sketch to a parametric Mechanical Desktop sketch, you resolve it. To complete the parametric sketch, modify it by applying appropriate geometric and dimension constraints.

PARAMETRIC SKETCHING CONCEPTS

When you begin to design, you would ordinarily concentrate on shapes rather than on dimensions. However, many conventional computer-aided design applications require you to input precise lengths and orientations of entities from the onset. In the preliminary design stage, such data are generally not available.

SKETCHING

With the Part module of Mechanical Desktop, use the computer as an electronic sketching pad to record your design idea. You start from rough sketches and determine the dimensions later. In the sketch, dimensions are not important, lines need not be absolutely horizontal or vertical, and objects need not be joined precisely together at their ends. In fact, you should concentrate only on forms and shapes. You will focus on the dimensions at a later design stage. Figure 2–1 shows a rough sketch.

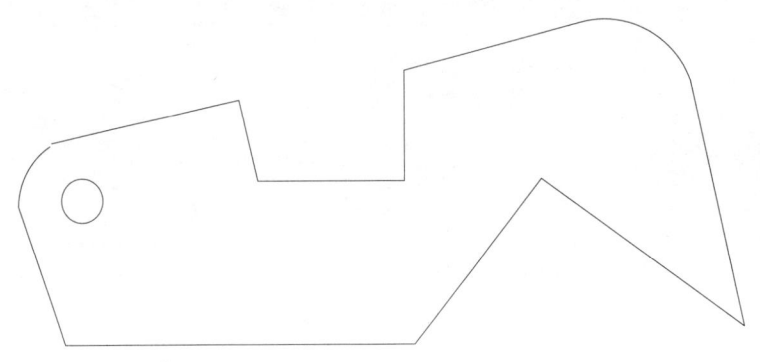

Figure 2–1 *Rough sketch*

Because Mechanical Desktop sits on top of AutoCAD, the sketch you construct, by default, is a non-parametric AutoCAD sketch. To convert it to a parametric sketch, let Mechanical Desktop resolve it. Depending on how you plan to use the parametric sketch subsequently, resolve it to one of various objects: profile, path, break line, cut line, or split line. Use of these objects is outlined below:

- Profile—a cross section for extruding, revolving, sweeping, and lofting
- Path—a guide rail for sweeping
- Break line—used to define a break out section drawing view
- Cut line—used to define a cutting plane for an offset or aligned section view
- Split line—used to split the face of a solid into two faces or split a solid part into two parts

Here in this chapter, you will learn how to resolve a sketch to a profile (see Figure 2–2) You will learn how to use break line and cut line in Chapter 10 and how to use split line in Chapter 9.

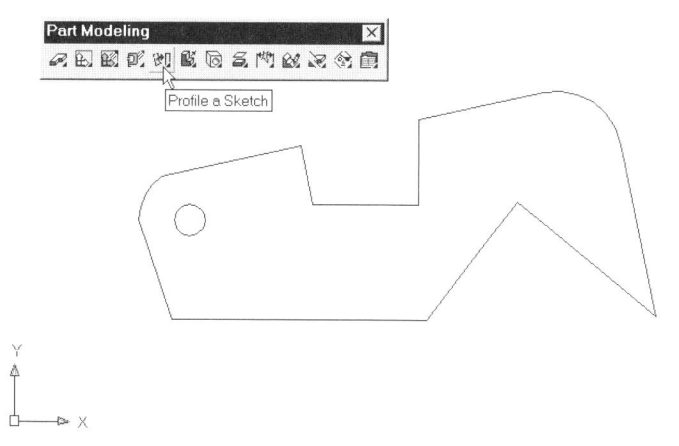

Figure 2–2 *Resolved sketch*

In the process of resolving a sketch, Mechanical Desktop applies two sets of rules below unless otherwise determined.

The first set of rules concerns geometric characteristics:

- Lines will become horizontal if they are nearly horizontal.
- Lines will become vertical if they are nearly vertical.
- Lines will become perpendicular if they are nearly perpendicular to each other.
- Lines will become parallel if they are nearly parallel to each other.
- Lines, arcs, and circles will become tangential if they are nearly tangential to each other.
- Lines will become collinear if they lie nearly on the same straight line.
- Arcs and circles become concentric if they are nearly concentric.
- Centers of arcs and circles will have the same X coordinate if their X coordinates are nearly the same.
- Centers of arcs and circles will have the same Y coordinate if their Y coordinates are nearly the same.
- End points of lines will have the same X coordinates if their X coordinates are nearly the same.
- End points of lines will have the same Y coordinates if their Y coordinates are nearly the same.
- Arcs and circles will have the same radius if they have nearly the same radius.
- Lines will have equal length if they have nearly the same length.

The second set of rules joins objects together or attaches objects to another object:

- Objects will join together at their end points if they are close enough.
- End points that are near to a line, an arc, or a circle are treated as attached to the line, arc, or circle.

Before you use a resolved sketch for any operation, you should fully constrain it. Otherwise, you may have unpredictable results later when you modify your solid part. Fully constraining a resolved sketch involves two stages.

MANIPULATING GEOMETRIC CONSTRAINTS

First, make necessary adjustments to the geometric constraints that Mechanical Desktop automatically applies to the sketch. Delete those constraints that you do not want, and add those constraints that Mechanical Desktop does not automatically apply for you. For example, delete the unwanted concentric constraints that Mechanical Desktop applied to two circles because you put their centers so close together, and add a horizontal constraint to a line that Mechanical Desktop does not apply for you because the line deviates largely from being horizontal. The symbols in Figure 2–3 show the geometric constraints applied to the sketch.

ADDING PARAMETRIC DIMENSIONS

Besides manipulating geometric constraints, you need to add parametric dimensions to complete the constraint requirement. To add dimensions, you may apply a numeric value or an expression as an equation. For example, you may set the width of a rectangle to be in a certain proportion to its length. You also establish a set of parameters that control the dimensions of a number of solid parts. Figure 2–4 shows the parametric dimensions applied to the sketch.

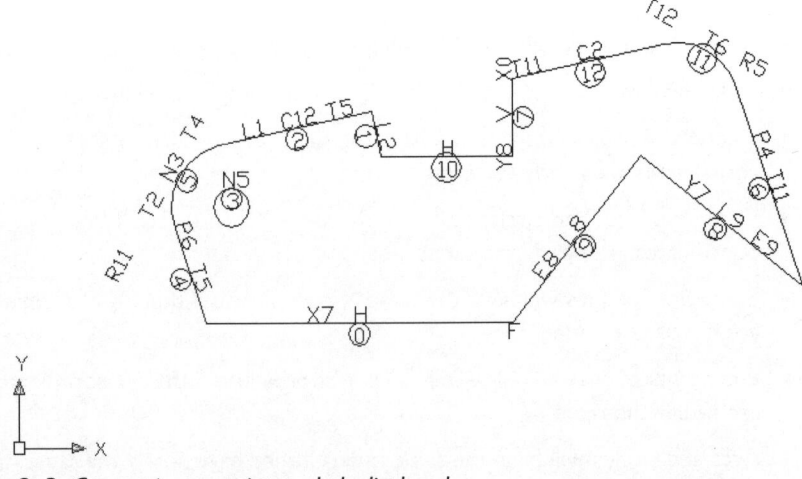

Figure 2–3 *Geometric constraint symbols displayed*

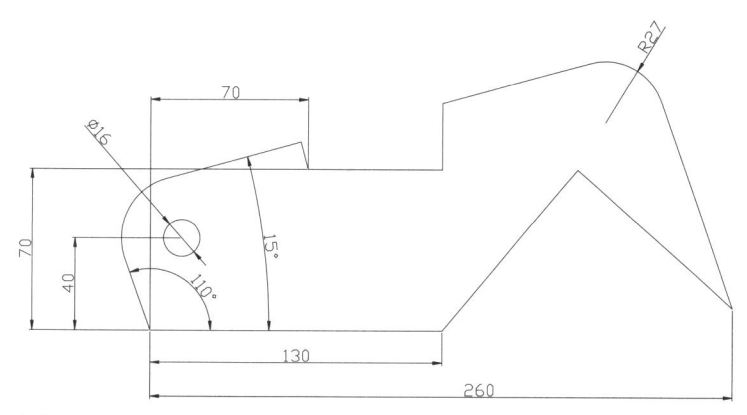

Figure 2–4 *Parametric dimensions added*

OPTIONS

Having explained earlier that sketches you construct need not be precise at all because Mechanical Desktop will apply geometric constraints to your sketch automatically, there are yet two questions to answer:

How rough can a sketch be?

How does Mechanical Desktop interpret your sketch?

DISTANCE AND ANGULAR TOLERANCES

To control how rough a sketch can be, you set distance and angular tolerances by using two system variables, PICKBOX and AMSKANGTOL.

Distance Tolerance

The PICKBOX variable determines the linear distance in terms of screen pixel size. If the pickbox size is set to five pixels, two end points will join together if you zoom the screen display such that the gap between them is smaller than five screen pixels. In other words, a gap that appears to be larger if you zoom close enough will not join together simply because the gap is larger than the pickbox size. At the other end of the spectrum, if you zoom very far away, a number of vertices that are close together within the specified pickbox size will join together as a single vertex. As a result, you may lose some line segments. Therefore, you should set the pickbox neither too large nor too small.

Now you will set the pickbox size.

    1. Start a new part file. Start from scratch and use Metric default.

    2. Select Assist>Options and select the Selection tab (see Figure 2–5.)

       Command: OPTIONS

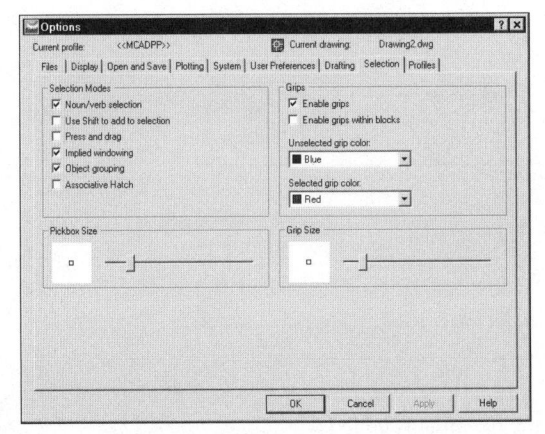

Figure 2–5 *Selection tab of the Options dialog box*

3. In the Selection tab, move the slider bar in the Pickbox Size area to manipulate the size of the pickbox.
4. Select the OK button.

Angular Tolerance

The AMSKANGTOL variable determines the angular distance. For example, if AMSKANGTOL is set to 4, lines inclined at less than 4 degree to the horizontal will be treated as a horizontal line. Now you will set these two system variables.

5. Select Part>Options or the Options button from the Part Modeling toolbar.

   Command: AMOPTIONS

6. In the Part tab of the Mechanical Desktop Options dialog box, set angular tolerance to 4 degree.
7. Do not close the dialog box.

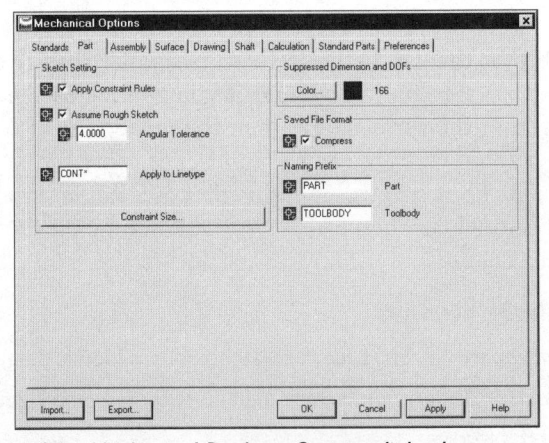

Figure 2–6 *Part tab of the Mechanical Desktop Options dialog box*

APPLY CONSTRAINT RULES AND ASSUME ROUGH SKETCH

To control how Mechanical Desktop interprets your sketch, you manipulate two system variables, AMRULEMODE and AMSKMODE.

As explained earlier, Mechanical Desktop applies two sets of rules on the sketch while resolving. However, there are times when you do not want to apply the first set of geometric characteristics to your sketch. Instead you want to attach the end points only. In this case, use the AMRULEMODE variable.

Sometimes you may want to resolve a line as inclined precisely to 3 degrees, despite setting the angular tolerance to 4 degrees. To override the angular tolerance and treat the sketch as precise, manipulate the AMSKMODE variable.

Now you will learn how to set the AMRULEMODE and AMSKMODE variables.

8. In the Part tab of the Mechanical Desktop Options dialog box, make sure that the Apply Constraint Rules (AMRULEMODE) and the Assume Rough Sketch (AMSKMODE) boxes are checked (see Figure 2–6)

OTHER PART MODELING OPTIONS

Apart from Apply Constraint Rules, Assume Rough Sketch, and Angular Tolerance, there are other settings in the Parts tab of the Mechanical Options dialog box:

Apply to Linetype	The linetype you select here is the linetype that will be used in the parametric sketch for making the solid feature. Any other kinds of linetype included in the sketch will be regarded as construction objects.
Constraint Size	This button enables you to change the height of the constraint symbols while displaying, editing, or deleting them.
Suppressed Dimensions and DOFs	You can change the color of dimensions that are referencing suppressed features. Otherwise the dimensions will take on the color properties of the current layer. This setting also enables you to control the color of the degree of freedom (DOF) symbol.
Save File Format	If the Compress box is checked, the model will be saved in a compressed file. A compressed file takes less memory space to store. However, it will take you some time to uncompress the file the next time you open it.
Naming Prefix	This area sets the default name prefix for the parts and toolbodies in a file.

9. Select the OK button.

CONSTRUCTING A PARAMETRIC SKETCH

There are four basic steps to constructing a parametric sketch: sketching, resolving, manipulating geometric constraints, and manipulating parametric dimensions.

SKETCHING

Before you construct a sketch, you need to establish a sketch plane on which to perform sketching. The sketch plane for the first feature of the solid part can simply be the XY plane of the current UCS. For subsequent sketches of a solid, you need to specify a sketch plane explicitly. You will learn more about sketch plane and related construction features later in this book.

To compose a sketch, use AutoCAD entities such as lines, polylines, arcs, circles, ellipses, and splines. Therefore you need to be familiar with using AutoCAD. However, you do not need to be proficient in AutoCAD because you do not have to construct precise sketches. You simply use AutoCAD entities to build up a rough sketch.

To give you a better perception of the size of the sketch that you are going to construct, zoom your screen display to an appropriate known size.

10. Select Assist>Format>Drawing Limits.

    Command: LIMITS

11. Type 0,0 at the command line area to specify the location of the lower-left corner.

12. Type 100,80 at the command line area to specify the location of the upper-right corner.

13. Select View>Zoom>All.

    Command: ZOOM

Your screen now displays an area of approximately 100 units times 80 units. Now you will construct a sketch.

14. With reference to Figure 2–7, construct a sketch.

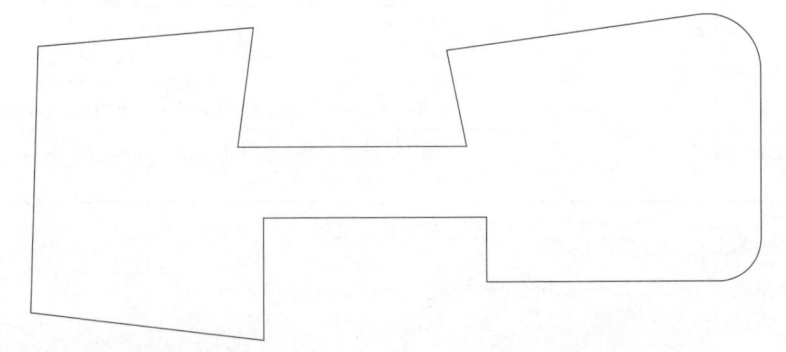

Figure 2–7 *Sketch constructed*

RESOLVING A SKETCH

By default, the sketch you construct is an AutoCAD sketch and it is non-parametric. To convert it to a parametric sketch, you resolve it. Depending on how you use the resolved sketch for subsequent operations, you resolve it into various kinds of objects. Now you will resolve your sketch to a profile.

To resolve a sketch to a profile, select either Single Profile (if there is only a single object: a circle, ellipse, or polyline) or the Profile from the Sketch Solving cascading menu of the Part pull-down menu (if there is more than one object).

15. Select Part>Sketch Solving>Profile or the Profile button from the Part toolbar (see Figure 2–8)

    Command: AMPROFILE

16. Select the lines and arcs and press the ENTER key.

Alternatively, steps 15 and 16 can be replaced by a single step below if the sketch is a single object, like a rectangle, circle, or polyline.

Select Part>Sketch Solving>Single Profile

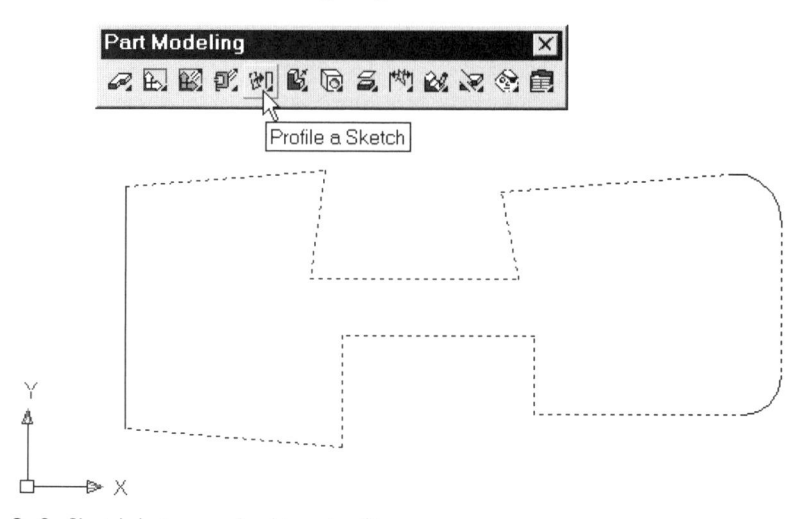

Figure 2–8 *Sketch being resolved to a profile*

After you resolve a sketch, you will find the prompt "Solved under constrained sketch requiring ?? dimensions or constraints" at the command line area.

The "??" is a numeric figure depicting the number of parametric dimensions or geometric constraints that you need to add to the resolved sketch in order to fully constrain it. Because the rough sketch that is shown in Figure 2–7 is not exactly the same as your sketch, the geometric constraints that Mechanical Desktop automatically applies to your sketch will be different. Consequently, the number of parametric dimensions or geometric constraints that you need to add will be different.

MANIPULATING GEOMETRIC CONSTRAINTS

To properly constrain a resolved sketch geometrically, you need to remove those constraints that you do not want and add those constraints that Mechanical Desktop does not apply for you. Before you remove and add constraints, you need to find out what constraints are already added.

Display Constraint Symbols

Now you will find out what geometric constraints are applied to your sketch. As we have said, the resolved sketch shown here may differ from yours. Therefore the kinds of constraints applied may be different. However, you will find a point on the sketch that is set as the fixed point. As the name implies, this point will not move.

17. Select Part>2D Constraints>Show Constraints or the Show Constraints button from the Part Modeling toolbar.

    Command: AMSHOWCON

18. Type A at the command line area to display all the constraints (see Figure 2–9)

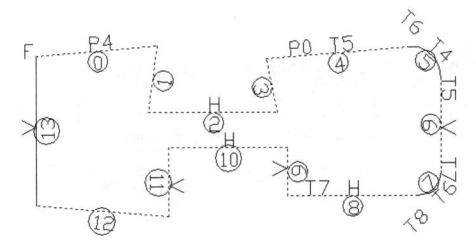

Figure 2–9 *Constraint symbols displayed*

In Figure 2–9, you see numbers in circles and alphabetic symbols. The numbers denote the entity name. The letters denote the geometric constraints. For example, T1 means that the entity is tangential to number 1 entity. The meanings of all the alphabetic constraint symbols are listed below:

H	Horizontal constraint
V	Vertical constraint
L	Perpendicular constraint
P	Parallelism constraint
T	Tangential constraint
C	Collinear constraint
N	Concentric constraint
J	Projected constraint
X	X-values constraints for center points and end points
Y	Y-values constraints for center points and end points
R	Same-radius constraint
E	Equal length constraint
M	Mirror constraint
F	Fixed point

Constraint Symbol Display Size

The display size of the constraint symbol is governed by the setting you made in the Part tab of the Mechanical Desktop Options dialog box.

19. Select Part>Part Options
20. In the Part tab of the Mechanical Desktop Options dialog box, select the Constraint Size button.
21. In the Constraint Display Size dialog box, move the slider bar to manipulate the display size, select Apply and Close, and then the OK buttons.

Delete and Add Geometric Constraints

Now let us suppose that we have the following requirements:

Lines 0, 2, 4, 8, 10, and 12 (Figure 2–9)	horizontal
Lines 1, 3, 6, 9, 11, and 13 (Figure 2–9)	vertical
Lines 3 and 6 (Figure 2–9)	collinear
Lines 1 and 11 (Figure 2–9)	not collinear
Lines 0 and 4 (Figure 2–9)	not collinear
Lines 8 and 12 (Figure 2–9)	not collinear
Arc 5 and line 4 (Figure 2–9)	tangent
Arc 5 and line 6 (Figure 2–9)	tangent
Arc 7 and line 6 (Figure 2–9)	tangent
Arc 7 and line 8 (Figure 2–9)	tangent
Arcs 5 and 7 (Figure 2–9)	equal radii

With reference to the table above, delete the unwanted constraints.

22. Select Part>2D Constraints>Delete Constraints or the Delete Constraint button from the Part Modeling toolbar.
    Command: AMDELCON
23. Select the constraints not listed on the above table and press the ENTER key.

To fulfill the constraint requirement, add constraints.

24. Select Part>2D Constraints and then select the appropriate constraint.
    Command: AMADDCON
25. Select the entity to apply the constraint.

Depending on whether the constraint has already been applied, you get one of the following prompts:

```
This constraint already exists. Enter an option.
Solved under constrained sketch requiring ?? dimensions or
   constraints.
```

```
This constraint cannot be added. Existing dimensions, con-
straints, or a fix constraint prevent the constraint from
being applied.
```

The "??" in the second prompt is a number that decreases by 1 each time a constraint is added. In accordance with the requirements mentioned earlier, continue to constrain the sketch. Figure 2–10 shows the sketched properly constrained.

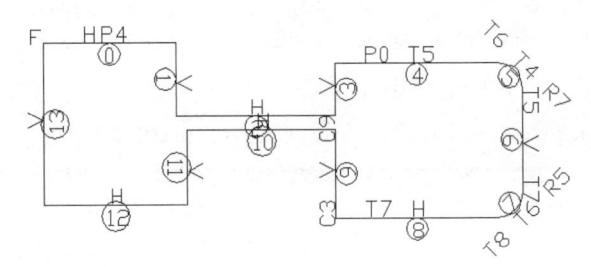

Figure 2–10 *Unwanted constraints removed and appropriate constraints added*

MANIPULATING PARAMETRIC DIMENSIONS

The final step in constructing a parametric sketch is to add parametric dimensions to it. Before you add parametric dimensions to a resolved sketch, you set the general dimension style and the parametric dimension display method.

Dimension Style

General dimension style concerns dimension formats such as dimension text height, arrow size, and so forth. Now you will set the general dimension style and select display method.

26. Select Annotate>Edit Dimensions>Dimension Style (See Figure 2–11.)

    Command: DDIM

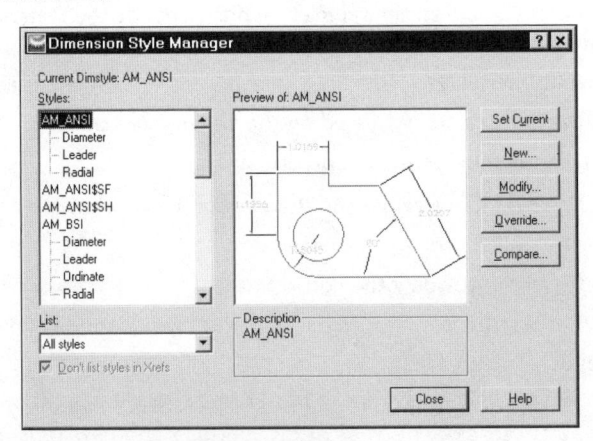

Figure 2–11 *Dimension Style Manager dialog box*

27. In the Dimension Style Manager dialog box, select the Modify button, make necessary adjustment to the dimension style settings, and select the OK buttons to exit.

Dimension Display Method

The parametric dimension display method concerns how the parametric dimensions are displayed. There are three display methods: numeric display, parameter display, and equation display. Numeric display will tell you the exact dimension value. Parameter display will tell you the parameter name assigned to the dimension. The first dimension that you add in a drawing is called d0, the second d1, and so on. Depending on the sequence of dimensioning, the parameter names that are displayed on your screen may not be the same as those shown in the illustration in this book. Therefore, you may want to write down the parameter names of your drawing on a separate piece of paper while you are working, if they are different from those in the illustration. In the third type of display, the equation display, the dimension value is expressed as an equation with the parameter name equal to an expression. Now you will select a dimension display method.

28. Select Part>Dimensioning>Dimensions As Equation to use equation display.

    Command: AMDIMDSP

Applying Parametric Dimensions

Now you will add dimensions to the resolved sketch. There are two commands: AMPARDIM and AMPOWERDIM. Both commands are similar, but the AMPOWERDIM command has additional capabilities to enable you to specify locations of extension lines. Figure 2–12 shows the dimensions that you need to add.

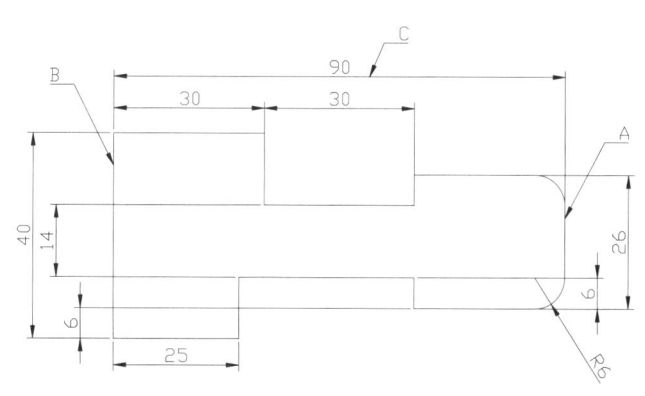

Figure 2–12 *Dimensions to be added*

Let us start with the outermost dimension.

29. Select Part>Dimensioning>New Dimension

    Command: AMPARDIM

30. Select A (Figure 2–12) to specify the first object.
31. Select B (Figure 2–12) to specify the second object.
32. Select C (Figure 2–12) to specify the location of the dimension.

After you select the objects and the dimension placement position, the measured dimension value will be displayed at the command line area. The dimension value "??" shown below is the exact size of the dimension.

```
Enter dimension value or [Undo/Hor/Ver/Align/Par/aNgle/Ord/
    Diameter/pLace] <??>:
```

To change it to 90 units as required, type **90** at the command line area. However, you should first check the measured value against the required value. If the measured value is much larger, say 200, and you type 90, the dimension will shrink from 200 units to 90 units. Because the other dimensions remain unchanged, the profile will be distorted and the result may resemble Figure 2–13. Therefore, do not type the value 90 if the measured value is much larger. Instead, type U to use the undo option.

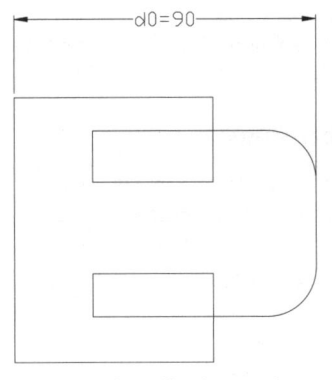

Figure 2–13 *Outer dimension shrunk and profile distorted*

On the other hand, if the measured dimension is less than 90 and you type in a value of 90, the outer dimension stretches. The result will resemble Figure 2–14. Here the shape is not changed. Therefore, you can continue to add the remaining dimensions.

Figure 2–14 *Outer dimension stretched, others unchanged*

As a rule of thumb, compare the measured dimension that is shown at the command line against the required dimension before inputting the required dimension value. If you start dimensioning the outermost dimension of the profile and the measured value is much larger than the required value, you should undo the command. Otherwise it may distort the profile and change the shape. To tackle sketches that are constructed much larger than required, you should start dimensioning the smallest dimension.

Continue to add dimensions until you get the prompt "Solved fully constrained sketch." To exit the command, press the ENTER key. Because your sequence of dimensioning may not be the same, the parameter names of the dimensions may not be the same as those shown in Figure 2–15.

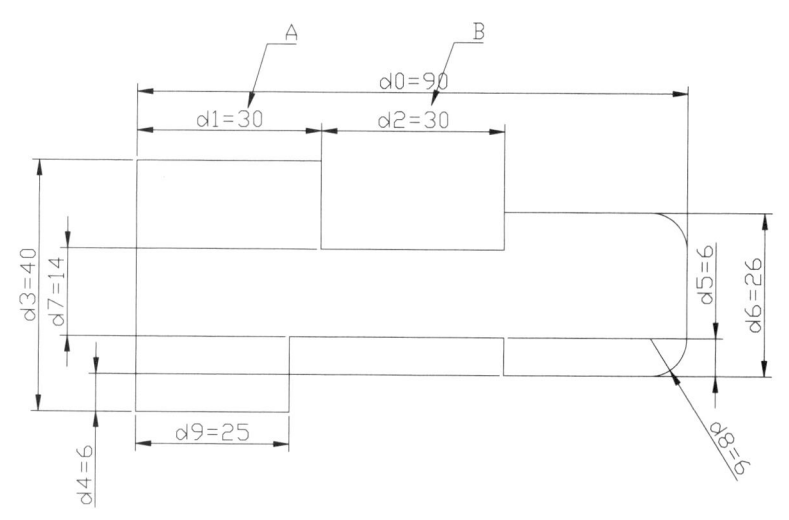

Figure 2–15 *Resolved sketch fully dimensioned*

As you can see, the dimensions A and B (Figure 2–15) are the same. Instead of specifying the same dimension value, you can set dimension A (Figure 2–15) equal to dimension B (Figure 2–15).

In Figure 2–15, the parameter name for dimension B is d2. However, the parameter name of dimension B in your drawing may not be d2; it can be d?, in which "?" is a number that depends on your dimensioning sequence. (Note that d0 is the first dimension in a drawing, d1 is the second dimension, and so forth.) You should replace d2, below, with the parametric name of dimension B in your drawing.

To modify a parametric dimension, you use either the AMMODDIM command or AMPOWEREDIT command. The AMMODDIM command enables you to modify the dimension value and the AMPOWEREDIT command enables you to edit various aspects of the dimension, including the dimension value, dimension format, dimension geometry, and units of measurement.

Now you will set dimension B equal to dimension A.

33. Select Part>Dimensioning>Edit Dimension
    Command: AMMODDIM
34. Select B.
35. Type **d1** at the command line area.

Dimension B is changed (see Figure 2–16). Now you will modify dimension A.

36. Select A (see Figure 2–16).
37. Type **20**.
38. Press the ENTER key to exit.

After you changed dimension A, dimension B changes as well because it is equal to dimension A (see Figure 2–17).

The parametric sketch is now complete. Save your file (file name: *Profile.dwg*).

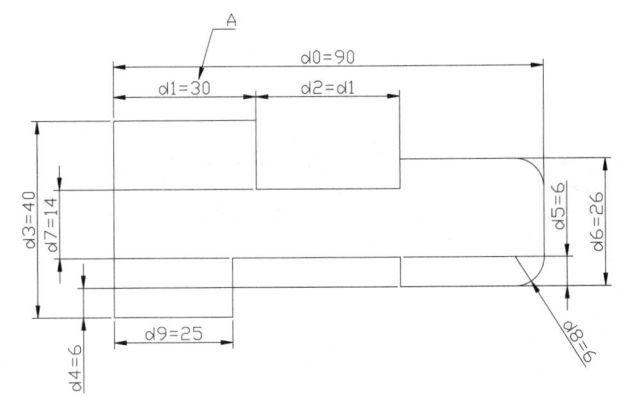

Figure 2–16 *Dimension d2 set equal to dimension d1*

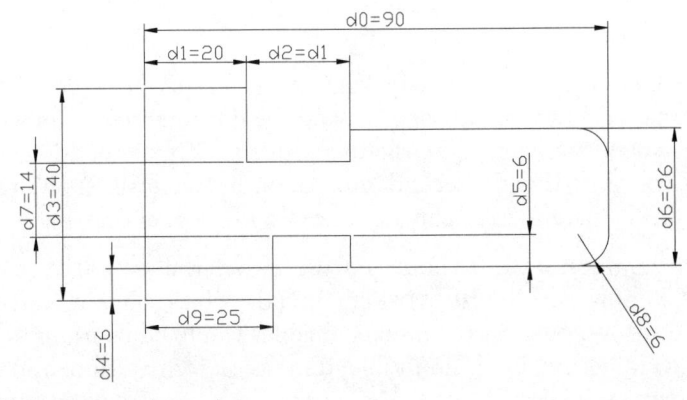

Figure 2–17 *Dimensions changed*

DESKTOP BROWSERS

Desktop browsers show the hierarchy of objects in a file. Sketches and features that you construct will display as objects in the hierarchy.

In addition to delineating a hierarchy of objects in the file, desktop browsers also provide command shortcuts and controls on the construction of parts, features, assemblies, and drawings. By right-clicking in the browser, you use the context-sensitive menu. By selecting different tabs of the browser, you switch among different working modes and activate the appropriate toolbars. Applicable commands are displayed and inappropriate commands are grayed out.

Now check your screen display. If the desktop browser is not shown in your screen, check View>Display>Desktop Browser.

TYPES OF BROWSERS

Because there are two kinds of files, there are two kinds of desktop browsers: the part drawing browser and the assembly drawing browser. Figure 2–18 shows the part file desktop browser (left) and the assembly file desktop browser (right).

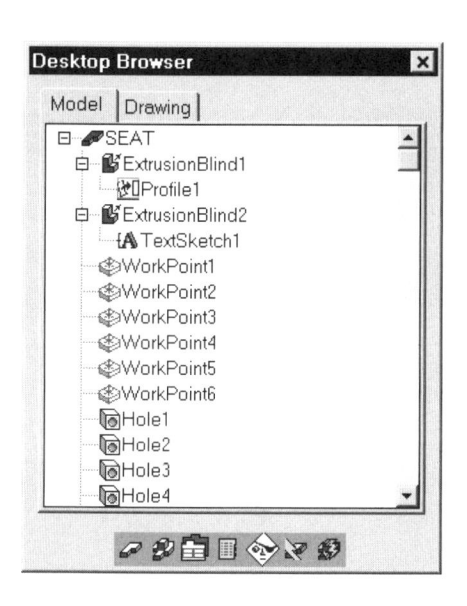

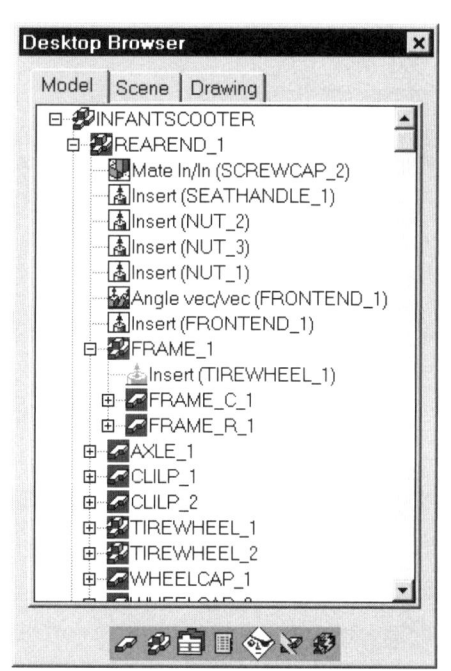

Figure 2–18 *Desktop browsers for a part file (left) and an assembly file (right)*

Part Browser

The part file browser has two tabs: Model and Drawing. The Model tab provides control for the construction of solid parts and features, while the Drawing tab provides control for the creation of drawings. The assembly drawing browser has three tabs: Model, Scene, and Drawing. The Scene tab is used for setting up assembly scenes. (The Scene tab will be discussed further in Chapter 6, and the Drawing tab will be discussed in greater detail in Chapter 10.)

At the bottom of the desktop browser, there are seven buttons (see the table below).

Part Filter toggle button	controls the display of parts and features in the browser
Assembly Filter toggle button	controls the display of assembly and constraints
Options button	displays the Mechanical Desktop dialog box
Catalog button	displays the Catalog dialog box
Visibility button	displays the Desktop Visibility dialog box
Update Part button	updates the solid part
Update Assembly button	updates the assembly

Minimizing the Browser

With the desktop browser in action, some working space on your screen is occupied. To maximize the working space, you can either close or collapse the browser. To close the browser, you select the [x] (close) button at the upper-right corner of the browser. To collapse the browser, you drag the browser to the central part of your screen, place the cursor on the frame of the browser, and double-click (see Figure 2–19). When the browser is collapsed, double-clicking while the cursor is on the browser frame expands the browser.

Figure 2–19 *Browser collapsed*

SKETCHING TECHNIQUES

To help establish a relationship among objects in a sketch, you use construction geometry. During the design process, we modify the sketch for various reasons. Therefore, you will add or remove entities from the resolved sketch. In a part file, you can construct multiple sketches before you use them for subsequent operations. Including text objects in a solid part is a common practice in design. To construct text entities, use textual sketch.

CONSTRUCTION GEOMETRY

Construction geometry consists of sketch entities included in the parametric sketch. You use construction entities to help apply geometric constraints and parametric dimensions, but the entities will not be used in subsequent operations. For example, you use a construction line to define an axis of symmetry or use a construction line to help align end points of lines. When you extrude the sketch, the construction geometry is not extruded. To differentiate construction geometry from the other entities, use a different linetype.

Now you will construct the sketch for making an extruded solid feature of the model shown in Figure 2–20.

1. Start a new part file from scratch. Use metric as the default.
2. Set the drawing limit to 1,000 units times 800 units and zoom the display to extent.

   Command: LIMITS

   Command: ZOOM
3. With reference to Figure 2–21, construct a set of line segments. Entity 2 is a hidden line. It will be treated as a construction line after the sketch is resolved.
4. Resolve the sketch (including the hidden line) to a profile.
5. Add horizontal, vertical, and mirror geometric constraints in accordance with Figure 2–21.

Figure 2–20 *Seat of the infant scooter*

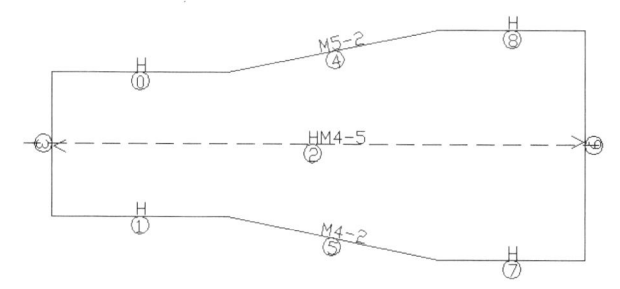

Figure 2–21 *Sketch resolved and geometric constraints applied*

6. Add parametric dimensions in accordance with Figure 2–22.

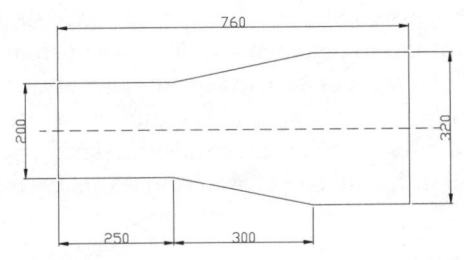

Figure 2–22 *Parametric dimensions added*

The sketch for a sketched solid feature of the seat of an infant scooter is complete. Save your file (file name: *Seat.dwg*).

Now you will construct the sketch for making the tire of the infant scooter. Figure 2–23 shows the completed model.

1. Start a new part file from scratch. Use metric as the default.
2. With reference to Figure 2–24, construct a sketch. Entities 0, 1, and 2 are hidden lines and will be used as construction lines.

Figure 2–23 *Tire of the infant scooter*

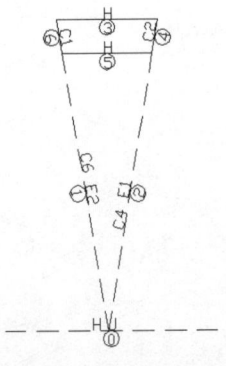

Figure 2–24 *Sketch resolved and geometric constraint applied*

3. Resolve the sketch to a profile and add geometric constraints.
4. Add parametric dimensions in accordance with Figure 2–25.

The sketch is complete. Save your file (file name: *Tire.dwg*).

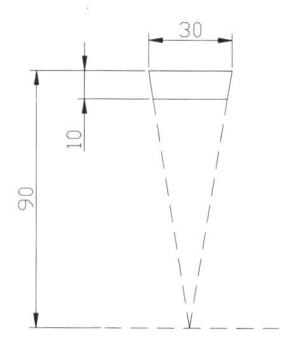

Figure 2–25 *Parametric dimensions added*

APPEND AND RE-SOLVE

To modify the shape or size of a parametric sketch, you manipulate its geometric constraints and parametric dimensions. If you wish to add additional AutoCAD entities to the parametric sketch, construct the entities and append them to the parametric sketch. If you wish to remove objects from the parametric sketch, delete the objects from the sketch and make any necessary modifications. After modifying, you re-solve the sketch.

Now you will modify the resolved sketch of the tire of the infant scooter.

1. Open the file *Tire.dwg* if you already closed it.
2. Delete a line and add an arc in accordance with Figure 2–26.

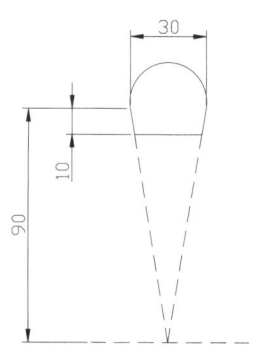

Figure 2–26 *Line deleted and arc constructed*

3. Select Part>Sketch Solving>Append or select the sketch in the Browser, right-click, and select Append.
4. Select the arc and press the ENTER key.

5. Select Part>Sketch Solving>Re-solve or select the sketch in the Browser, right-click, and select Re-solve.
6. With reference to Figure 2–27, add geometric constraints (Tangent) to the arc.

The modification is complete. Save your file.

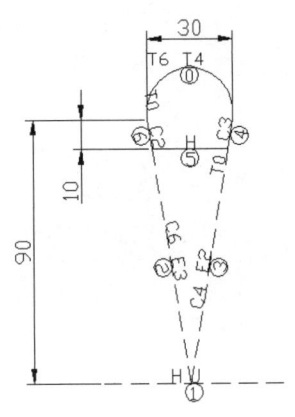

Figure 2–27 *Geometric constraints and parametric dimension added*

MULTIPLE SKETCHES

Using a resolved sketch, you construct various kinds of sketched solid features. After you use a sketch in an operation, the sketch is said to be consumed. To prepare for several operations, you may construct multiple sketches.

Now you will construct a second sketch in the file *Tire.dwg*.

1. Open the file *Tire.dwg*, if you had already closed it.
2. With reference to Figure 2–28, construct three lines and an arc, resolve the lines and arc to a second profile, and add geometric constraints.

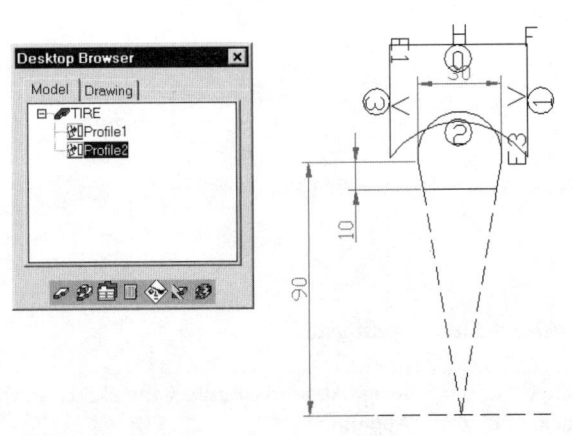

Figure 2–28 *Second profile resolved*

Note that there should be two profile objects in the browser.

3. Add parametric dimensions in accordance with Figure 2–29.

The second sketch is complete. Save and close your file.

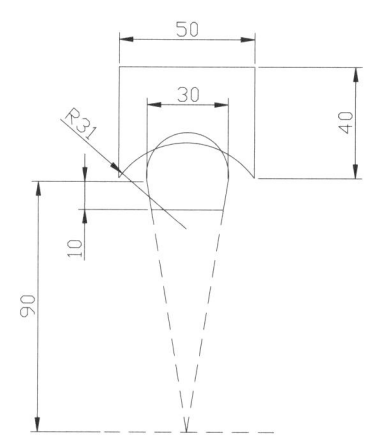

Figure 2–29 *Parametric dimensions added*

TEXT SKETCH

A text object is a special kind of sketch. You use it to construct solid parts that resemble the shape of a text string.

Now you will construct a text sketch in the file *Seat.dwg*.

1. Open the file *Seat.dwg*, if you had already closed it.
2. Select Part>Sketch Solving>Text Sketch

    Command: AMTEXTSK
3. In the Text Sketch dialog box, select font type and style, type a text string, and select the OK button (see Figure 2–30).

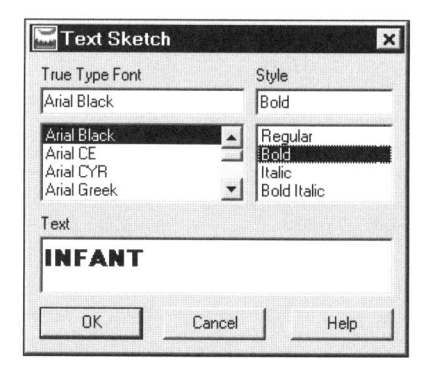

Figure 2–30 *Text Sketch dialog box*

4. With reference to Figure 2–31, select two points to indicate the location of the text sketch.
5. Double-click the dimension to modify the height.

The second sketch, a text sketch, is complete. Save and close your file.

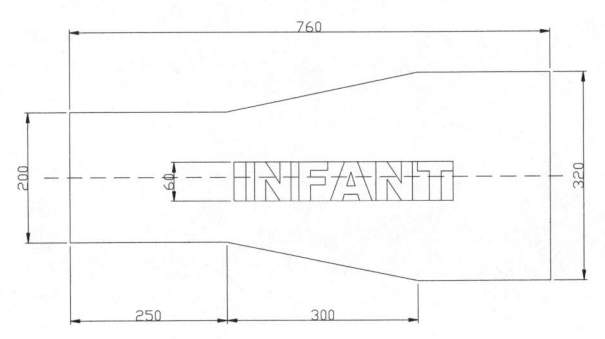

Figure 2–31 *Text sketch constructed*

OTHER SKETCHING TECHNIQUES

You need a sketch plane for each sketch. By default, the sketch plane lies on the XY plane of the world coordinate system. In addition to the default plane, you may use existing solid faces if you already constructed a solid. (Note that establishing a sketch plane in Mechanical Desktop is equivalent to setting up a User Coordinate System (UCS) in AutoCAD.

COPY SKETCH

If you have already constructed a sketch or a feature, you can construct a sketch by copying the sketch or the sketch of the feature. Re-using sketches you have already constructed saves you time in sketch construction.

COPY EDGE

Edges of solid features on the sketch plane can also be re-used. You copy the edges to become a portion of the new sketch.

You will learn how to establish a sketch plane on a face of a solid and copy a sketch from a feature in Chapter 3, and how to copy edges of a solid in Chapter 4.

SUMMARY

Construction of a parametric solid part starts from making a parametric sketch. Making a parametric sketch involves four basic steps. To construct a sketch, you use AutoCAD curve construction tools to construct a sketch on a sketch plane. Because the sketch is, by default, an AutoCAD non-parametric sketch, you let Mechanical Desktop resolve it to a parametric sketch. Depending on how you will use the parametric sketch in subsequent operations, you resolve the sketch to a profile, path, break line, cut line, or split line. (In this chapter, you learned how to resolve the sketch to a profile.) After resolving, you examine the geometric constraints and add or remove constraints as may be necessary to meet your design intent. Finally, you add parametric dimensions to control the size of the resolved sketch. While making a sketch, you may include construction geometry to help establish geometric constraints and parametric dimensions. Construction geometry of the parametric sketch will not be used in subsequent operations on the sketch. After you have constructed a parametric sketch, you can modify it by appending and resolving.

PROJECTS

To enhance your sketching skills, work on the following sketching projects. These sketches, together with the other sketches in this chapter, will be used in subsequent chapters for making the assembly of the infant scooter shown in Figure 2–32.

Figure 2–32 *Infant scooter*

Now set up a folder, *C:\Projects\InfantScooter*, in your computer and put all the files that you created in this chapter there.

STEERING SHAFT

Now you will construct the sketch for making an extruded solid feature of the steering shaft of the infant scooter. Figure 2–33 shows the completed model.

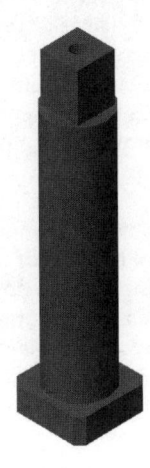

Figure 2–33 *Steering shaft*

1. Start a new part file from scratch using the metric default.
2. Construct a rectangle.
3. Resolve the rectangle to a profile.
4. With reference to Figure 2–34, add equal length, vertical, and horizontal geometric constraints to the profile and add a parametric dimension.

The sketch for the extruded solid feature is complete. Save and close your file (file name: *SteeringShaft.dwg*).

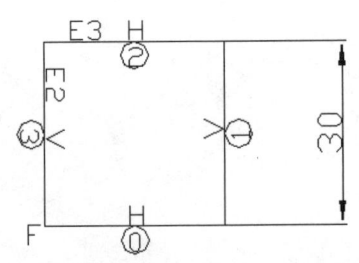

Figure 2–34 *Geometric constraints and parametric dimensions*

CLIP

Now you will construct the sketch for making the clip of the infant scooter shown in Figure 2–35.

1. Start a new part file from scratch using the metric default.
2. With reference to Figure 2–36, construct a sketch.

3. Resolve the sketch to a profile.
4. Add equal length and concentric geometric constraints and parametric dimensions to the profile, in accordance with Figure 2–37.

Figure 2–35 *Clip*

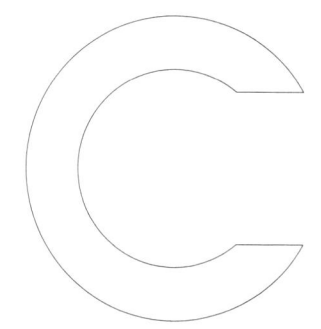

Figure 2–36 *Sketch*

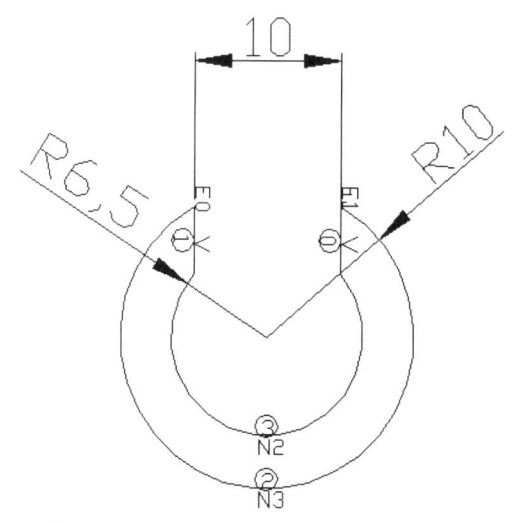

Figure 2–37 *Parametric dimensions and geometric constraints*

The sketch is complete. Save and close your file (file name: *Clip.dwg*).

AXLE
Now you will construct the sketch for making the axle shown in Figure 2–38.

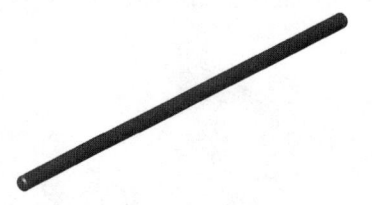

Figure 2–38 *Axle*

1. Start a new part file from scratch using the metric default.
2. With reference to the dimensions shown in Figure 2–39, construct a sketch.
3. Resolve the sketch to a profile.
4. Add collinear, equal length, horizontal, and vertical constraints and parametric dimensions to the profile.

The sketch is complete. Save and close your file (file name: *Axle.dwg*).

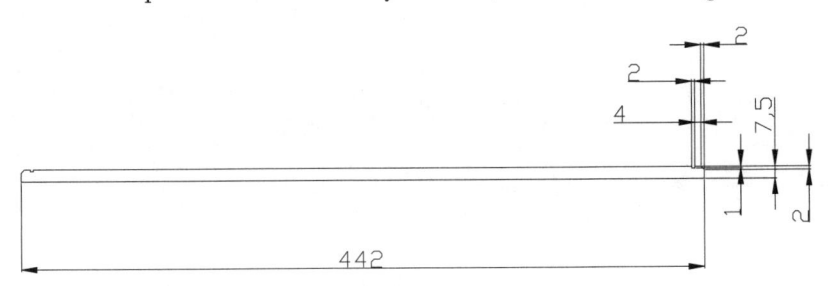

Figure 2–39 *Sketch*

SCREW CAP
Now you will construct the sketch for making the screw cap shown in Figure 2–40.

Figure 2–40 *Screw cap*

1. Start a new part file from scratch using the metric default.
2. With reference to the dimensions shown in Figure 2–41, construct a sketch.

3. Resolve the sketch to a profile.
4. Add horizontal and vertical geometric constraints and parametric dimensions.

The sketch is complete. Save and close your file (file name: *ScrewCap.dwg*).

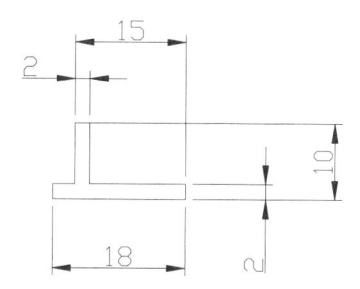

Figure 2–41 *Sketch*

WHEEL CAP

Now you will construct the sketch for making a wheel cap. Figure 2–42 shows the completed model.

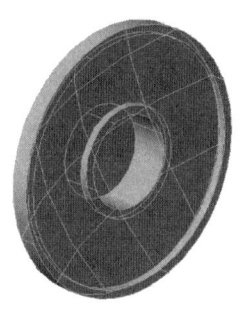

Figure 2–42 *Wheel cap*

1. Start a new part file from scratch using the metric default.
2. With reference to Figure 2–43, construct a sketch, resolve it to a profile, and add geometric constraints.

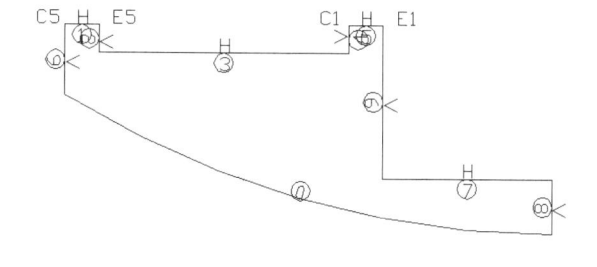

Figure 2–43 *Geometric constraints*

3. Add parametric dimensions in accordance with Figure 2–44. The zero size dimension specifies the horizontal distance between the center of the arc and the vertical line.

The sketch is complete. Save and close your file (file name: *WheelCap.dwg*).

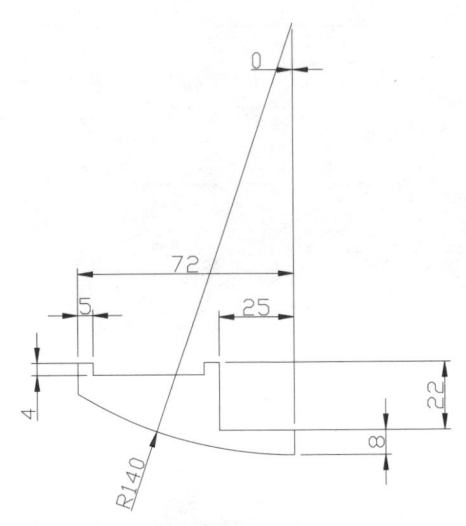

Figure 2–44 *Parametric dimensions*

NUT

Now you will construct the sketch for making the hexagonal body of a nut. The completed model is shown in Figure 2–45.

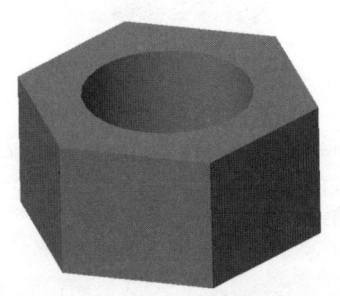

Figure 2–45 *Nut*

1. Start a new part file from scratch using the metric default.
2. With reference to Figure 2–46, construct a circle and a hexagon (a polygon with six sides).
3. Change the linetype of the circle to hidden so that it will be treated as construction geometry in a resolved profile.

4. Resolve the sketch to a profile.
5. Add a parametric dimension and geometric constraints to the profile in accordance with Figure 2–47.

The sketch is complete. Save and close your file (file name: *Nut.dwg*).

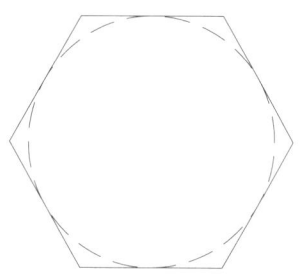

Figure 2–46 *Circle and polygon*

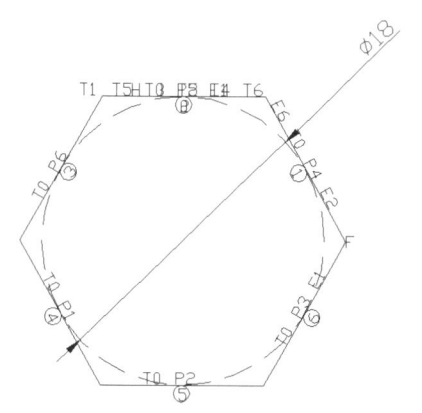

Figure 2–47 *Geometric constraints and parametric dimension*

WHEEL

Now you will construct the sketch for making the wheel shown in Figure 2–48.

Figure 2–48 *Wheel*

1. Start a new part file from scratch using the metric default.

2. With reference to Figure 2–49, construct a set of line segments (including two hidden lines to be used as construction geometry) and resolve the sketch to a profile. (The two construction lines serve the purpose of guiding the two inclined lines of the sketch.)
3. With reference to Figure 2–50, add equal length, collinear, vertical, and horizontal constraints to the profile.
4. Add parametric dimensions.

The sketch is complete. Save and close your file (file name: *Wheel.dwg*).

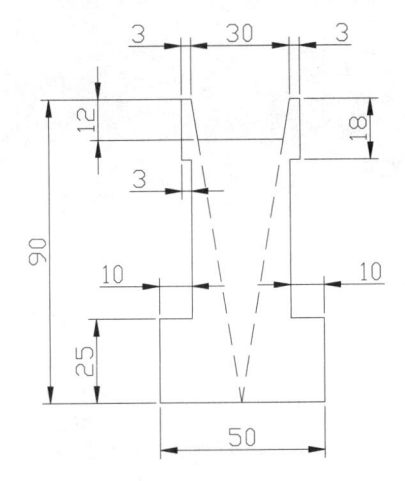

Figure 2–49 *Sketch*

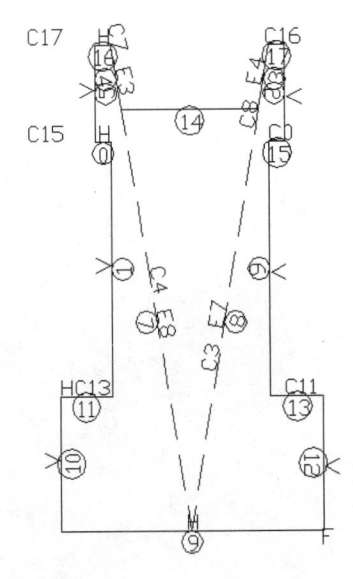

Figure 2–50 *Geometric constraints*

FRONTAL FRAME

Now you will construct the sketch for making an extruded solid feature of the frontal frame of the infant scooter. Figure 2–51 shows the completed model.

Figure 2–51 *Frontal frame*

1. Start a new part file from scratch using the metric default.
2. With reference to Figure 2–52, construct a sketch and resolve it to a profile.
3. Add geometric constraints in accordance with Figure 2–53.
4. Add parametric dimensions.

The sketch is complete. Save and close your file (file name: *Frame_F.dwg*).

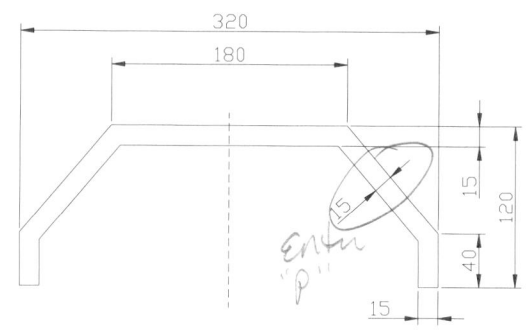

Figure 2–52 *Sketch*

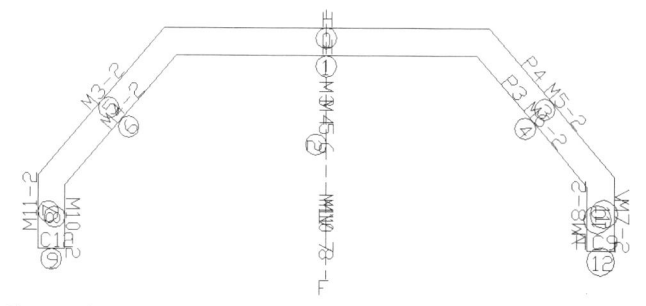

Figure 2–53 *Geometric constraints*

REVIEW QUESTIONS

1. List four items that can be set in the Part tab of the Mechanical Options dialog box that affect profiling a sketch.

2. When constructing a sketch for making a profile, there can be a gap in the sketch. How big can the gap be?

3. If a sketch is constructed to exact size and shape, does it need to be resolved?

4. Can a geometric constraint be removed? If so, how?

5. When an angled dimension is required in a profile, a horizontal dimension appears. How can this be changed to an angle dimension?

6. How many ways can you resolve a sketch? What are they?

7. Delineate the steps to construct a parametric sketch.

8. How would you remove and include entities to a parametric sketch?

9. Explain the meaning of construction geometry, multiple sketches, and text sketch.

10. A profile does not need to be fully constrained. True or false?

11. There can be multiple fix constraints on a sketch. True or false?

12. After a profile is fully constrained, dimension's value cannot be changed. True or false?

CHAPTER 3

Part Modeling I

OBJECTIVES

The goals of this chapter are to introduce the Mechanical Desktop visualization tools; to explain how to construct extruded, revolved, and bend features; to illustrate the ways to combine sketched solid features; and to depict various ways to modify a parametric solid and to evaluate mass properties. After studying this chapter, you should be able to do the following:

- Manipulate display visualization tools
- Construct extruded, revolved, and bend features
- Apply Boolean operations to combine sketched solid features
- Modify parametric solid parts
- Evaluate mass properties of a solid part
- Include attributes in a solid part

OVERVIEW

Before you use a sketch to make a three-dimensional solid feature, you need to know how to manipulate the screen display in order to gain a better perception of the objects you construct in 3D. A solid feature starts from a parametric sketch. Two basic ways to use a sketch are to extrude it in a direction perpendicular to the sketch and revolve it about an axis. If you already have a solid, you construct a simple line sketch and bend the solid along the line. The solid that results from bending is called a bend solid feature. To build a more complex solid object, you construct more sketches in various ways, construct additional sketched solid features, and combine the solid features. Continual improvement and change are necessary in the design development process. With a parametric solid, you can change the parameters any time you like and retrieve mass properties from the solid. To include textual information to the solid, you add attributes.

VISUALIZATION

The graphic window of the computer display can be regarded as a viewing camera through which you see the 3D virtual working space where you construct the 3D objects. By default, you view the working space in the top view direction. Thus, you see the XY plane lying parallel to the screen. To see the object from other directions, you orbit the display. A 3D solid can be displayed in either wireframe or shaded mode. The default display mode is wireframe. Tools that facilitate 3D visualization are available from the Mechanical View toolbar (see Figure 3–1).

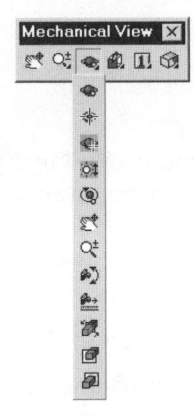

Figure 3–1 *Mechanical View toolbar*

ORBIT

Using the orbit tool, you view the 3D object from different angles.

1. Select the 3D Orbit button from the Mechanical View toolbar (see Figure 3–2).

Figure 3–2 *Orbiting*

You will find a large circle and four small circles at the quadrant position of the large circle.

2. Click anywhere, hold down the left mouse button, and drag. The display orbits.
3. Click one of the four small circles, hold down the left mouse button, and drag. The display orbits either vertically or horizontally.
4. Right-click and select Exit on the screen.

DISPLAY MODE

Because a solid has edges and surface data as well as volume information, you may view objects you have constructed in various kinds of display mode. Figure 3–3 shows wireframe, flat shaded, Gouraud shaded, flat-shaded (edges on), and Gouraud shaded (edges on) modes(from top left to lower right).

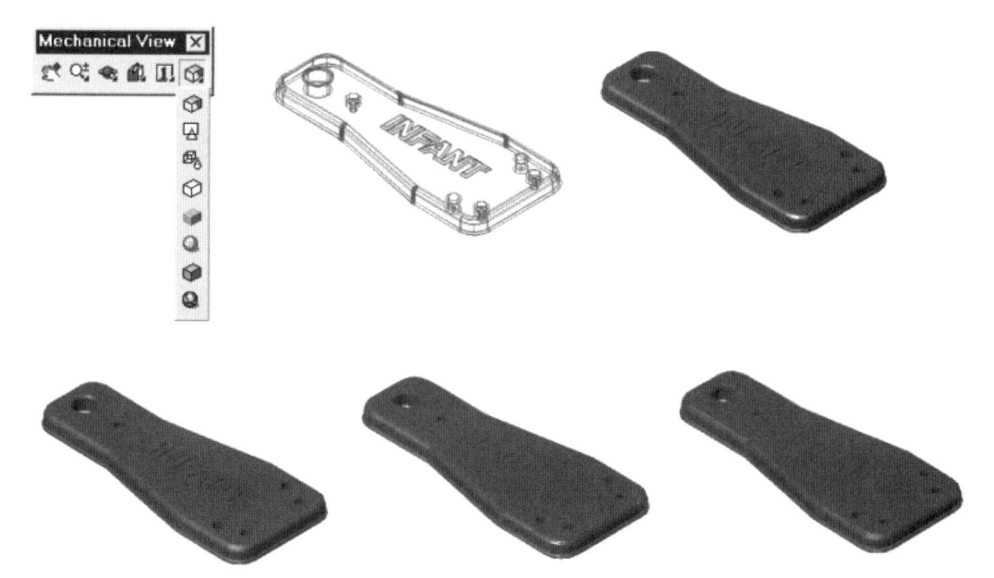

Figure 3–3 *Various display modes*

In general, shaded mode gives a better idea of the shape of the object. However, details behind the outer faces of the object are hidden. Therefore you may need to switch from wireframe mode to shaded mode from time to time.

LIGHTS

The color brightness of the faces of a model in shaded display mode is determined by three elements: intensity of ambient light, intensity of direct light, and location of the target point. To adjust these parameters, select the Lighting Control button from the Mechanical View toolbar.

Figure 3–4 (top) shows the effect of increasing the ambient light intensity and reducing the direct light intensity. Figure 3–4 (left) shows the effect of reducing ambient light intensity and increasing direct light intensity. Figure 3–4 (right) shows the effect of putting the target location on the top face of the model.

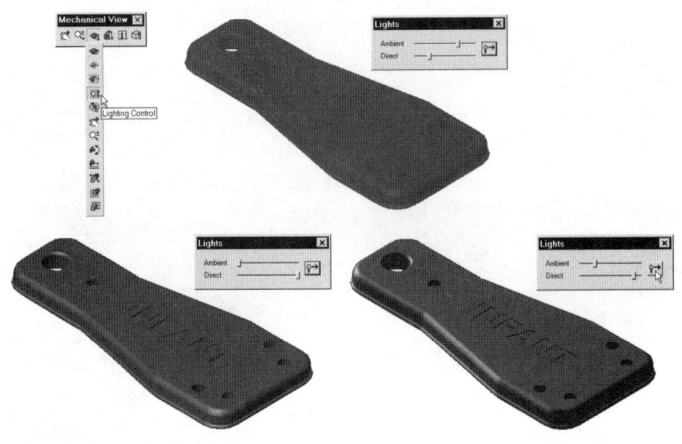

Figure 3–4 *Lighting control*

SKETCHED FEATURES: EXTRUDE, REVOLVE, AND BEND

Using the parametric feature-based approach, you can construct a 3D solid part by making individual solid features of the 3D object and combining the features to form the final complex object. The first step to making a 3D solid part is to construct parametric sketches. Using the sketches, you construct solid features. Because this kind of solid feature derives from sketches, we call them sketched solid features.

Using Mechanical Desktop, you can construct seven kinds of solid features from sketches: extruded, revolved, bend, rib, sweep, loft, and face split. Among them, extruded, revolved, sweep, and loft features are basic independent solid shapes; you can use them as individual solid objects. Bend, rib, and face split features are special kinds of sketched solid features that you build on existing solids; you bend a solid to form a bend feature, you construct a rib on an existing solid, and you split a face of an existing solid. Because sketches for making rib, sweep, and loft solid features require special construction tools and face split feature concerns when making face drafts, you will learn rib, sweep, and loft features in the following chapter, after learning how to use work features as construction tools and learning face split and face draft in Chapter 9. Now you will learn how to construct extruded, revolved, and bend solid features.

EXTRUDED SOLID FEATURE

To construct an extruded solid, you construct a sketch and extrude the sketch in a direction perpendicular to the plane of the sketch. You can extrude the first sketch in the solid part in either direction, from mid-plane, to a plane, or from a plane to another plane.

Now you will construct an extruded solid for making the main body of the seat of the infant scooter. Because there are two resolved profiles in this file, you need to select one of them to extrude, if you use the command from the Part pull-down menu. Alternatively, you may select the sketch to be extruded in the browser and use the right-click menu.

1. Open the file *Seat.dwg* that you constructed in Chapter 2.
2. Type **8** at the command line area to set the display to an isometric view.
3. Referring to Figure 3–5, hit the + sign to expand the browser, select the sketch, right-click, and select Extrude.

    Command: AMEXTRUDE

4. In the Extrusion dialog box (Figure 3–6), set the distance of extrusion to 40 units.
5. Accept the default draft angle of 0 degree.
6. Set termination type to blind.
7. Select the Flip button so that the direction of extrusion is in accordance with Figure 3–6.
8. Select the OK button.

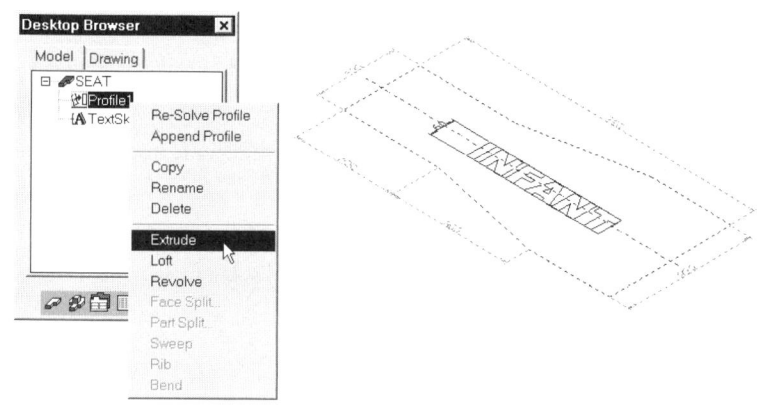

Figure 3–5 *Selecting Extrude from the right-click menu in the browser*

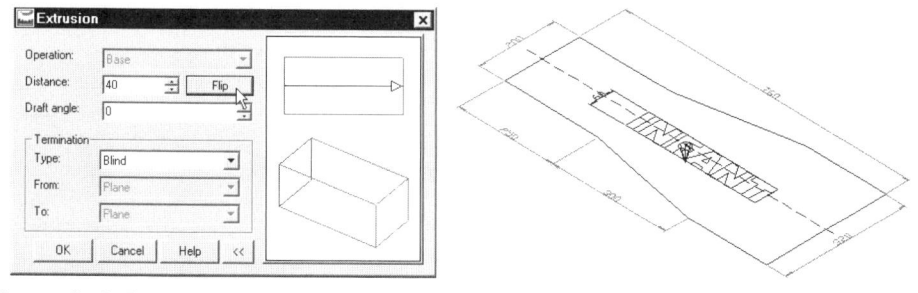

Figure 3–6 *Extrusion dialog box*

The selected profile is extruded to a solid feature (see Figure 3–7). Because this solid feature is the first solid feature in the solid part, it is called the base solid feature. Now save your file. Note that there is still an unconsumed sketch (Text sketch). You will extrude it later.

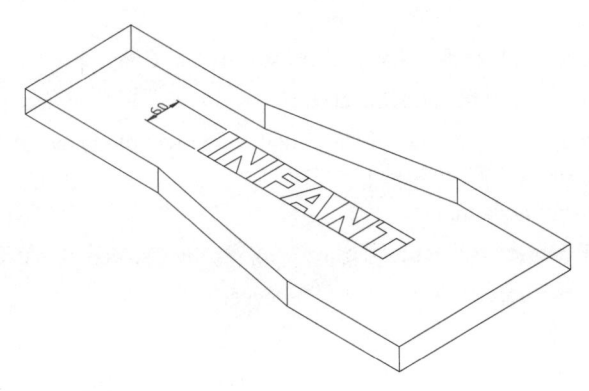

Figure 3–7 *Profile extruded*

REVOLVED SOLID FEATURE

To construct a revolved solid, you revolve a sketch about an axis. You can revolve the first sketch of the solid part in either direction, from mid-plane, to a plane, or from a plane to another plane.

Now you will construct a revolved solid for making the main body of the tire of the infant scooter.

1. Open the file *Tire.dwg* that you constructed in Chapter 2.
2. Type **8** at the command line area to set the display to isometric view.

There are two sketches in this file. You will revolve one of them.

3. Select the first profile in the browser, right-click, and select Revolve (see Figure 3–8).

   Command: AMREVOLVE

4. Select the construction line (hidden line) indicated in Figure 3–9 as the revolution axis.
5. In the Revolution dialog box, accept the default if the angle of revolution is 360 degrees, and select the OK button.

A revolved solid feature is constructed from the selected profile (see Figure 3–10). Save and close your file. Note that there is an unconsumed sketch in the file.

Part Modeling I 67

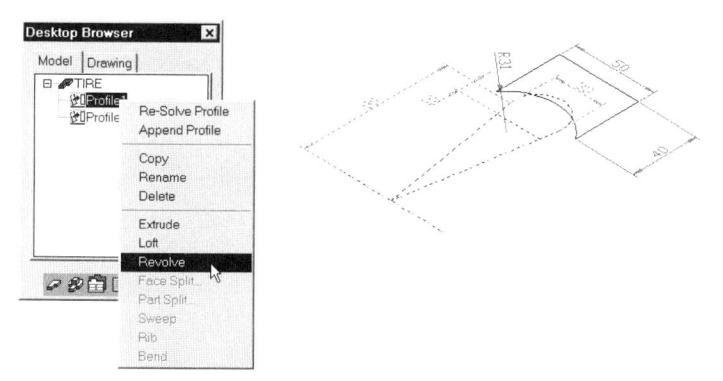

Figure 3-8 *Selecting Revolve from the right-click menu in the browser*

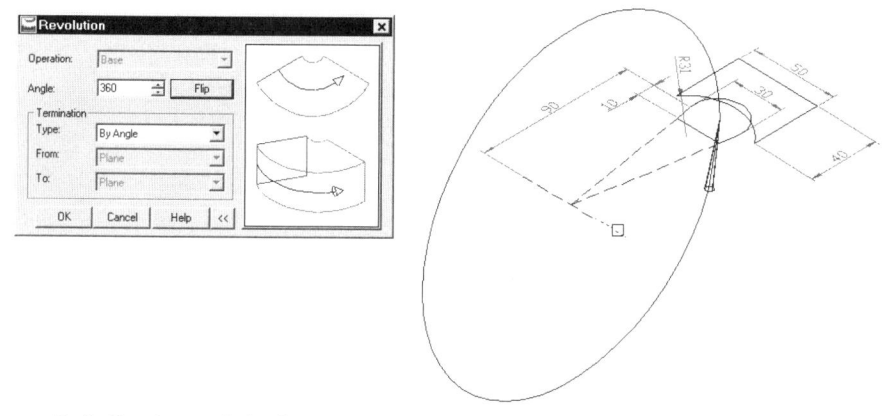

Figure 3-9 *Revolution dialog box*

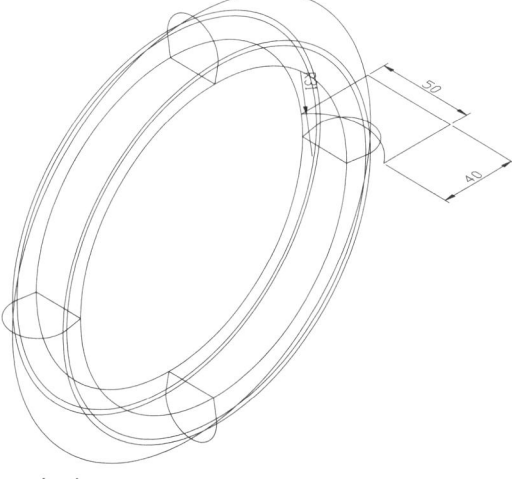

Figure 3-10 *Profile revolved*

BEND FEATURE

As the name implies, a bend feature bends a solid around a folding line. You construct a parametric folding line on a face of the solid and bend the solid about the line.

Now you will use the bend feature to construct the main body of the rear frame of the infant scooter. Figure 3–11 shows the completed model. It is an extruded solid and the solid is bent along two profile lines.

Figure 3–11 *Rear frame*

1. Start a new part file. Use metric as the default.
2. Referring to Figure 3–12, construct a rectangle, resolve it to a profile, and add parametric dimensions.
3. Set the display to an isometric view and extrude the profile to an extruded solid for a distance of 20 mm.

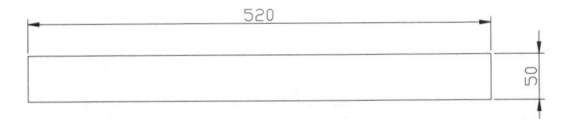

Figure 3–12 *Sketch*

4. Referring to Figure 3–13, construct a line.
5. Select Part>Sketch Solving>Single Profile.
6. Press the ENTER key to resolve the line to an open profile.
7. Add a vertical constraint and a parametric dimension.
8. Select the open profile from the browser, right-click, and select Bend (see Figure 3–14).
9. In the Bend dialog box, select a combination of angle+radius, set radius to 10 mm and angle to 90 degree, select the Flip Direction button and the Flip Bend Side button, and select the OK button.

Part Modeling I 69

The solid is bent along the profile (see Figure 3–15).

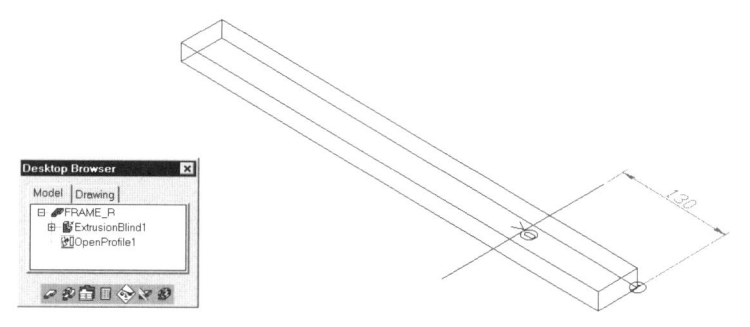

Figure 3–13 *Open profile constructed*

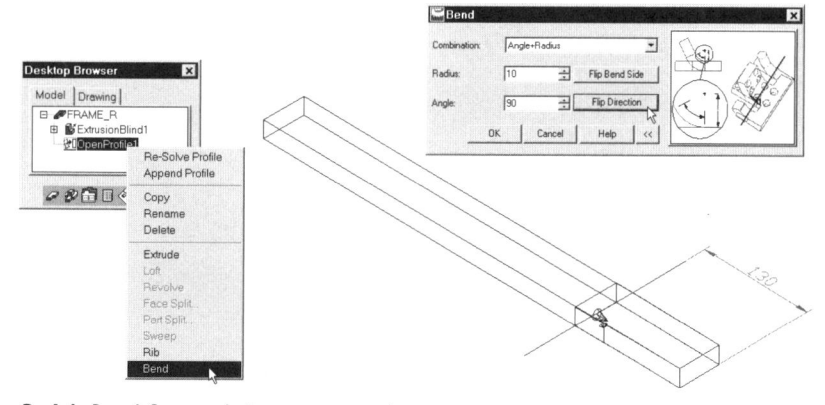

Figure 3–14 *Bend feature being constructed*

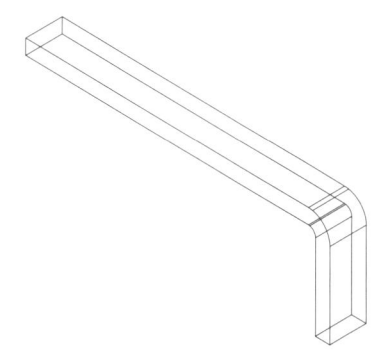

Figure 3–15 *Bend feature constructed*

   10. Refer to Figures 3–16 and 3–17 to construct another bend feature.

Two bend features are complete. Save and close your file. (file name: *Frame_R.dwg*)

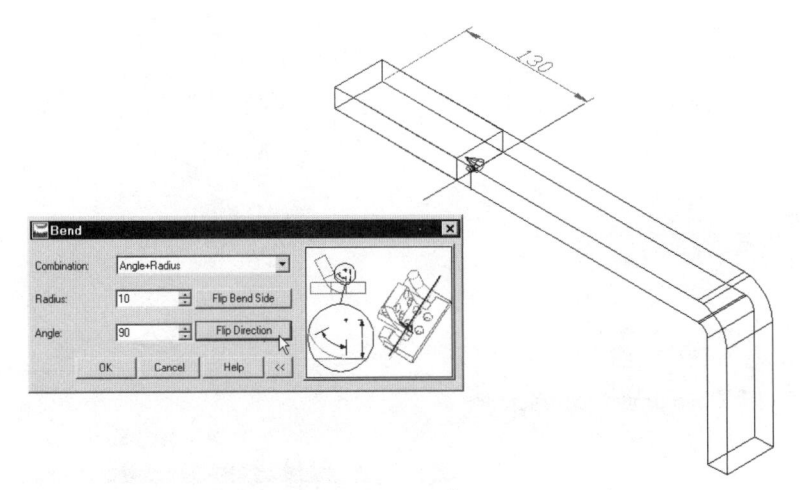

Figure 3–16 *Second bend feature being constructed*

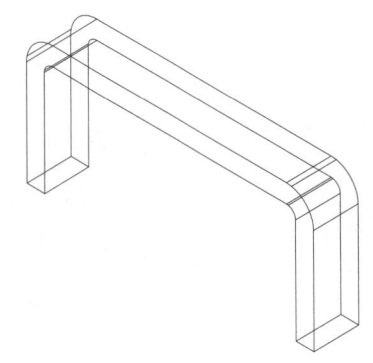

Figure 3–17 *Second bend feature constructed*

BOOLEAN OPERATIONS

To make a solid part of a complex shape, you make a number of features and combine them together one by one. Among the features you will choose one to make first. This is the base solid feature. Subsequent features you make will join to, cut from, intersect with, or split with the existing solid.

JOIN

Joining a sketched solid feature to an existing solid forms a solid that consists of all the volume enclosed by the newly formed sketched feature and the existing solid. Now you will work on the steering shaft of the infant scooter. You will construct three extruded solid features and join them together. While making the extruded solid features, you will also learn how to establish sketch planes, copy a sketch from a feature, and append objects to a profile.

Base Solid Feature

1. Open the file *SteeringShaft.dwg* that you constructed in Chapter 2.
2. Set the display to an isometric view by using the shortcut key 8.
3. Referring to Figure 3–18, extrude the profile a distance of 10 mm.

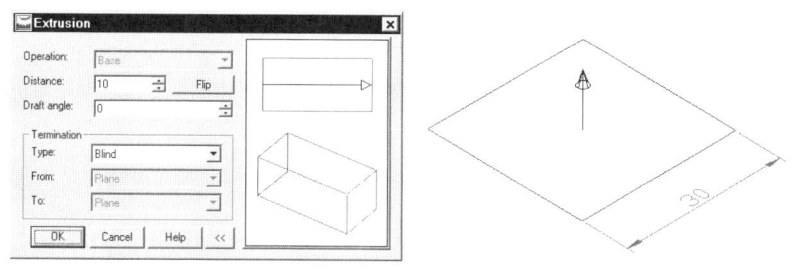

Figure 3–18 *Profile being extruded*

An extruded solid feature is constructed. This, being the first solid feature of a solid part, is called the base solid feature.

New Sketch Plane and Second Sketch

Now you will establish a sketch plane and construct a sketch on the sketch plane.

4. Select Part>New Sketch plane

    Command: AMSKPLN

5. Select the top face of the solid.

One of the faces of the solid is highlighted. Because the highlighted face may not be the one you want, the cursor changes to the shape of a mouse with a blinking red (left) button and a rotating arrow to tell you that you can change to other faces by pressing the left mouse button (see Figure 3–19).

6. Left-click (press the left mouse button) to cycle through the possible faces.
7. Right-click (press the right mouse button) to accepted.

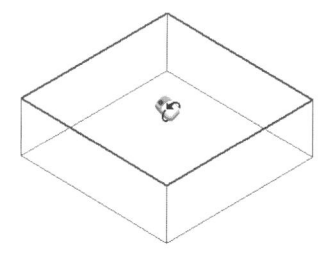

Figure 3–19 *Sketching plane being selected*

Now you will specify the orientation of the X-axis of the sketch plane. A plane together with a tripod depicting the X, Y, and Z orientation of the sketch plane are displayed, and the rotating arrow of the blinking mouse changes to a rotating L-shaped symbol (see Figure 3–20).

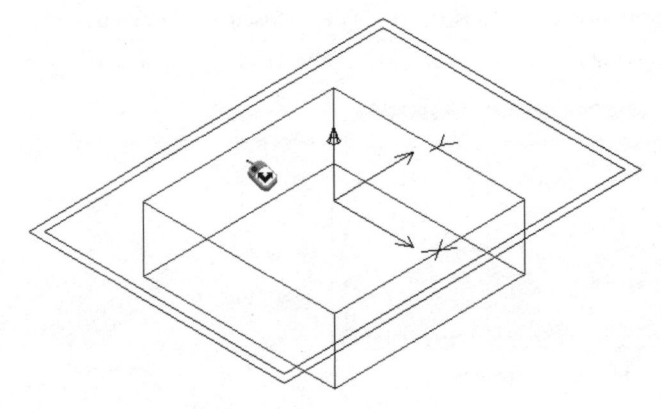

Figure 3–20 *X-axis orientation being specified*

8. Left-click to cycle through the possible X-axis orientations.
9. Right-click to accept an orientation.
10. Referring to Figure 3–21, construct two circles and change the linetype of the outer circle to hidden.

Note: You can double-click the outer circle and change the linetype in the Properties dialog box.

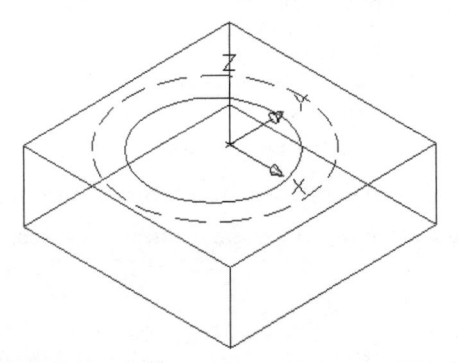

Figure 3–21 *Circles constructed*

11. Type **9** (shortcut key) at the command line area to set the display to the top view of the current sketch plane.
12. Resolve the sketch to a profile.
13. Add concentric constraint to the circles.
14. Add tangent constraint to the circle (hidden line) and the edges of the solid part.
15. Add a parametric dimension (see Figure 3–22).

Part Modeling I 73

A sketch is constructed on a sketch plane residing on a face of the solid faces.

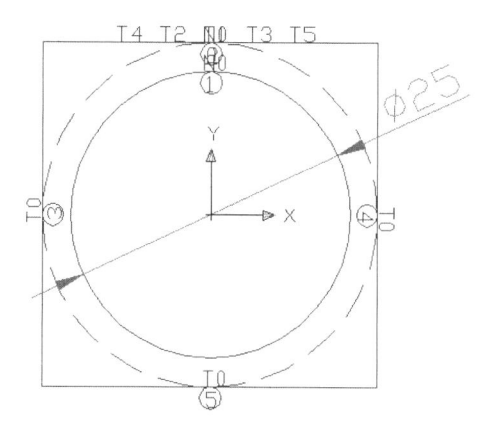

Figure 3–22 *Geometric constraints and parametric dimensions added*

Extrude and Join

Now you will construct an extruded solid feature and join it to the base solid feature.

16. Set the display to an isometric view by using the shortcut key 8.

17. Select the profile in the browser, right-click, and select Extrude (see Figure 3–23).

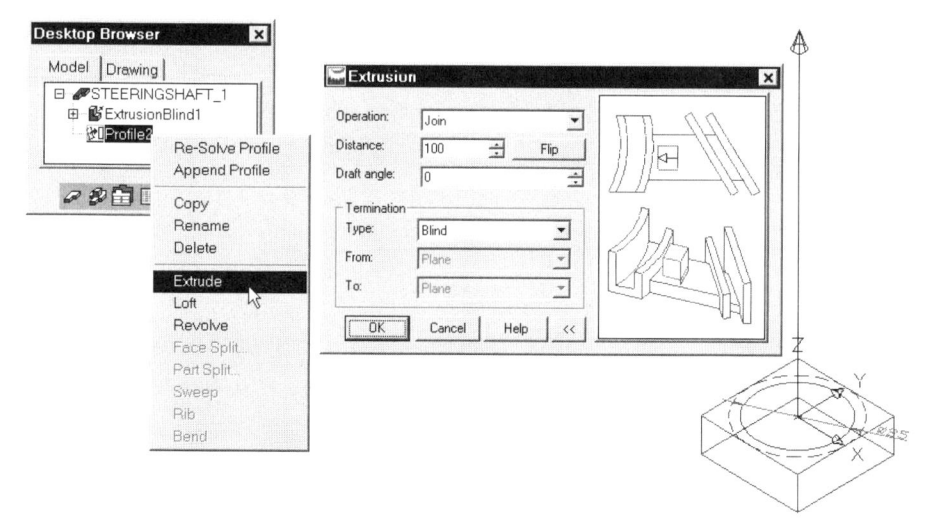

Figure 3–23 *Sketch being extruded*

18. In the Extrusion dialog box, select Join operation, set the distance to 100 and the draft angle to 0, select Blind termination, and select the OK button.

The sketch is extruded and the extruded solid feature is joined to the solid part (see Figure 3–24).

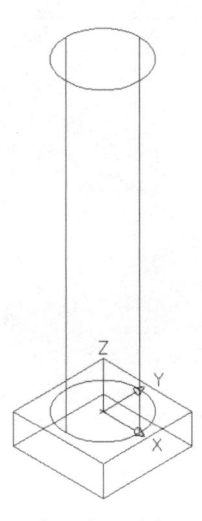

Figure 3–24 *Sketch extruded and joined to the solid part*

Copy Sketch and Append

Now you will learn how to copy a sketch and append sketch objects to a sketch.

19. Referring to Figure 3–25, establish a sketch plane.

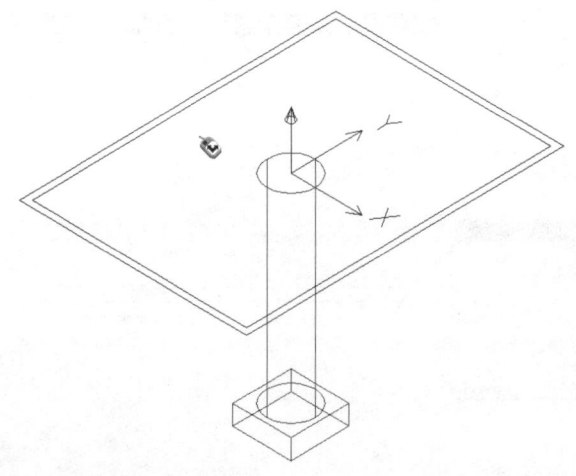

Figure 3–25 *Sketch plane being established*

20. Select Part>Sketch Solving>Copy Sketch
    Command: AMCOPYSKETCH

21. Type **F** at the command line area to use the Feature option. (Using the feature option, you copy a sketch from a feature.)
22. Referring to Figure 3–26, select the base feature, select a location to specify the sketch center, and press the ENTER key.

The sketch of a feature is copied. Now you will append sketch objects to a sketch.

23. Referring to Figure 3–27, construct a circle and change its linetype to hidden.
24. Select Part>Sketch Solving>Append
25. Select the circle and press the ENTER key.
26. Delete the Fix constraint so that you can relocate the sketch.
27. Add tangent constraints to the circle and the square and change the dimension value to 16 mm.

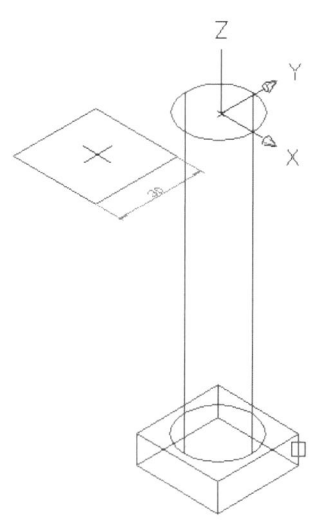

Figure 3–26 *Sketch being copied*

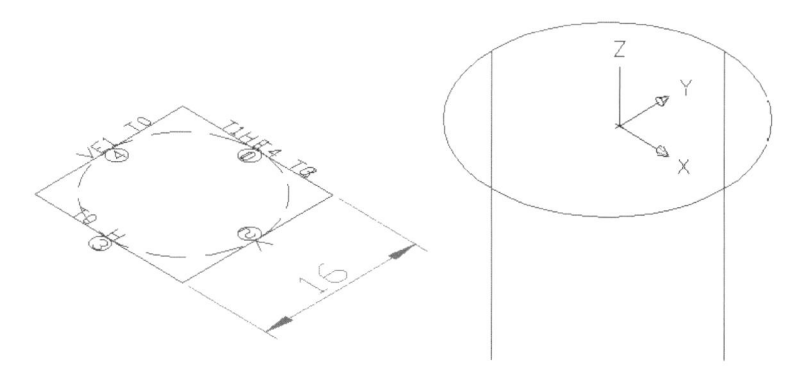

Figure 3–27 *Circle constructed and appended to the sketch*

28. Referring to Figure 3–28, add a concentric constraint to the circle and the circular edge of the solid.

The sketch is appended and properly constrained.

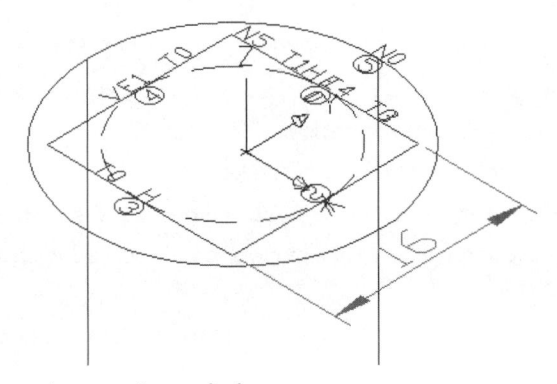

Figure 3–28 *Concentric constraint applied*

Extrude and Join

Now you will extrude the sketch and join it to the solid.

29. Select the profile in the browser, right-click, and select **Extrude**.
30. In the Extrusion dialog box, select Join operation, set the distance to 20 and the draft angle to 0, select Blind termination, and select the OK button (see Figure 3–29).

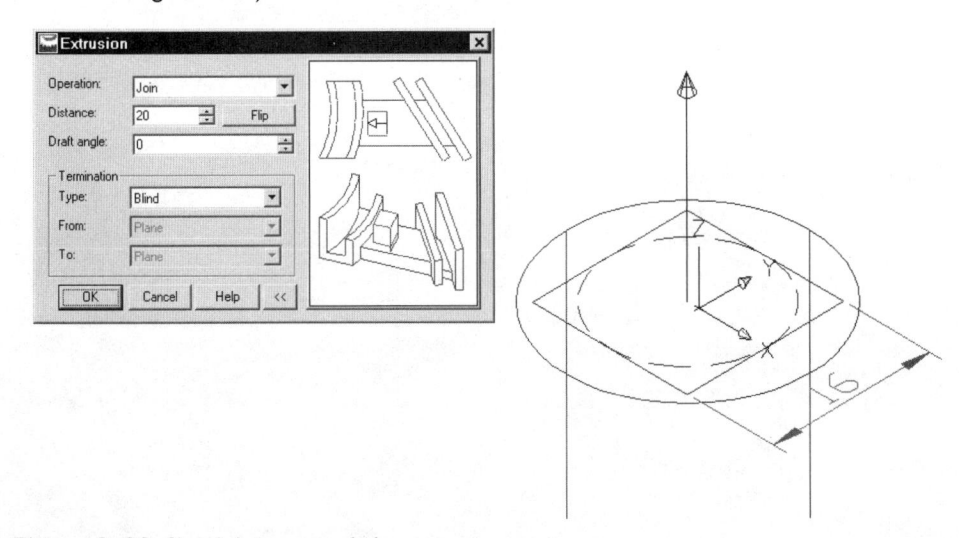

Figure 3–29 *Sketch being extruded*

The profile is extruded and joined to the solid. Shade the display to see the result (see Figure 3–30). Save and close the file.

Figure 3–30 *Extruded solid joined to the solid part*

CUT

Cutting a sketched solid feature from an existing solid forms a solid that encloses the volume enclosed by the original solid but not by the newly formed sketched feature. Now you will construct an extruded solid feature and cut it from the solid.

1. Open the file *Seat.dwg*, if you have already closed it.
2. Set the display to the top view by using the shortcut key 9.
3. Delete the fix constraint of the text sketch.
4. Referring to Figure 3–31, add two parametric dimensions.

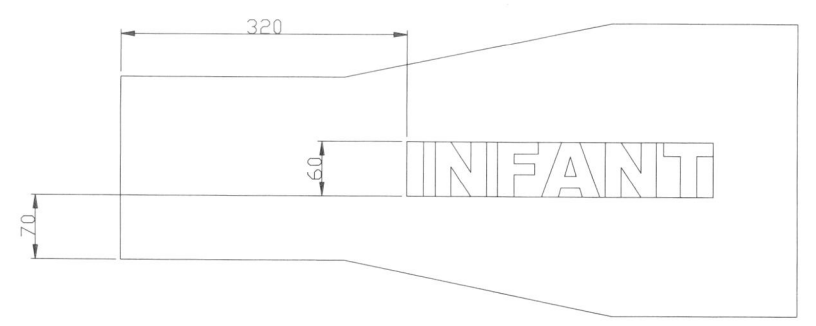

Figure 3–31 *Parametric dimensions added*

5. Set the display to an isometric view.

6. Select the text profile in the browser, right-click, and select Extrude (see Figure 3–32).
7. In the Extrusion dialog box, select Cut operation, set the distance to 3 and the draft angle to 0, select Blind termination, select the Flip button to change the direction of extrusion, and select the OK button.

The extruded solid is constructed and cut from the solid (see Figure 3–33). Save and close your file.

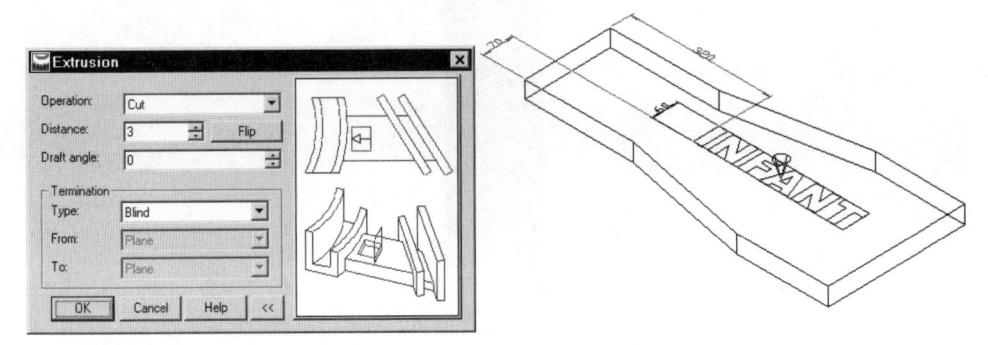

Figure 3–32 *Text profile being extruded*

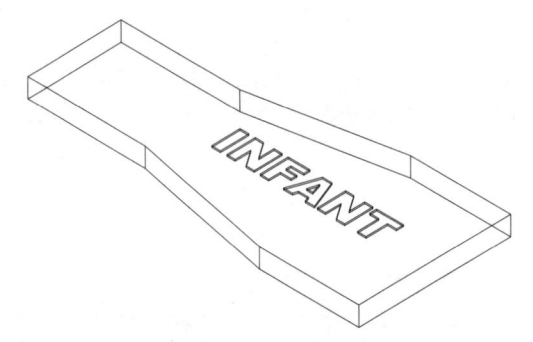

Figure 3–33 *Extruded solid feature constructed and cut from the solid*

INTERSECT

Intersecting a solid feature with an existing solid forms a solid that encloses the volume of the portion of the newly formed sketched solid feature that is also contained in the original solid. Now you will construct two solid features and intersect the second solid feature with the base solid feature.

Base Solid Feature

1. Open the file *Frame_F.dwg*, that you constructed in Chapter 2.
2. Set the display to an isometric view.
3. Select the profile from the browser, right-click, and select Extrude.

4. In the Extrusion dialog box, set the distance to 60, the draft angle to 0, and blind termination, and select the OK button.

A base solid feature is constructed.

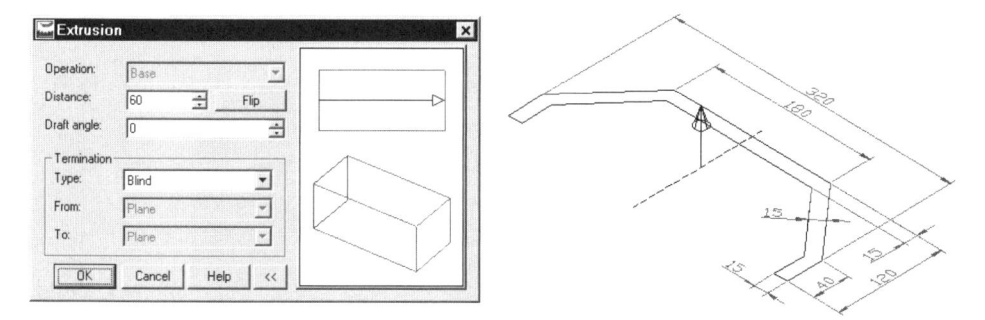

Figure 3–34 *Profile being extruded*

Sketch Plane and Sketching

Now you will establish a sketch plane and construct a sketch.

5. Referring to Figure 3–35, establish a sketch plane.

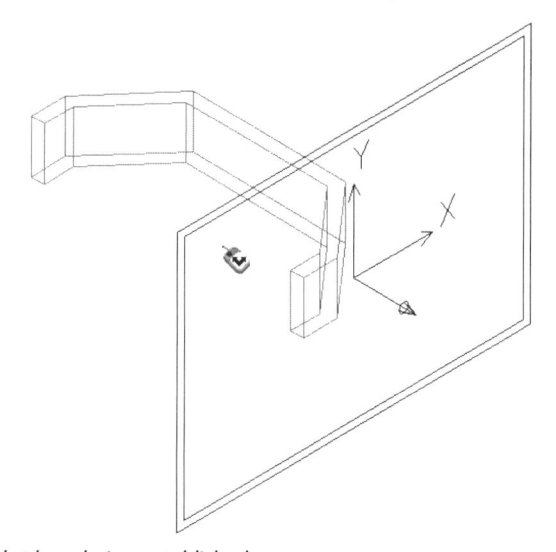

Figure 3–35 *Sketch plane being established*

6. Use the shortcut key to set the display to the top view of the sketch plane.
7. Referring to Figure 3–36, construct a sketch and resolve it to a profile.
8. Add geometric constraints and parametric dimensions in accordance with Figure 3–37.

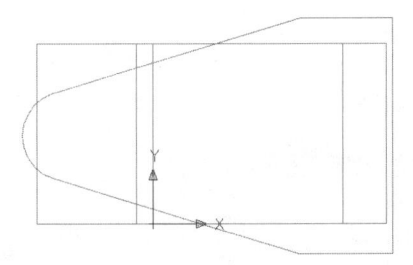

Figure 3-36 Sketch

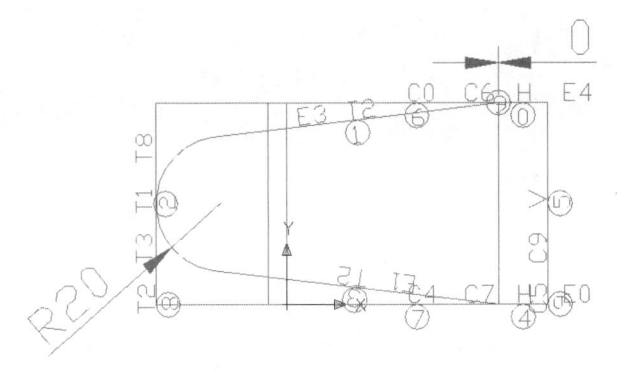

Figure 3-37 Geometric constraints and parametric dimensions

Extrude and Intersect

Now you will construct an extruded solid from the sketch and intersect the solid with the base solid.

9. Set the display to an isometric view.
10. Select the profile from the browser, right-click, and select Extrude.
11. In the Extrusion dialog box, select Intersect operation, set termination to Through, and select the OK button (see Figure 3-38).

The profile is extruded and the extruded solid is intersected with the solid (see Figure 3-39). Save and close your file.

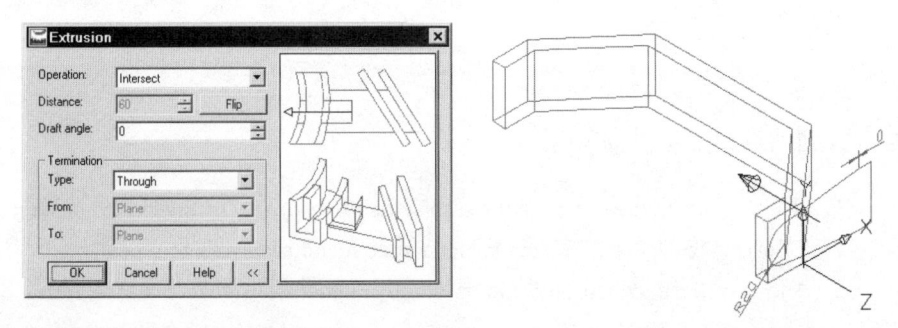

Figure 3-38 Sketch being extruded

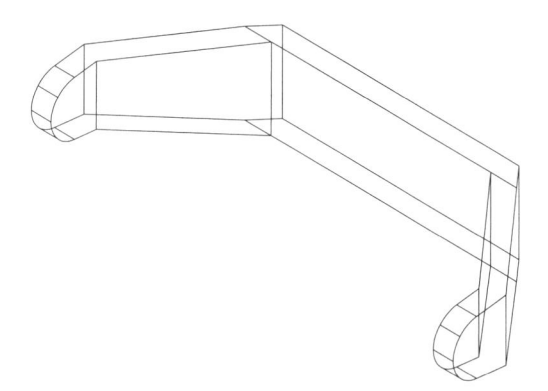

Figure 3–39 *Sketch extruded and solids intersected*

SPLIT

Splitting is a special kind of operation in which two solid parts are derived from the original solid part: One part is formed by subtracting the newly formed sketched solid feature from the original solid and the other is formed by intersecting the newly formed sketched solid feature with the original solid.

Now you will construct two profiles, extrude them one by one, and combine the second extruded solid to the first solid by splitting.

1. Start a new part file. Use metric as the default.
2. Construct two sketches and resolve them one by one into two profiles (see Figure 3–40).
3. Set the display to an isometric view.
4. Select the profile (rectangle) from the browser, right-click, and select Extrude to extrude the rectangle a distance of 30 mm (see Figure 3–41).
5. Extrude the other profile. In the Extrusion dialog box, select Split operation, set termination to Through, and select the OK button (see Figure 3–42).
6. Type Split1 at the command line area to specify the name of the split solid part.

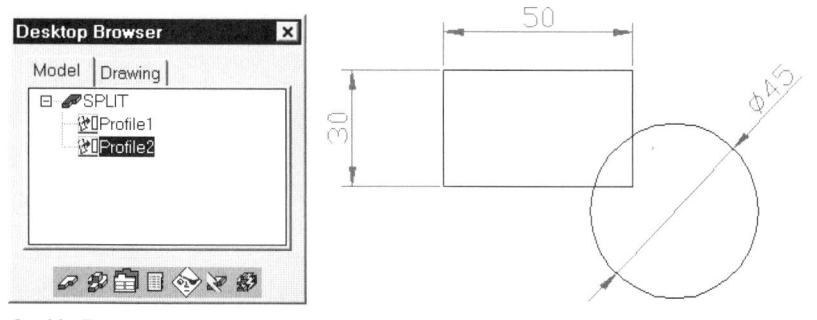

Figure 3–40 *Two profiles constructed*

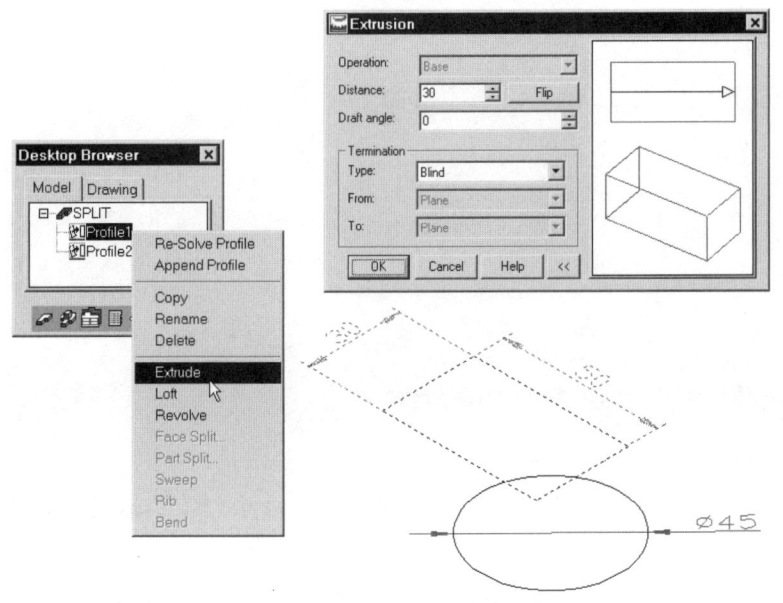

Figure 3–41 *Rectangle being extruded to an extruded solid*

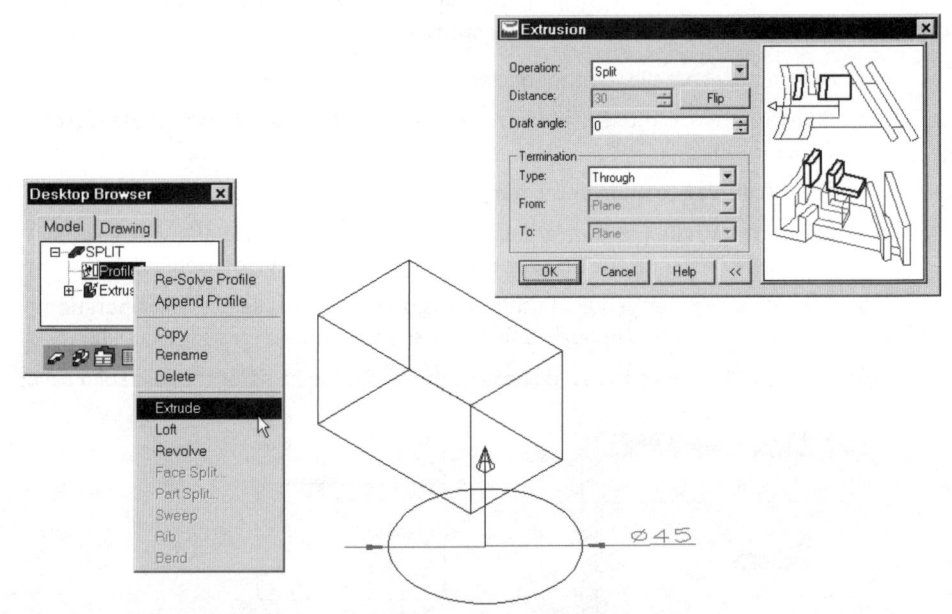

Figure 3–42 *Second profile being extruded and split solid being formed*

Now you have two solid parts in a part file (see Figure 3–43). Save your file (file name: *Split.dwg*). You will learn more about multiple solid parts in Chapter 9.

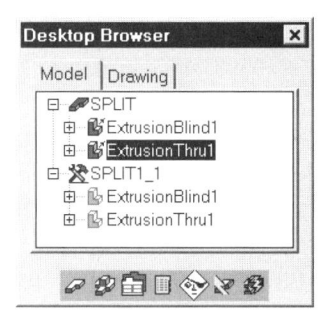

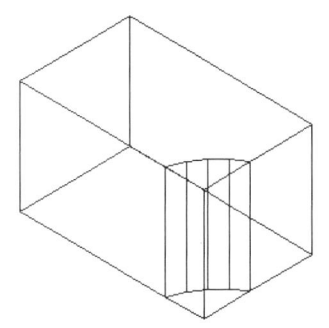

Figure 3–43 *Split solid constructed*

EDITING

The most significant characteristic of a parametric feature-based solid is that you can change the solid's parameter in various ways to edit the solid part. You can modify the sketch of a feature and the parameters of a feature to change the sketched solid feature's size and shape. To remove a feature from a solid part, you delete or suppress the feature. To change the sequence of solid modeling construction, you reorder the features in the feature hierarchy. Besides, you can replay and truncate the modeling process.

EDIT SKETCH OF A FEATURE

The parameter of a sketch can be edited even after a solid feature is constructed from it. You can modify the geometric constraints and parametric dimensions, append objects, and resolve the sketch. There are three ways to edit a sketch:

- Select the sketch from the browser, right-click, and select Edit Sketch
- Select the sketch from the browser and double-click
- Select the feature of the solid part and double-click

Among the three methods, the third method is the most convenient but it may take you some time to find out the appropriate selection point of the solid part. If the solid is a complex solid consisting of many features, selecting the wrong location and double-clicking will lead you to the wrong feature of the solid.

Now you will edit a sketch of a solid feature.

1. Open the file *Split.dwg*, if you have already closed it.
2. Referring to Figure 3–44, select an extruded solid feature from the browser, right-click, and select Edit Sketch or select the sketch from the browser and double-click.
3. Change the dimension of the circle to 60 in accordance with Figure 3–45, right-click, and select enter.

The selected profile is modified.

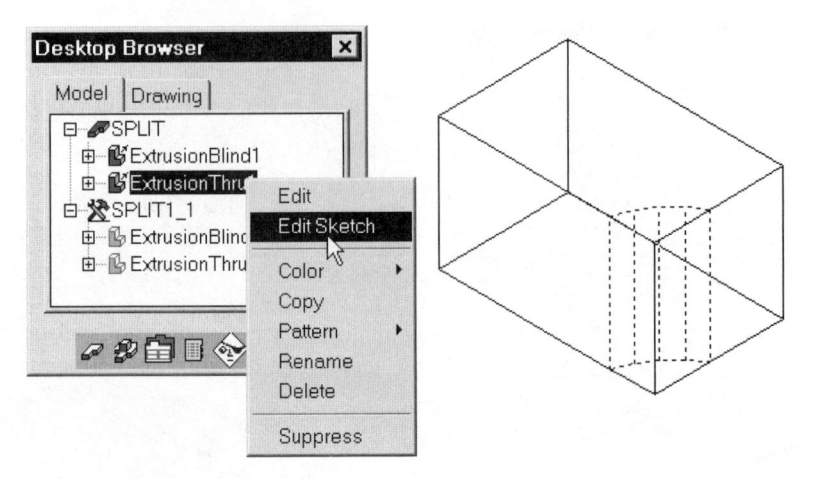

Figure 3-44 *Solid feature selected from the browser*

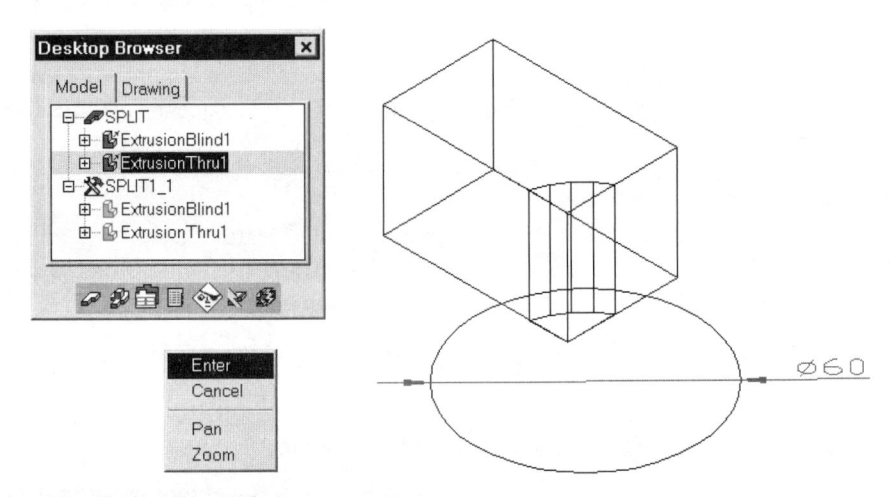

Figure 3-45 *Sketch modified*

UPDATE

After you edit a solid part, the parameters change. You need to update the model because updating is not automatic. The reason why updating is not automatic is because it may take considerable time, depending on the complexity of the object. If updating is automatic, you will have to wait many times as you change the parameters of a solid. To update the solid, do it manually.

4. Select the Update button from the browser (see Figure 3-46).

The selected solid is updated.

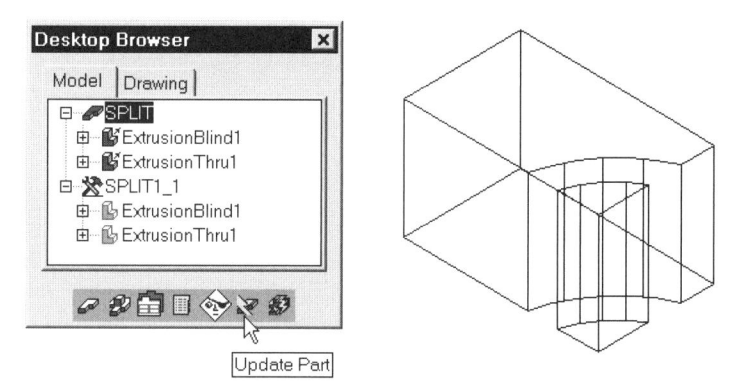

Figure 3–46 *Solid being updated*

EDIT FEATURE

Similar to editing the sketch of a feature, you can edit a feature in three ways:

- Select the feature from the browser, right-click, and select Edit
- Select the feature from the browser and double-click
- Select the feature of the solid part and double-click

The original dialog box that you use to construct the feature reappears. Modify the parameters and select the OK button. After that, select the Update button from the browser to update the changes.

5. Referring to Figure 3–47, double-click Split1 from the browser to activate it, select the extruded feature, right-click, and select Edit. (You may select the feature from the browser and double-click.)

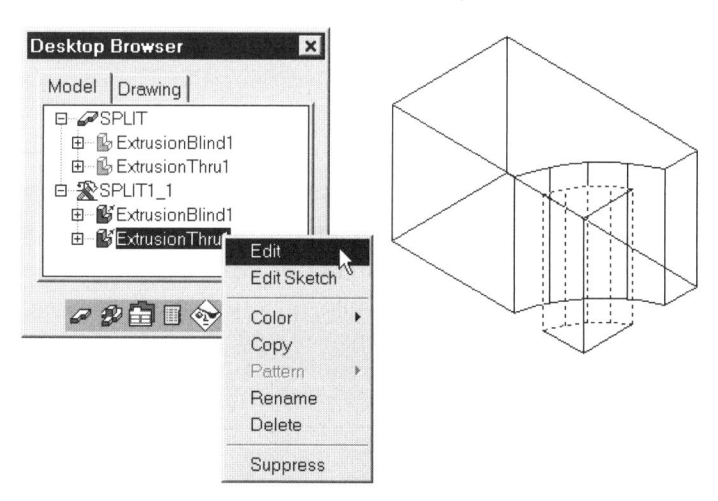

Figure 3–47 *Feature selected*

6. In the Extrusion dialog box, change termination type to Blind, set the distance to 10, and select the OK button (see Figure 3–48).
7. Right-click and select ENTER.

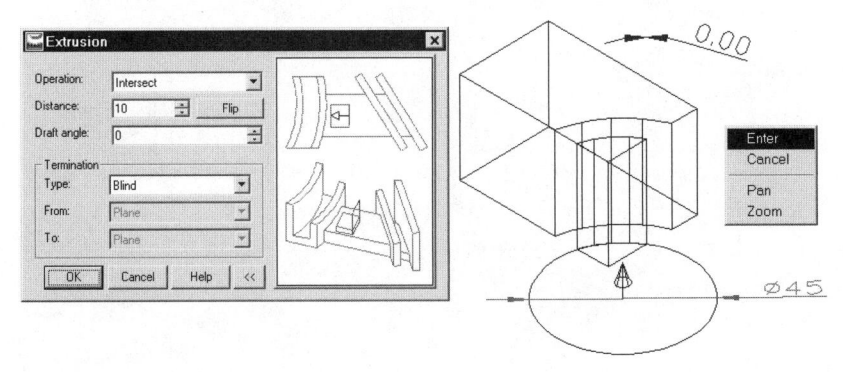

Figure 3–48 *Feature being modified*

8. Select the Update button from the browser to update the change.

The selected solid is modified and updated (see Figure 3–49).

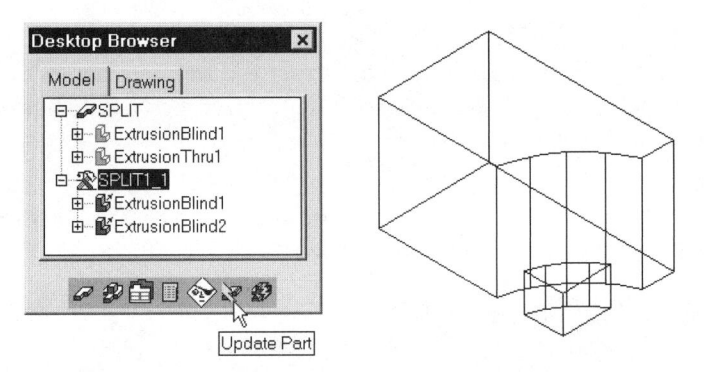

Figure 3–49 *Solid updated*

REORDER FEATURES

To construct a solid with a number of sketched solid features, you construct them one by one sequentially. While you are modeling, you may construct the features in the wrong sequence or you may want to add one operation before another that has already been performed. To illustrate how the working sequence affects the modeling outcome and to reorder the sequence, you will construct three extruded solid features and combine them by joining and cutting.

1. Start a new part file. Use metric as the default.
2. Referring to Figure 3–50, construct a rectangle of 50 mm times 40 mm, resolve it to a profile, and extrude it a distance of 30 mm.

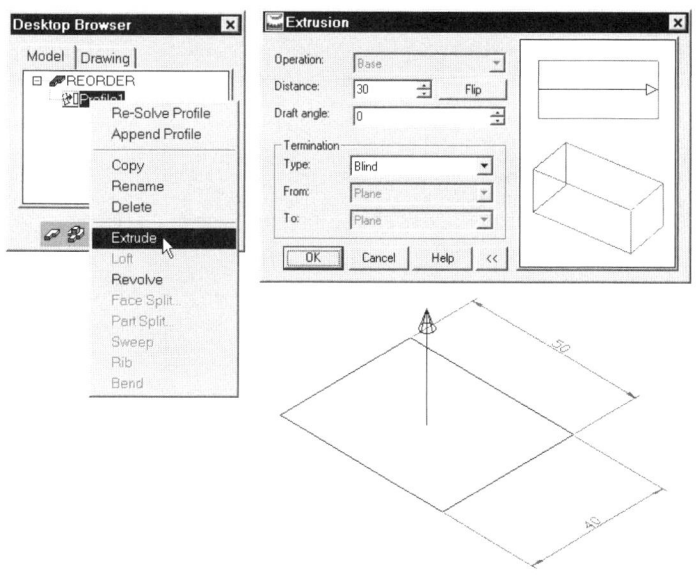

Figure 3–50 *Rectangle constructed and being extruded*

3. Construct a circle of 20 mm diameter, resolve it to a profile, extrude it a distance of 60 mm, and join the extruded solid to the base solid (see Figure 3–51).
4. Referring to Figure 3–52, construct a circle of 20 mm diameter, resolve it to a profile, extrude it through, and cut the extruded solid from the solid.

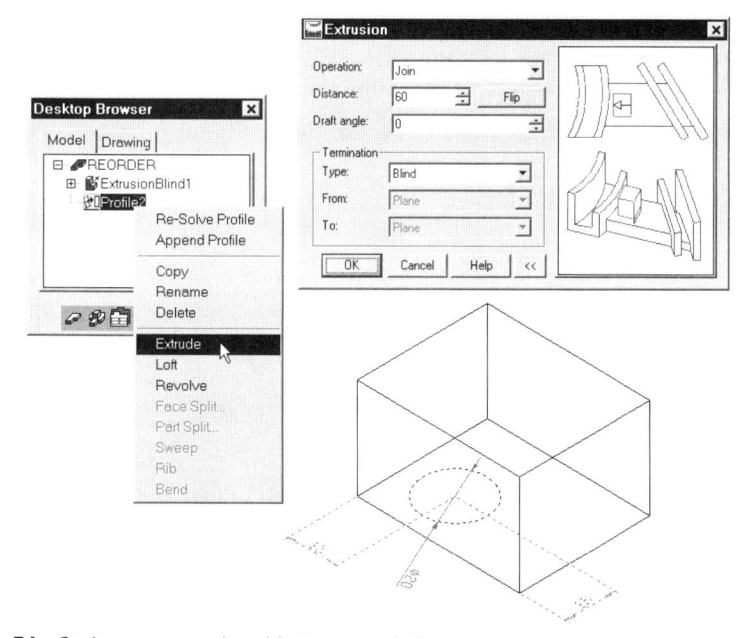

Figure 3–51 *Circle constructed and being extruded*

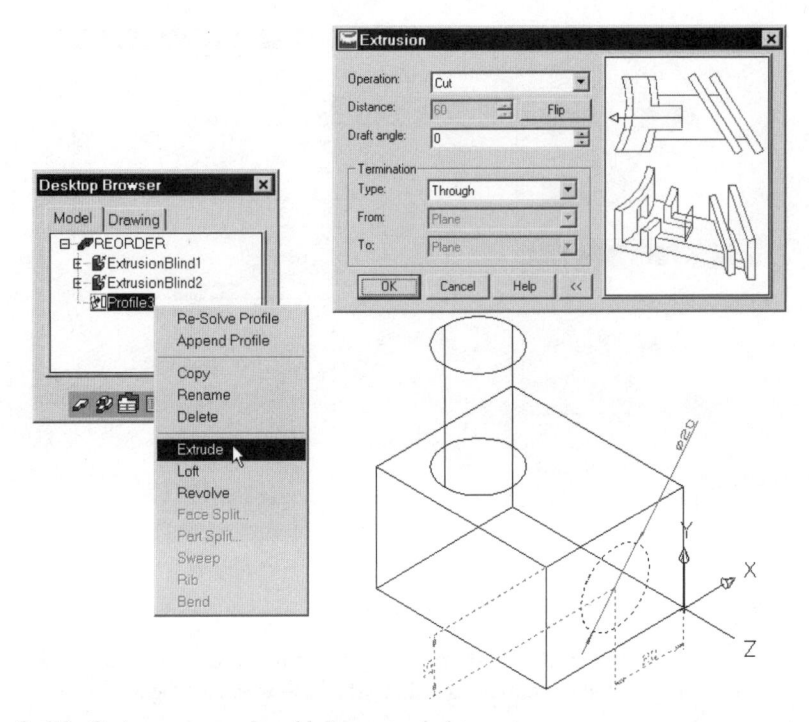

Figure 3–52 *Circle constructed and being extruded*

Now you will reorder the operations to change the outcome.

    5. Referring to Figure 3–53, select the third extrusion feature from the browser, drag, and move it above the second extrusion feature.

The sequence of the operation is reordered. Compare Figure 3–54 with Figure 3–53.

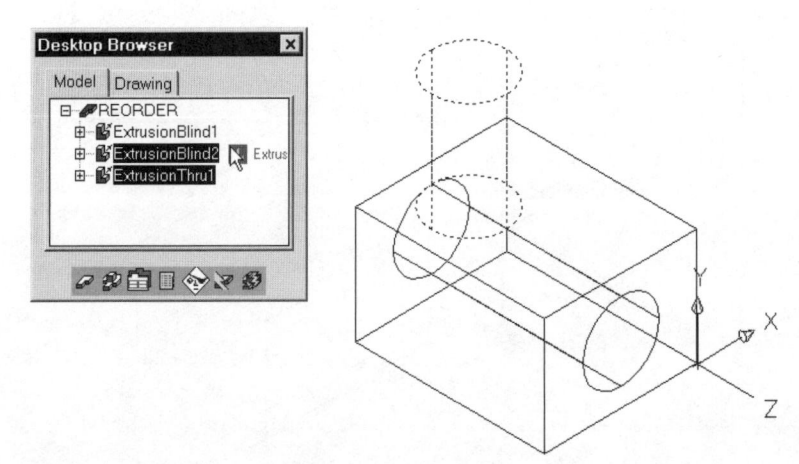

Figure 3–53 *Sequence being reordered*

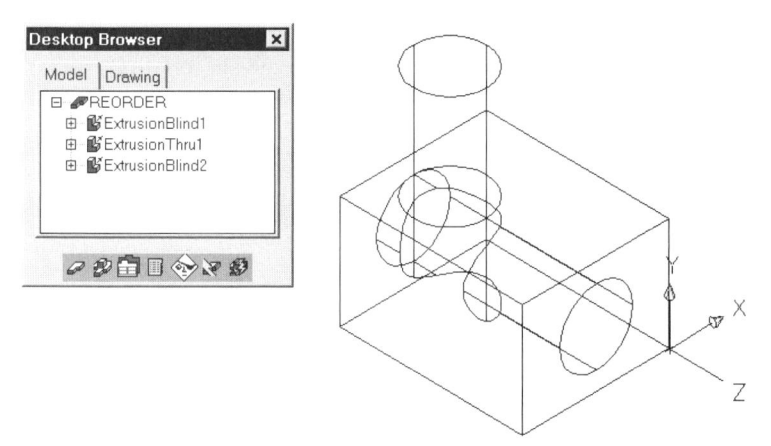

Figure 3–54 *Sequence reordered*

The model is complete. Save your file (file name: *ReOrder.dwg*).

FEATURE SUPPRESSION

To remove a feature temporarily, you can suppress it. After a feature is suppressed, it is disregarded in the solid model as if it had been deleted. However, you can recover a suppressed feature by unsuppressing it. To suppress a feature, you select it from the browser, right-click, and select Suppress.

6. Referring to Figure 3–55, select an extruded solid feature, right-click, and select Suppress.
7. Press the ENTER key to confirm.

The selected feature is suppressed

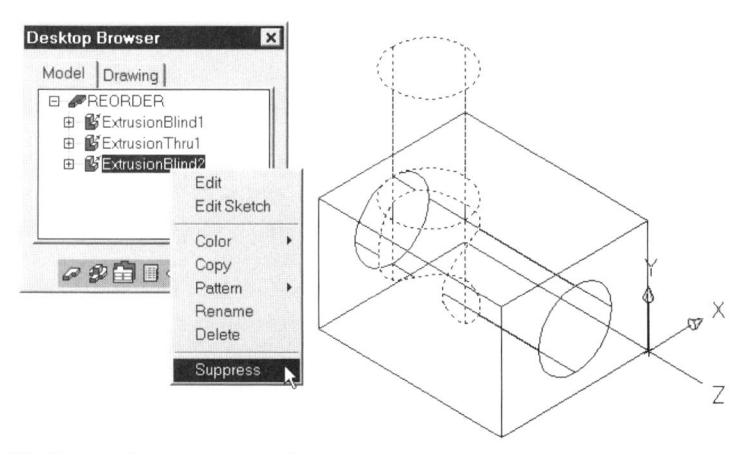

Figure 3–55 *Feature being suppressed*

8. Select the suppressed feature from the browser, right-click, and select Unsuppress.

The suppressed feature is unsuppressed.

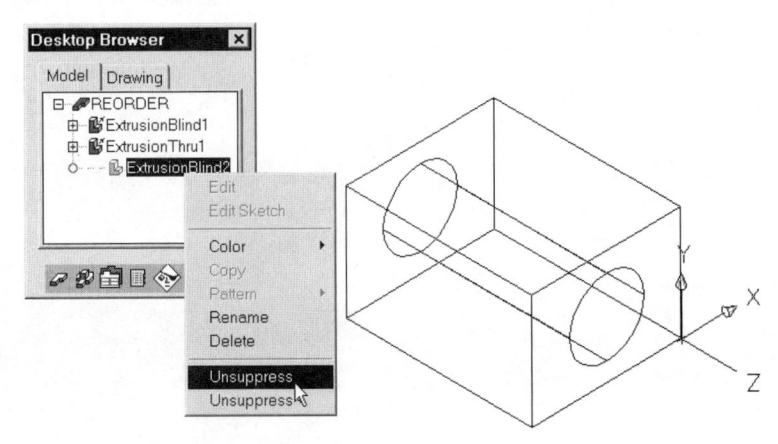

Figure 3–56 *Feature being unsuppressed*

DELETE FEATURE

To remove a feature permanently, you delete the feature. You delete the unwanted feature by selecting the feature in the browser, right-clicking, and selecting Delete. Once a feature is removed and the file saved and reopened, you can no longer retrieve the deleted feature any more.

9. Referring to Figure 3–57, select the feature to be deleted from the browser, right-click, and select Delete.
10. Press the ENTER key to confirm.
11. To recover a deleted feature, undo the command immediately by typing **U** at the command line area.

    Command: U

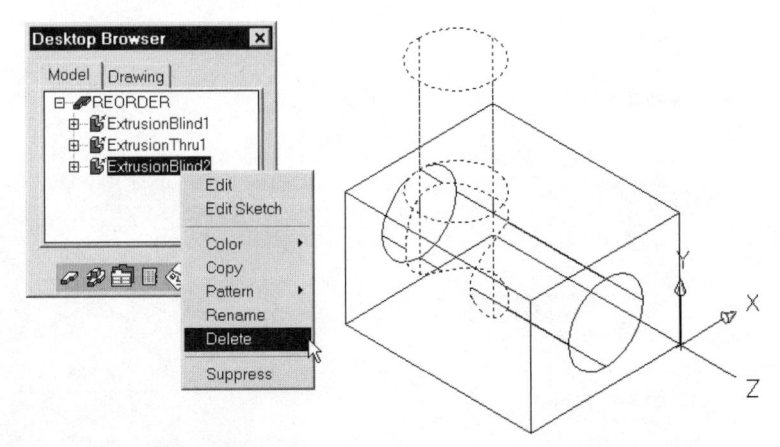

Figure 3–57 *Feature being deleted*

REPLAY

The entire process of making the solid model can be replayed in a step-by-step fashion. While replaying, you can stop the replay and remove the remaining operations by truncating. After truncating, all subsequent operations will be deleted from memory.

12. Select Part>Part>Replay

    Command: AMREPLAY

13. Press ENTER to cycle to the next operation.

To remove the remaining features, type T to truncate.

14. Undo the AMREPLAY command.

BASE PART

A base part is a solid part without any history in the browser hierarchy. It cannot be modified. To change a parametric solid to a base solid, select Part>Part>Make Base Part (see Figure 3–58). You will learn more about base solid editing in Chapter 9.

Now close the file without saving it.

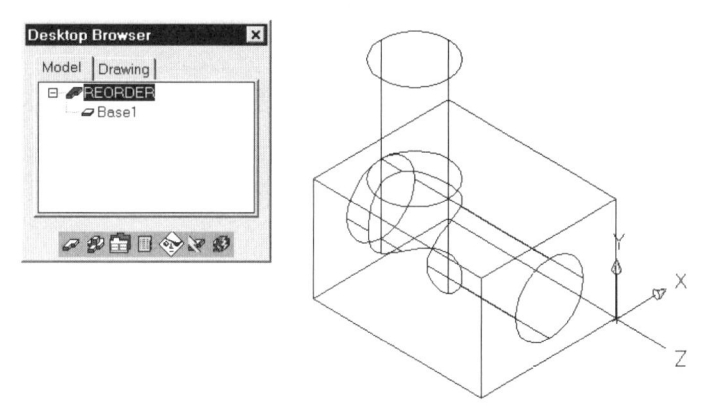

Figure 3–58 *Parametric part converted to a base part*

MASS PROPERTIES

One very important attribute of a solid model is that it has all the information of a 3D object, including vertices, edges, surfaces, and volume. Therefore, with a solid model, you evaluate its mass properties information: mass, volume, surface area, centroid, mass moments of inertia, mass products of inertia, radii of gyration, principal mass moments, and principal axes. Now you will evaluate the mass properties of a solid part.

1. Open the file *Frame_F.dwg*.
2. Select Part>Part>Mass Properties

    Command: AMPARTPROP

3. Select the solid part and press the ENTER key (see Figure 3–59).

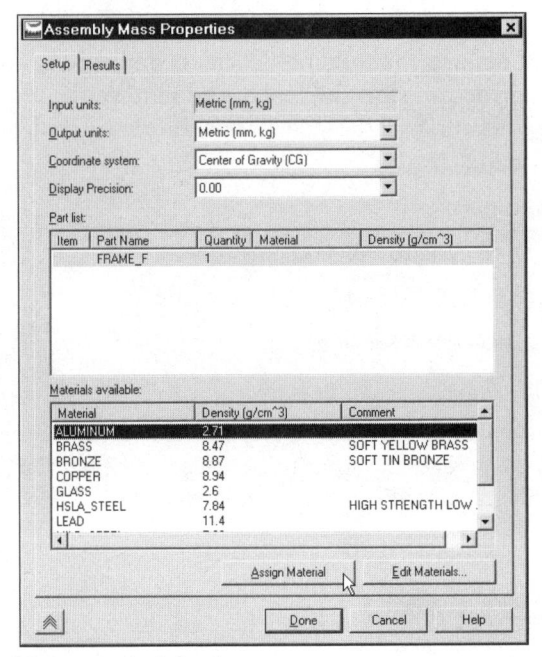

Figure 3-59 *Assembly Properties dialog box*

Because this command can be used for evaluating a set of solid parts in a part or assembly file, you have to select objects and then press the ENTER key after you have finished object selection.

   4. Select the Setup tab, if it is not already selected.
   5. Select Metric from the output units box.
   6. Select Center of Gravity from the coordinate system box.
   7. Set display precision to 0.00.
   8. Select Frame_F from the part list.

The solid part is highlighted. (If you selected more than one solid part for evaluation, highlighting a part helps you confirm which object is being evaluated.)

   9. Select Aluminum from the Material available box and select the Assign Material button.
   10. Select the OK button to confirm (see Figure 3-60).

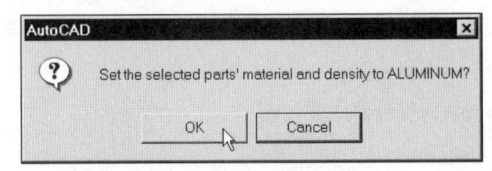

Figure 3-60 *Confirmation dialog box*

11. Select the Edit Materials button.
12. Check the data in the Physical Materials List dialog box (see Figure 3–61) against your standard materials handbook, make adjustments to the values if necessary, and select the OK button.
13. Select the Results tab and select the Calculate button (see Figure 3–62).

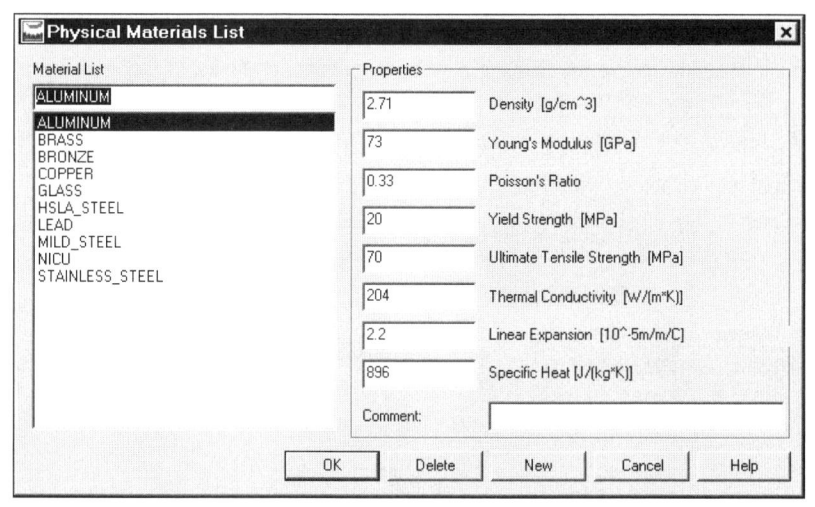

Figure 3–61 *Physical Materials List dialog box*

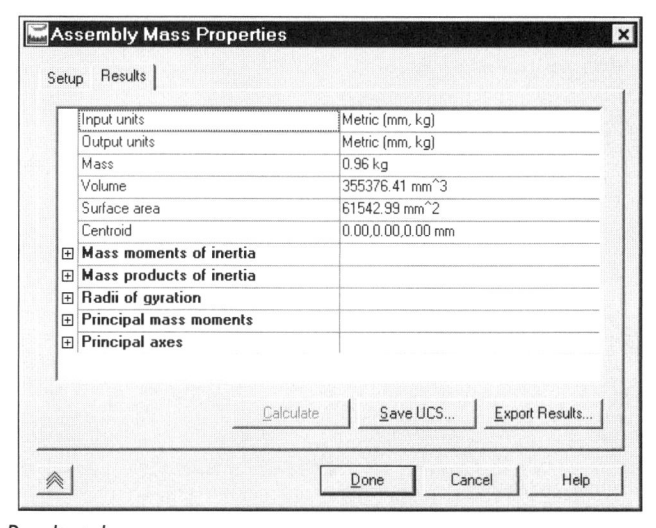

Figure 3–62 *Results tab*

14. If you want to export the result, select the Export Results button and specify a file name.
15. On returning to the Assembly Mass Properties dialog box, select the Done button.

ATTRIBUTES

Attributes are information attached to each component part. The information can be a text string, a whole number, or real (decimal) number. You assign attributes in a part file or in an assembly file. Now you will assign attributes to a solid part.

1. Select Part>Part>Attributes.
2. In the Assign Attributes dialog box, select the Add button (see Figure 3–63).
3. In the Add New Attribute dialog box, specify an attribute name and attribute value, select a data type, and select the OK button (see Figure 3–64).
4. Select the Done button. Use the Cancel button if you did not add any attributes.

An attribute is assigned. Save and close your file.

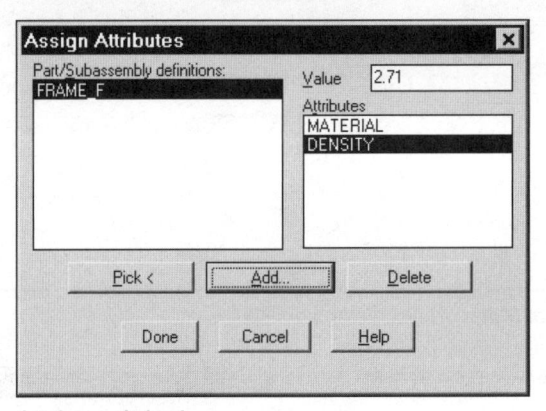

Figure 3–63 *Assign Attributes dialog box*

Figure 3–64 *Add new Attribute dialog box*

SUMMARY

Working in 3D, you will need to view the object from different directions and use different kinds of display modes. To view the 3D object from different directions, orbit the display. To see the faces of the 3D object clearly, use shaded display mode.

The fundamental way to construct parametric solids is to construct a parametric sketch and extrude the sketch in a direction perpendicular to the sketch, or revolve the sketch about an axis to form a solid feature. A bend feature is a special kind of sketched solid feature. You bend an existing solid along a bending line.

To create complex 3D objects, construct a number of solid features one by one and combine them together by joining, cutting, intersecting, or splitting. Features of a parametric solid are listed as objects in the hierarchy in the desktop browser. To modify the solid, select an object from the hierarchy and change the parameters. To delete a feature, select and remove it from the hierarchy. You can also modify the solid by reordering the sequence of features in the hierarchy. To review the process of model construction, replay the sequence. To remove all the features from any designated point of the sequence, truncate the replay. Because a solid has all the information on the 3D object, you can retrieve mass properties. To include additional information to the solid, you add attributes.

PROJECTS

The following projects have been designed for you to practice on and enhance your newly acquired knowledge.

CLIP

The clip of the infant scooter is an extruded solid feature.

1. Open the file *Clip.dwg* that you constructed in Chapter 2.
2. Set the display to an isometric view.
3. Select the profile from the browser, right-click, and select Extrude.
4. In the Extrusion dialog box, set the distance to 2 and draft angle to 0, select Blind termination, and select the OK button (see Figure 3–65).

The solid part is complete (see Figure 3–66). Save and close your file.

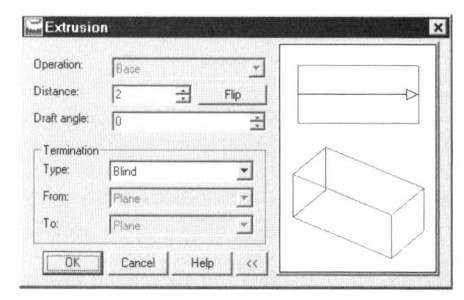

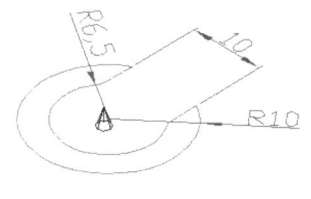

Figure 3–65 *Profile being extruded*

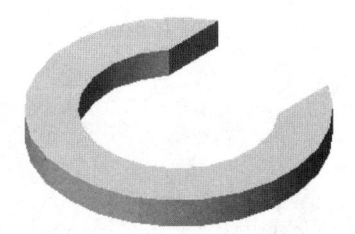

Figure 3–66 *Clip*

AXLE

The axle of the infant scooter is a revolved solid.

1. Open the file *Axle.dwg* that you constructed in Chapter 2.
2. Select the profile from the browser, right-click, and select Revolve.
3. Referring to Figure 3–67, select the lower edge of the profile as the revolution axis.
4. In the Revolution dialog box, set the angle to 360 and select the OK button.

The solid part is complete (see Figure 3–68). Save and close your file.

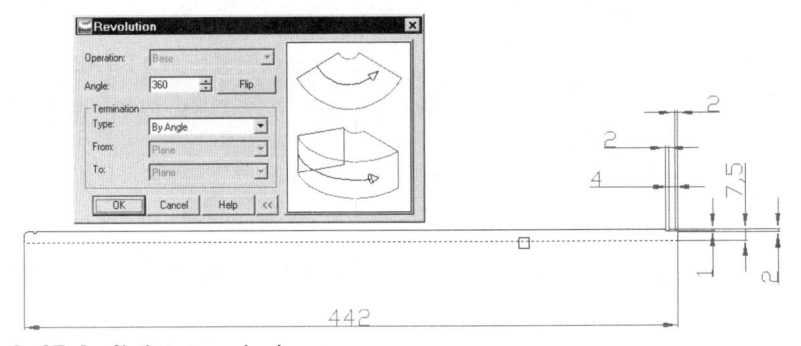

Figure 3–67 *Profile being revolved*

Figure 3–68 *Axle*

SCREW CAP

The screw cap is a revolved solid.

1. Open the file *ScrewCap.dwg* that you constructed in Chapter 2.
2. Set the display to an isometric view.
3. Select the profile from the browser, right-click, and select Revolve.
4. Referring to Figure 3–69, select the edge indicated as the revolution axis.
5. In the Revolution dialog box, set the angle to 360 and select the OK button.

The solid part is complete (see Figure 3–70). Save and close the file.

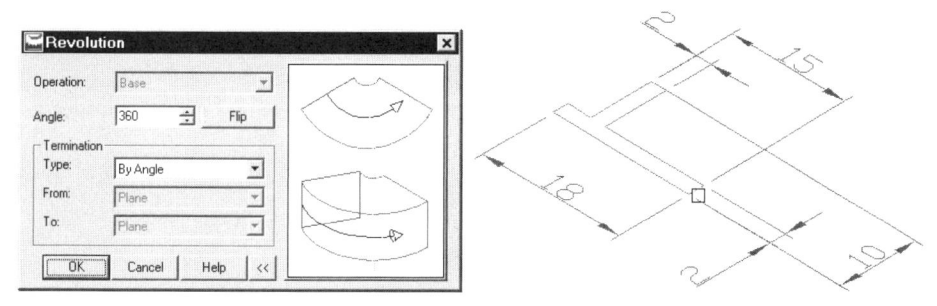

Figure 3–69 *Profile being revolved*

Figure 3–70 *Screw cap*

WHEEL CAP

The wheel cap is a revolved solid.

1. Open the file *WheelCap.dwg*, that you constructed in Chapter 2.
2. Set the display to an isometric view.
3. Select the profile from the browser, right-click, and select Revolve.
4. Referring to Figure 3–71, select the edge indicated as the revolution axis.
5. In the Revolution dialog box, set the angle to 360 and select the OK button.

The solid part is complete (see Figure 3–72). Save and close the file.

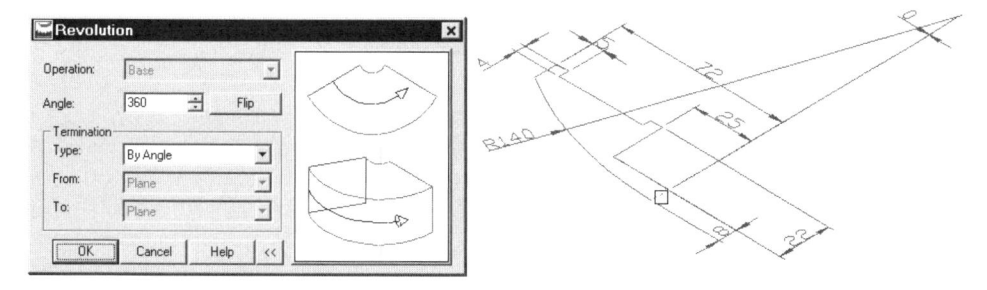

Figure 3–71 *Profile being revolved*

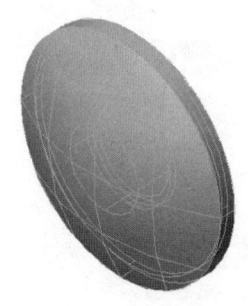

Figure 3–72 *Wheel cap*

WHEEL

The wheel is a revolved solid.

1. Open the file *Wheel.dwg*, that you constructed in Chapter 2.
2. Set the display to an isometric view.
3. Select the profile from the browser, right-click, and select Revolve.
4. Select the line indicated in Figure 3–73 as the revolution axis.
5. In the Revolution dialog box, set the angle to 360 and select the OK button.

The solid part is complete (see Figure 3–74). Save and close the file.

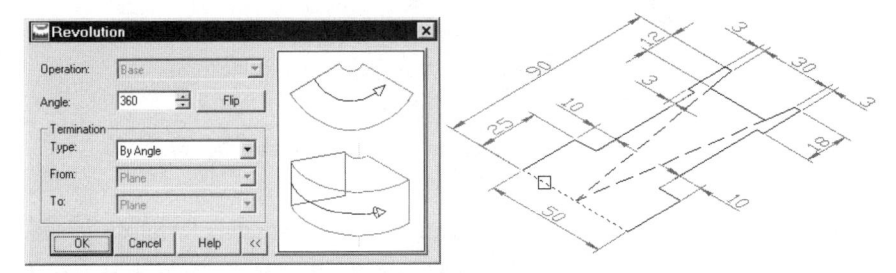

Figure 3–73 *Profile being revolved*

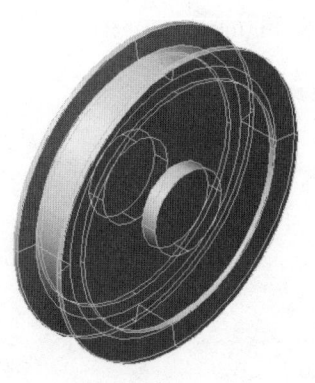

Figure 3–74 *Wheel*

NUT

The nut has two extruded solid features. You cut the second extruded solid feature from the first solid feature.

1. Open the file *Nut.dwg,* that you constructed in Chapter 2.
2. Set the display to isometric view.
3. Select the profile from the browser, right-click, and select Extrude.
4. In the Extrusion dialog box, set the extrusion distance to 8 and select the OK button (see Figure 3–75).

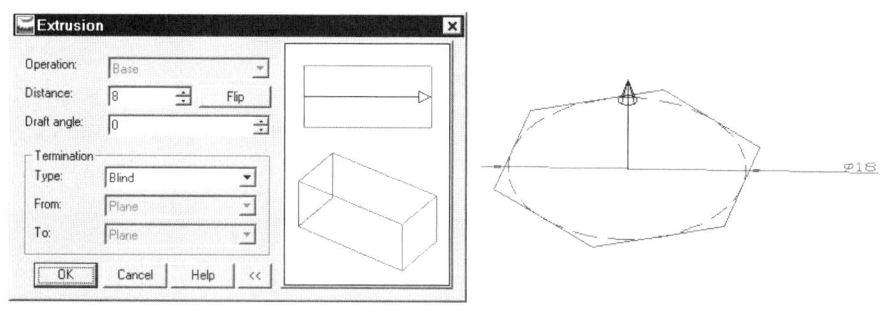

Figure 3–75 *Profile being extruded*

The profile is extruded (see Figure 3–76).

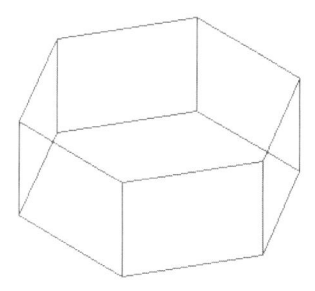

Figure 3–76 *Profile extruded*

5. Use the shortcut key 9 to set the display to top view.
6. Construct two circles, change the linetype of the larger circle to hidden, add geometric constraints, and add parametric dimensions. (see Figure 3–77).
7. Set the display to isometric view.
8. Select the second profile from the browser, right-click, and select Extrude.
9. In the Extrusion dialog box, select Cut operation, set termination to Through, and select the OK button (see Figure 3–78).

The second solid feature is constructed and is cut from the solid (see Figure 3–79). The model is complete. Save your file.

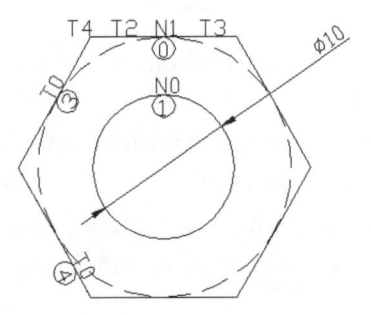

Figure 3-77 *Sketch constructed*

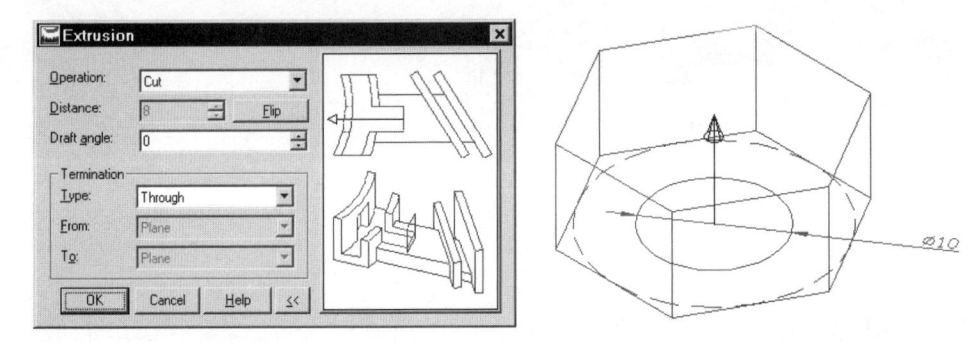

Figure 3-78 *Sketch being extruded*

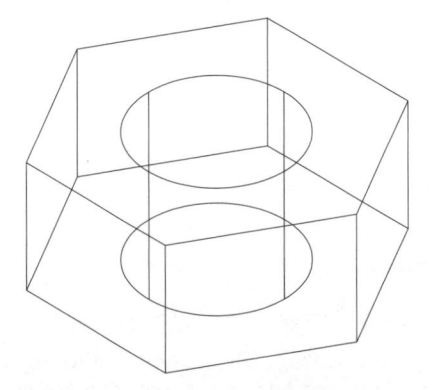

Figure 3-79 *Extruded solid cut*

BOLT

The bolt, like the nut, has two extruded solids. You join the second extruded solid to the first solid.

    1. Open the file *Nut.dwg* if you have already closed it.

    2. Select File>Save As and specify a new file name (file name: *Bolt.dwg*).

3. Referring to Figure 3–80, select the extruded feature, right-click, and select Edit.
4. In the Extrusion dialog box, set operation to Join, termination to Blind, and distance to 60, select the Flip button to flip the direction of extrusion, and select the OK button.
5. Right-click and select ENTER (see Figure 3–81).
6. Select the Update button from the browser.

The solid is modified (see Figure 3–82). Save and close your file.

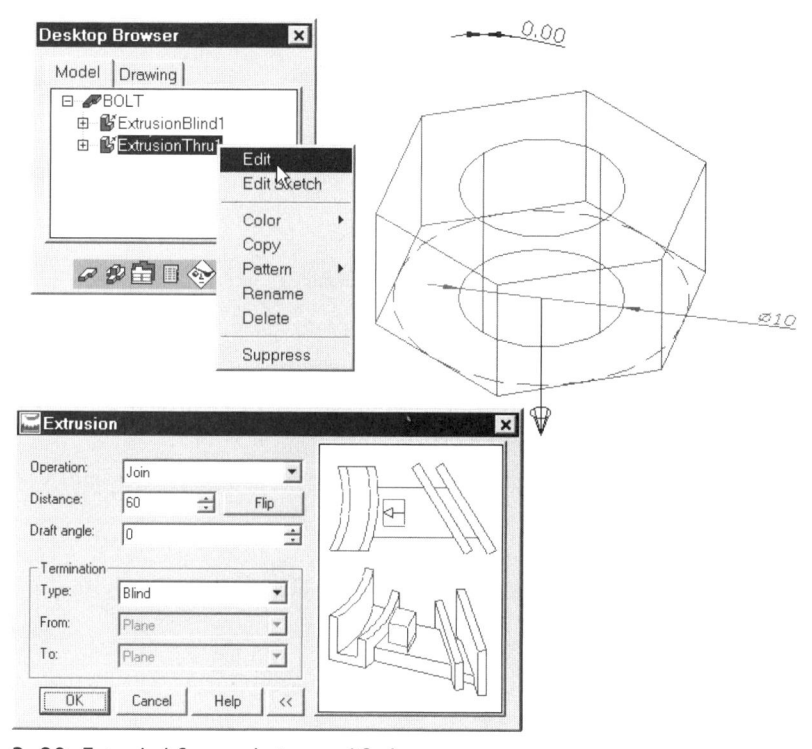

Figure 3–80 *Extruded feature being modified*

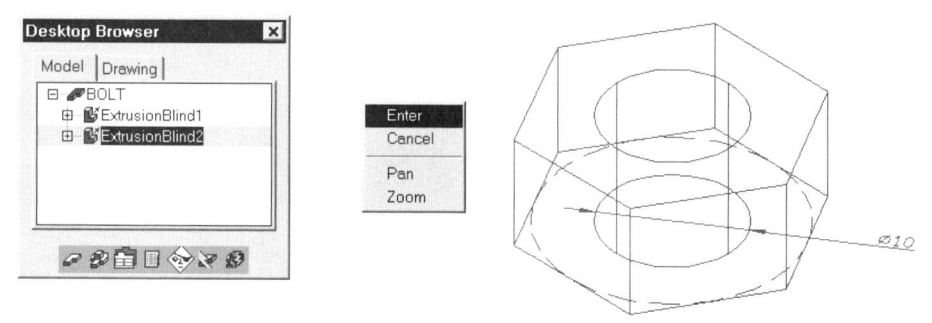

Figure 3–81 *Solid pending update*

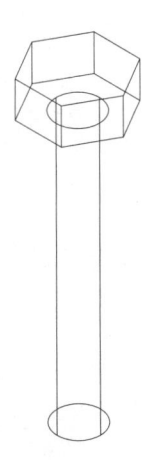

Figure 3–82 *Solid updated*

FRONTAL FRAME

Now you will construct two extruded solid features.

1. Open the file *Frame_F.dwg* if you have already closed it.
2. Referring to Figure 3–83, establish a sketch plane.
3. Construct a circle on the sketch plane, resolve it to a profile, and add parametric dimensions in accordance with Figure 3–84.
4. Select the profile from the browser, right-click, and select Extrude.

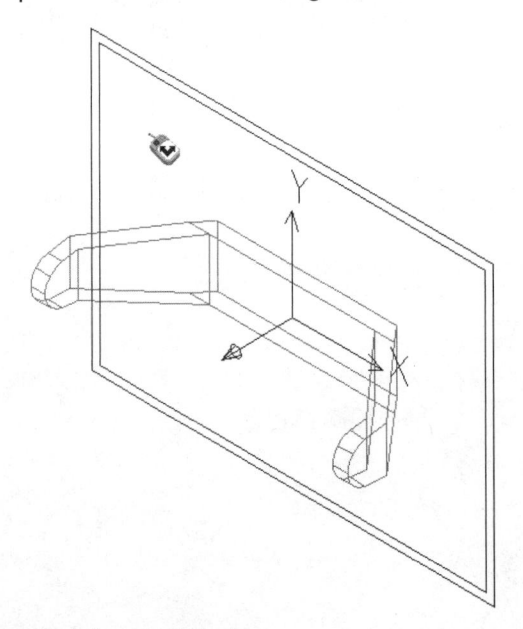

Figure 3–83 *Sketch plane being established*

Part Modeling I   103

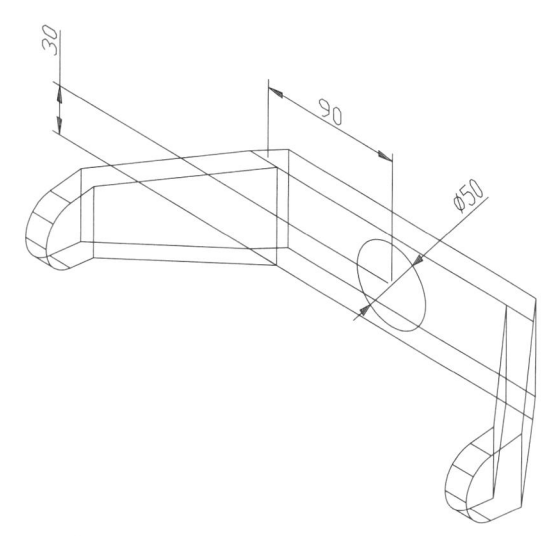

Figure 3–84 *Parametric dimensions added*

5. In the Extrusion dialog box, select Join operation, set termination to MidPlane and distance to 40, and select the OK button (see Figure 3–85).

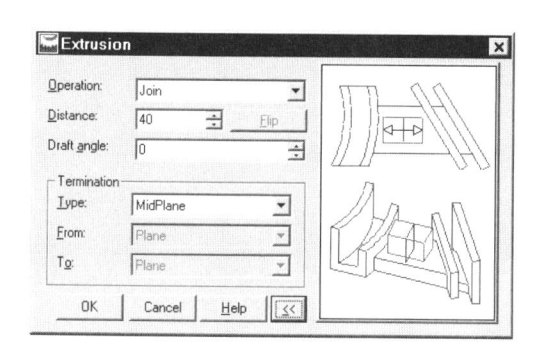

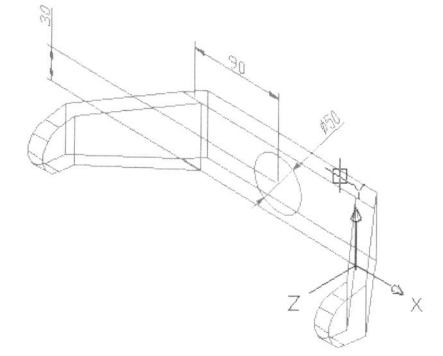

Figure 3–85 *Profile being extruded*

6. Referring to Figure 3–86, establish a new sketch plane.
7. Construct a circle (hidden line) and a rectangle.
8. Resolve the circle and rectangle to a profile.
9. Add tangent constraints to the circle and the rectangle.
10. Add concentric constraints to the circle and the cylindrical solid feature.
11. Add parametric dimension (30 mm) (see Figure 3–87).

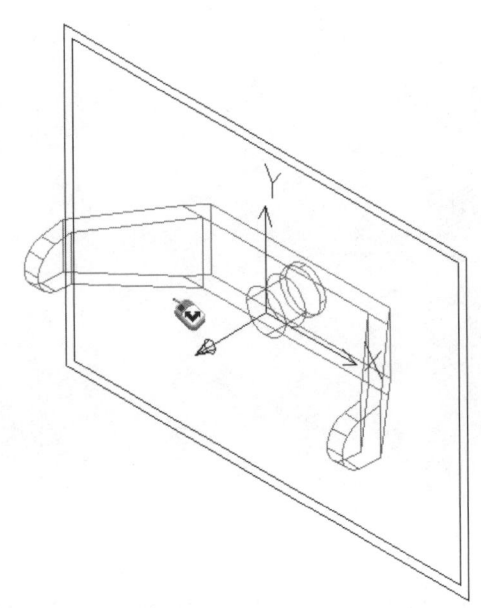

Figure 3–86 *New sketch plane being established*

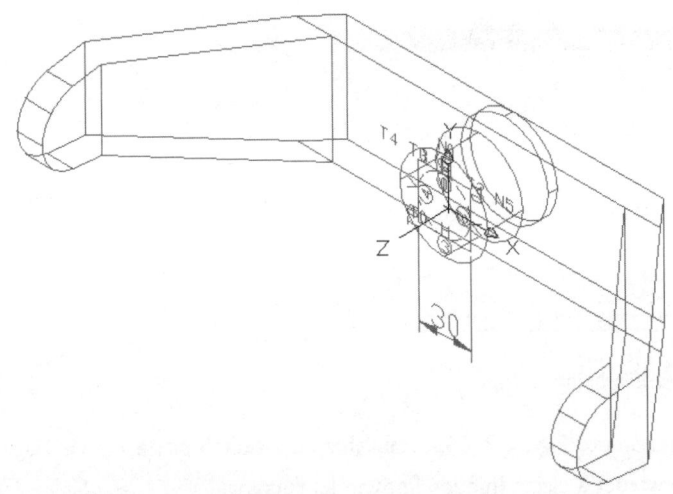

Figure 3–87 *Rectangle and circle constructed*

12. Select the profile from the browser, right-click, and select Extrude.
13. In the Extrusion dialog box, select Cut operation and Blind termination, set distance to 10, and select the OK button (see Figure 3–88).

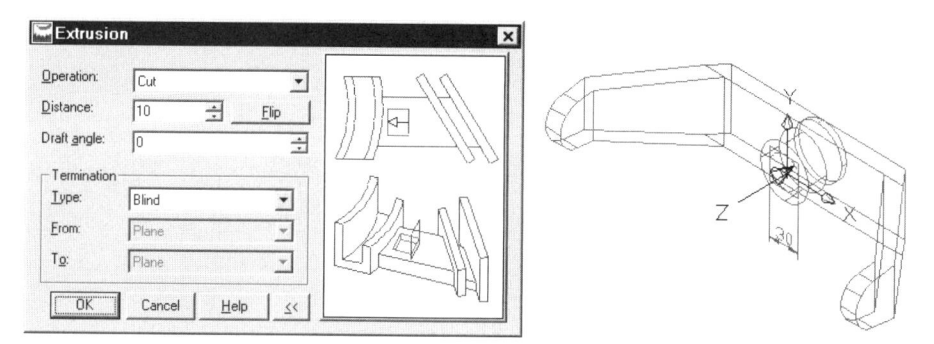

Figure 3-88 *Profile being extruded*

The solid is cut (see Figure 3-89). Save and close your file.

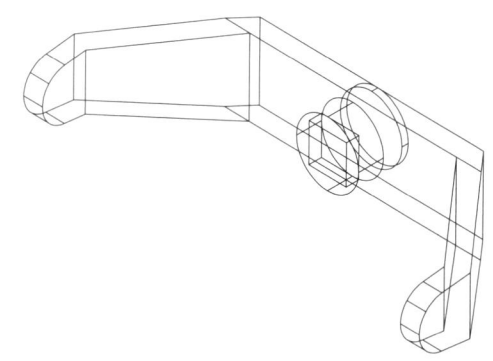

Figure 3-89 *Profile extruded*

PARTS FOR THE INFANT SCOOTER

In this chapter, you constructed a number of parts for the infant scooter. Some of these parts are completed, and some will be completed by adding more features in later chapters. Figure 3-90 shows the completed parts and Figure 3-91 shows the parts to be completed in later chapters.

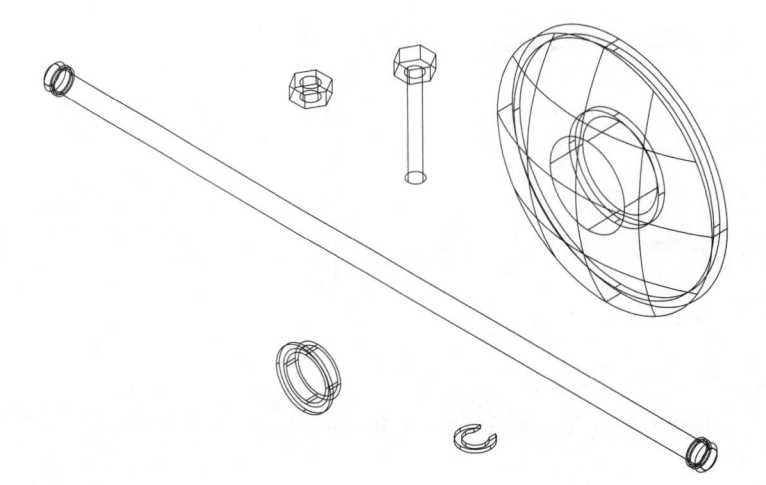

Figure 3–90 Completed component parts

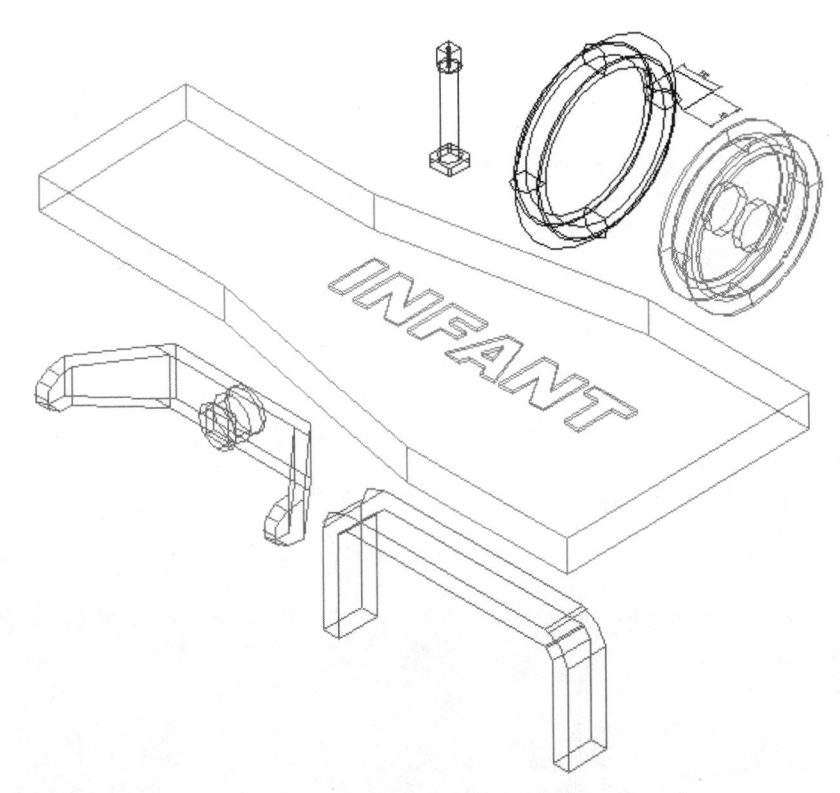

Figure 3–91 Component parts to be completed in later chapters

REVIEW QUESTIONS

1. Describe, with the aid of sketches, the process of making extruded, revolved, and bend solid features.

2. Explain the Boolean operations that you use to combine sketched solid features.

3. Delineate the ways to edit a solid part.

4. What mass properties can you retrieve from a solid part?

5. What attributes can you include in the part?

6. A cut operation can be performed without having a base solid feature. True or false?

CHAPTER 4

Part Modeling II

OBJECTIVES

The aims of this chapter are to delineate ways to construct work features and use work features in solid modeling; to practice constructing rib feature, various kinds of sweep features, and loft features; to introduce the use of placed features in design; and to depict various ways to construct hole, thread, fillet, chamfer, shell, and pattern features. After studying this chapter, you should be able to do the following:

- Construct work features
- Use work features to help make solid features
- Construct rib, sweep, and loft features
- State the use of placed solid features in solid modeling
- Construct hole, thread, fillet, chamfer, shell, and pattern features in a solid part

OVERVIEW

The fundamental way to make a parametric solid model is to construct parametric sketches, use the sketches to construct sketched solid features, and combine the solid features to form a complex solid. Besides extruded, revolved, and bend features that you learned about in the last chapter, you will learn how to construct three other kinds of sketched solid features: rib, sweep, and loft. Among them, sweep and loft solid features require multiple sketches. You need two sketches for a sweep solid and two or more sketches for a loft solid. To help establish sketch planes and other construction objects, you use work planes, work axes, and work points. Some common engineering features like hole, thread, fillet, and chamfer are repeatedly used in our design. To save time to construct these and other common features, they are provided as pre-constructed solid features in such a way that you select them from the menu, specify the parameters, and indicate the locations. Because you do not have to construct parametric sketches and you have only to place them in your solid, these pre-constructed solid features are called placed solid features. In this chapter, you will learn how to construct six basic kinds of placed solid features: Hole, thread, fillet, chamfer, shell, and array.

WORK FEATURES

Work axes, work planes, and work points are called work features. They are integral parts of the parametric solid part. Using them, you define and maintain parametric relationships among the solid features.

WORK PLANES

The work plane is a construction plane that you place in the solid part. You use a work plane to define a sketch plane and establish an artificial feature face for feature construction. To construct a work plane, select Part>Work Features>Work Plane. Figure 4–1 shows the Work Plane dialog box.

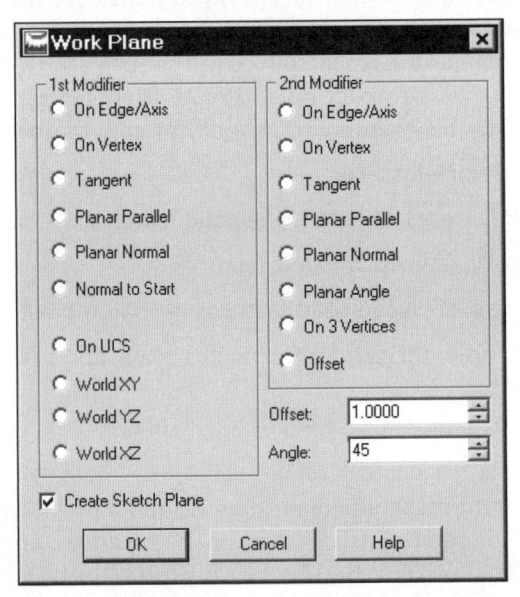

Figure 4–1 *Work Plane Feature dialog box*

You construct work planes in any of the following ways:

- Through two parallel edges/axes
- Through an edge/axis and a vertex
- Through an edge/axis and tangent to a cylindrical face
- Through an edge/axis and parallel to a plane
- Through an edge/axis and normal to a plane

Part Modeling II   111

- Through an edge/axis and at an angle to a plane
- Through a vertex and parallel to a plane
- Through three vertices
- Tangent to two cylindrical faces
- Tangent to a cylindrical face and parallel to a plane
- Tangent to a cylindrical face and normal to a plane
- Parallel to a plane and offset at a specified distance
- Normal to the start point of a path
- On the current UCS (User Coordinate System)
- On the XY plane of WCS (World Coordinate System)
- On the YZ plane of WCS
- On the XZ plane of WCS

Now you will learn how to use work planes in solid part modeling.

1. Start a new part file. Use metric as the default.
2. Construct a rectangle that measures 50 mm times 40 mm.
3. Resolve the rectangle to a profile, add parametric dimensions, and extrude the profile a distance of 30 mm (see Figure 4–2).
4. Select Part>Work Features>Work Plane

   Command: AMWORKPLN

5. In the Work Plane dialog box, select On Edge/Axis (1st Modifier) and On Edge/Axis (2nd Modifier), uncheck Create Sketch Plane, and select the OK button.
6. Select the edges indicated in Figure 4–3.

A work plane passing through two selected edges is constructed.

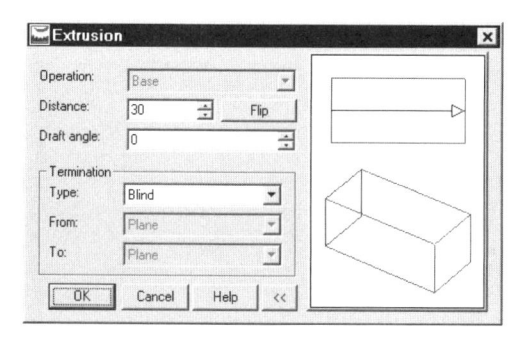

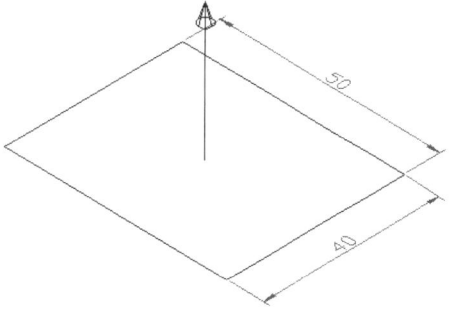

Figure 4–2 *Profile being extruded*

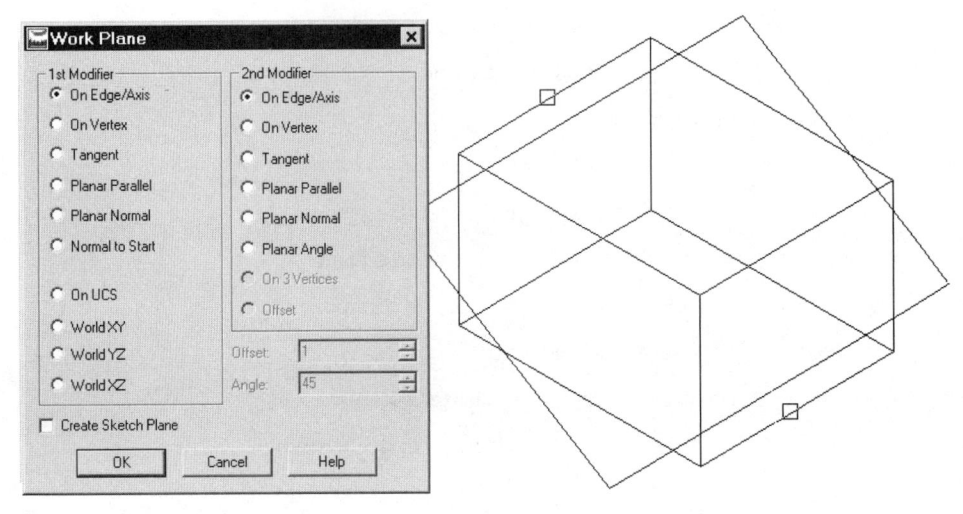

Figure 4–3 *Work plane passing through two edges being constructed*

7. Repeat the AMWORKPLN command.
8. In the Work Plane dialog box, select On Vertex (1st Modifier) and Planar Parallel (2nd Modifier), and select the OK button.
9. Select the vertex and work plane indicated in Figure 4–4.

A work plane parallel to the select plane and passing through the selected vertex is constructed.

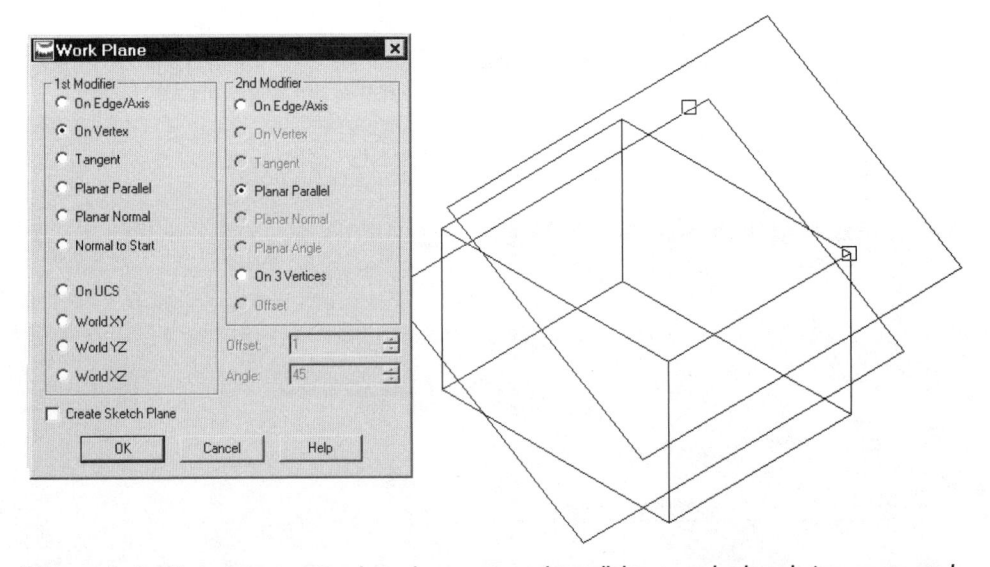

Figure 4–4 *Work plane passing through a vertex and parallel to a work plane being constructed*

Now you will construct an extruded solid feature from a work plane to another work plane.

10. Construct a new sketch by selecting the work plane indicated in Figure 4–5.

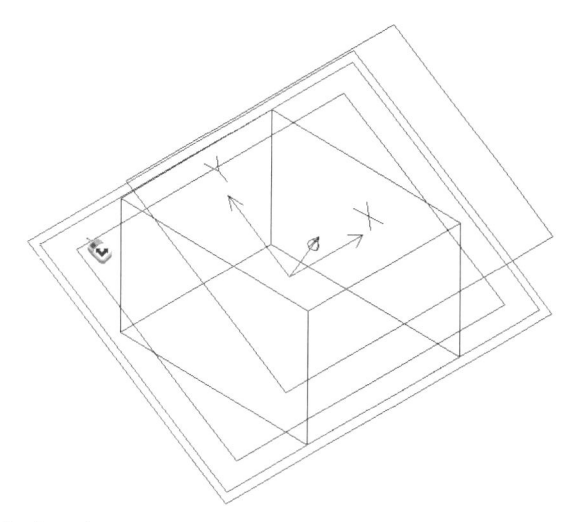

Figure 4–5 *Sketch plane being established on a work plane*

11. Construct a circle, resolve it to a profile, and add parametric dimensions as shown in Figure 4–6.
12. Select the profile from the browser, right-click, and select Extrude.
13. In the Extrusion dialog box, select Join operation and Plane termination and select the OK button.
14. Select the work plane indicated in Figure 4–6.

The profile is extruded to a plane.

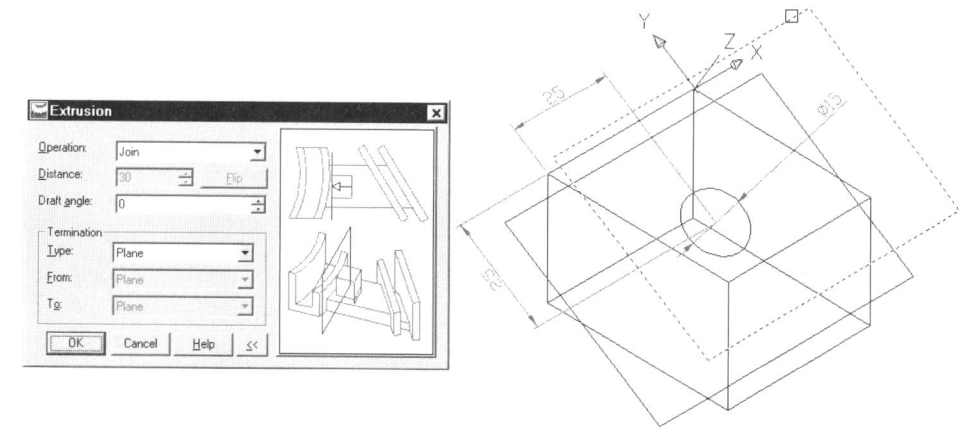

Figure 4–6 *Profile being extruded to the plane*

Now you will construct a work plane passing through three vertices and an extruded solid on the plane.

15. Select Part>Work Features>Work Plane
16. In the Work Plane dialog box, select On Vertex (1st Modifier) and On 3 Vertices (2nd Modifier), check the Create Sketch Plane button, and select the OK button.
17. Select the three vertices indicated in Figure 4–7.

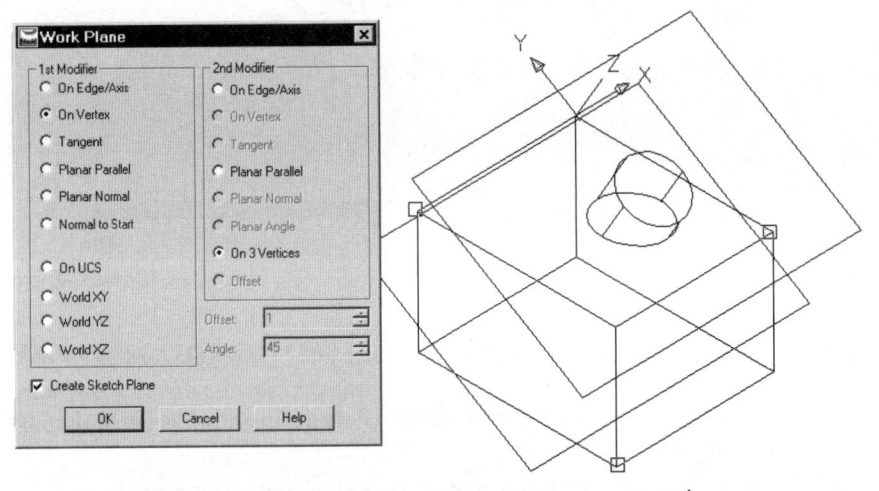

Figure 4–7 *Work plane passing through three vertices being constructed*

18. Right-click to accept (see Figure 4–8).

A work plane is constructed and a sketch plane is established on the work plane.

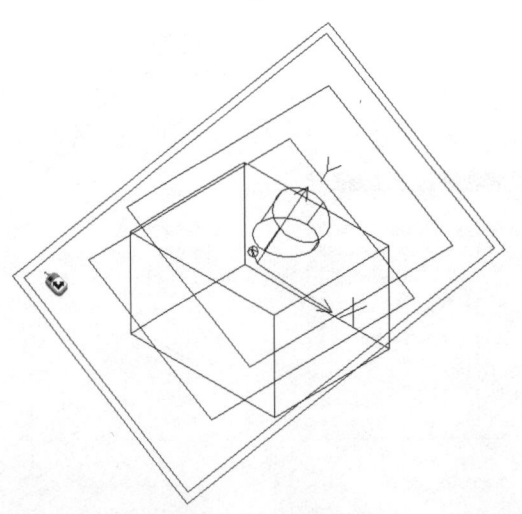

Figure 4–8 *Sketch plane being established*

Now you will construct a profile and extrude the profile to cut through the solid.

19. Construct three line segments, resolve them to a profile, and add project constraints to the end points of the profile in accordance with Figure 4–9. Using the Endpoint osnap will help in the selection.

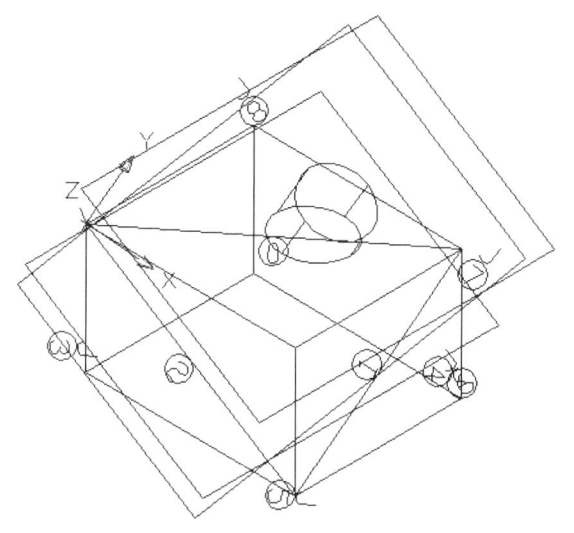

Figure 4–9 *Profile constructed*

20. Select the profile in the browser, right-click, and select Extrude.
21. In the Extrusion dialog box, select Cut operation and Through termination, select the Flip button so that the extrusion direction is in accordance with Figure 4–10, and select the OK button.

The profile is extruded to cut through the solid.

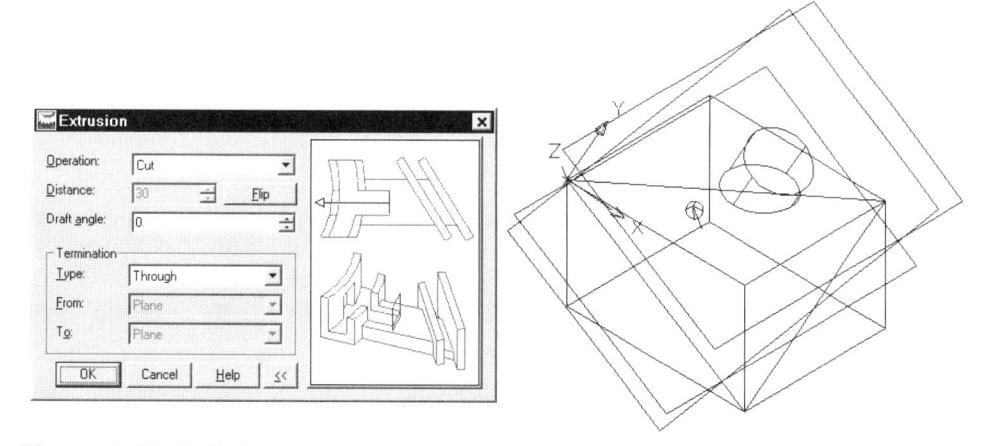

Figure 4–10 *Profile being extruded*

Now you will hide two work planes and construct an extruded solid.

22. Select the work planes highlighted in Figure 4–11 one by one, right-click, and deselect Visible.

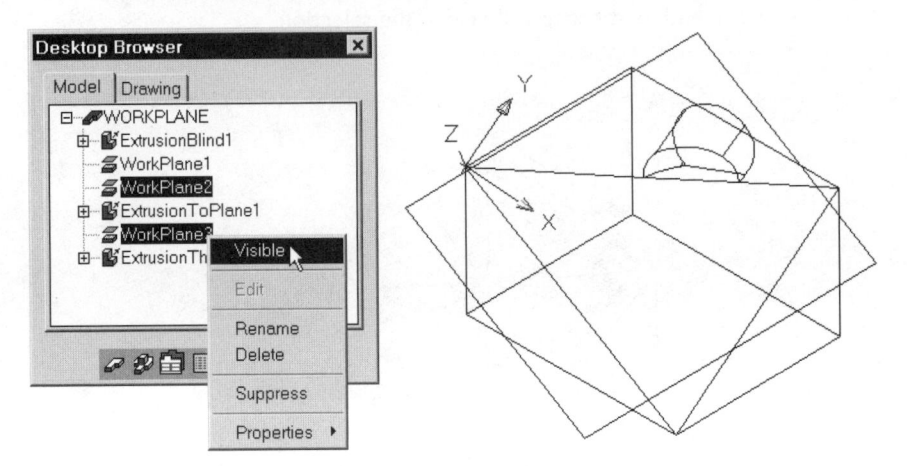

Figure 4–11 *Work plane hidden*

23. Referring to Figure 4–12, establish a sketch plane.
24. Construct a profile in accordance with Figure 4–13.

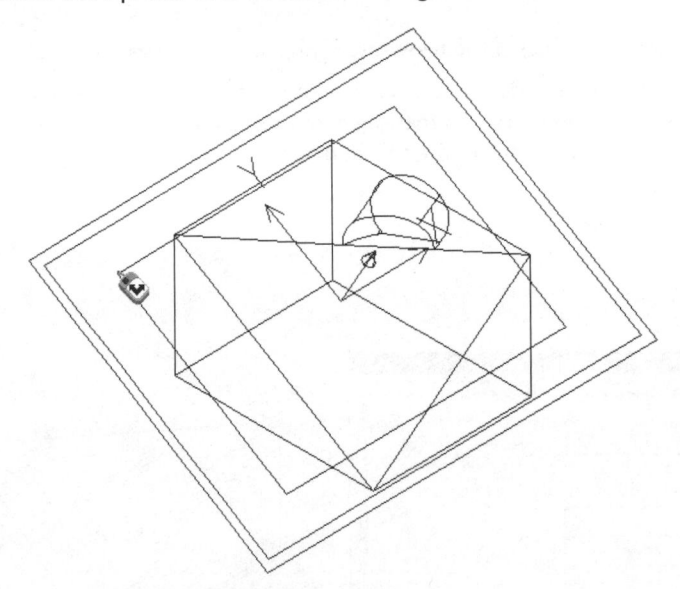

Figure 4–12 *Sketch plane being established*

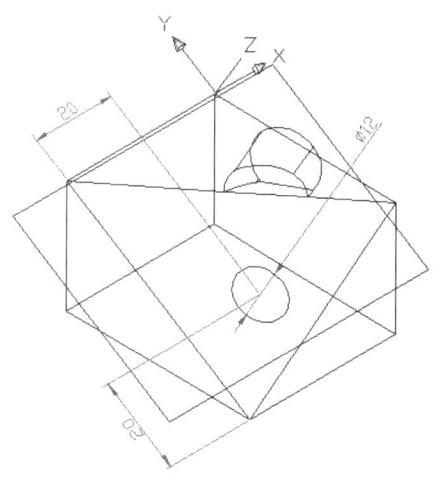

Figure 4-13 *Profile constructed*

25. Select the profile from the browser, right-click, and select Extrude.
26. In the Extrusion dialog box, select Join operation and Blind termination, set distance to 50, and select the OK button (see Figure 4-14).

Now you will hide all the work planes, construct a new work plane, and construct an extruded solid on the new work plane.

27. Referring to Figure 4-15, select the work plane from the browser, right-click, and deselect Visible.
28. Select Part>Work Features>Work Plane
29. In the Work Plane dialog box, select Tangent (1st Modifier) and Tangent (2nd Modifier), and select the OK button.
30. Select the cylindrical features indicated in Figure 4-16.

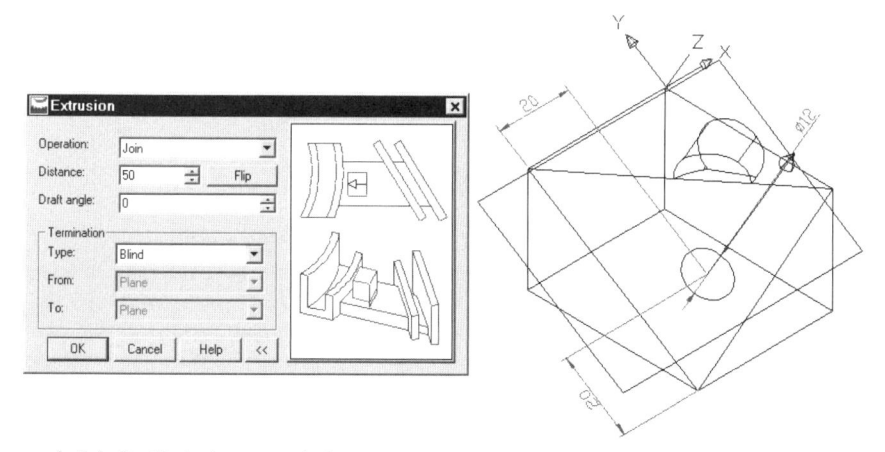

Figure 4-14 *Profile being extruded*

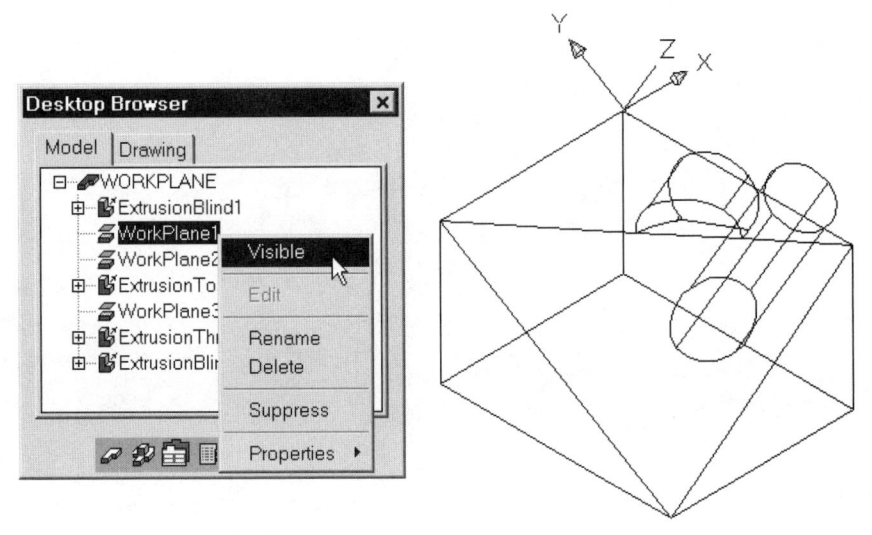

Figure 4–15 *Work plane hidden*

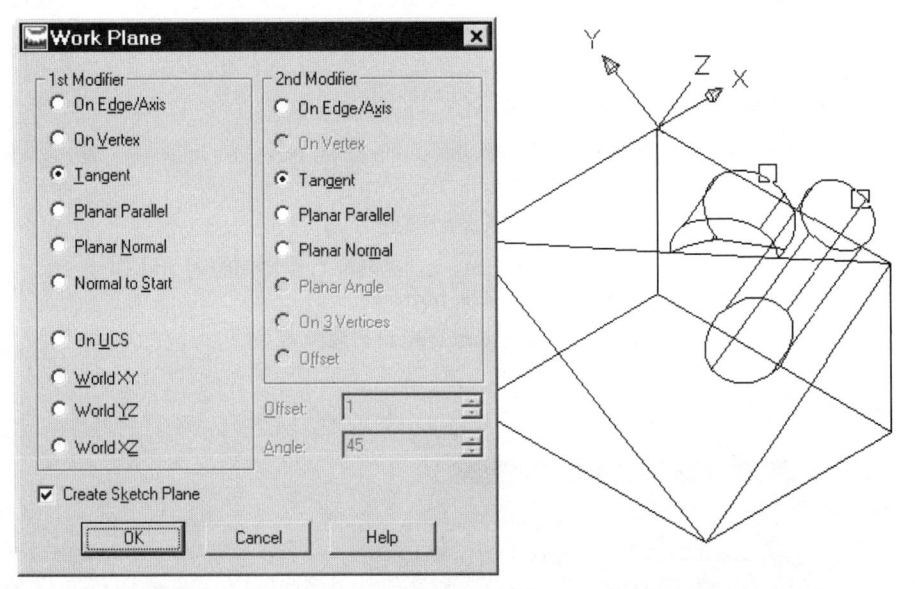

Figure 4–16 *Work plane tangential to two cylindrical features being constructed*

    31. Referring to Figure 4–17, establish a sketch plane.

Part Modeling II   119

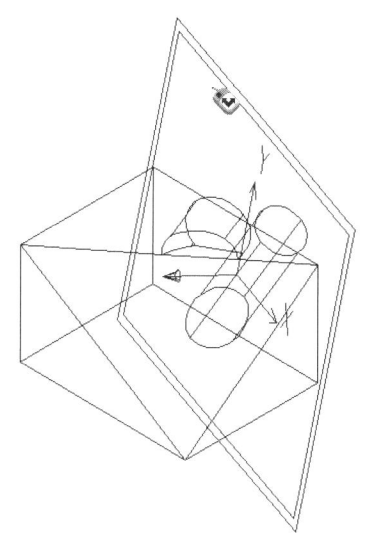

Figure 4–17 *Sketch plane being established*

32. Referring to Figure 4–18, construct a rectangle, resolve it to a profile, and extrude it to cut through the solid.
33. Hide the work plane.

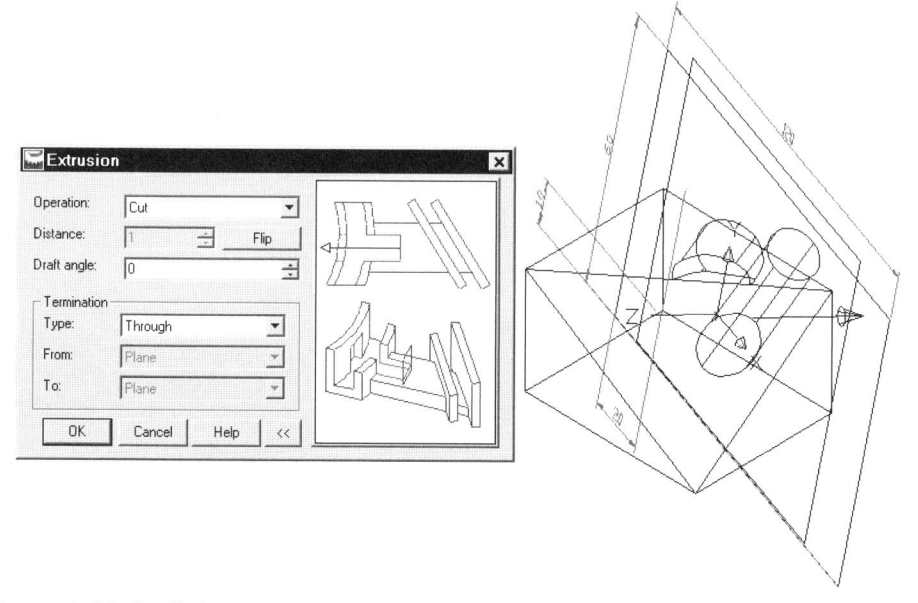

Figure 4–18 *Profile being extruded*

The solid is complete (See Figure 4–19). Save and close your file (File name: *WorkPlane.dwg*).

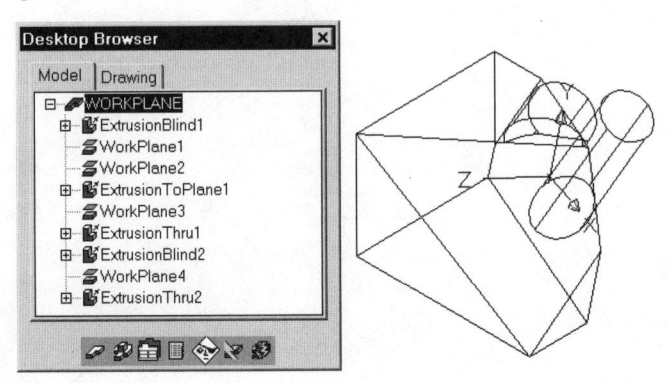

Figure 4–19 *Completed solid part*

Basic 3D Work Planes

From the onset of making a solid part, you can define three work planes parallel to the XY, YZ, and XZ of the UCS by specifying the location of the origin. The three work planes are called basic work planes. If you construct these basic work planes after you have already constructed a solid part, a second solid part is activated automatically. This solid part is called the tool body. You combine the tool body with a solid part to form a more complex solid. You will learn how to use the tool body in Chapter 9.

1. Start a new part file. Use metric as the default.
2. Set the display to an isometric view.
3. Select Part>Work Features>Basic 3D Work Features

   Command: AMBASICPLANES
4. Select a point in your screen (see Figure 4–20).

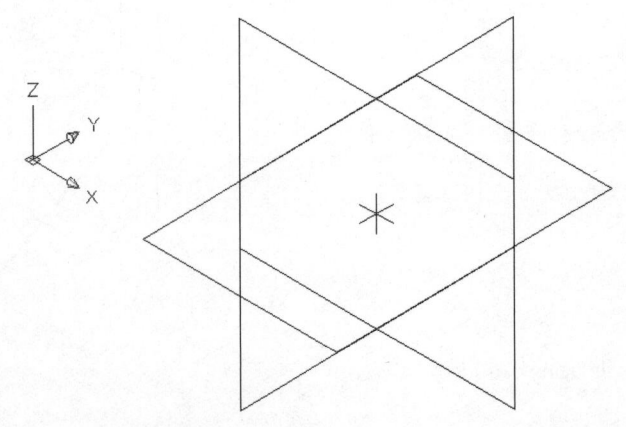

Figure 4–20 *Three basic work planes*

Three work planes, together with a work point at the intersection of the work planes, are constructed. Now save and close your file (file name: *BasicWorkPlanes.dwg*).

WORK AXIS

A work axis is a parametric center line. You construct a work axis in two ways:

- Specify two points on a sketch plane
- Select a cylindrical or elliptical object

The work axis helps you place parametric dimensions; construct a work plane through the center of cylindrical, elliptical, or torus objects; define the center of polar array; and define the axis of a helical sweep feature.

Work Axis on a Sketch Plane

A work axis can be constructed by specifying two points on a sketch plane. Now you will construct a work axis on a sketch plane. You will use this work axis for making a helical spring later in this chapter.

1. Start a new part file. Use metric as the default.
2. Select Part>Work Features>Work Axis

    Command: AMWORKAXIS

3. Select two points on the screen to indicate the orientation of the work axis.

The work axis is complete. Save your file (file name: *Spring.dwg*).

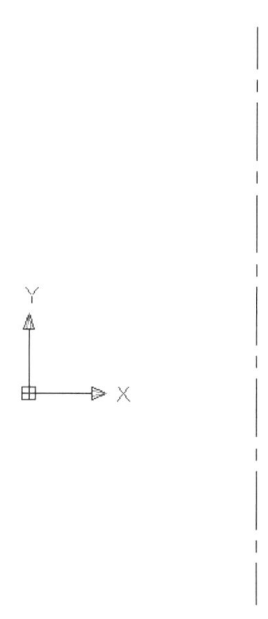

Figure 4–21 *Work axis constructed*

Work Axis on Cylindrical/Elliptical Feature

A work axis can be constructed by selecting a cylindrical or elliptical feature. Now you will construct an elliptical extruded solid with a work axis passing through the center of the elliptical solid.

1. Start a new part file. Use metric as the default.
2. Construct an ellipse.
3. Resolve the ellipse to a profile.

Now you will add two dimensions to the profile. To dimension an ellipse, you need to use the QUA object snap.

4. Select Part>Dimensioning>New Dimension
   Command: AMPARDIM
5. Type QUA at the command line area to use the QUA object snap.
6. Referring to Figure 4–22, select the quadrant of the ellipse, a point on the ellipse (to use the center), and a location to specify the dimension location.
7. Type **50** to specify a dimension.

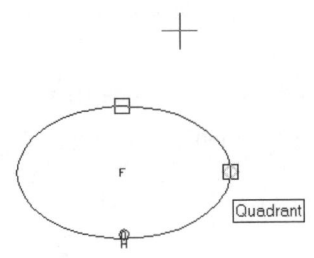

Figure 4–22 *Quadrant and center of the ellipse and dimension location selected*

8. Repeat the AMPARDIM command.
9. Type QUA at the command line area.
10. Referring to Figure 4–23, select the quadrant and center of the ellipse, and a point to indicate the dimension location.
11. Type **30** to specify the dimension.
12. Press the ENTER key.

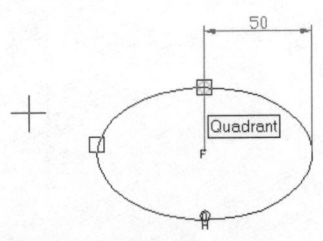

Figure 4–23 *Second dimension being constructed*

Part Modeling II   123

Now you will extrude the profile and construct a work axis.

13. Set the display to an isometric view.
14. Select the profile from the browser, right-click, and select Extrude to extrude the profile a distance of 120 mm (see Figure 4–24).

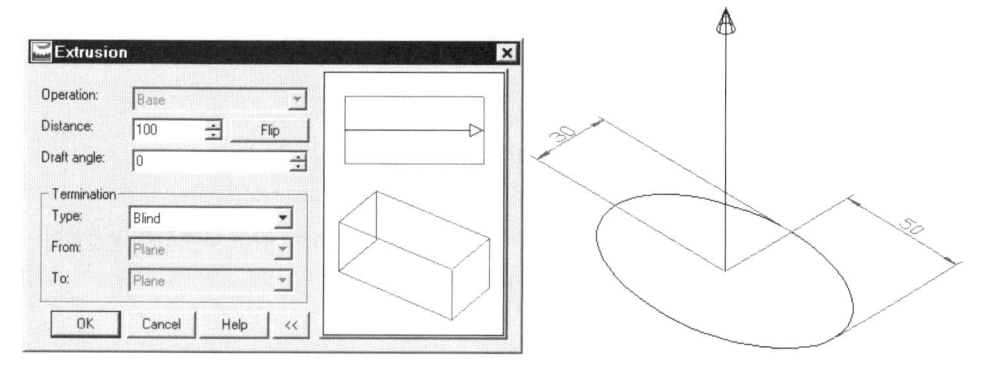

Figure 4–24 *Elliptical solid feature being constructed*

15. Select Part>Work Features>Work Axis
16. Select the elliptical solid feature.

A work axis is constructed at the center of the elliptical solid feature (see Figure 4–25). Save your file (file name: *Ellipse.dwg*).

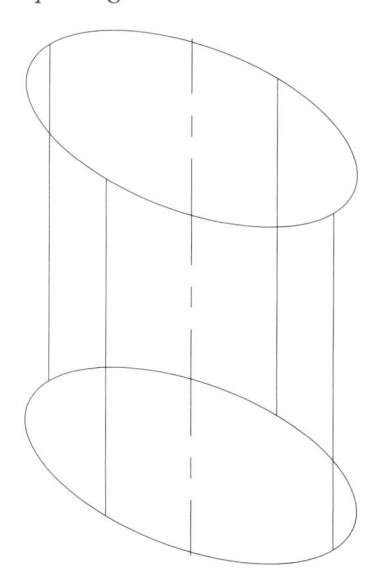

Figure 4–25 *Work axis constructed*

WORK POINTS

A work point is a construction point you place on a sketch plane. Using a work point, you locate the position of a hole and the center of polar array.

Now you will construct the central frame of the infant scooter. You will construct an extruded solid, establish a sketch plane on the solid, construct a work point on the sketch plane, and use the work point to construct a hole later. Figure 4–26 shows the completed model.

1. Start a new part file. Use metric as the default.
2. Construct a rectangle that measures 650 mm by 60 mm.
3. Referring to Figure 4–27, extrude the profile a distance of 20 mm.
4. Establish a sketch plane on the top face of the extruded solid (see Figure 4–28).
5. Select Part>Work Features>Work Point

   Command: AMWORKPT
6. Select a point on the sketch plane.
7. Referring to Figure 4–29, add two parametric dimensions to locate the work point.

The extruded solid and the work point are constructed. Save your file (file name: *Frame_C.dwg*).

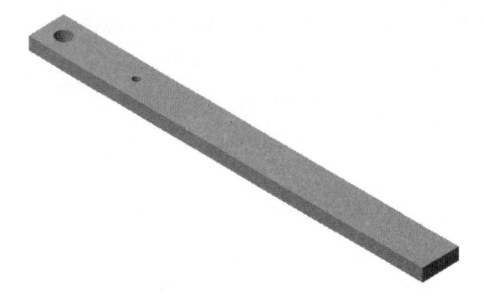

Figure 4–26 *Central frame of the infant scooter*

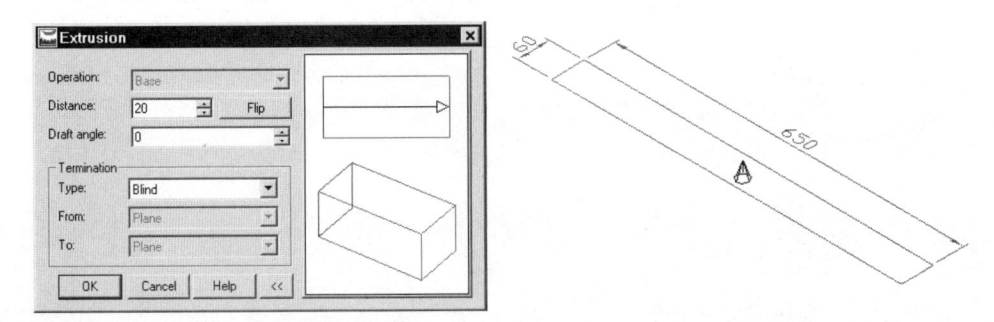

Figure 4–27 *Profile being extruded*

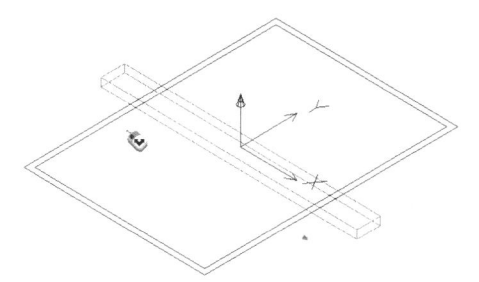

Figure 4–28 *Sketch plane being established*

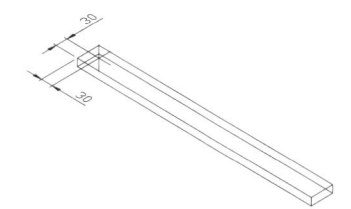

Figure 4–29 *Parametric dimensions constructed*

VISIBILITY OF WORK FEATURES

Work features are essential construction objects that are required in the modeling process. Once they are used in constructing solid features, you must not discard them. If you delete a work feature, all its dependent features will be deleted as well. If you do not want to see them in your model, you can hide them.

By selecting Part>Part Visibility, you use the Desktop Visibility dialog box (shown in Figure 4–30) to control the visibility of work planes, work axes, work points, cutting lines, and threads. (You will learn how to use cutting lines in Chapter 10 and how to construct threads later in this chapter.)

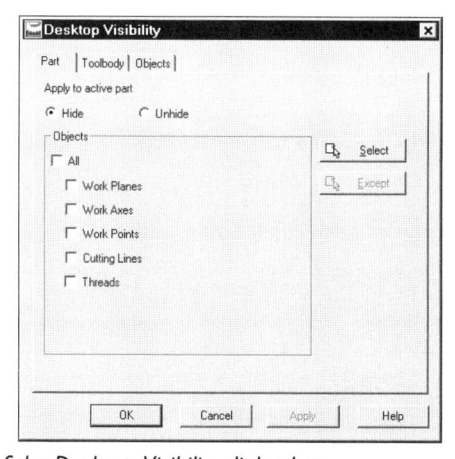

Figure 4–30 *Part tab of the Desktop Visibility dialog box*

SKETCHED FEATURES: RIB, SWEEP, AND LOFT

After introducing you to the three kinds of work features, you will now learn how to construct rib, sweep, and loft features. Common to the work features, you use sketch features as construction tools and as a means to maintain parametric relation with other features in the solid part.

RIB FEATURE

In engineering design, a rib feature in a component serves as a reinforcing object. Figure 4–31 shows an angle piece with a rib reinforcement. Set up a work plane that cuts across an existing solid part, construct an open loop profile to depict the outline of the rib feature, and use the open loop profile to construct the rib feature.

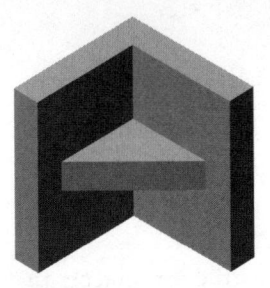

Figure 4–31 *Angle bracket*

Now you will construct an extruded solid.

1. Start a new part file. Use metric as the default.
2. Referring to Figure 4–32, construct a sketch, resolve it to a profile, and add geometric constraints and parametric dimensions.

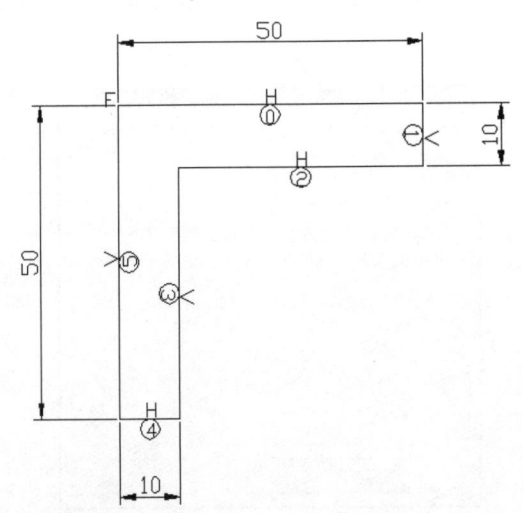

Figure 4–32 *Profile constructed*

3. Extrude the profile a distance of 60 mm (see Figure 4–33).
4. Construct a work plane parallel to and offset a distance of 30 mm from the face indicated in Figure 4–33. In the Work Plane dialog box, check the Create Sketch Plane box to establish a sketch plane on the new work plane.
5. Select the face indicated in Figure 4–34. Press the right mouse button to cycle until the face is selected and left-click to accept.
6. Left-click to flip the direction (if required) and right-click to accept.

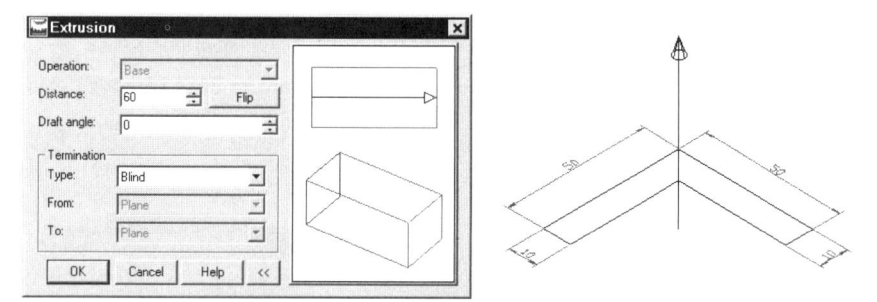

Figure 4–33 *Profile being extruded*

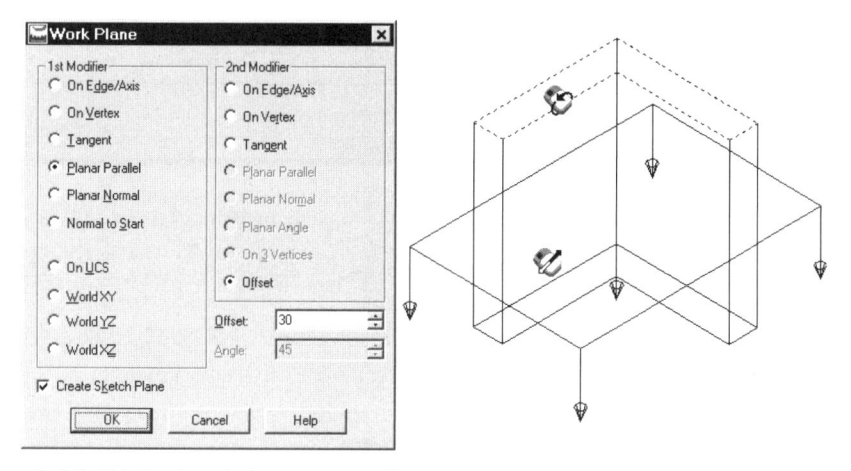

Figure 4–34 *Work plane being constructed*

A work plane is constructed and a sketch plane is established on the work plane (see Figure 4–35).

7. Use the shortcut key 9 to set the display to the top view of the sketch plane.
8. Construct a line.
9. Select Part>Sketch Solving>Profile
10. Select the line and press the ENTER key to resolve it to an open profile.

11. Add two project constraints and two parametric dimensions in accordance with Figure 4-36.
12. Set the display to an isometric view.
13. Select the open profile from the browser, right-click, and select Rib.
14. In the Rib dialog box, select Two Directions, set Thickness 1 to 4 and Thickness 2 to 6, and select the OK button (see Figure 4-37).
15. Select the work plane from the browser, right-click, and deselect Visible.

The rib feature is constructed and the work plane is hidden (see Figure 4-31). Save and close your file (file name: *Rib.dwg*).

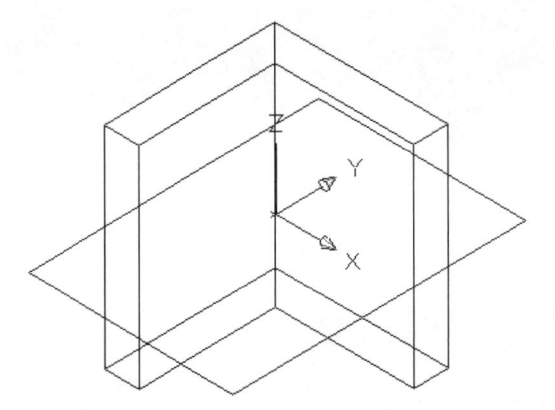

Figure 4-35 *Sketch plane established on the work plane*

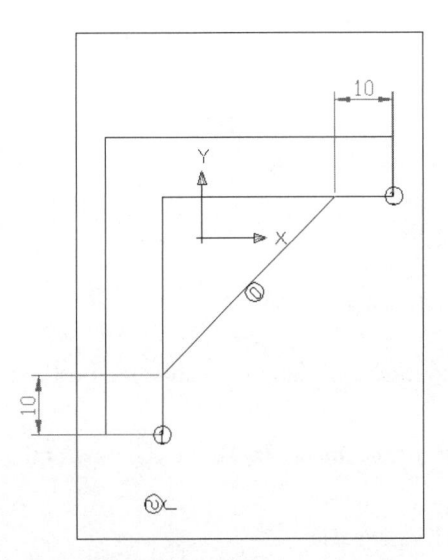

Figure 4-36 *Open profile constructed*

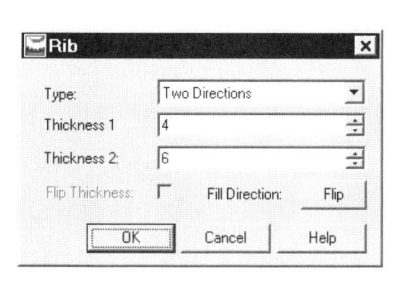

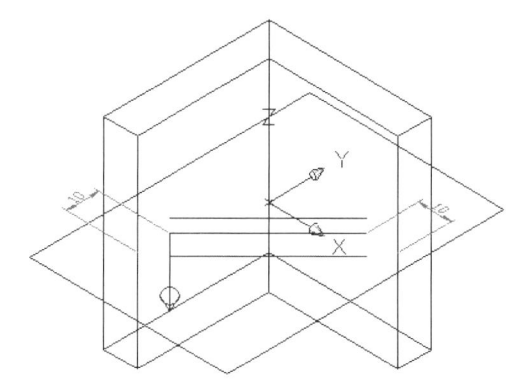

Figure 4-37 *Rib being constructed*

SWEEP FEATURES

To make a sweep solid, you construct two parametric sketches: a path sketch and a profile sketch normal to the start point of the path, and sweep the profile along the path. There are five kinds of paths: 2D path, 3D helix path, 3D edge path, 3D pipe path, and 3D spline path.

Sweeping 2D Path

A 2D path is a parametric sketch that lies on a sketch plane. Construction of a 2D path is similar to making a profile. You construct a sketch, resolve the sketch to a path, manipulate geometric constraints, and add parametric dimensions. After you have constructed a path, construct a profile on a plane normal to the start of the path.

Now you will construct a 2D path.

1. Start a new part file. Use metric as the default.
2. Referring to Figure 4-38, construct two lines and two arcs.

Figure 4-38 *Sketch*

3. Select Part>Sketch Solving>2D Path

   Command: AM2DPATH

4. Select the lines and arcs and press the ENTER key.
5. Select the left end of the sketch to indicate the start point of the path.

Now you have to decide whether or not you want a work plane perpendicular to the start point of the path.

6. Type **N** at the command line area.
7. Apply equal radius constraint to the arcs.
8. Apply equal length constraint to the lines.
9. Apply tangent constraints to the arcs and lines.
10. Add three parametric dimensions (see Figure 4–39).

Now you will construct a work plane normal to the start of the path and construct a profile.

11. Set the display to an isometric view.
12. Select Part>Work Features>Work Plane
13. In the Work Plane dialog box, select Normal to Start, check the Create Sketch Plane box, and select the OK button.
14. Referring to Figure 4–40, right-click to accept.

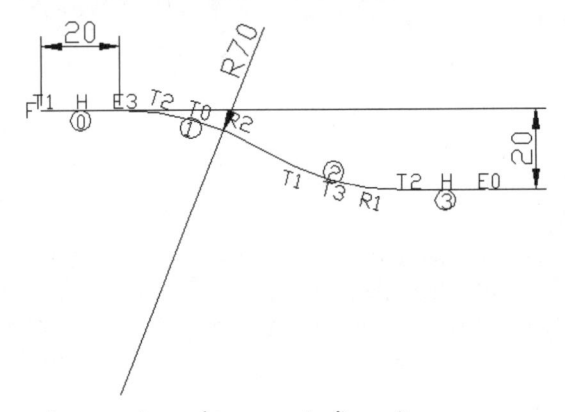

Figure 4–39 *Geometric constraint and parametric dimensions*

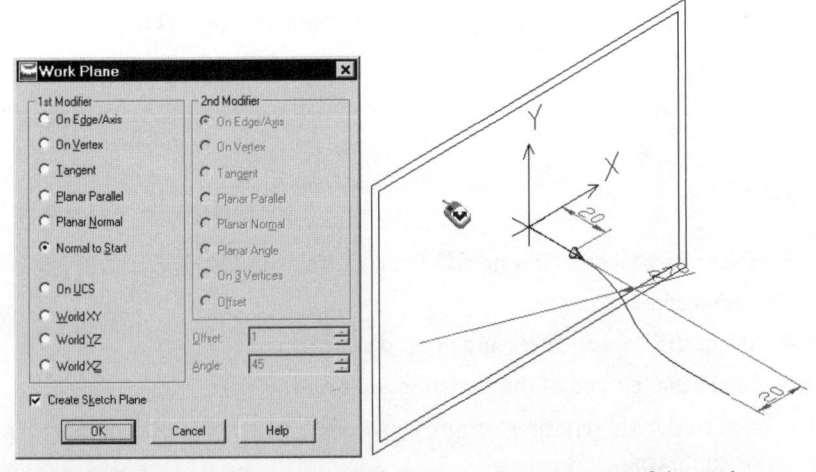

Figure 4–40 *Work plane being constructed normal to the start point of the path*

15. Construct a circle and resolve it to a profile.
16. Delete the fix constraint so that you can reposition it.
17. Referring to Figure 4–41, add three parametric dimensions. (The zero dimensions locate the center of the profile in relation to the work point at the start point of the 2D path.)

Now you will construct a sweep solid.

18. Select Part>Sketched Features>Sweep

    Command: AMSWEEP

19. In the Sweep dialog box, select the OK button (see Figure 4–42).
20. Select the work plane in the browser, right-click, and deselect Visible (see Figure 4–43).

A sweep solid is constructed and the work plane is hidden. Save and close your file (file name: *2DSweep.dwg*).

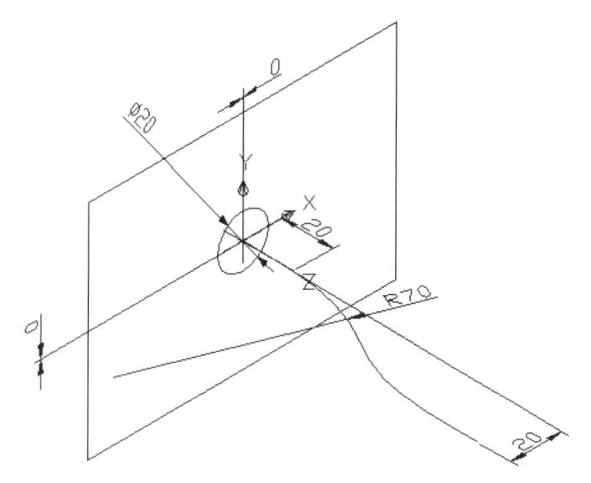

Figure 4–41 *Profile constructed*

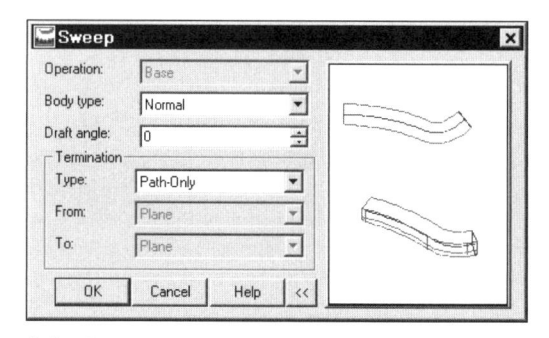

Figure 4–42 *Sweep dialog box*

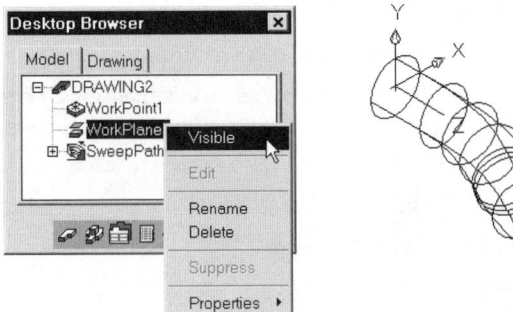

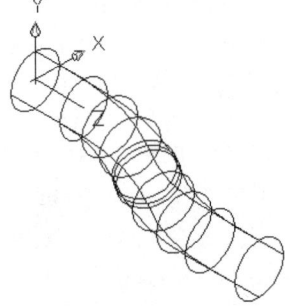

Figure 4–43 *Sweep solid constructed and work plane hidden*

Sweeping 3D Helix Path

A helix path is a parametric helical curve. To construct a helical path, you select an axis, a cylindrical object, or an elliptical object and specify the parameters of the helix. Prior to making the helical path, you have to set up the sketch plane perpendicular to the axis of the helix.

Now you will construct a helical path.

1. Open the file *Spring.dwg* that you constructed earlier in this chapter.
2. Set the display to an isometric view.
3. Select Part>Work Features>Work Plane
4. In the Work Plane dialog box, select Normal to Start, check the Create Sketch Plane box, and select the OK button.
5. Select the end point of the work axis.
6. Referring to Figure 4–44, right-click to accept.

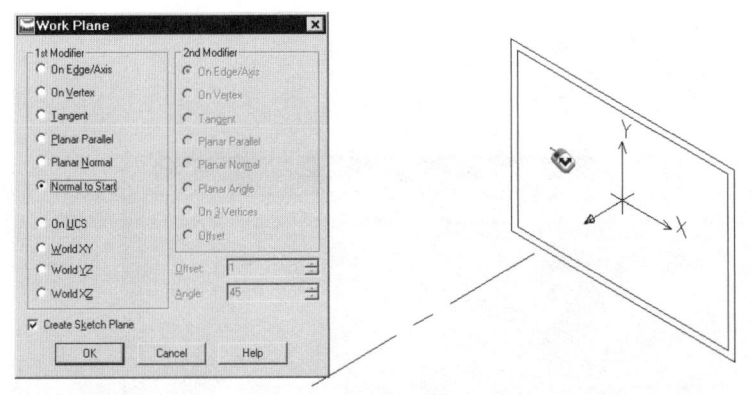

Figure 4–44 *Work plane being constructed normal to the start of the work axis*

A work plane is constructed and a sketch plane is established normal to the start point of the work axis. Now you will construct a helical path.

7. Select Part>Sketch Solving>3D Helix Path
8. Select the work axis as the helical center.
9. In the Helix dialog box, select Height and Pitch in the Type box, set pitch to 12 mm and height to 48 mm, select Circle in the shape box, set the diameter to 30 mm, and select the OK button (see Figure 4–45).

A helical path is complete. Now you will construct a work plane at the start point of the helical path and construct a profile.

10. Select Part>Work Features>Work Plane
11. In the Work Plane dialog box, select Normal to Start, check the Create Sketch Plane box, and select the OK button.
12. Select the helical path.
13. Referring to Figure 4–46, right-click to accept.
14. Construct a circle, resolve it to a profile, and delete the fix constraint.
15. Add three parametric dimensions (see Figure 4–47).

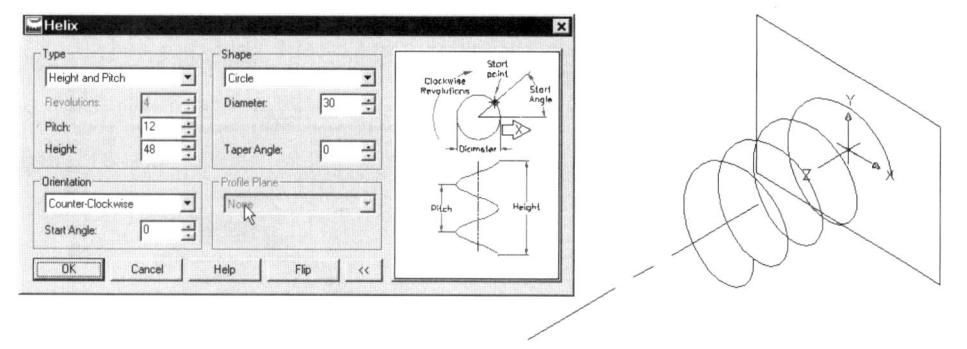

Figure 4–45 *Helical path being constructed*

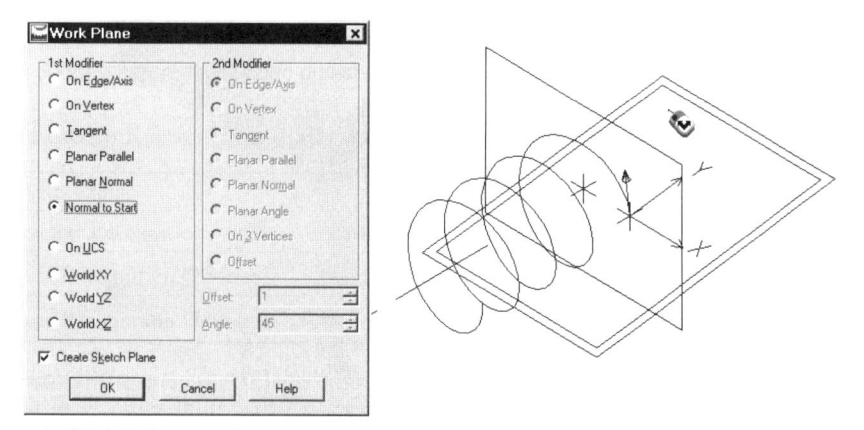

Figure 4–46 *Work plane being constructed at the start of the helical path*

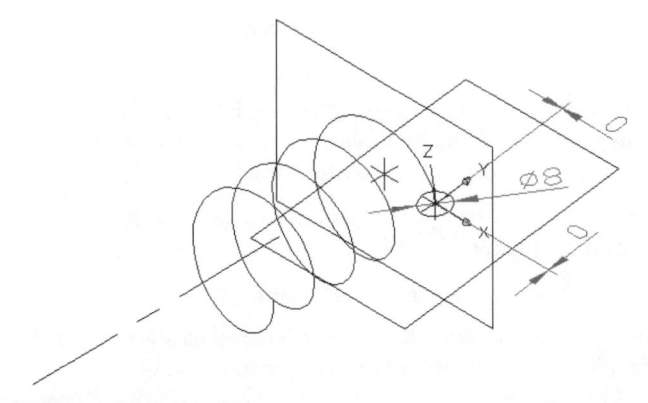

Figure 4–47 *Profile constructed*

Now you will sweep the profile along the path.

16. Select Part>Sketched Features>Sweep
17. In the Sweep dialog box, select the OK button.
18. Select the work planes from the browser one by one, right-click, and deselect Visible.

The sweep solid is constructed and the work planes are hidden (see Figure 4–48). Save and close your file.

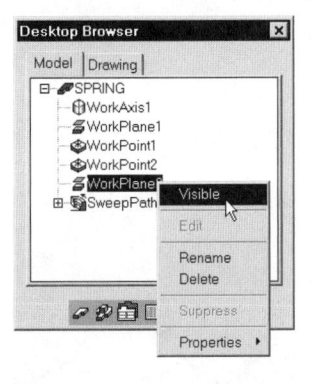

Figure 4–48 *Sweep solid constructed*

Sweeping 3D Edge Path

A 3D edge path derives from the edges of an existing solid part. To construct a 3D edge path, select edges from a solid part. Because an edge path has a parametric relation with the edges from which the path is derived, the edge path will change automatically if the solid part's edges change. Now you will construct an extruded solid, construct an edge path by selecting edges of the solid, construct a profile at the start point of the path, and sweep the profile along the path to construct a sweep solid.

1. Start a new part file. Use metric as the default.
2. Construct a rectangle, resolve it to a profile, and add a parametric dimension (50 mm times 40 mm).
3. Referring to Figure 4–49, extrude the profile a distance of 30 mm.

Now you will construct a 3D edge path.

4. Select Part>Sketch Solving>3D Edge Path
5. Referring to Figure 4–50, select three edges and press the ENTER key.
6. Select the lower-right end of the path to indicate the start of the 3D edge path.
7. Type **Y** at the command line area to construct a work plane at the start point of the path.
8. Referring to Figure 4–51, right-click to accept.

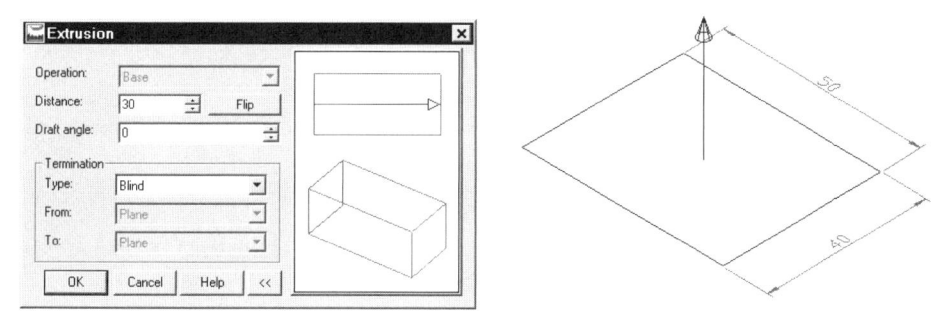

Figure 4–49 *Profile being extruded*

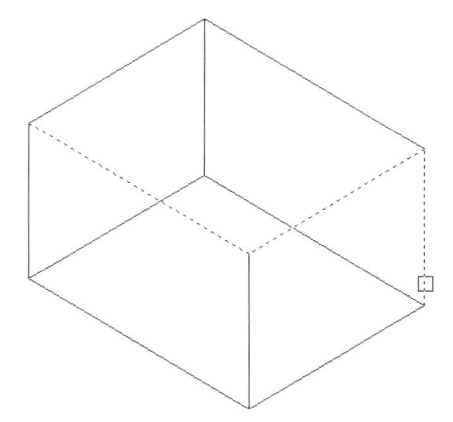

Figure 4–50 *Edges of solid part selected*

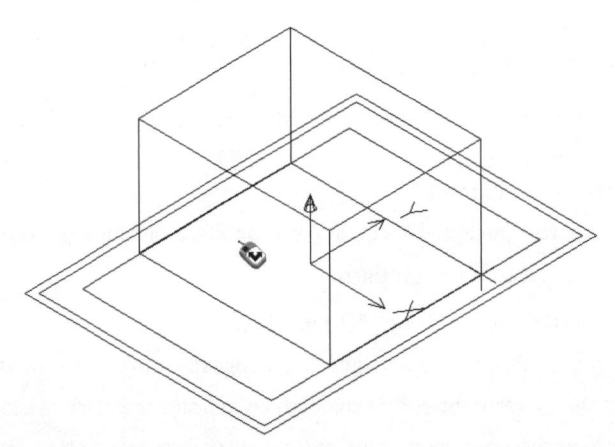

Figure 4–51 *Work plane being constructed at the start point of the 3D edge path*

The 3D edge path is complete. Now you will construct a profile on a work plane normal to the start point of the path.

    9. Construct a circle, resolve it to a profile, and add three parametric dimensions (see Figure 4–52).

Now you will sweep the profile along the 3D edge path and combine the sweep solid to the solid part by cutting.

    10. Select Part>Sketched Features>Sweep

    11. In the Sweep dialog box, select Cut operation and select the OK button (see Figure 4–53).

    12. Select the work plane and the work point from the browser one by one, right-click, and deselect Visible.

The sweep solid is constructed and the work features are hidden (see Figure 4–54). Save and close your file (file name: *EdgeSweep.dwg*).

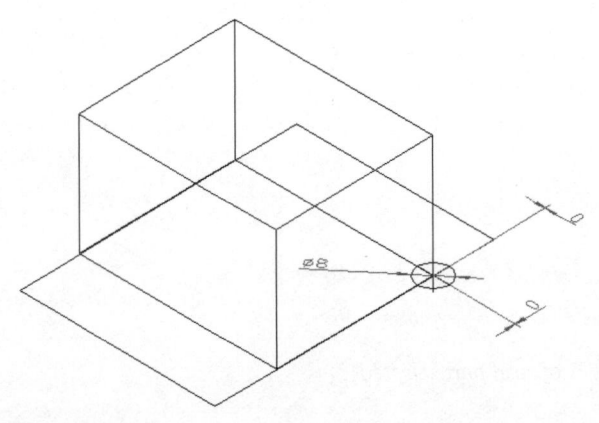

Figure 4–52 *Profile constructed*

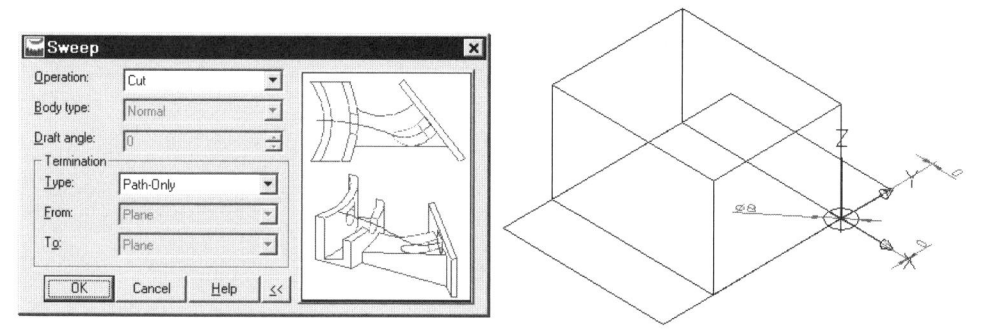

Figure 4–53 *Sweep solid being constructed*

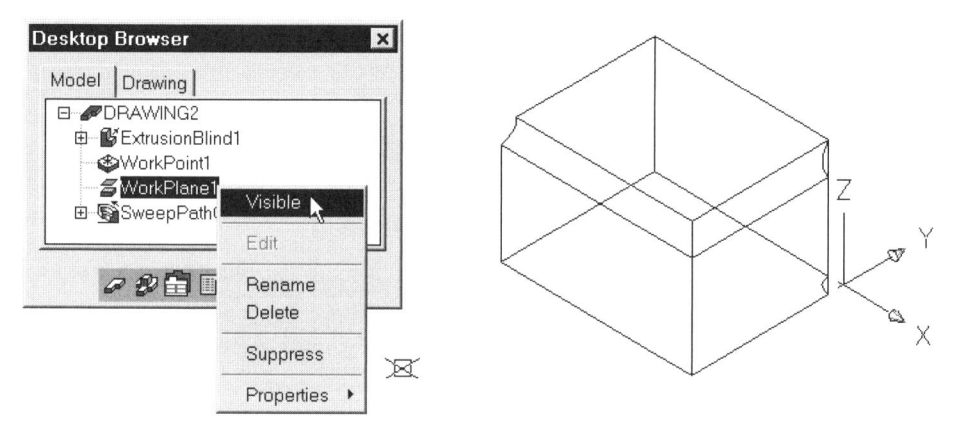

Figure 4–54 *Sweep solid cut*

Sweeping 3D Pipe Path

To construct a 3D pipe path, you construct a 2D polyline, 3D polyline, or a set of line segments and resolve the polyline or line segments. To incorporate a parametric relationship with other solid features, you construct work points and use object snap to snap to the work points while constructing the polyline or line segments.

Now you will construct a set of work points, construct a 3D polyline that passes through the work points, resolve the polyline to a 3D pipe path, construct a profile at the start point of the path, and construct a sweep solid by sweeping the profile along the path.

1. Start a new part file. Use metric as the default.
2. Set the display to isometric view.
3. Construct three basic work planes.
4. Referring to Figure 4–55, construct two work points on the horizontal work plane and add four parametric dimensions.

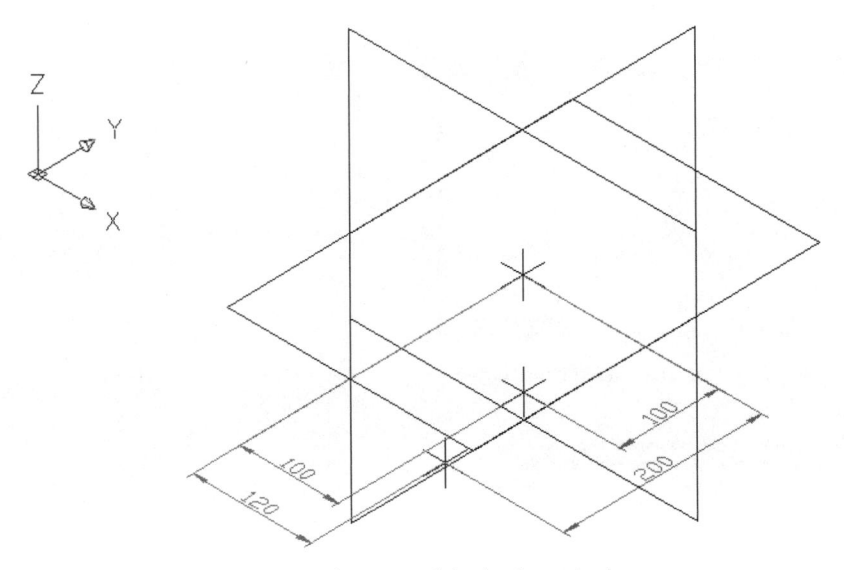

Figure 4-55 Work points constructed on one of the basic work planes

5. Referring to Figure 4–56, establish a sketch plane.
6. Construct a work point and add two parametric dimensions (see Figure 4–57).

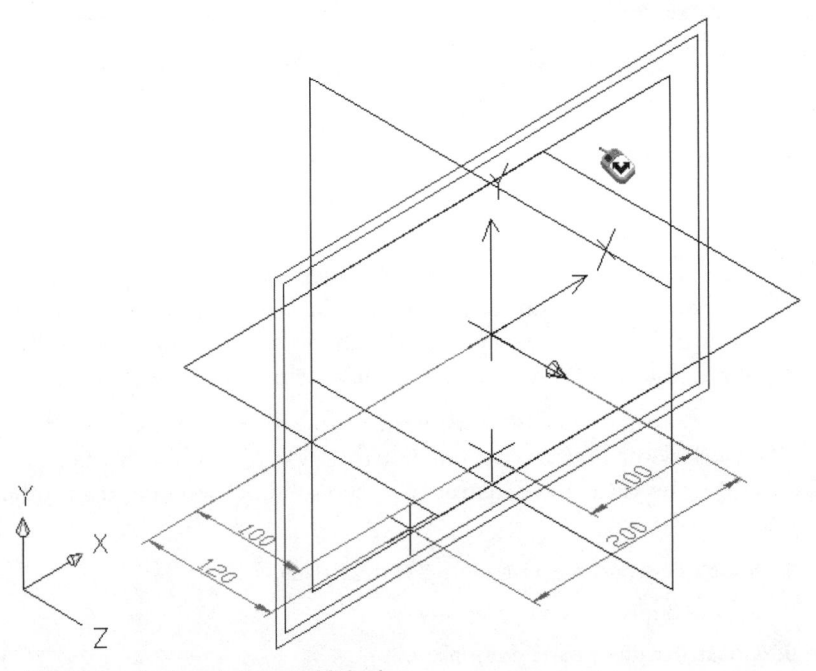

Figure 4-56 Sketch plane being established

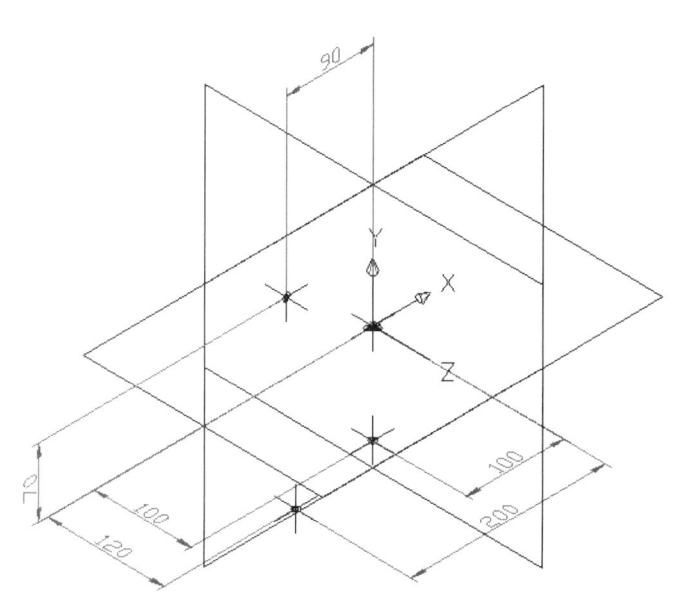

Figure 4-57 Work point constructed

7. Establish a sketch plane on the third work plane shown in Figure 4-58.
8. Referring to Figure 4-59, construct a work point and add two parametric dimensions.

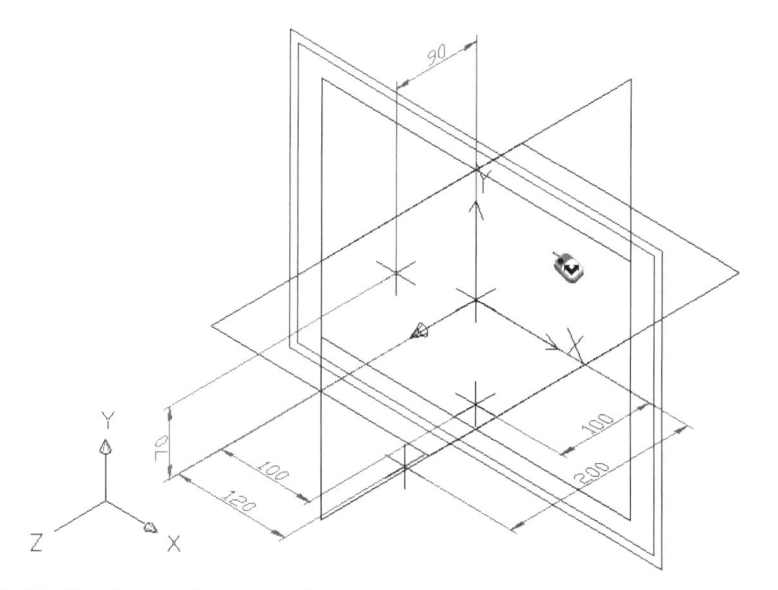

Figure 4-58 Sketch plane being established

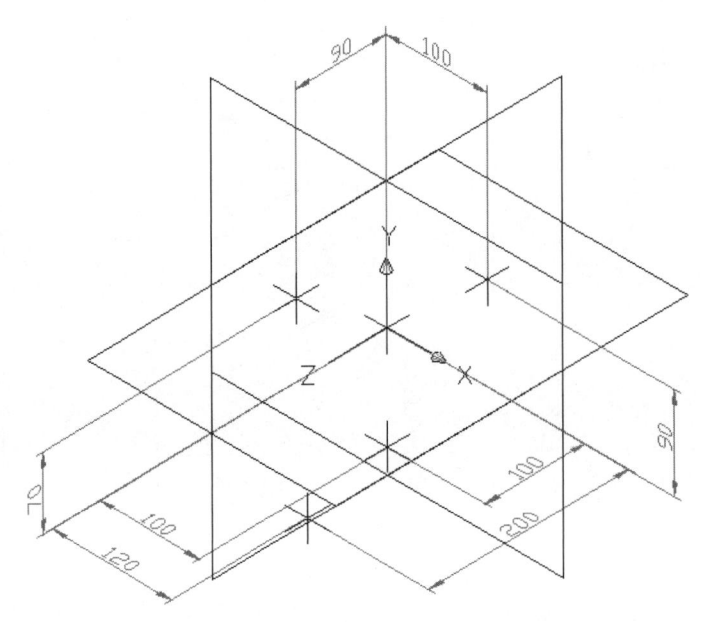

Figure 4–59 *Work point constructed*

A set of work points are constructed. Now you will construct a 3D polyline and resolve it to a 3D pipe path. A 3D pipe path resolved this way is parametric and is constrained to the work points automatically.

    9. Select Assist>Drafting Settings>Drafting Settings

       Command: DSETTINGS

   10. In the Drafting Settings dialog box, select the Object Snap tab.

   11. In the Object Snap tab, select the Clear All button, check the Node box, check Object Snap on, and select the OK button.

   12. Select Design>3D Polyline

       Command: 3DPOLY

   13. Referring to Figure 4–60, select the work points one by one and then press the ENTER key.

   14. Select Part>Sketch Solving>3D Pipe Path

   15. Select the polyline and press the ENTER key.

   16. Select the lower-left end of the polyline to indicate the start point.

After you select the start point of the pipe path, the 3D Pipe Path dialog box displays (see Figure 4–61).

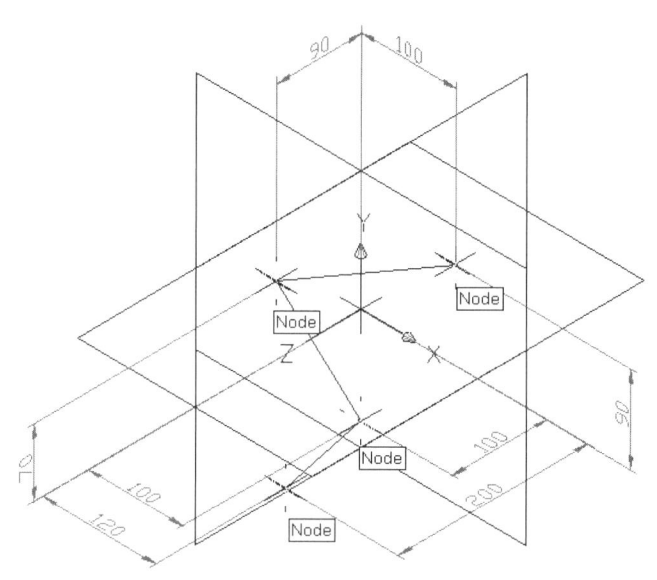

Figure 4–60 *3D polyline being constructed*

Figure 4–61 *3D Pipe Path dialog box*

In the 3D Pipe Path dialog box, there are four rows (because there are four fit points) and ten columns. Each row gives the information about the fit points of the pipe path. Here all the columns, except the Radius column, are grayed out. You are not allowed to change the grayed out columns because the pipe path is constrained to the work points. There are two ways to modify the pipe path:

- You select the 3D Pipe Path from the browser, right-click, and select Edit Sketch to modify the dimensions of the work points from which the pipe path is derived.

- You select an item in column C (C stands for constraint) in the 3D Pipe Path dialog box and right-click (see Figure 4–62). In the right-click menu, you select Constraint to Path Point. After that, the fit point will no longer be constrained to the work point and you can change the data in the dialog box.

3D Pipe Path								
No.	C	From	Delta X	Delta Y	Delta Z	Length	Angle XY	Angle Z	Radius
1			120	0	200	101.98	180	-78.69	
2			100	0	100	122.47	145.01	-4.68	0
3		Constrain to Work Point			90	136.01	11.31	-41.43	0
4		Constrain to Path Point			0				
5		Unconstrain							
		Insert							
		Delete							

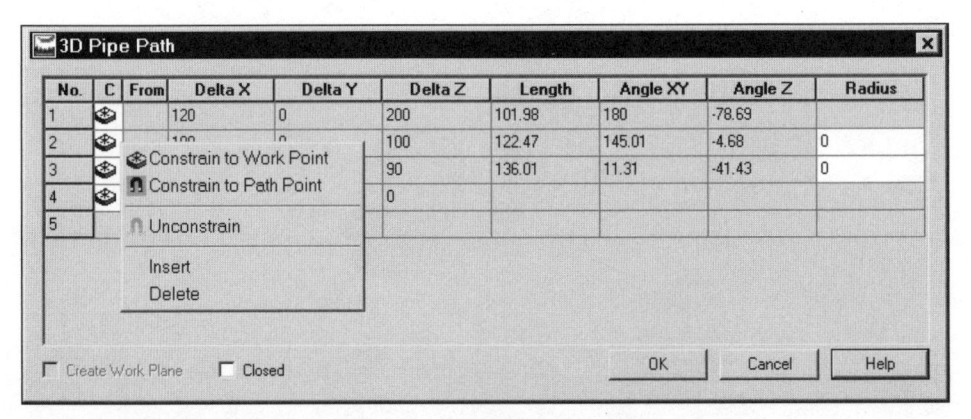

☐ Create Work Plane ☐ Closed   OK   Cancel   Help

Figure 4–62 *Right-click menu for column C of the 3D Pipe Path dialog box*

Now you will keep the parametric relationship with the work points by leaving the fit points to constrain to the work points, and you will add a radius bend at fit points 2 and 3.

17. Set the radius in the Radius column for fit points 2 and 3 to 20 mm and select the OK button (see Figure 4–63).

3D Pipe Path								
No.	C	From	Delta X	Delta Y	Delta Z	Length	Angle XY	Angle Z	Radius
1			120	0	200	101.98	180	-78.69	
2			100	0	100	122.47	145.01	-4.68	20
3			0	70	90	136.01	11.31	-41.43	20
4			100	90	0				
5									

☑ Create Work Plane ☐ Closed   OK   Cancel   Help

Figure 4–63 *Corner radius changed*

18. Referring to Figure 4–64, right-click to accept the sketch plane orientation.

The 3D pipe path and a work plane normal to the start point of the path are constructed and a sketch plane is established on the work plane. Now you will construct a profile on the sketch plane.

19. Construct a circle, resolve it to a profile, and delete the fix constraint.
20. Add three parametric dimensions to specify the size of the circle and its location in relation to the start point of the 3D path (see Figure 4–65).

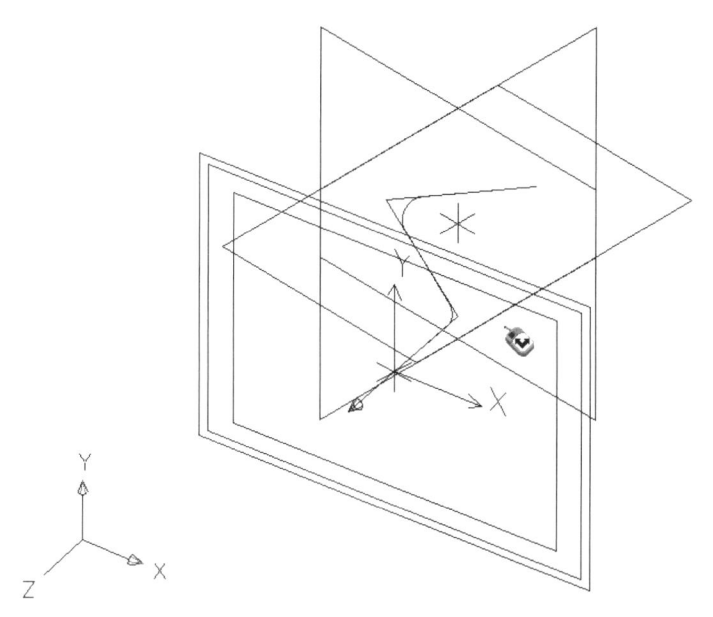

Figure 4–64 *Sketch plane being established on the work plane at the start point of the 3D path*

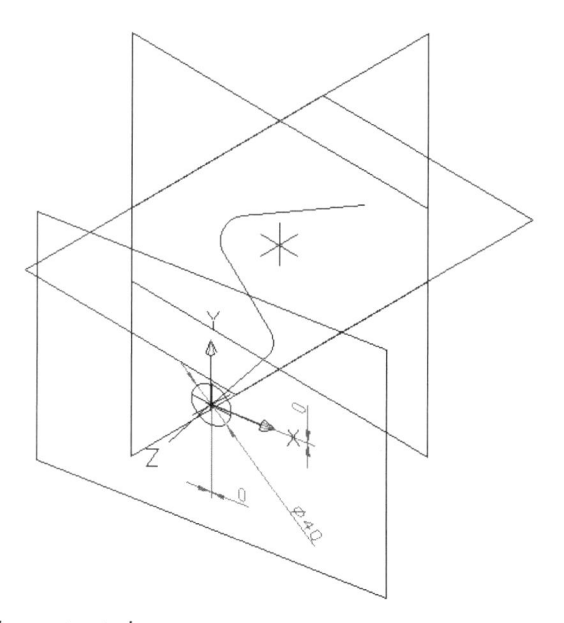

Figure 4–65 *Profile constructed*

The profile is complete. Now you will construct a sweep solid.

21. Select the profile from the browser, right-click, and select Sweep.

22. In the Sweep dialog box, select the OK button (see Figure 4–66).
23. Select the work features from the browser one by one, right-click, and deselect Visible (see Figure 4–67).

The sweep solid is complete and the work features are hidden. Save and close your file (file name: *PipeSweep.dwg*).

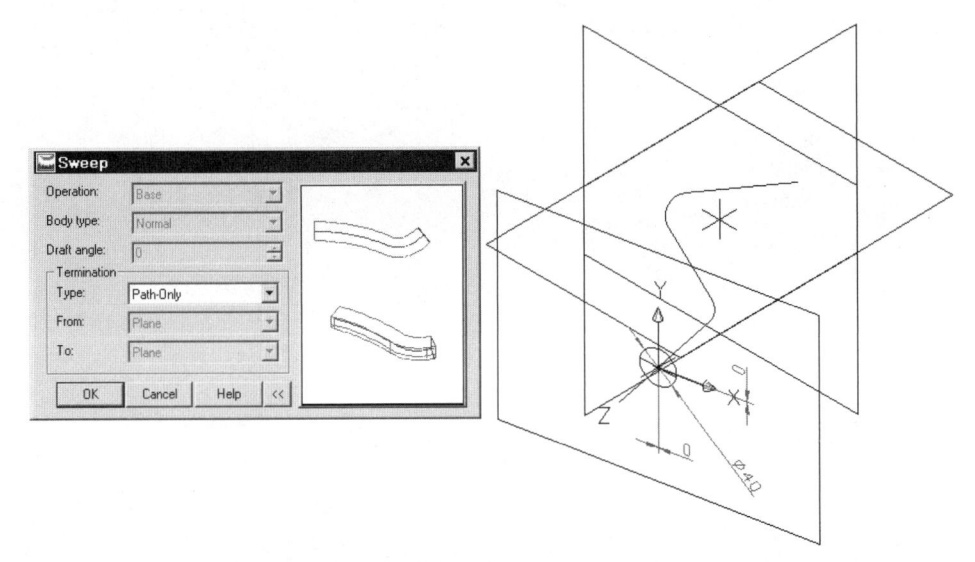

Figure 4–66 *Profile being swept along the 3D pipe path*

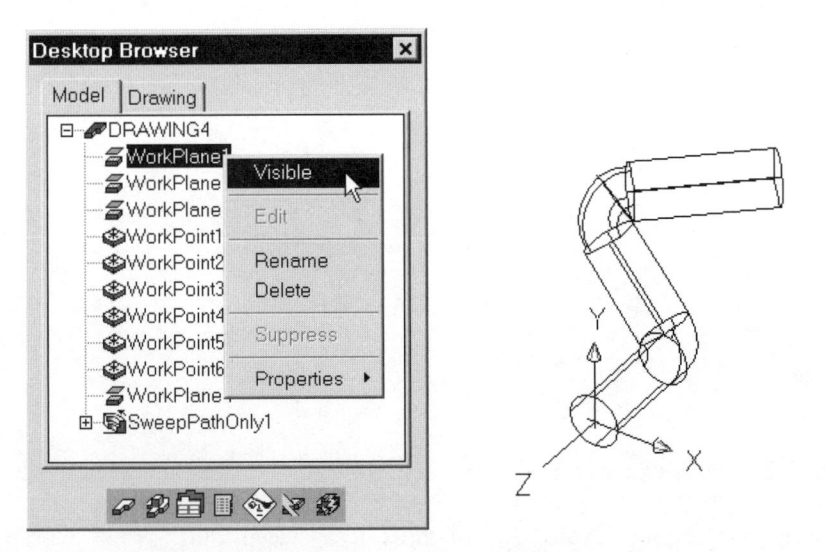

Figure 4–67 *Sweep solid constructed and work features hidden*

Sweeping 3D Spline Path

A spline path derives from a spline; you construct a spline and resolve it to a spline path. Similar to making a pipe path, you construct work points to establish a parametric relationship with other features in the solid. Now you will construct a set of work points, construct a spline to pass through the work points, resolve the spline to a spline path, construct a profile at the start point of the path, and sweep the profile along the path.

1. Start a new part file. Use metric as the default.
2. Set the display to an isometric view.
3. Construct three basic work planes
4. Select work planes 2 and 3 from the browser one by one, right-click, and deselect Visible.
5. Referring to Figure 4–68, establish a sketch plane on work plane 1, construct two work points, and add four parametric dimensions.

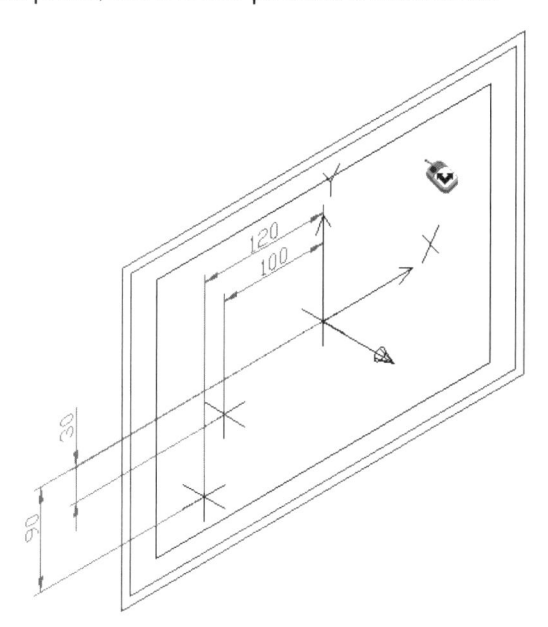

Figure 4–68 *Work points constructed on work plane 1*

Now you will construct an offset work plane and construct two work points on the new work plane.

6. Construct a work plane parallel to work plane 1 and offset at a distance of 200 mm (see Figure 4–69).
7. Establish a sketch plane on the new work plane, construct two work points, and add four parametric dimensions (see Figure 4–70).

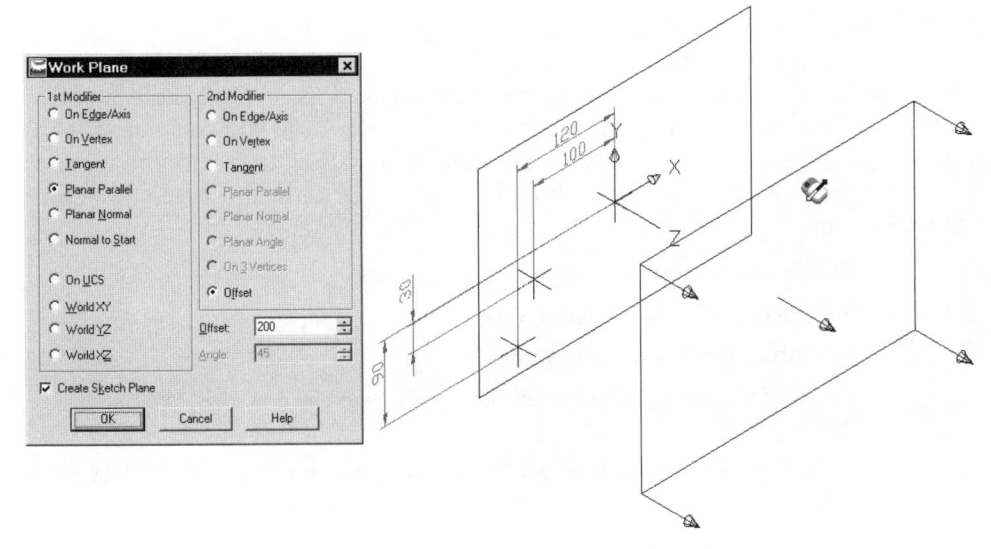

Figure 4–69 Work plane parallel to work plane 1 being constructed

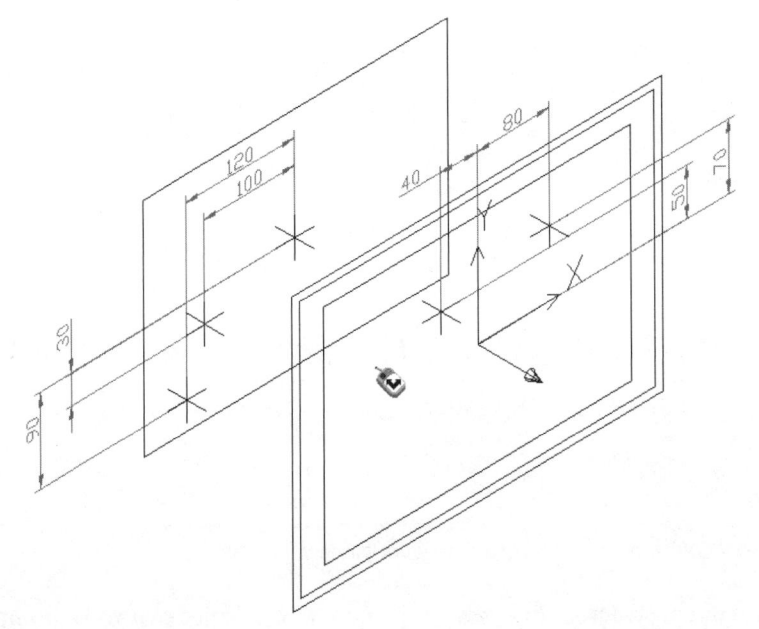

Figure 4–70 Work points constructed

Now you will construct a spline.

    8. Set object snap mode to NODE and construct a spline to pass through the four work points indicated in Figure 4–71.

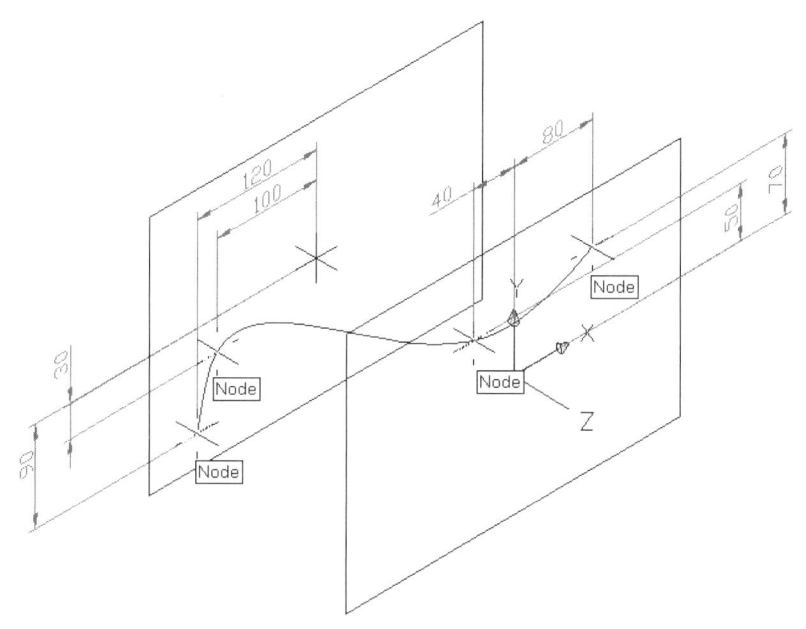

Figure 4-71 *Spline constructed*

The spline is complete. Now you will resolve it to a spline path.

    9. Select Part>Sketch Solving>3D Spline Path

   10. Select the spline.

   11. Select the lower-left end of the spline to indicate the start point of the path.

After you select the start point of the pipe path, the 3D Pipe Path dialog box displays (see Figure 4–72).

Figure 4-72 *3D Spline Path dialog box*

The 3D Spline Path dialog box has 11 columns depicting the information of the spline at the fit points. Among them, the i, j, and k columns delineate the tangency vector at the fit points. Except the start point and the end point, these three columns are grayed out because the spline is constrained to the work points. Similar to modifying the pipe path, you modify the work points or unconstrain the fit points to modify the parameters in the 3D Spline Path dialog box.

12. In the 3D Spline Path dialog box, select the OK button.
13. Referring to Figure 4–73, right-click to accept.

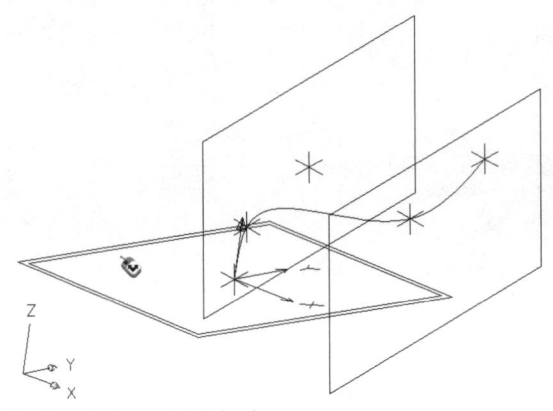

Figure 4–73 *Sketch plane being established*

The spline path is constructed and a sketch plane is established on the work plane normal to the start of the path. Now you will construct a profile and sweep the profile along the path.

14. Construct a circle, resolve it to a profile, and delete the fix constraint.
15. Add three parametric dimensions (see Figure 4–74).

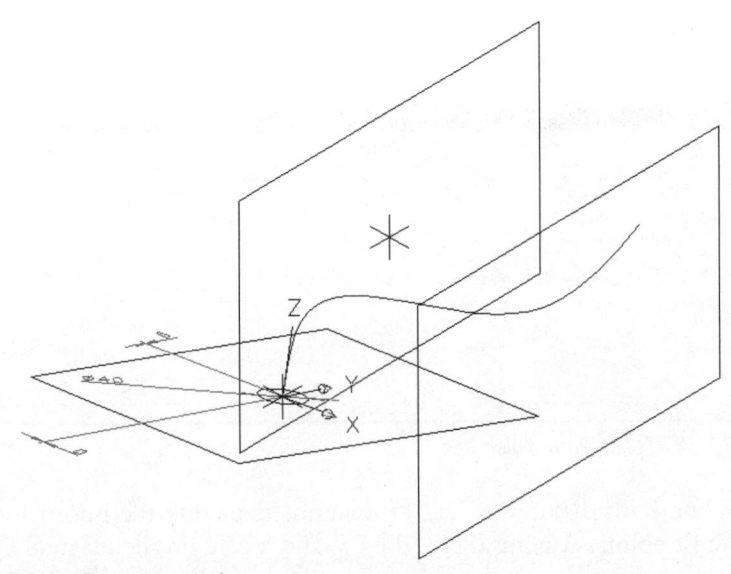

Figure 4–74 *Profile constructed*

15. Select the profile from the browser, right-click, and select Sweep.
16. In the Sweep dialog box, select the OK button (see Figure 4–75).
17. Shade the display.
18. Select the work features from the browser bar one by one, right-click, and deselect Visible (see Figure 4–76).

The sweep solid is constructed and the work features are hidden. Save and close your file (file name: *SplineSweep.dwg*).

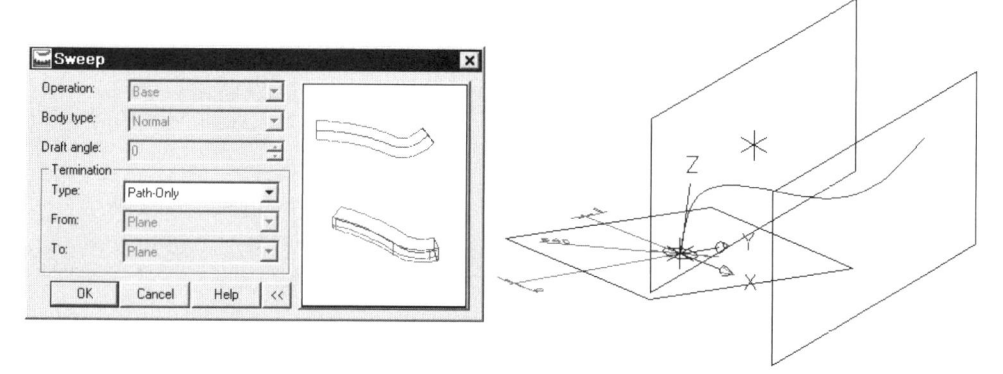

Figure 4–75 *Sweep solid being constructed*

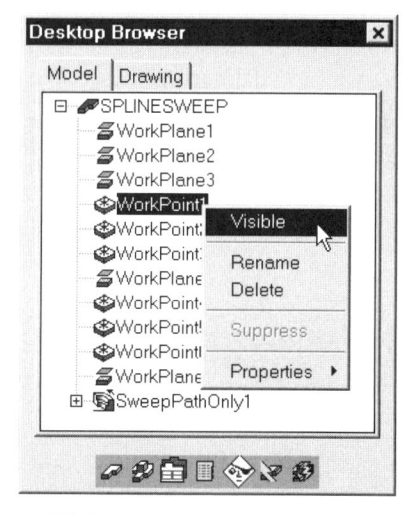

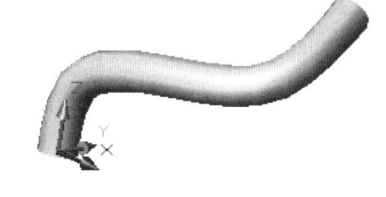

Figure 4–76 *Work features hidden*

COPY EDGE, 3D EDGE PATH, AND 3D PIPE PATH

As mentioned in Chapter 2, you can copy edges of a solid part to become individual entities. However, you must not confuse edge copying with constructing an edge path.

To illustrate the difference between the two operations, you will construct a sweep solid by using a 3D pipe path constructed from edges copied from a solid part and compare the outcome with a sweep solid by using an edge path.

1. Open the file *EdgeSweep.dwg* that you constructed earlier in this chapter.
2. Select File>Save As and specify a new file name (file name: *SweepCompare.dwg*).
3. Select Part>Sketch Solving>Copy Edge
4. Select the edges highlighted in Figure 4–77 and press the ENTER key.

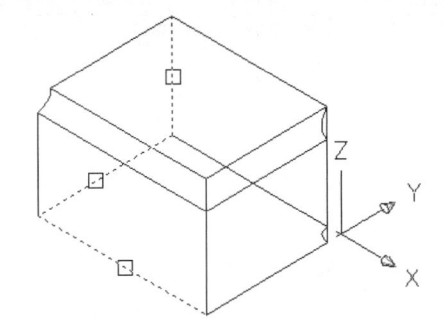

Figure 4–77 *Edges copied*

Three edges are copied. Now you will construct a pipe path from the copied edges.

5. Select Part>Sketch Solving>3D Pipe Path
6. Select the copied edges described by the rectangular window selection box indicated in Figure 4–78 and press the ENTER key.
7. Select the lower corner to specify the start point of the path.

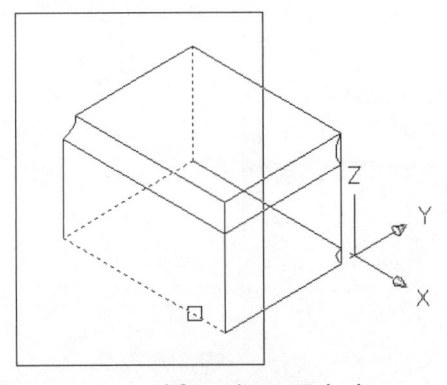

Figure 4–78 *Pipe path being constructed from the copied edges*

After you select the start point, the 3D Pipe Path dialog box displays (see Figure 4–79). Note that the points are unconstrained to any work point.

8. Select the OK button.
9. Referring to Figure 4–80, right-click to accept the sketch plane.

No.	C	From	Delta X	Delta Y	Delta Z	Length	Angle XY	Angle Z	Radius
1			0	-40	0	50	180	0	
2		1	-50	0	0	40	90	0	0
3		2	0	40	0	30	0	90	0
4		3	0	0	30				
5									

☑ Create Work Plane   ☐ Closed   OK   Cancel   Help

Figure 4–79 *3D Pipe Path dialog box*

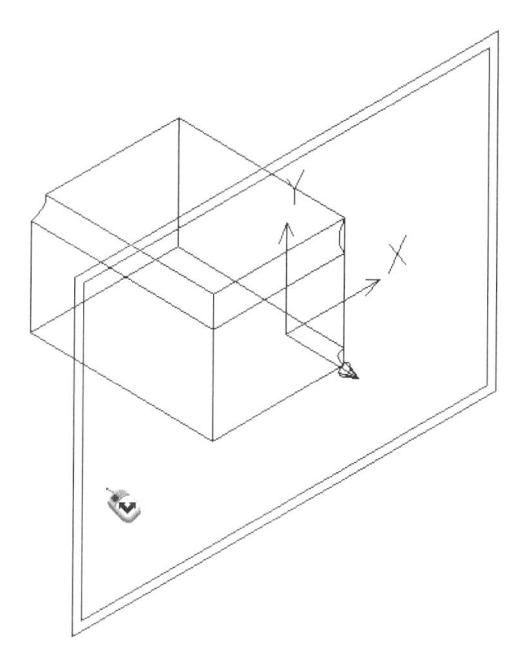

Figure 4–80 *Sketch plane being established normal to the start of the path*

The pipe path is complete. Now you will construct a profile and construct a sweep solid.

10. Construct a circle, resolve it to a profile, and add three parametric dimensions (see Figure 4–81).
11. Select the profile from the browser, right-click, and select Sweep.

12. In the Sweep dialog box, select Join operation and select the OK button (see Figure 4–82).
13. Select the work plane from the browser, right-click, and deselect Visible (see Figure 4–83).

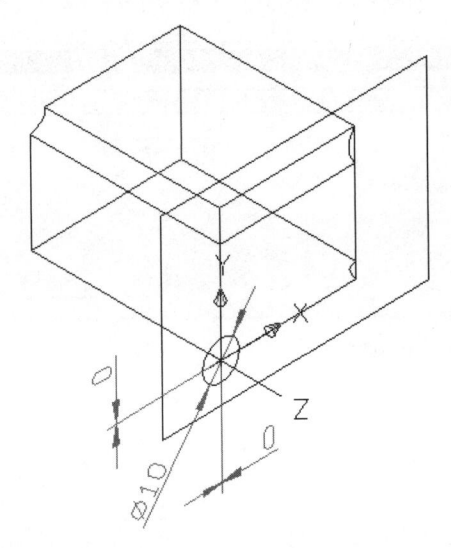

Figure 4–81 *Profile constructed*

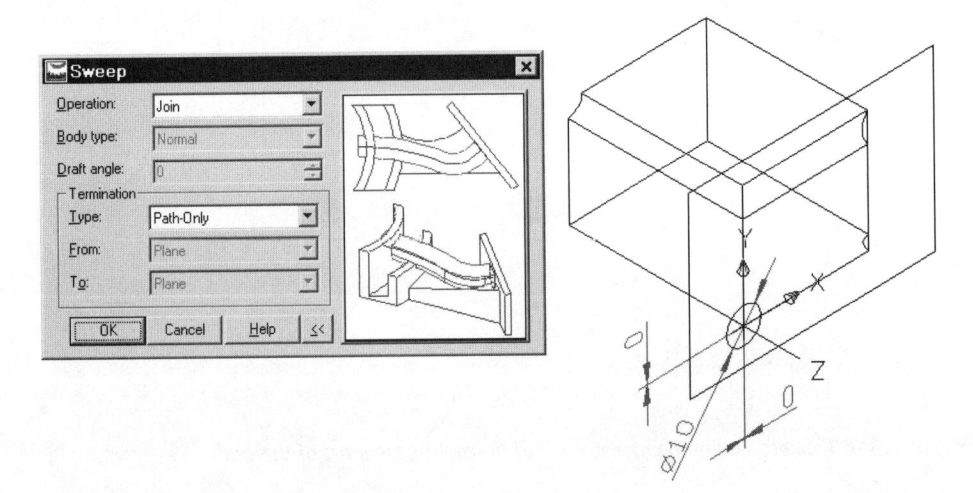

Figure 4–82 *Sweep feature being constructed*

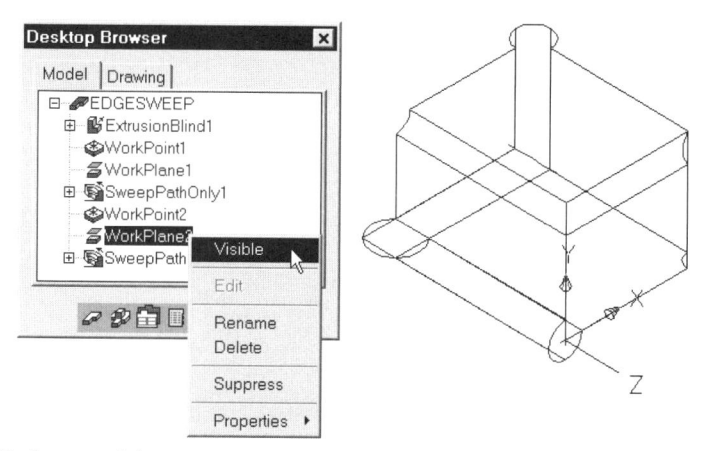

Figure 4–83 *Sweep solid constructed and work plane hidden*

The sweep solid feature is complete. Now you will modify the solid part.

   14. Select the first extruded feature from the browser, right-click, and select Edit Sketch. (You may select the sketch of the feature from the browser and double-click.)
   15. Modify the dimensions of the rectangle to 80 mm by 60 mm (see Figure 4–84).
   16. Right-click and select ENTER.
   17. Select the Update button from the browser.

The solid part is updated (see Figure 4–85). Because the sweep solid from the edge path has a parametric relation with the edges of the extruded solid, it changes automatically. However, the sweep solid from the pipe edge created by using the copied edges does not change because there is no parametric relationship between the copied edges and the solid part. If you want to establish a parametric relationship between the copied edges and the solid's edges, you may add geometric constraints or parametric dimensions. Now save and close the file.

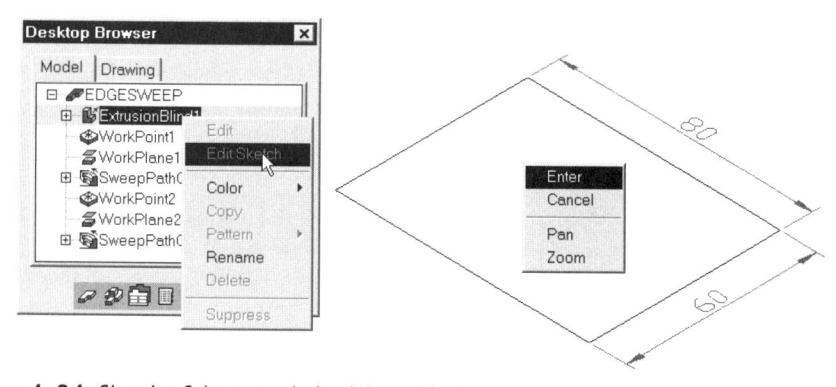

Figure 4–84 *Sketch of the extruded solid modified*

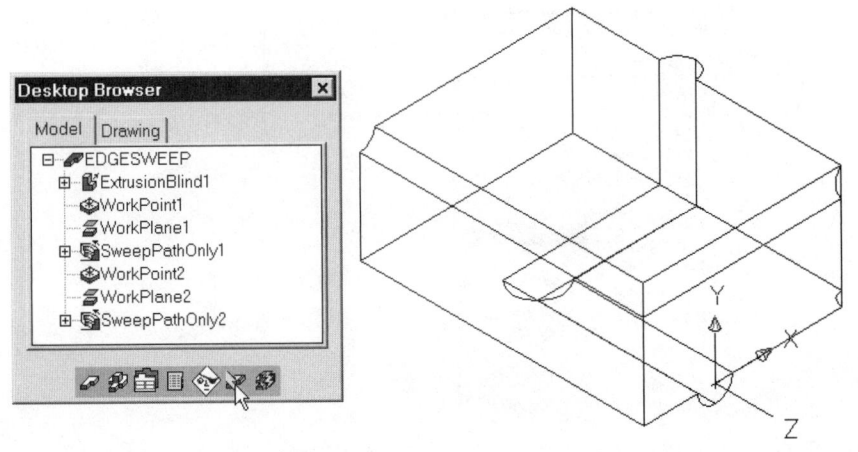

Figure 4–85 *Solid part updated*

LOFT FEATURE

A loft feature is constructed by lofting along a series of profiles. Unlike a sweep feature that has uniform cross sections, a loft feature has a variable cross section that changes from one profile to another.

Now you will construct three profiles on different sketch planes and loft along these profiles.

1. Start a new part file. Use metric as the default.
2. Set the display to an isometric view.
3. Construct three basic work planes.
4. Referring to Figure 4–86, construct a rectangle, resolve it to a profile, delete the fix constraint, and add four parametric dimensions.

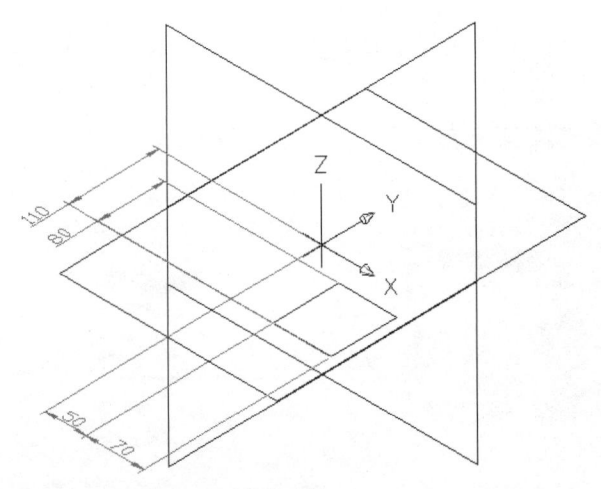

Figure 4–86 *Profile constructed on work plane 3*

5. Establish a sketch plane on work plane 1, construct a circle, resolve it to a profile, delete the fix constraint, and add three parametric dimensions (see Figure 4–87).

6. Referring to Figure 4–88, construct a work plane parallel to work plane 1 and offset at a distance of 120.

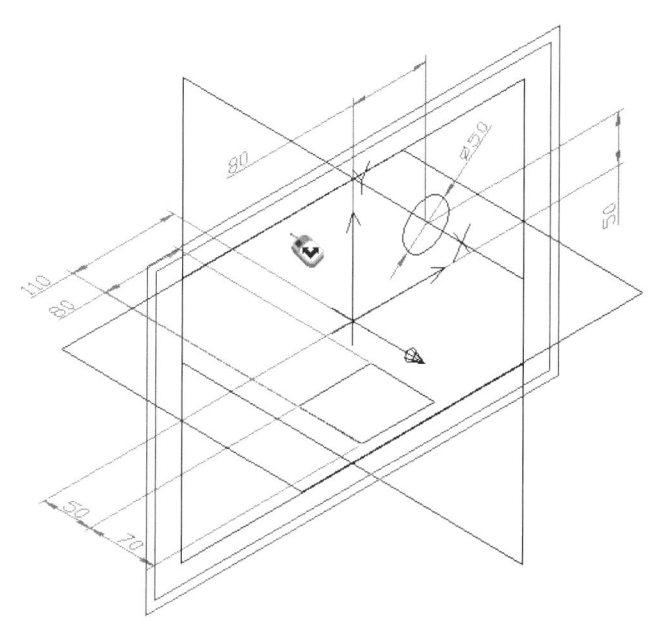

Figure 4–87 *Profile constructed on work plane 1*

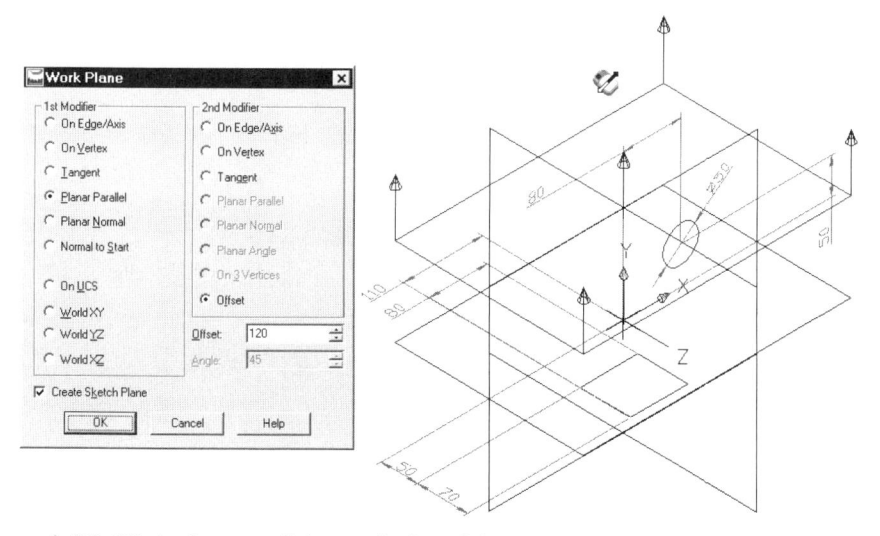

Figure 4–88 *Work plane parallel to work plane 3 being constructed*

7. Construct a rectangle on the work plane, resolve it to a profile, delete the fix constraint, and add four parametric dimensions (see Figure 4–89).

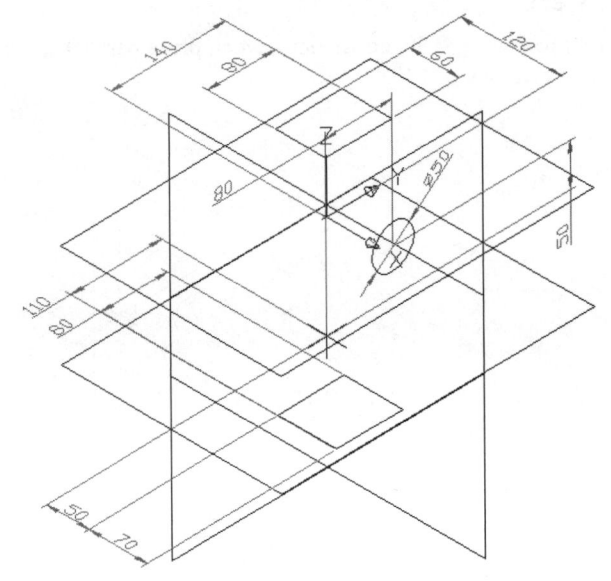

Figure 4–89 *Profile constructed*

Three profiles are complete. Now you will loft along these profiles.

8. Select a profile (any one from the three profiles) from the browser, right-click, and select Loft.
9. Select the profiles one by one, from bottom to top, and press the ENTER key.
10. In the Loft dialog box, select the OK button (see Figure 4–90).
11. Select the work features from the browser one by one, right-click, and deselect Visible.

The loft solid is constructed and the work features are hidden (see Figure 4–91). Save and close your file (file name: *Loft.dwg*).

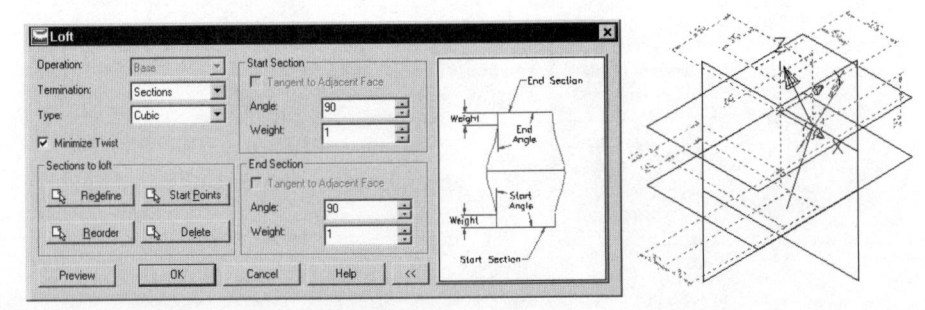

Figure 4–90 *Loft solid being constructed*

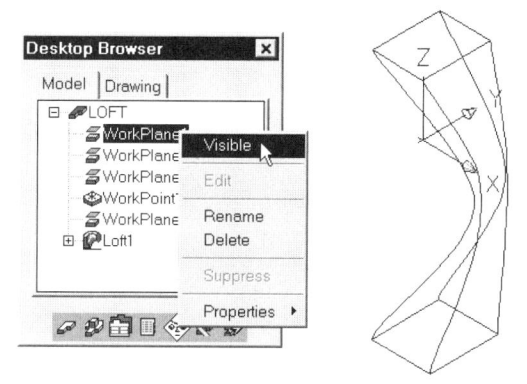

Figure 4–91 *Loft solid constructed and work features hidden*

PLACED FEATURES: HOLE, THREAD, FILLET, CHAMFER, SHELL, AND PATTERN

Placed solid features are preconstructed solid features. To make a placed feature, you select a feature from the menu, specify the parameters, and indicate a location. Now you will learn hole, thread, fillet, chamfer, shell, and pattern (rectangular, polar, and axial) features. In Chapter 9, you will learn surface cut, combine, part split, and face draft features.

HOLE FEATURE

The hole feature is very common in engineering design. To make a hole in a real object, you use a drill. To remove additional material from the object to allow for the head of screws and bolts, you use counter-bore and counter-sink operations. Sometimes, you tap the hole to fit a screw. Now you will construct an extruded solid and put a number of holes in it.

1. Start a new part file. Use metric as the default.
2. Construct a rectangle, resolve it to a profile, add parametric dimensions, and extrude it to an extruded solid (see Figure 4–92).

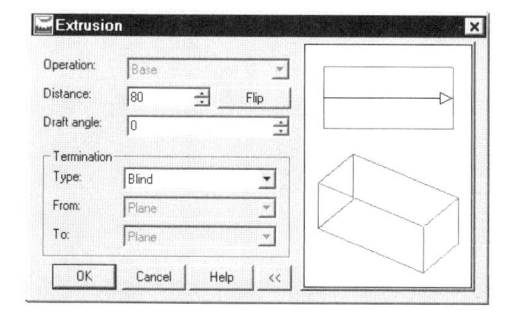

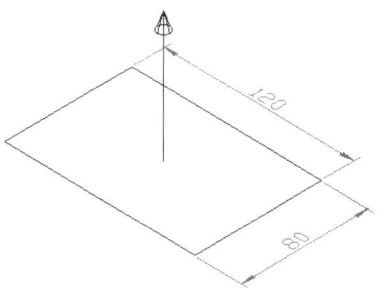

Figure 4–92 *Rectangle being extruded*

3. Referring to Figure 4–93, establish a sketch plane on the top face of the extruded solid, construct a work point, and add two parametric dimensions to position the work point.

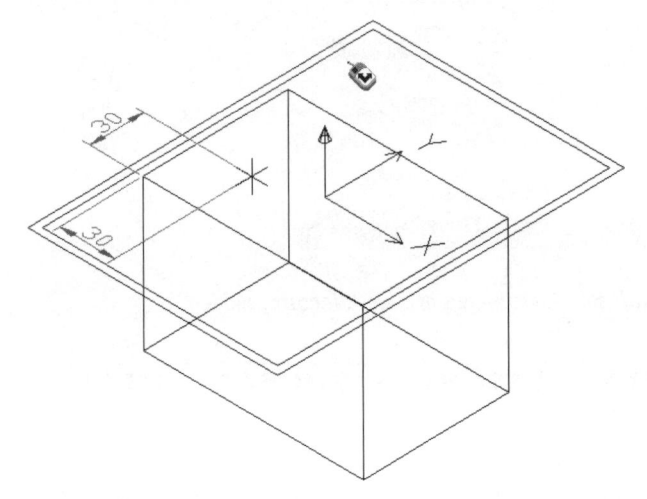

Figure 4–93 *Work point constructed*

Now you will place a tapped hole on the work point.

    4. Select Part>Placed Features>Hole

       Command: AMHOLE

    5. Select the Threads tab of the Hole dialog box (see Figure 4–94).

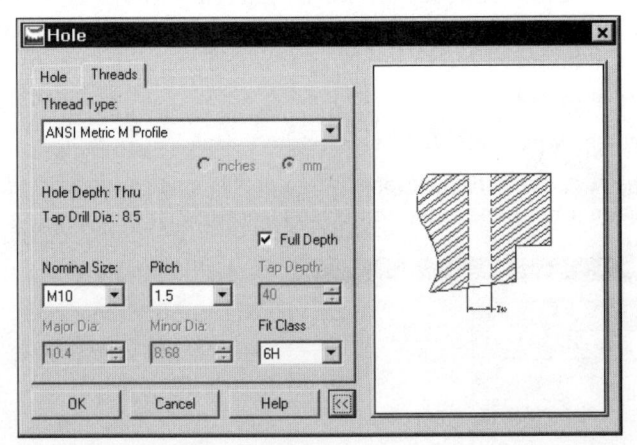

Figure 4–94 *Threads tab of the Hole dialog box*

    6. In the Threads tab, select ANSI Metric M Profile thread type, select M10 nominal size, and check the Full Depth box. (*Note:* Accept the default pitch and fit class.)

7. Select the Hole tab.
8. In the Hole tab, select the Drilled button, select Blind termination, select On Point placement, set depth to 40, and select the OK button. (*Note:* Accept the default Pt angle. Pt angle means the drill point angle.)
9. Select the work point and press the ENTER key (see Figure 4–95).

A blind tapped drilled hole is placed on the work point. Note that the thread is not shown in the solid part. However, threaded hole conventions will be placed in the drawing view, if you output an engineering drawing from the solid part.

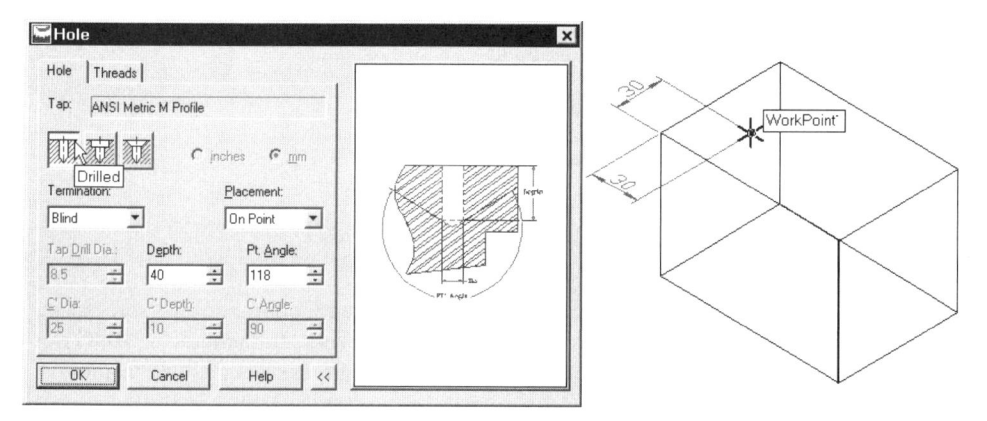

Figure 4–95 *Hole being placed*

Now you will place a hole by specifying two edges.

10. Repeat the AMHOLE command.
11. Select the Threads tab from the Hole dialog box.
12. In the Threads tab, select None thread type.
13. Select the Hole tab.
14. In the Hole tab, select the Counterbore button, select Through termination, select 2 Edges placement, set diameter to 10, C'Dia (counterbore diameter) to 20, and C'Depth (counterbore depth) to 8, and select the OK button (see Figure 4–96).
15. Select the edges highlighted in Figure 4–96 one by one (from right to left) as the reference edges and select a location.
16. Type **30** at the command line area to specify the distance from the first reference edge.
17. Type **50** at the command line area to specify the distance from the second reference edge.
18. Press the ENTER key.

A counter bore hole is placed on the solid.

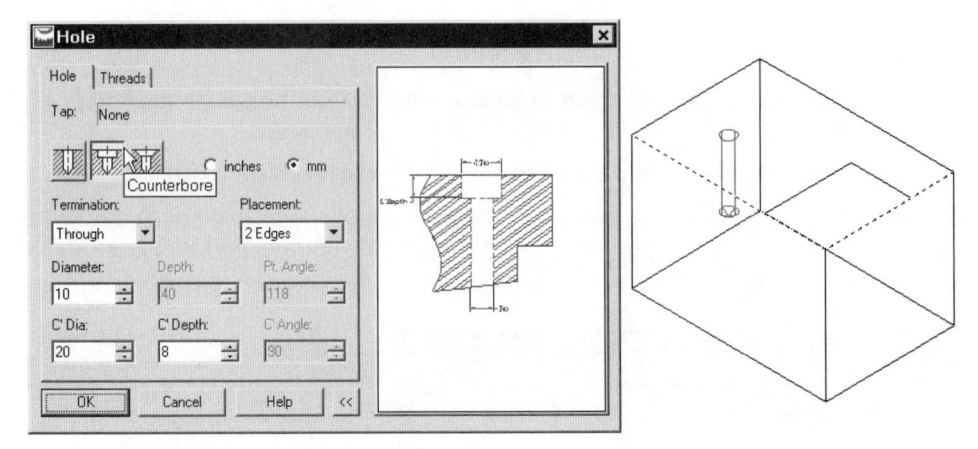

Figure 4-96 *Reference edges being selected*

Now you will place a hole by specifying distances from two holes.

19. Repeat the AMHOLE command.
20. In the Hole tab of the Hole dialog box, select the Countersink button, select Through termination, select From Hole placement, set diameter to 10, C'Dia (countersink diameter) to 16, and C'Angle (countersink angle) to 90, and select the OK button.
21. Select the top face of the solid part (see Figure 4-97).

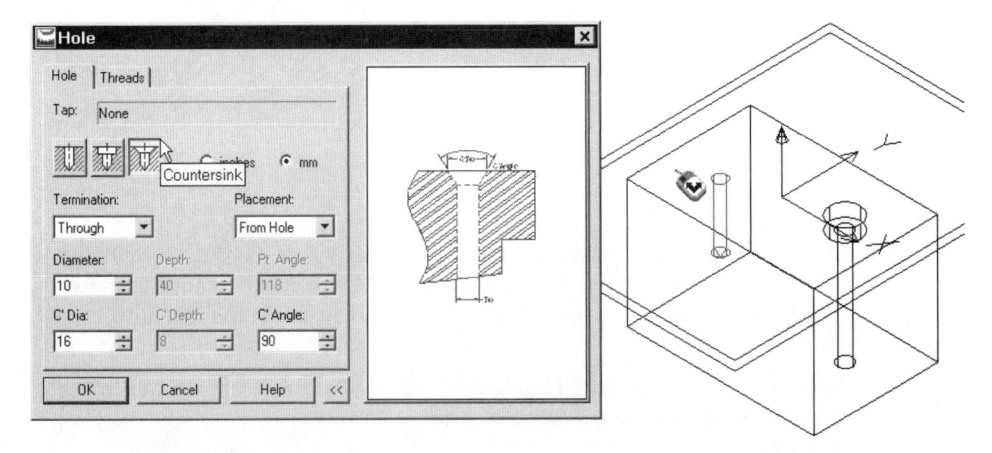

Figure 4-97 *Top face selected*

22. Select Hole 1 as the X reference and select Hole 2 as the Y reference (see Figure 4-98).
23. Referring to Figure 4-99, indicate the direction from the references.

24. Type **20** at the command line area to specify the distance from the first reference in the X direction.
25. Type **20** at the command line area to specify the distance from the first reference in the Y direction.
26. Press the ENTER key.

A countersink hole is placed.

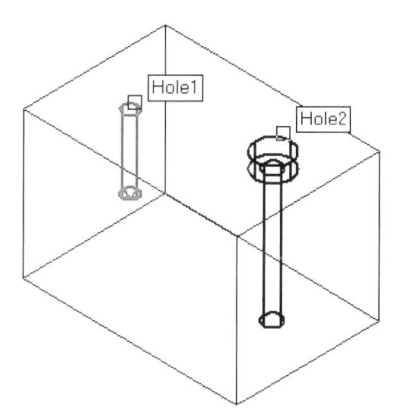

Figure 4–98 *Reference holes selected*

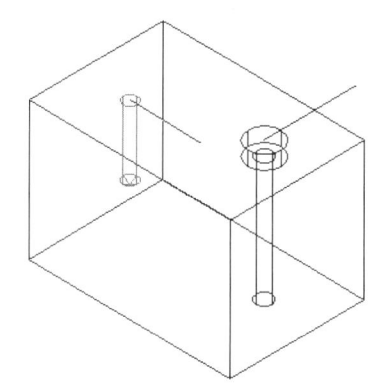

Figure 4–99 *Reference directions specified*

Now you will place a concentric hole.

27. Repeat the AMHOLE command.
28. In the Hole tab of the Hole dialog box, select the Drilled button, select Blind termination, select Concentric placement, set diameter to 20, depth to 6, and PT angle to 180, and select the OK button.
29. Select the bottom plane and the hole highlighted in Figure 4–100.
30. Press the ENTER key.

A blind concentric hole is placed (see Figure 4–101). Save and close your file (file name: *Holes.dwg*).

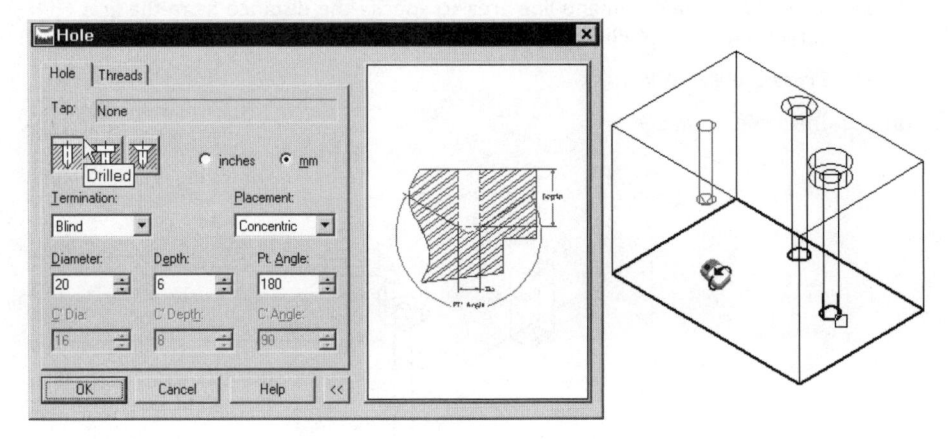

Figure 4–100 *Plane and concentric edge selected*

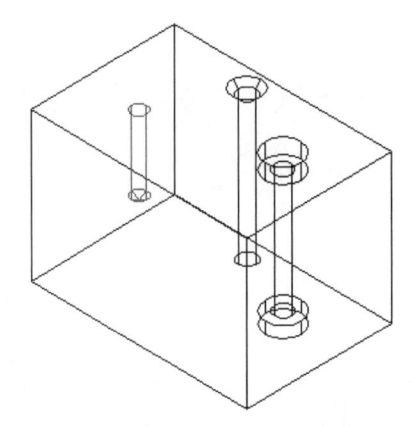

Figure 4–101 *Blind concentric hole placed*

THREAD FEATURE

A thread feature can be considered a cosmetic feature that places a helical curve on a cylindrical face of a solid. It is not, as you may expect, a helical groove cut on the solid part, because threads are often used in a solid part and a lot of helical sweep features in a solid may take up considerable memory space. However, threaded conventions will be placed on the drawing. Now you will construct a revolved solid and place thread features on it.

 1. Start a new part file. Use metric as the default.
 2. Referring to Figure 4–102, construct a sketch, resolve it to a profile, and add parametric dimensions.

Part Modeling II   163

3. Set the display to an isometric view.
4. Revolve the sketch about the hidden line (see Figure 4–103).

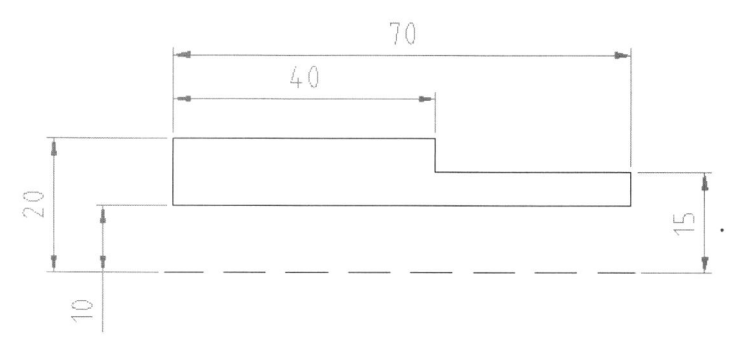

Figure 4–102 *Sketch constructed*

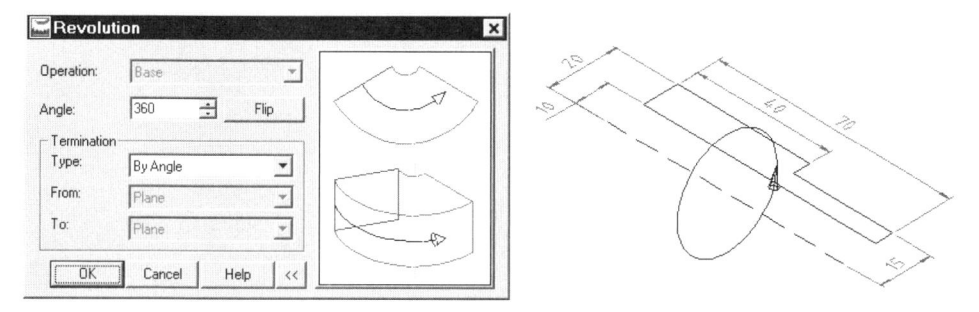

Figure 4–103 *Revolved solid constructed*

5. Select Part>Placed Features>Thread

   Command: AMTHREAD

6. Referring to Figure 4–104, select an edge.

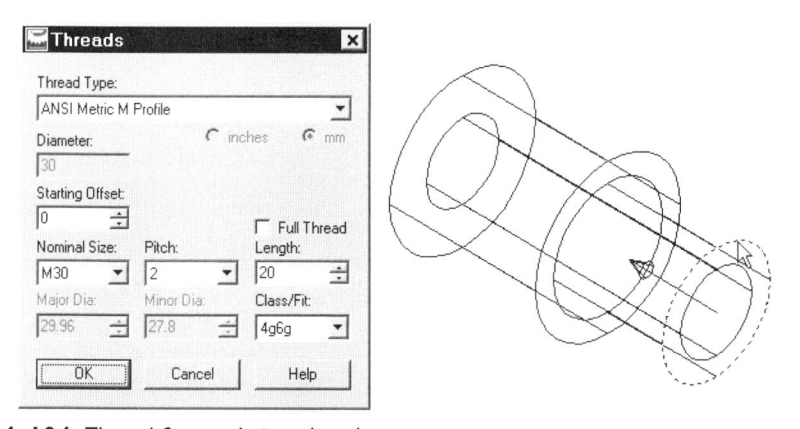

Figure 4–104 *Thread feature being placed*

7. In the Threads dialog box, select ANSI Metric M Profile thread type, uncheck the Full Thread box (if it is checked), set starting offset to 0, nominal size to M30, length to 20, and pitch to 2, and select the OK button.
8. Repeat the AMTHREAD command.
9. Select the edge indicated in Figure 4–105.
10. In the Threads dialog box, select ANSI Metric M Profile thread type, check the Full Thread box, set starting offset to 0, nominal size to M20 and pitch to 1, and select the OK button.

Thread features are placed (see Figure 4–106). Save and close your file (file name: *ThreadedBush.dwg*).

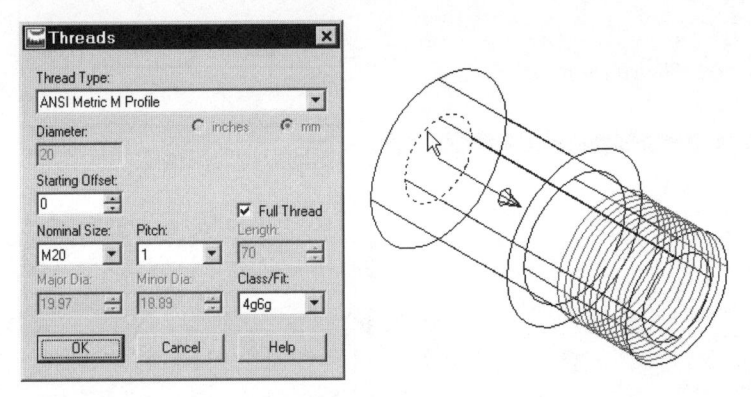

Figure 4–105 *Second thread feature being placed*

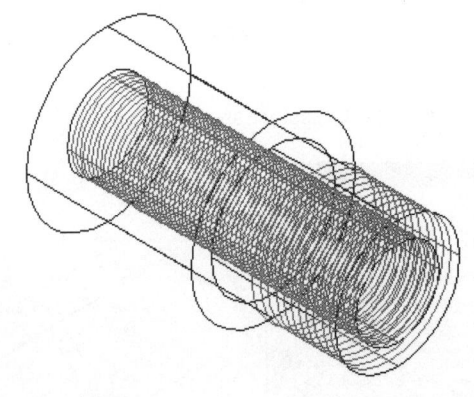

Figure 4–106 *Thread features placed*

FILLET FEATURE

A fillet is a rounded edge with a circular cross section. The radius of the circular cross section can be constant or variable. There are three ways to specify a variable radius. The first method is to specify a constant chord width. The second and third methods are to specify two radii and let the fillet radius change from the first radius to the second radius either cubically or linearly. Now you will place fillet features to a solid part.

1. Open the file *Holes.dwg*, select File>Save As, and specify a file name (file name: *Fillet.dwg*).
2. Select Part>Placed Features>Fillet

   Command: AMFILLET
3. In the Fillet dialog box, select Cubic and the OK button.
4. Select the edge indicated in Figure 4–107 and select the R symbols one by one to change the radius values.
5. Press the ENTER key.

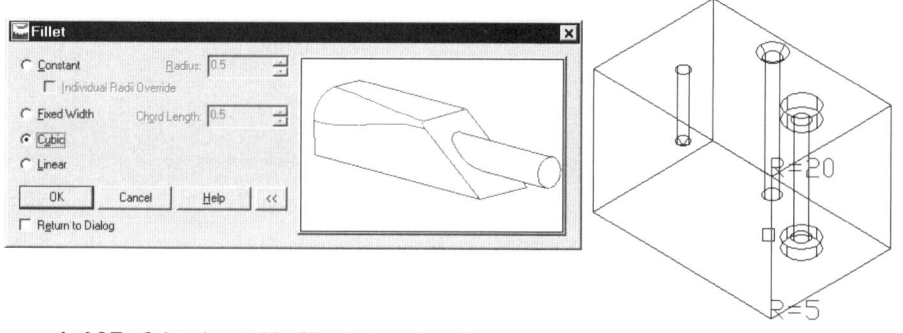

Figure 4–107 *Cubical variable fillet being placed*

A cubical variable fillet is placed. Now you will place another fillet.

6. Repeat the AMFILLET command.
7. In the Fillet dialog box, select Linear and the OK button.
8. Referring to Figure 4–108, select an edge and select the R symbols one by one to change the radius value.
9. Press the ENTER key.

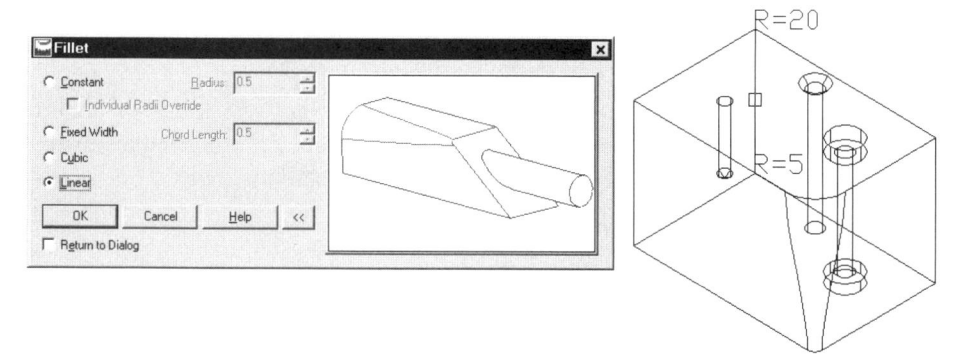

Figure 4–108 *Linear variable fillet being placed*

A linear variable fillet is placed. Now you will place a constant fillet.

10. Repeat the AMFILLET command.
11. In the Fillet dialog box, select Constant, check the Individual Radii Override box, and select the OK button.
12. Referring to Figure 4–109, select the edges and press the ENTER key.
13. Select the radius values one by one and change their value.
14. Press the ENTER key.

Constant radius fillets are constructed (see Figure 4–110). Save and close your file.

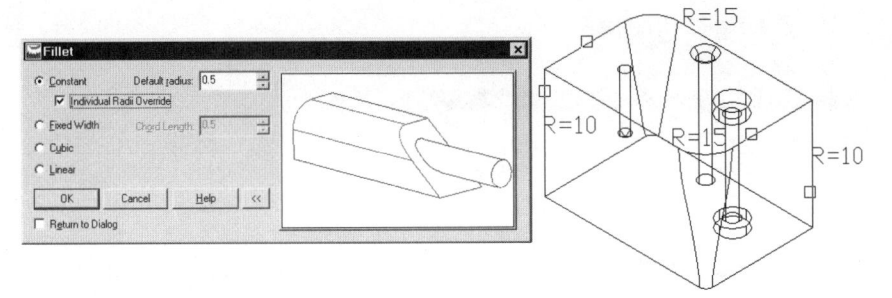

Figure 4–109 *Constant fillets with individual radii override being constructed*

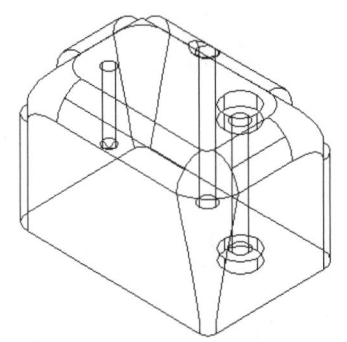

Figure 4–110 *Fillets constructed*

CHAMFER FEATURE

A chamfer is a bevel along the edge of a solid. There are two kinds of chamfer: chamfer of equal distances or non-equal distances. To place a non-equal distance chamfer feature, you specify two distances or a distance and an angle.

Now you will place chamfer features in a solid part.

1. Open the file *Holes.dwg*, select File>Save As, and specify a file name (file name: *Chamfer.dwg*).
2. Select Part>Placed Features>Chamfer
   Command: AMCHAMFER

3. In the Chamfer dialog box, select Equal Distance operation, set Distance 1 to 10, and select the OK button.
4. Select the edge indicated in Figure 4–111 and press the ENTER key.

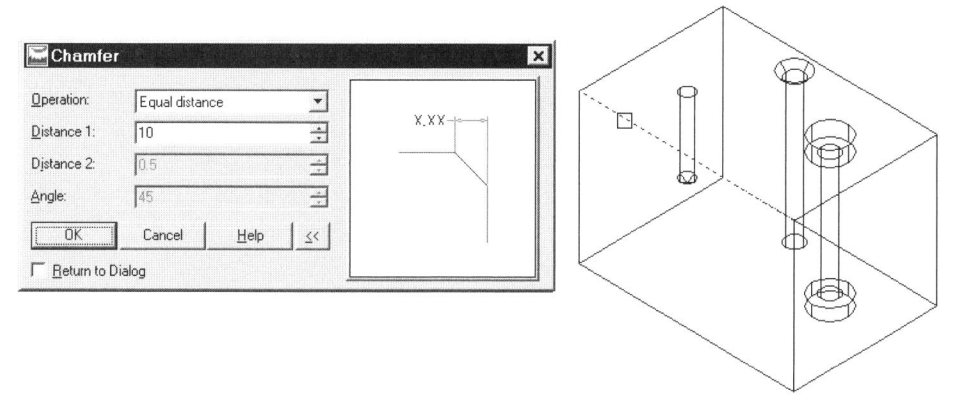

Figure 4–111 *Equal distance chamfer being placed*

An equal distance chamfer feature is placed. Now you will place another chamfer.

5. Repeat the AMCHAMFER command.
6. In the Chamfer dialog box, select Two Distance operation, set Distance 1 to 10 and Distance 2 to 20, and select the OK button.
7. Select the edge indicated in Figure 4–112 and press the ENTER key.
8. Select the face highlighted in Figure 4–112 and right-click to accept.

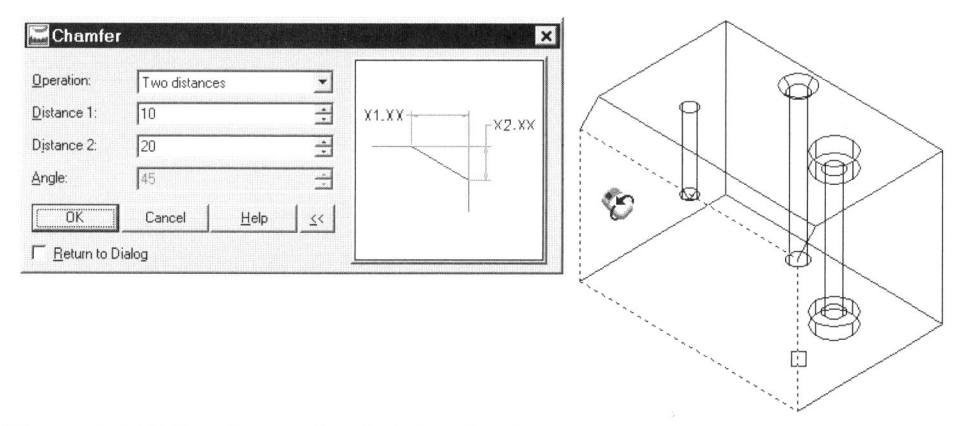

Figure 4–112 *Two-distance chamfer being placed*

A two-distance chamfer is placed. Now you will place another chamfer.

9. Repeat the AMCHAMFER command.

10. In the Chamfer dialog box, select Distance and Angle operation, set Distance 1 to 10 and Angle to 15, and select the OK button.
11. Referring to Figure 4–113, select an edge and press the ENTER key.
12. Select the highlighted face and right-click to accept.

Chamfer features are complete (see Figure 4–114). Save and close your file.

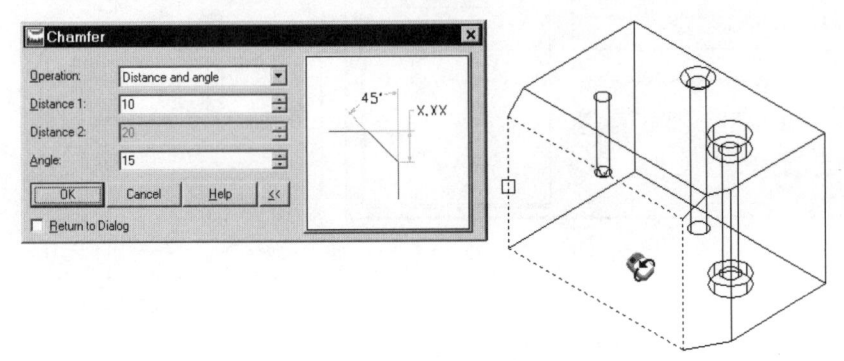

Figure 4–113 *Distance and angle chamfer being placed*

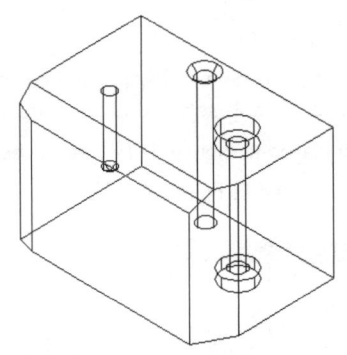

Figure 4–114 *Chamfer features placed*

SHELL FEATURE

A shell feature makes an object hollow. The concept of shelling is to derive a solid by offsetting the faces of the existing solid and cut the derived solid from the existing solid. This process is automatic. You need only to specify the offset distance and any faces that you want to remove after shelling.

Now you will make a solid part hollow.

1. Open the file *Fillet.dwg*, select File>Save As, and specify a file name (file name: *Shell.dwg*).
2. Select the hole feature (highlighted in Figure 4–115) from the browser, right-click, and select Delete.
3. Press the ENTER key to accept.

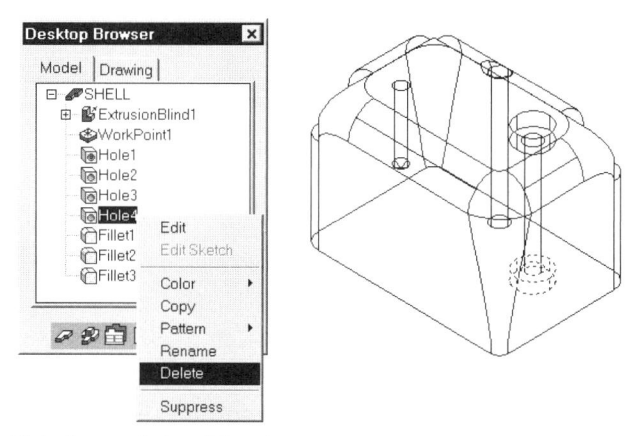

Figure 4–115 *Hole feature being deleted*

The selected hole feature is deleted. Now you will place a shell feature.

4. Select Part>Placed Features>Shell

   Command: AMSHELL

5. In the Shell dialog box, set inside thickness to 1 and select the Add button.
6. Select the face highlighted in Figure 4–116 and right-click to accept.
7. Press the ENTER key.
8. On return to the Shell dialog box, select the OK button.

The solid part is made hollow with a shell thickness of 1 mm and a face removed (see Figure 4–117). Save and close your file.

Because shelling involves the making of a solid offsetting from the existing solid, it is important to realize that shelling may not be successful if the radius of the curvature of a face of the solid is equal to or smaller than the shell thickness.

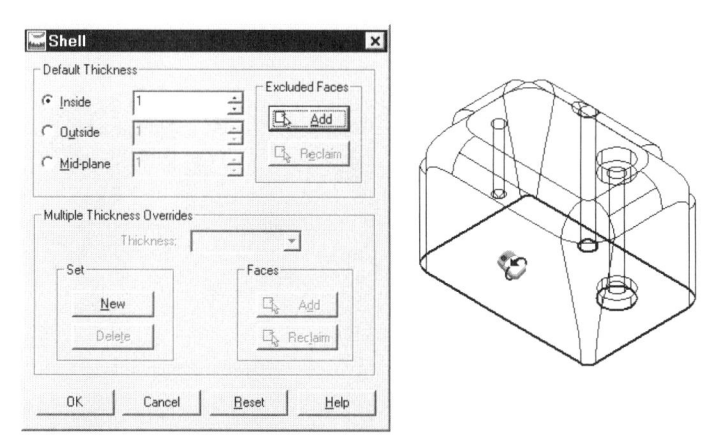

Figure 4–116 *Shell feature being placed*

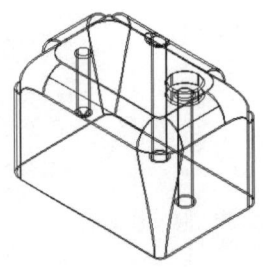

Figure 4–117 *Shell feature placed*

PATTERN FEATURE

A pattern feature repeats any existing solid features (sketched solid feature or placed solid features) in three ways. You select a source feature and repeat it in a rectangular pattern, a polar pattern, or an axial pattern.

Rectangular Pattern

A rectangular pattern repeats selected source features along specified rows and columns. Now you will place a rectangular pattern of features.

1. Open the file *Holes.dwg,* select File>Save As, and specify a file name (file name: *Pattern.dwg*).
2. Select Part>Placed Features>Rectangular Pattern

    Command: AMPATTERN

3. Select the hole feature indicated in Figure 4–118 and press the ENTER key.
4. In the Pattern dialog box, select Incremental Spacing and set instances to 2 and spacing to 12 in the Column Placement box, select Incremental Spacing and set instances to 2 and spacing to 20 in the Row Placement box, and select the OK button.

A rectangular pattern of the selected feature is placed (see Figure 4–119).

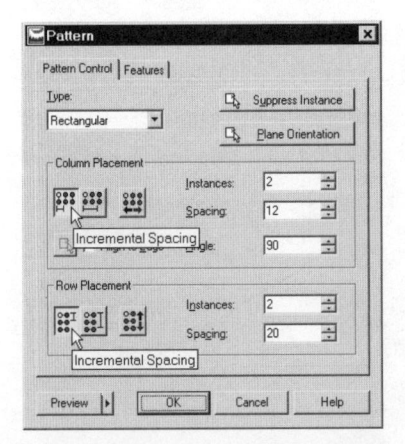

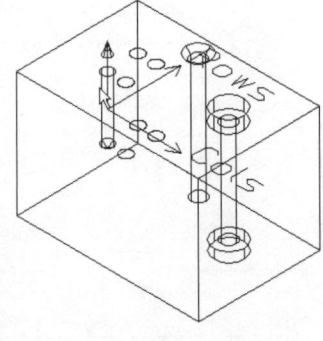

Figure 4–118 *Rectangular pattern being placed*

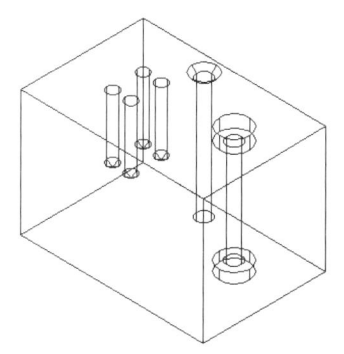

Figure 4–119 *Rectangular pattern placed*

Polar Pattern

A polar pattern repeats selected source features about a rotation center. The rotation center can be a work point, a work axis, or a cylindrical object.

Now you will place a polar pattern of a hole about a cylindrical object.

5. Select Part>Placed Features>Polar Pattern
6. Select Hole2 highlighted in Figure 4–120 and press the ENTER key.
7. Select the countersink hole as the rotational center.
8. In the Pattern dialog box, select the Incremental Angle button, select the Flip Rotation Direction button, set instances to 3 and spacing angle to 28, and select the OK button.

A polar pattern of the selected feature is placed (see Figure 4–121). Note that a work axis is automatically created. Now save and close your file.

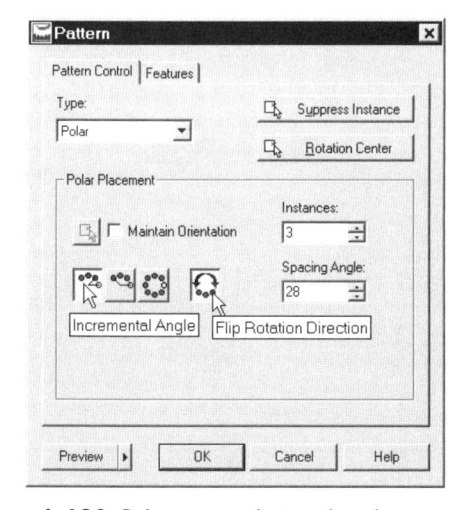

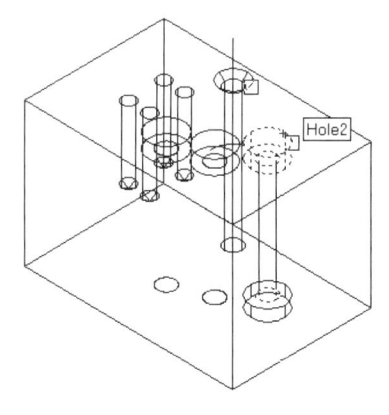

Figure 4–120 *Polar pattern being placed*

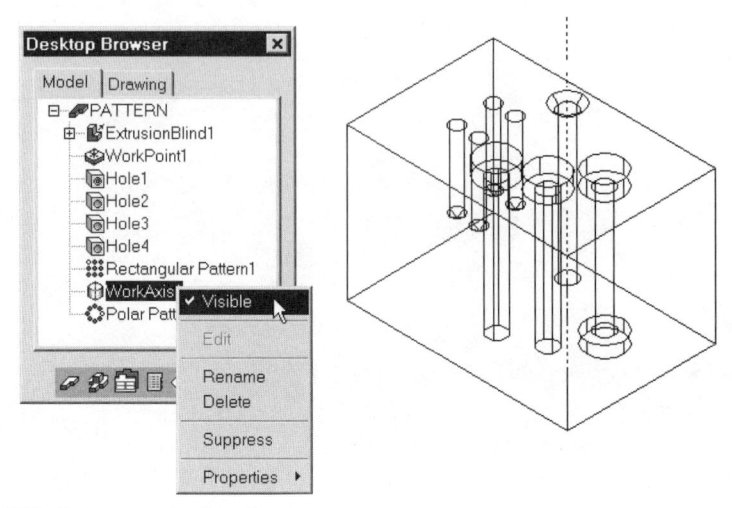

Figure 4–121 *Polar pattern placed*

Axial Pattern

An axial pattern repeats selected source features along a helix path. Now you will construct two revolved solid features and place an axial pattern of one of the revolved features around the other revolved feature.

1. Start a new part file. Use metric as the default.
2. Construct a rectangle, resolve it to a profile, and add parametric dimensions.
3. Set the display to an isometric view.
4. Construct a revolved solid (see Figure 4–122).
5. Referring to Figure 4–123, construct a rectangle, resolve it to a profile, add parametric dimensions, and construct a revolved solid to join the solid part.

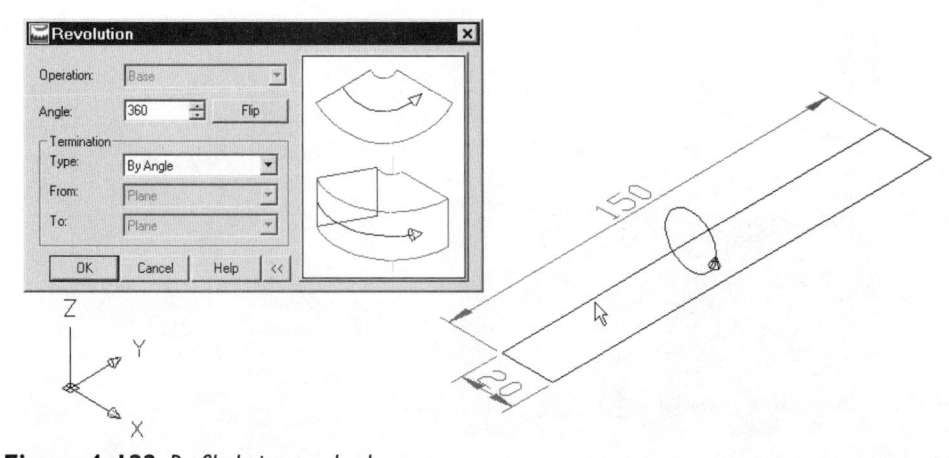

Figure 4–122 *Profile being revolved*

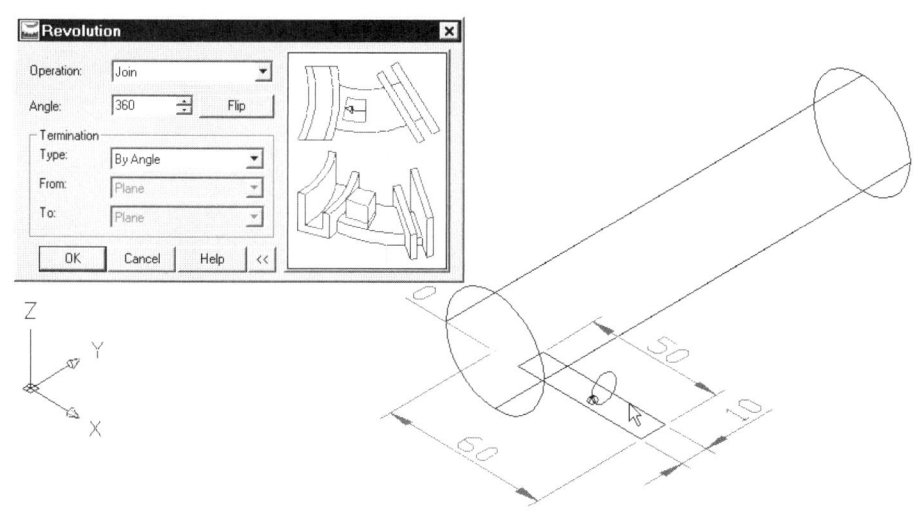

Figure 4–123 *Second profile being revolved*

Two revolved solid features are constructed. Now you will place an axial pattern.

6. Select Part>Placed Feature>Axial Pattern
7. Select the revolved solid feature indicated in Figure 4–124 and press the ENTER key.
8. Select the base solid feature as the rotational center.
9. In the Pattern dialog box, set instances to 6, select Incremental Angle, set spacing angle to 30, select Incremental Offset, set offset height to 25, and select the OK button. You might need to flip the axial pattern (see Figure 4–124).

An axial pattern is placed (see Figure 4–125). Save and close your file (file name: *AxialPattern.dwg*).

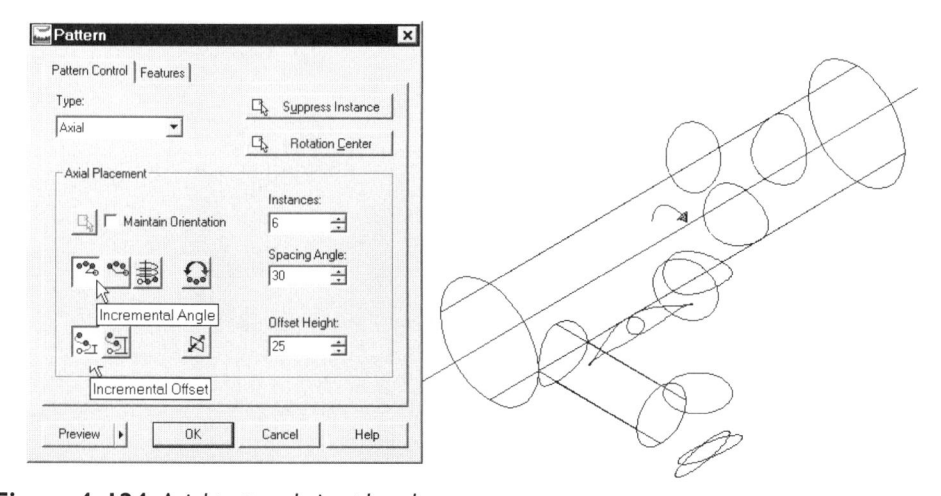

Figure 4–124 *Axial pattern being placed*

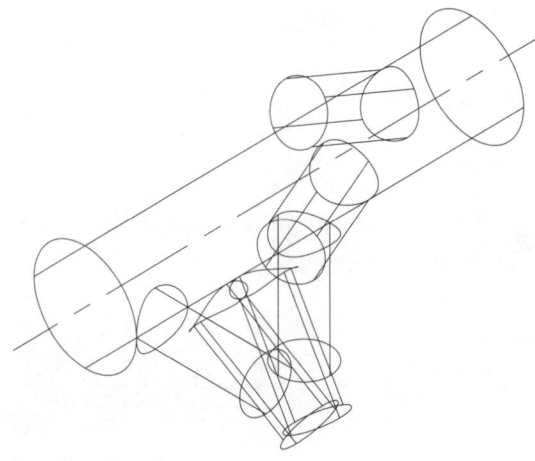

Figure 4–125 *Axial pattern placed*

Independent Pattern Instances

Each individual object in the rectangular, polar, or axial pattern is an instance of the source feature. By default, all the instances in the pattern will change if the source solid changes. However, you can make an instance of a pattern independent and modify it individually.

1. Open the file *Pattern.dwg*.
2. Select the rectangular pattern from the browser, right-click, and select Independent Instance (see Figure 4–126).
3. Select the instance indicated in Figure 4–126.

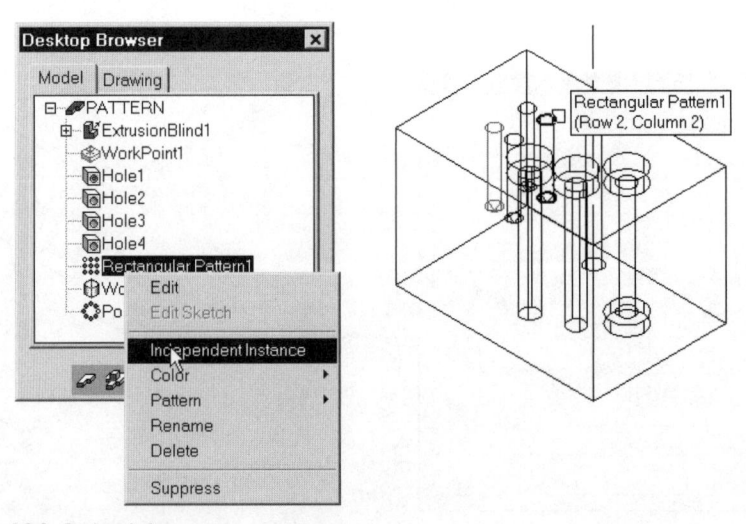

Figure 4–126 *Right-click menu*

Now the selected instance of the pattern is suppressed and a new feature is copied from the selected instance. There is not much change in the appearance of the solid part. However, you will find a new object in the browser (see Figure 4–127).

4. Select the independent feature from the browser (or the solid part) and double-click.
5. In the Hole dialog box, select the Threads tab.
6. In the Thread tab, change the thread size to M16 (see Figure 4–128).
7. Select the Hole tab.
8. In the Hole tab, select Through type and select the OK button (see Figure 4–129).
9. Press the ENTER key.
10. If needed, select the Update button from the browser.

The independent feature is modified (see Figure 4–130). Save your file.

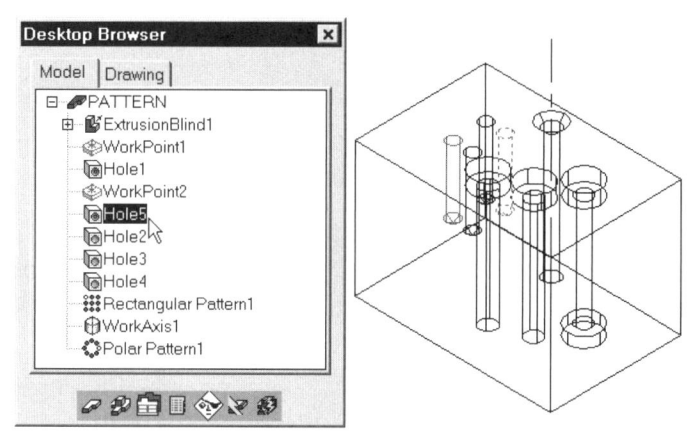

Figure 4–127 *New object in the browser*

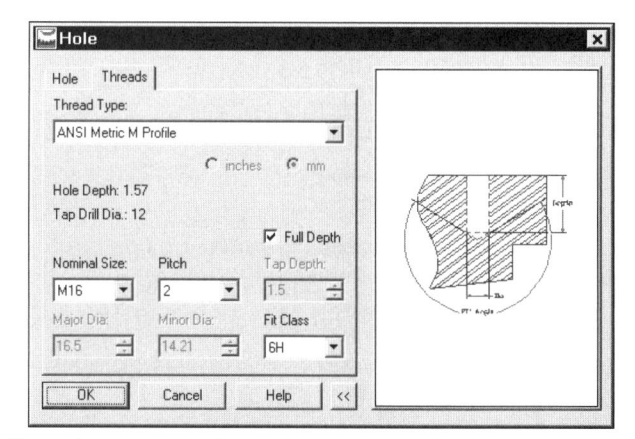

Figure 4–128 *Thread parameters changed*

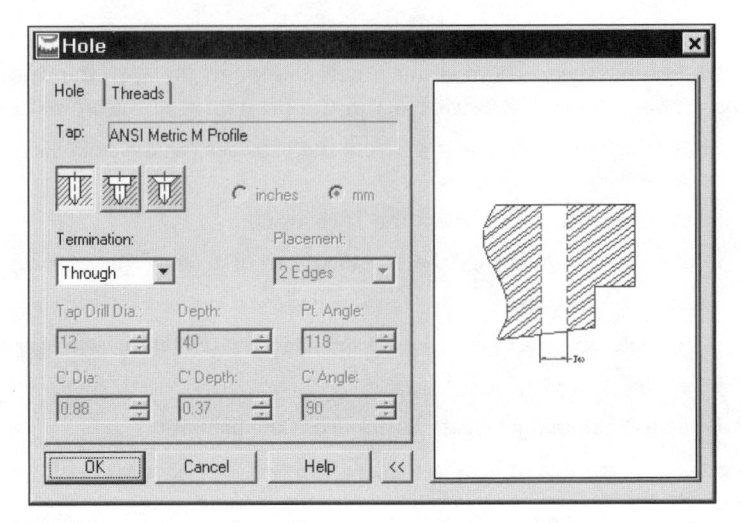

Figure 4–129 *Hole parameter changed*

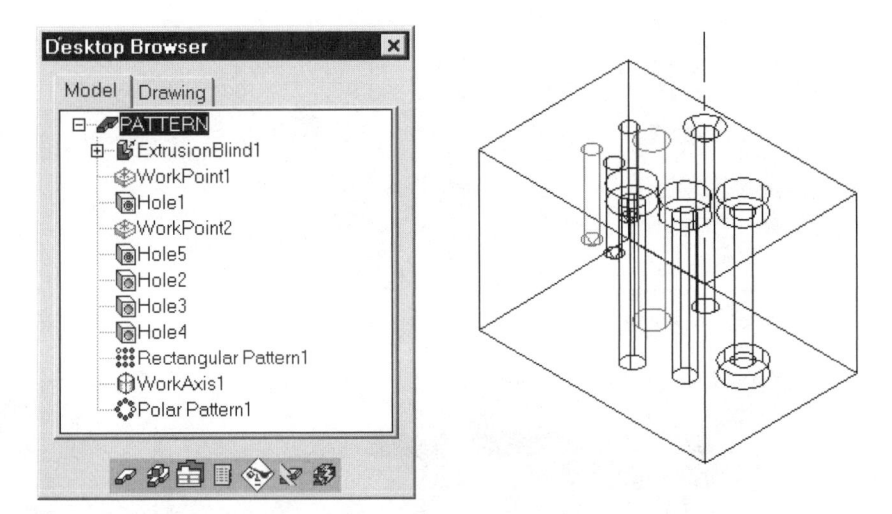

Figure 4–130 *Independent instance modified*

Suppression of Individual Instances

As mentioned in the last section, individual instances of a pattern can be suppressed. Now you will suppress an instance.

    11. Select the polar array from the browser and double-click (see Figure 4–131).

    12. In the Pattern dialog box, select the Suppress Instances button.

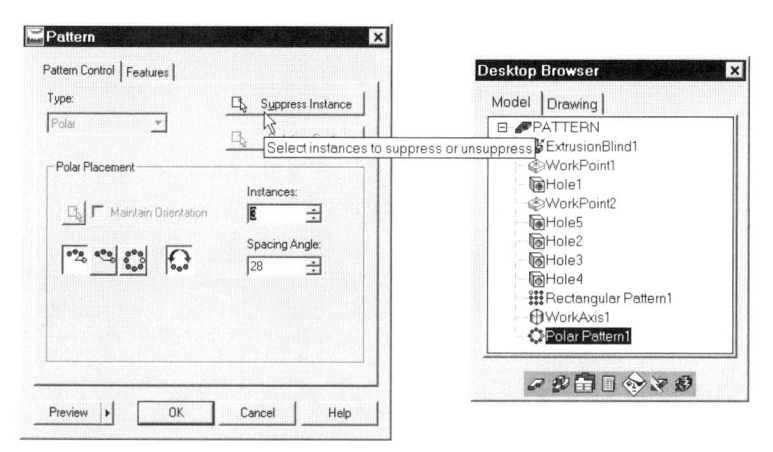

Figure 4–131 *Polar pattern selected*

13. Select the instance indicated in Figure 4–132 and press the ENTER key.
14. On returning to the Pattern dialog box, select the OK button.
15. Press the ENTER key.

The selected instance is suppressed (see Figure 4–133). Save and close your file.

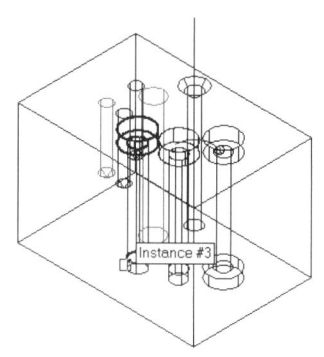

Figure 4–132 *Instance selected*

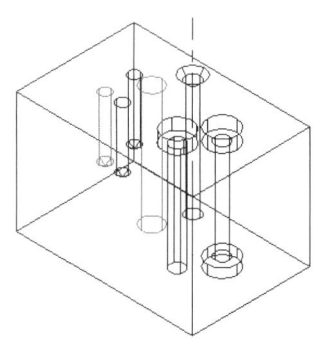

Figure 4–133 *Selected instance suppressed*

SUMMARY

To construct a parametric solid, you use two kinds of solid features: sketched solid features and placed solid features. In addition, you also use work features as construction objects. Hence, a solid part can have three kinds of features: sketched solid features, placed solid features, and work features.

There are three kinds of work features: work plane, work axis, and work point. You use them to help establish sketch plane and construction objects for making solid features in a solid part. Once a work feature is used, you must not discard them even after you have finished your solid part. You hide them if they are no longer needed.

Chapters 3 and 4 taught you about six kinds of sketched solid features: extruded, revolved, sweep, loft, rib, and bend. Both extruded and revolved solid features need only a single parametric sketch. A sweep solid feature needs two sketches: a profile sketch and a path sketch. You sweep the profile along the path. There are five kinds of path sketches: 2D path, 3D edge path, 3D helix path, 3D pipe path, and 3D spline path. To make a loft solid, you need two or more profile sketches to loft along the profiles. A rib solid can be considered a special kind of extruded feature. You need an open-loop sketch and to use the body of the solid to form a close loop for making the rib. A bend feature is, in fact, a kind of operation through which you bend a solid about a profile sketch which is a straight line.

Placed solid features are pre-constructed solid features available from the command menu. You select a feature and place them in your solid part. The distinction between a sketched solid feature and a placed solid feature is that a sketched solid feature starts from a parametric sketch, whereas a placed solid feature is added to the solid part directly. In this chapter, you learned about six basic kinds of placed solid features: hole, thread, fillet, chamfer, shell, and array. In Chapter 9, you will learn about the other kinds of placed solid features.

PROJECTS

Now you will work on the following projects to enhance your comprehension thus far.

WHEEL

You will complete the wheel of the infant scooter by adding a hole.

1. Open the file *Wheel.dwg* that you constructed in Chapter 3.
2. Select Part>Placed Features>Hole
3. In the Hole dialog box, select Drilled operation, Through termination, and Concentric placement, set drill diameter to 15, and select the OK button.
4. Select the circular face highlighted in Figure 4–134 to indicate a plane.

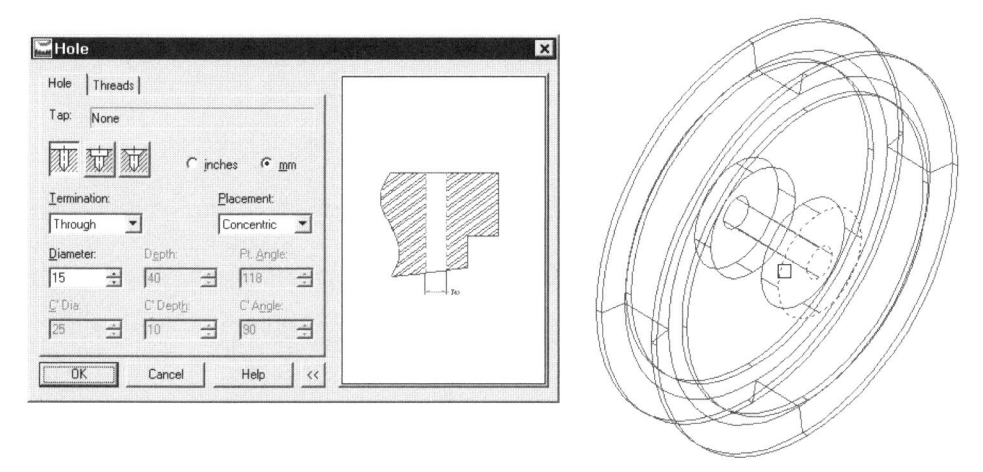

Figure 4-134 *Hole being placed*

5. Select the circular face again to indicate a concentric feature.
6. Press the ENTER key.

A drilled hole is placed. The model is complete. Save and close your file.

CENTRAL FRAME

You will now complete the central frame of the infant scooter by adding two holes.

1. Open the file *Frame_C.dwg* that you constructed earlier in this chapter.
2. Select Part>Placed Features>Hole
3. In the Hole dialog box, select Drilled operation, Through termination, and On Point placement, set drill diameter to 25, and select the OK button.
4. Select the work point indicated in Figure 4–135 and press the ENTER key.

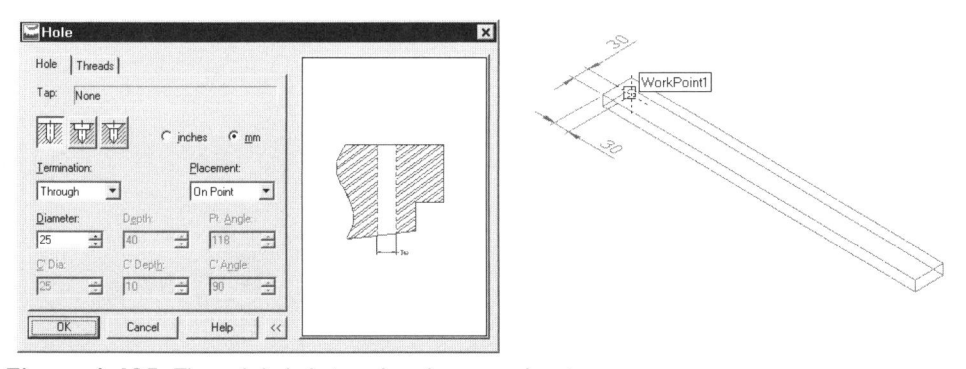

Figure 4-135 *Through hole being placed on a work point*

5. Repeat the AMHOLE command.

6. In the Hole dialog box, select Drilled operation, Through termination, and 2 Edges placement, set drill diameter to 11, and select the ENTER key.
7. Referring to Figure 4–136, select two edges and specify two distances (150 mm and 30 mm).
8. Press the ENTER key.

The central frame is complete. Save and close your file.

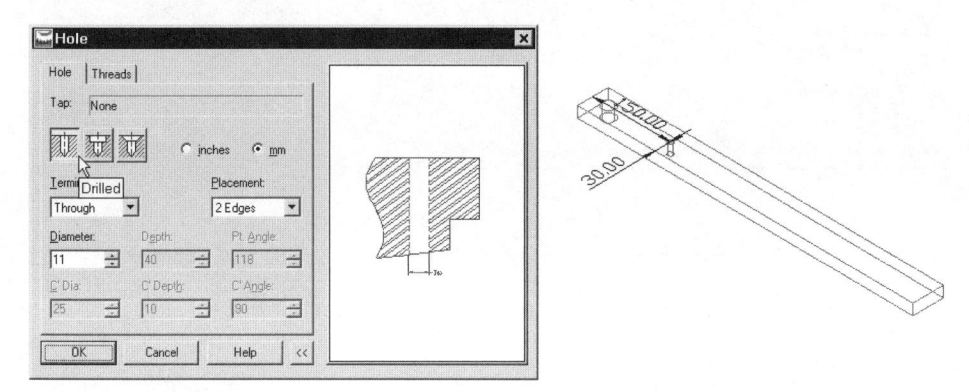

Figure 4–136 *Second hole being placed*

FRONTAL FRAME

Now complete the frontal frame of the infant scooter by adding a rib and holes.

1. Open the file *Frame_F.dwg* that you constructed in Chapter 3.
2. Referring to Figure 4–137, construct a work plane parallel to the selected face and offset a distance of 30.

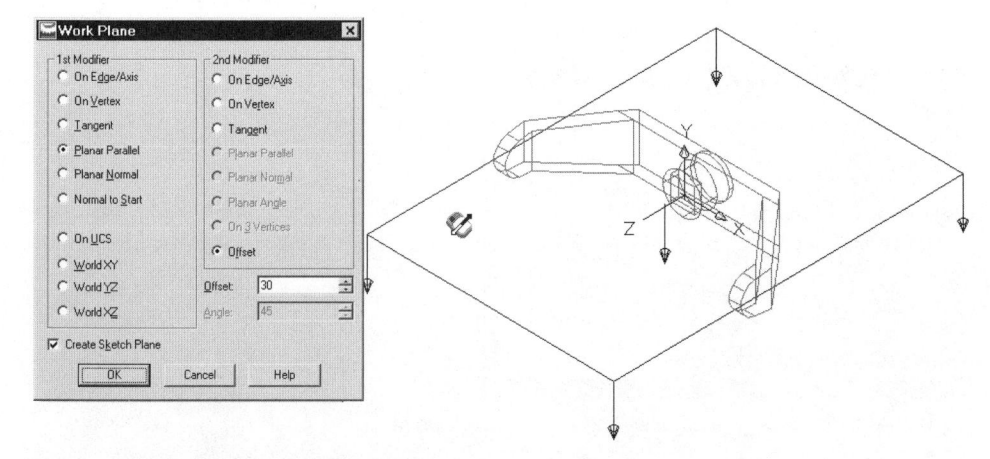

Figure 4–137 *Work plane being constructed*

Part Modeling II 181

3. Use the shortcut key 9 to set the display to the top view.
4. Construct a line and resolve it to an open profile (see Figure 4–138).
5. Add a horizontal constraint and a parametric dimension.
6. Set the display to an isometric view.
7. Referring to Figure 4–139, construct a rib with a thickness of 5 mm.

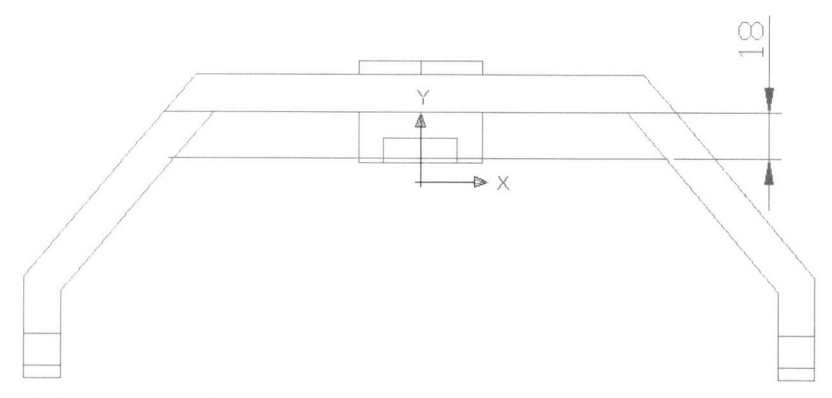

Figure 4–138 *Open profile constructed*

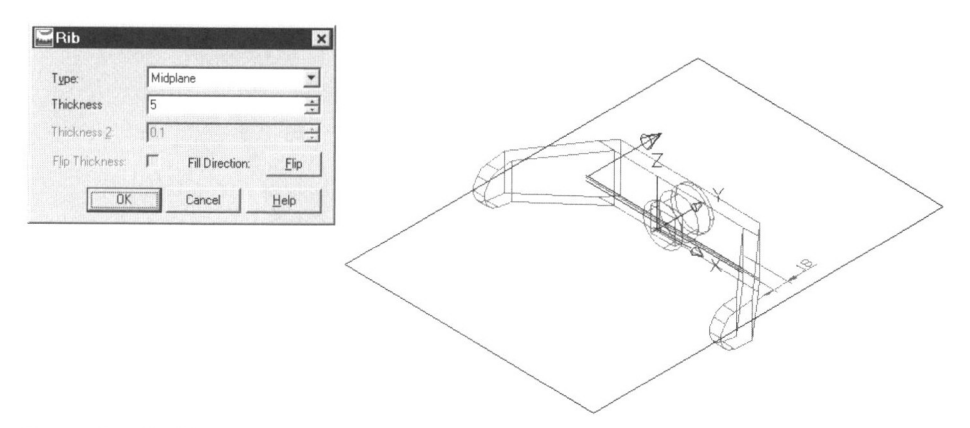

Figure 4–139 *Rib being constructed*

8. Hide the work plane.
9. Select the rib feature from the browser and reorder it (see Figure 4–140).
10. Referring to Figure 4–141, rotate the display and place a concentric through-hole with a diameter of 25 mm.
11. Set the display to an isometric view.
12. Referring to Figure 4–142, place a concentric through-hole of 15 mm diameter.

The model is complete. Save and close your file.

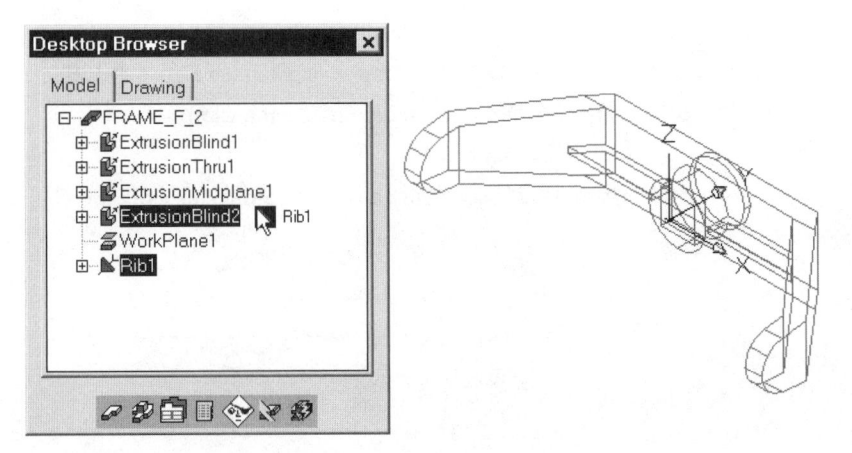

Figure 4–140 *Rib reordered*

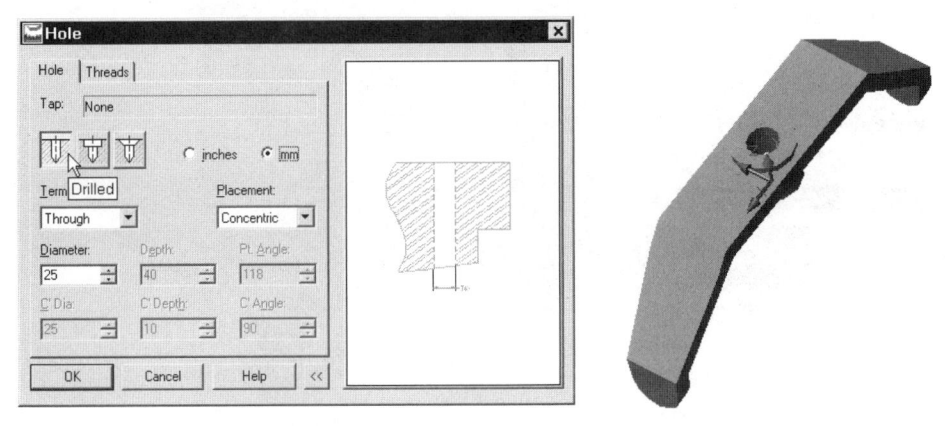

Figure 4–141 *Through-hole placed*

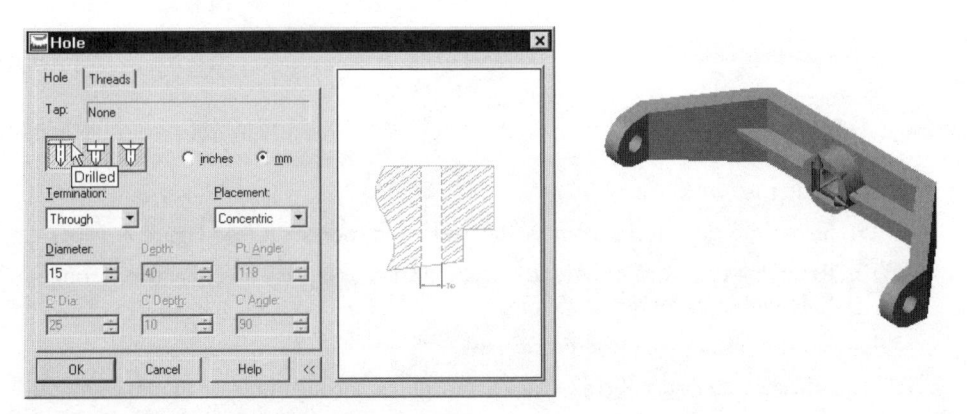

Figure 4–142 *Second through-hole placed*

REAR FRAME

Now complete the rear frame of the infant scooter by adding holes and chamfer features.

1. Open the file *Frame_R.dwg* that you constructed in Chapter 3.
2. Referring to Figure 4–143, set up a sketch plane and construct two work points.

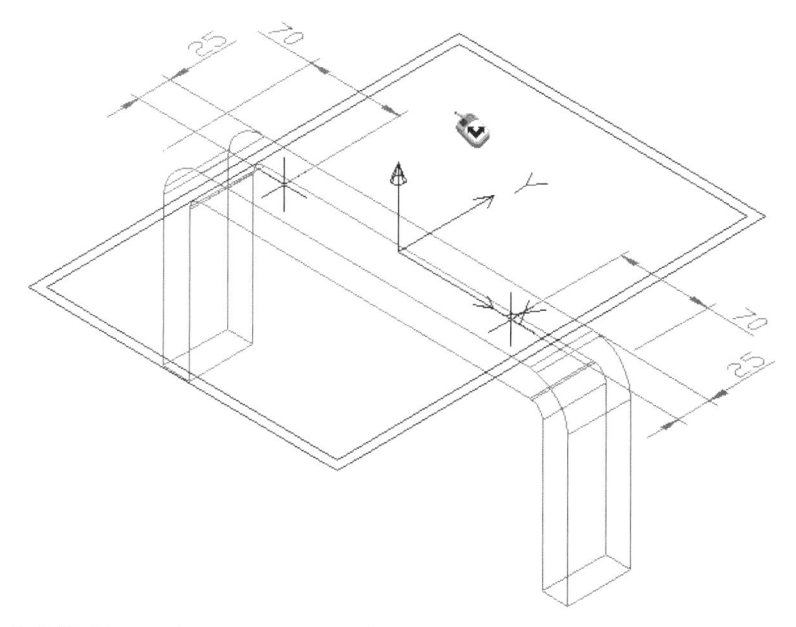

Figure 4–143 *Two work points constructed*

3. Place two through-holes of 11 mm diameter on the work points indicated in Figure 4–144.
4. Set up a sketch plane and construct a work point in accordance with Figure 4–145.

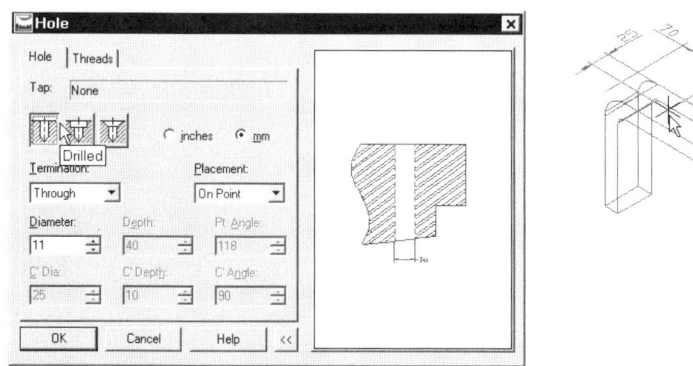

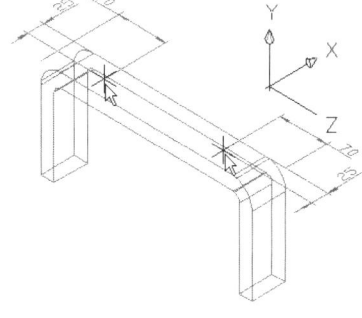

Figure 4–144 *Two holes being placed*

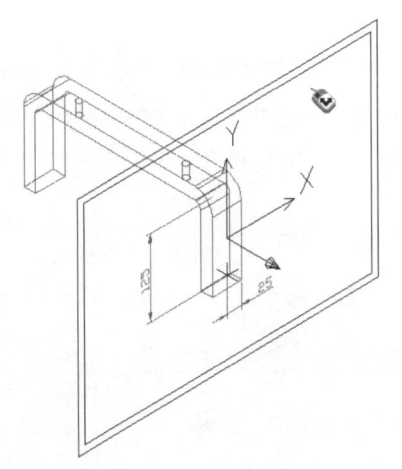

Figure 4–145 *Work point constructed*

5. Place a through-hole of 15 mm diameter on the work point (see Figure 4–146).
6. Referring to Figure 4–147, place four equal-distance chamfer features that measure 10 mm distance.

The rear frame is complete. Save and close your file.

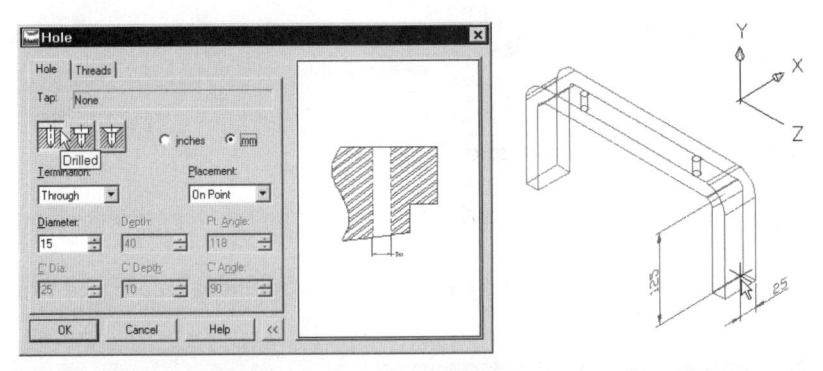

Figure 4–146 *Hole being placed*

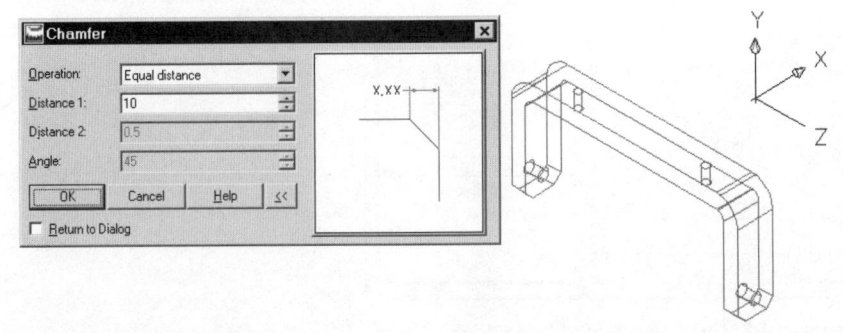

Figure 4–147 *Chamfer placed*

STEERING SHAFT

Complete the steering shaft of the infant scooter by adding hole and chamfer features.

1. Open the file *SteeringShaft.dwg* that you constructed in Chapter 3.
2. Place a drilled tapped blind hole with major diameter of 8 mm, drill size of 6 mm, and depth of 20 mm on the highlighted square face and concentric to a cylindrical feature as indicated in Figure 4–148.
3. Referring to Figure 4–149, place four equal-distance (3 mm) chamfer edges.

The steering shaft is complete. Save and close your file.

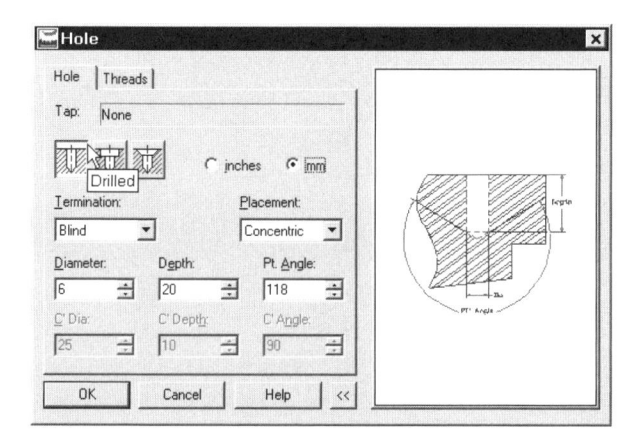

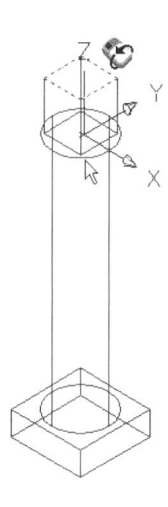

Figure 4–148 *Drilled tapped blind hole being placed*

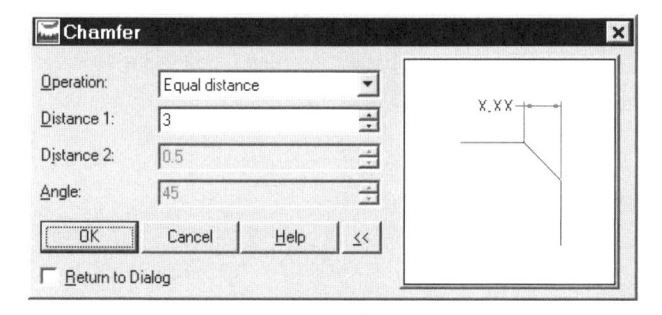

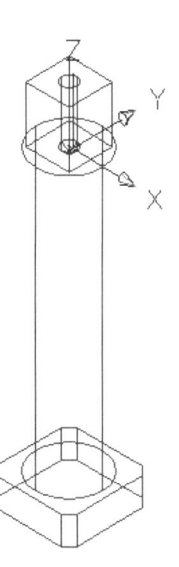

Figure 4–149 *Chamfer feature placed*

TIRE

Now complete the tire of the infant scooter by constructing a polar pattern feature.

1. Open the file *Tire.dwg* that you constructed in Chapter 3.
2. Delete the fix constraint from the unconsumed sketch.
3. Referring to Figure 4–150, add two parametric dimensions and construct a work axis.

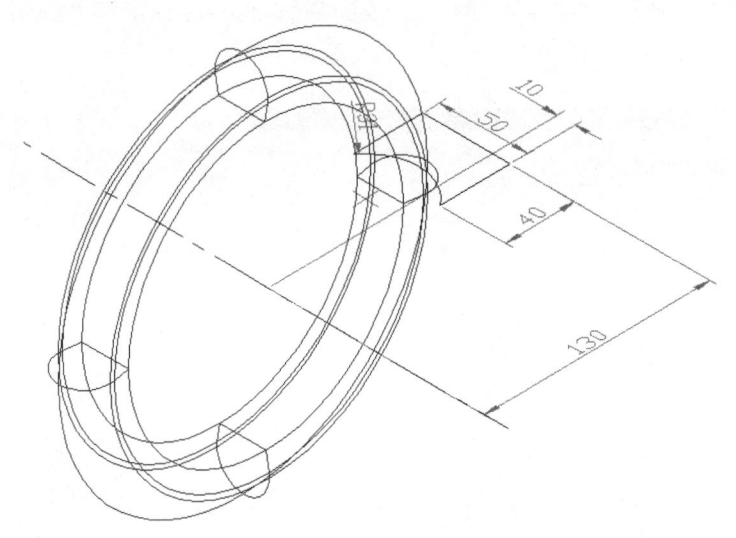

Figure 4–150 *Fix constraint deleted, parametric dimensions added, and work axis constructed*

4. Revolve the profile about the work axis from mid-plane for an angle of 10 degrees to cut the solid part (see Figure 4–151).
5. Construct a polar pattern of the revolve cut feature (see Figure 4–152).
6. Hide the work axis.

The tire is complete. Save and close your file.

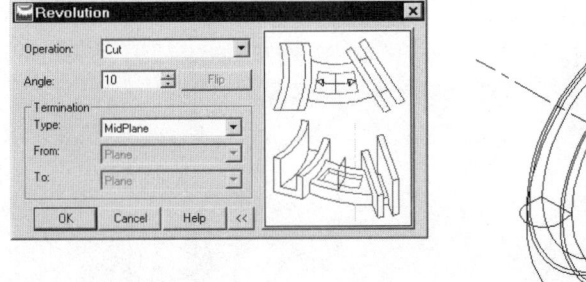

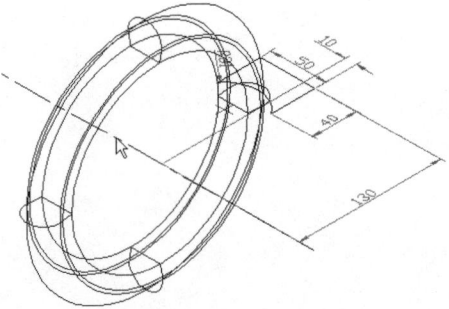

Figure 4–151 *Profile revolved to cut the solid part*

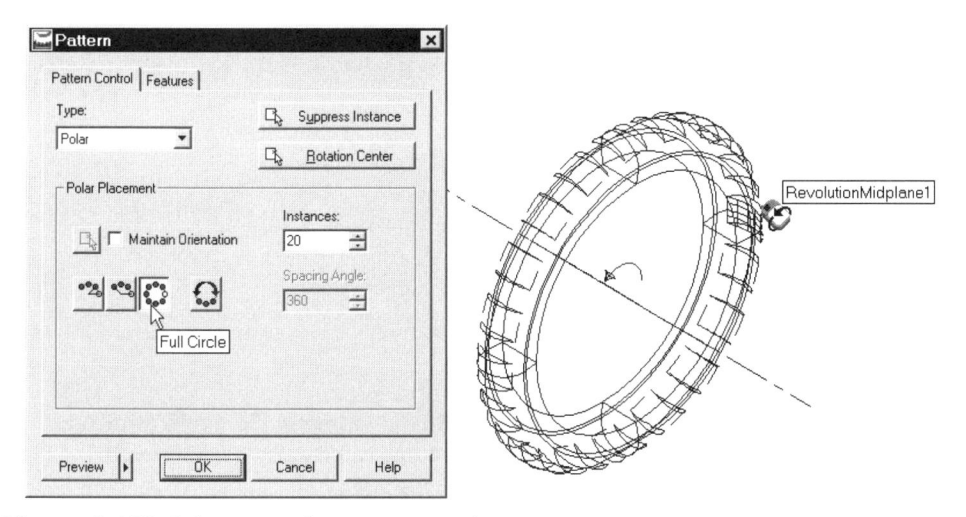

Figure 4-152 *Polar pattern being constructed*

SEAT

Now complete the model of the seat of the infant scooter by adding hole, pattern, shell, and sweep features.

1. Open the file *Seat.dwg* that you constructed in Chapter 3.
2. Referring to Figure 4–153, construct six work points.

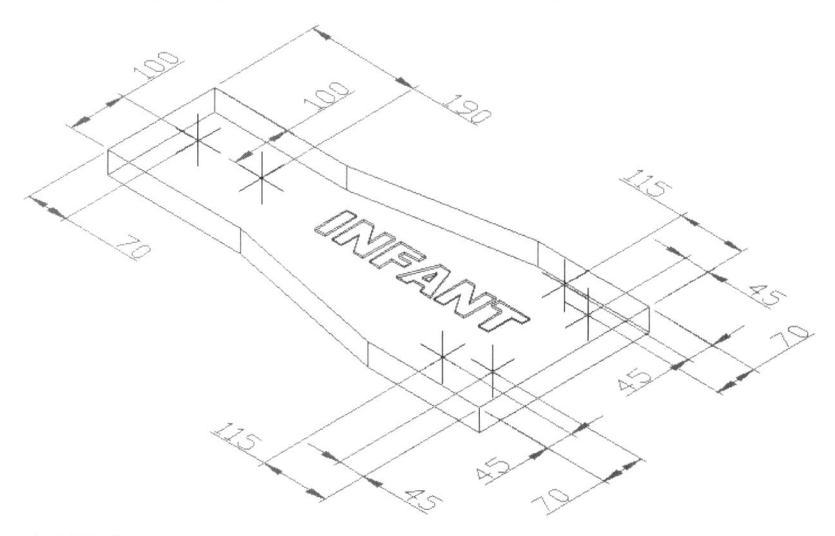

Figure 4-153 *Six work points constructed*

3. Place a drilled through-hole of 62 mm diameter on the work point indicated in Figure 4–154.

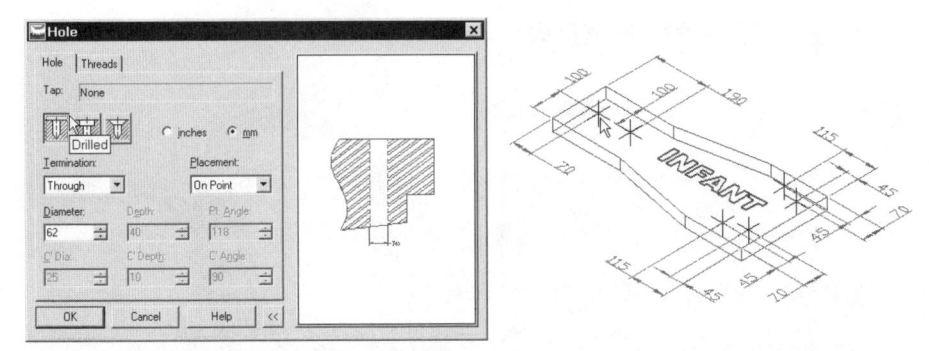

Figure 4–154 *Drilled through-hole being placed*

4. Place three C'bore through-holes of 10 mm diameter, 30 mm C'diameter, and 15 mm C' depth on the work point indicated in Figure 4–155.

5. Place two drilled blind-holes of 30 mm diameter, 30 mm depth, and 180 PT angle on the work point indicated in Figure 4–156.

6. Place a fillet feature of 50 mm radius at the vertical edges indicated in Figure 4–157.

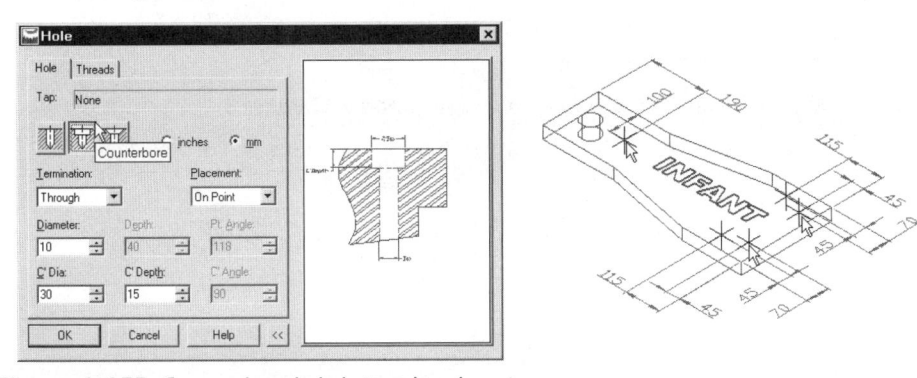

Figure 4–155 *Counter bore hole being placed*

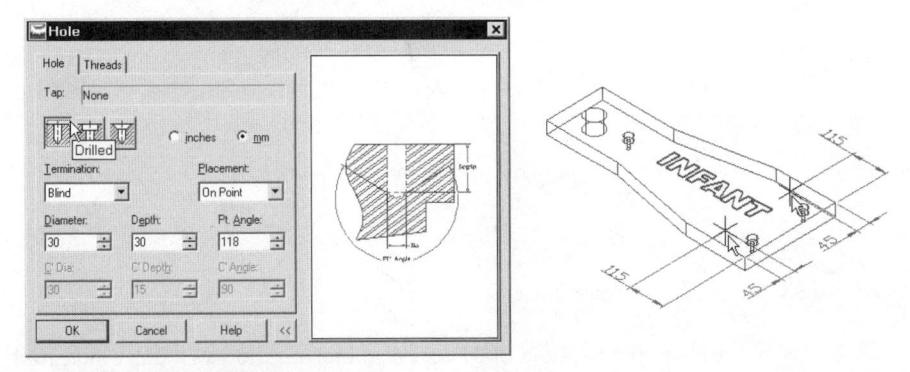

Figure 4–156 *Blind-hole being placed*

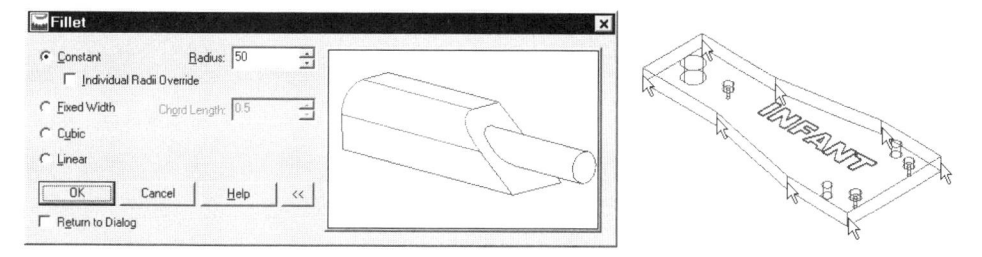

Figure 4–157 *Fillet feature being placed on eight edges*

7. Place fillet feature of 15 mm radius at two edges indicated in Figure 4–158.
8. Place a shell feature of 5 mm thick with the bottom face removed (see Figure 4–159).
9. Referring to Figure 4–160, construct a 3D edge path along the lower edge of the solid part and establish a sketch plane at the start point of the path.

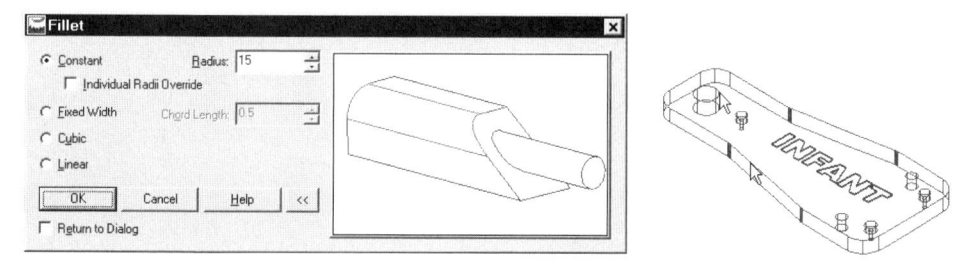

Figure 4–158 *Fillet feature being placed on two edges*

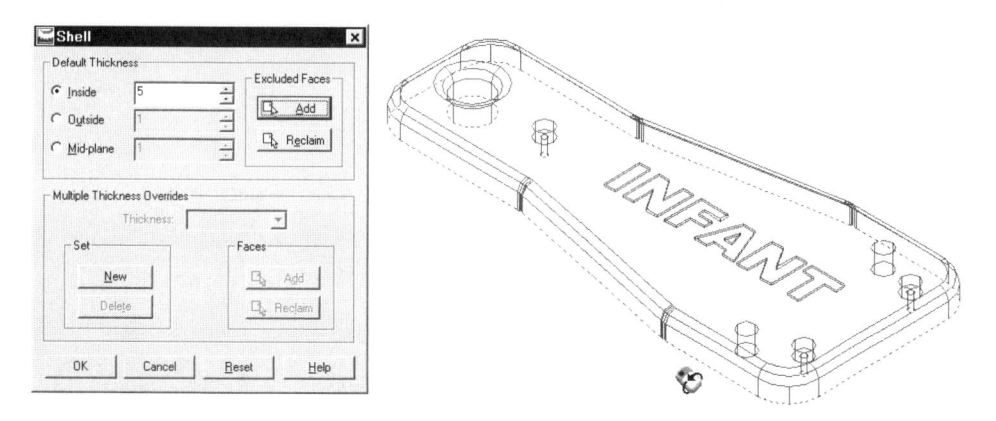

Figure 4–159 *Shell pattern placed*

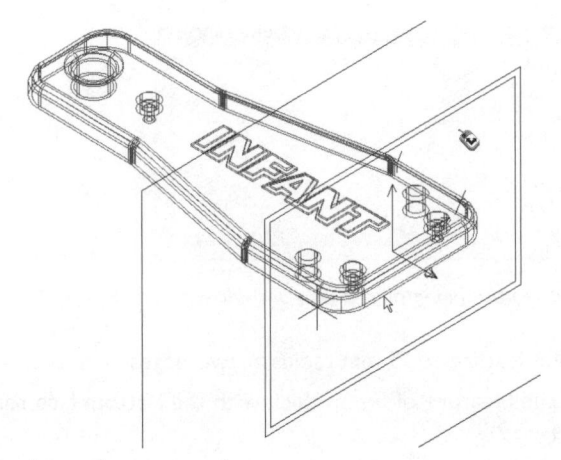

Figure 4-160 *3D edge path constructed*

10. Construct a profile in accordance to Figure 4-161.
11. Sweep the profile along the edge path to join the solid part.
12. Hide the work plane and shade the display (see Figure 4-162).

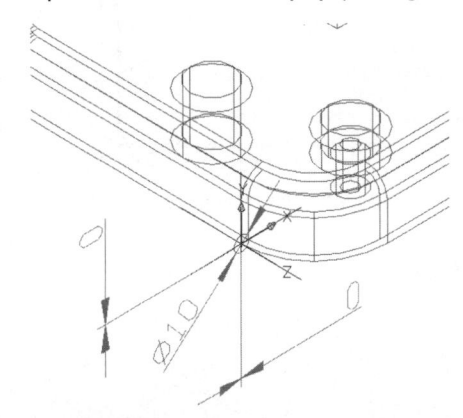

Figure 4-161 *Profile constructed on plane normal to start of the edge path*

Figure 4-162 *Sweep solid feature constructed*

13. Rotate the display and place two concentric holes of 10 mm diameter in accordance with Figure 4–163.

The solid part is complete. Save and close your file.

Figure 4–163 *Holes placed*

THREADED BOLT

Now construct two bolts, one with a cosmetic thread placed on the shank and one with a helical groove cut on the shank.

1. Open the file *Bolt.dwg* that you constructed in Chapter 3.
2. Select File>Save As and specify a new file name (file name: *Thread.dwg*).
3. Double-click the extruded solids from the browser one by one to edit the dimensions (see Figures 4–164 and 4–165). You will also be editing their sketches.
4. Save your file.
5. Select File>Save As and specify a new file name (file name: *ThreadBolt.dwg*).

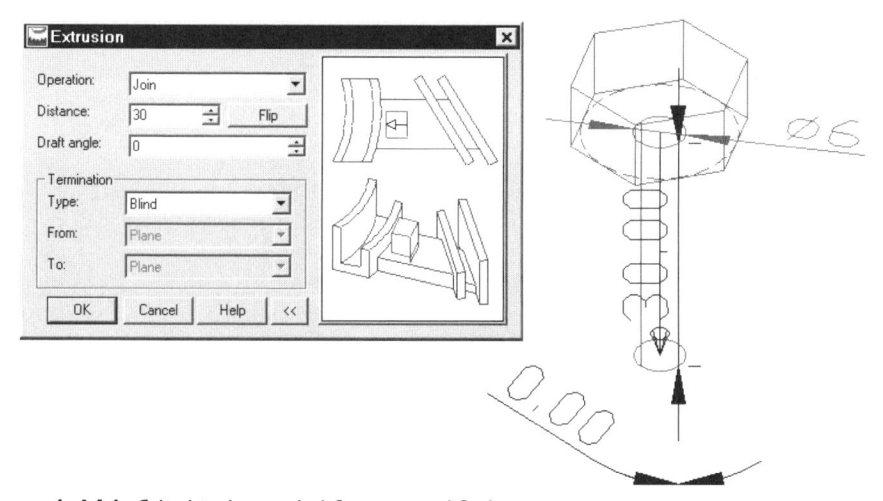

Figure 4–164 *Cylindrical extruded feature modified*

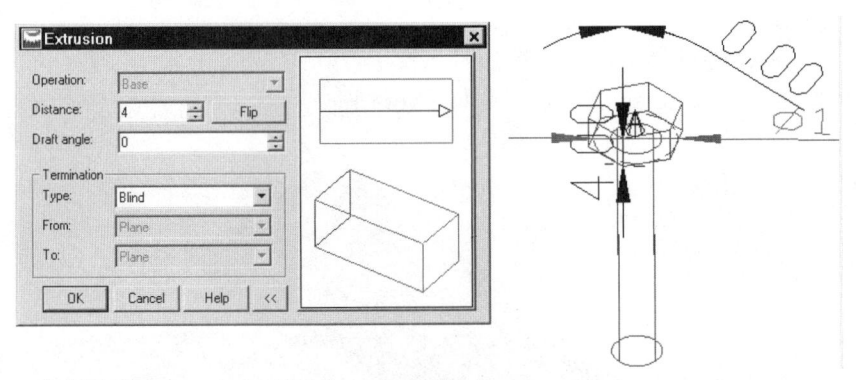

Figure 4-165 *Hexagonal extruded feature modified*

      6. Referring to Figure 4-166, place a thread feature on the shank of the bolt.
      7. Select the OK button.

Thread feature is placed (see Figure 4-167). Save and close your file.

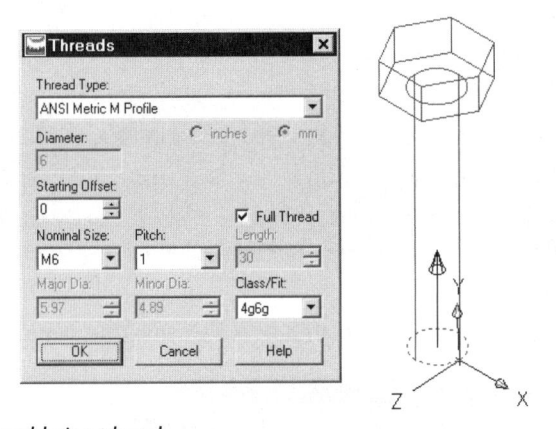

Figure 4-166 *Thread being placed*

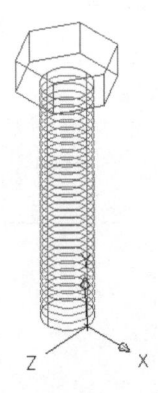

Figure 4-167 *Thread placed*

Now construct a bolt with a helical thread cut on it.

1. Open the file *Thread.dwg*.
2. Referring to Figure 4–168, construct a work axis on the cylindrical feature and establish a sketch plane on the bottom face.
3. Referring to Figure 4–169, construct a helical path with a pitch of 1 mm, 15 revolutions, and 6 mm diameter.

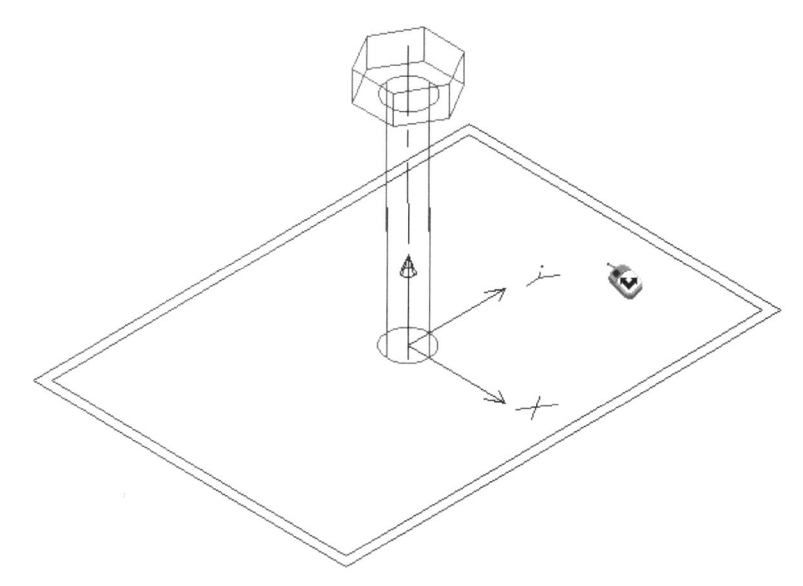

Figure 4–168 *Work axis constructed and sketch plane established*

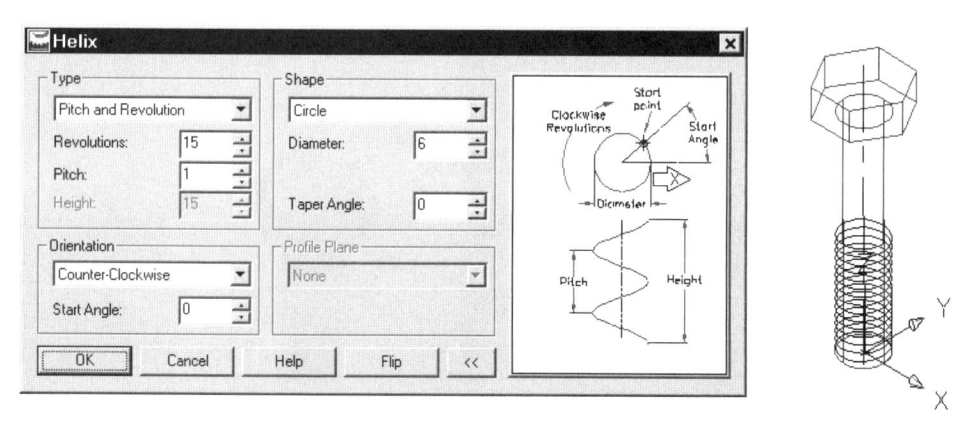

Figure 4–169 *Helical path being constructed*

4. Construct a work plane normal to the start of the helical path and establish a sketch plane on the work plane (see Figure 4–170).

5. Referring to Figure 4–171, construct a profile on the sketch plane.
6. Construct a sweep solid to cut the solid part.
7. Shade the display and hide the work features (see Figure 4–172).

The screw thread is complete. Save and close your file.

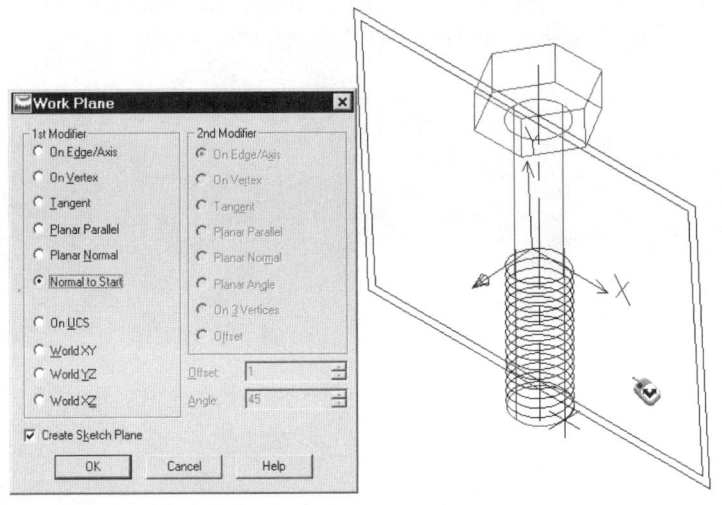

Figure 4–170 *Work plane being constructed*

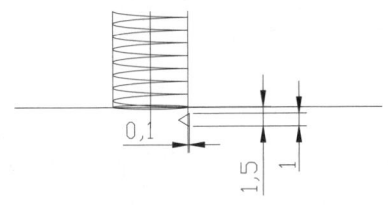

Figure 4–171 *Profile constructed*

Figure 4–172 *Helical groove constructed*

HANDLE

Now construct the handle of the infant scooter. Its main body is a sweep solid. You will construct a 3D pipe path and a circular profile and sweep the profile along the 3D pipe path.

1. Start a new part file. Use metric as the default.
2. Construct three basic work planes.
3. Referring to Figure 4–173, establish a sketch plane on Work Plane 1 and construct three work points.

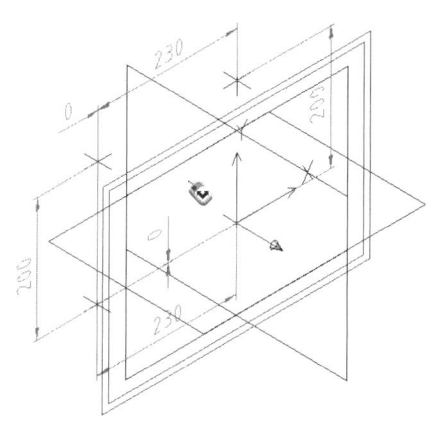

Figure 4–173 *Work points constructed on a work plane*

Together with the work point at the origin, you have four work points.

4. Construct a work plane that offsets a distance of 30 mm from Work Plane 1 (see Figure 4–174).
5. Set up a sketch plane on the offset work plane and construct two work points (see Figure 4–175).

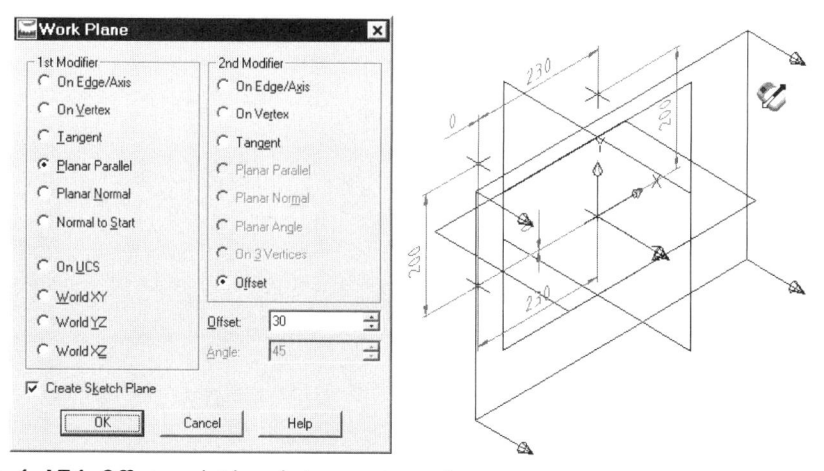

Figure 4–174 *Offset work plane being constructed*

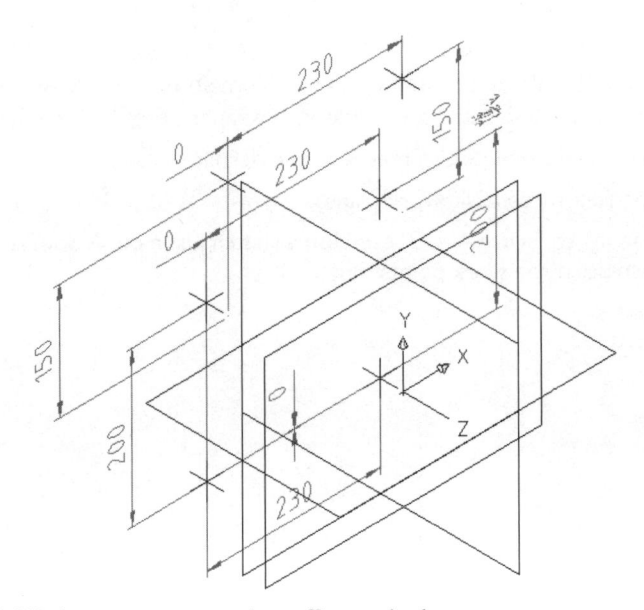

Figure 4–175 Work points constructed on offset work plane

6. Referring to Figure 4–176, construct a 3D polyline that passes through the work points.

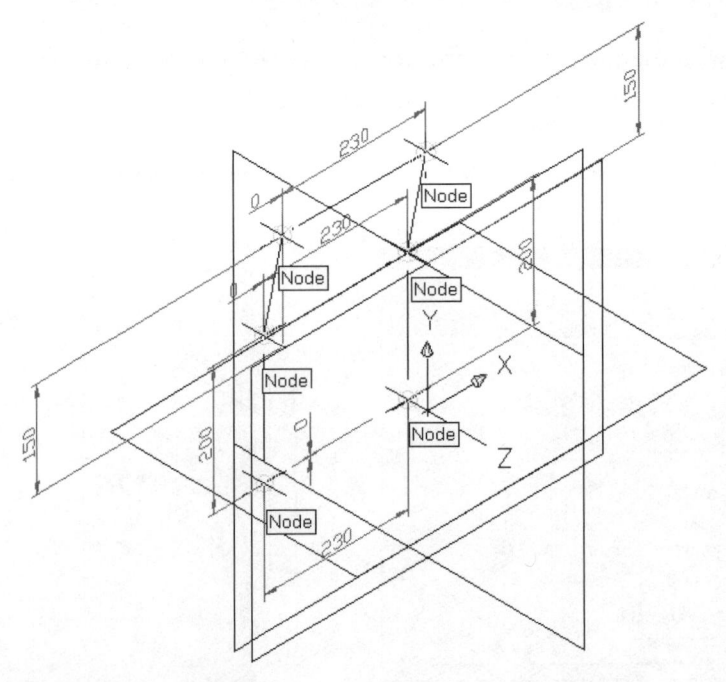

Figure 4–176 Pipe path constructed

Part Modeling II 197

7. Select the work features from the browser one by one, right-click, and deselect Visible.
8. Resolve the 3D polyline to a pipe path. Select the lower-left end as the start point.
9. In the 3D Pipe Path dialog box, set four corner radii to 30 (see Figure 4–177).

No.	C	From	Delta X	Delta Y	Delta Z	Length	Angle XY	Angle Z	Radius
1			-230	0	-30	200	90	0	
2			-230	200	-30	152.97	90	11.31	30
3			-230	350	0	230	0	0	30
4			0	350	0	152.97	-90	-11.31	30
5			0	200	-30	200	-90	0	30
6			0	0	-30				
7									

Figure 4–177 *Profile constructed*

10. Construct a work plane at the start of the path and construct a profile (see Figure 4–178).
11. Sweep the profile along the pipe path, hide the work plane, and add two blind drilled tapped holes (see Figure 4–179).

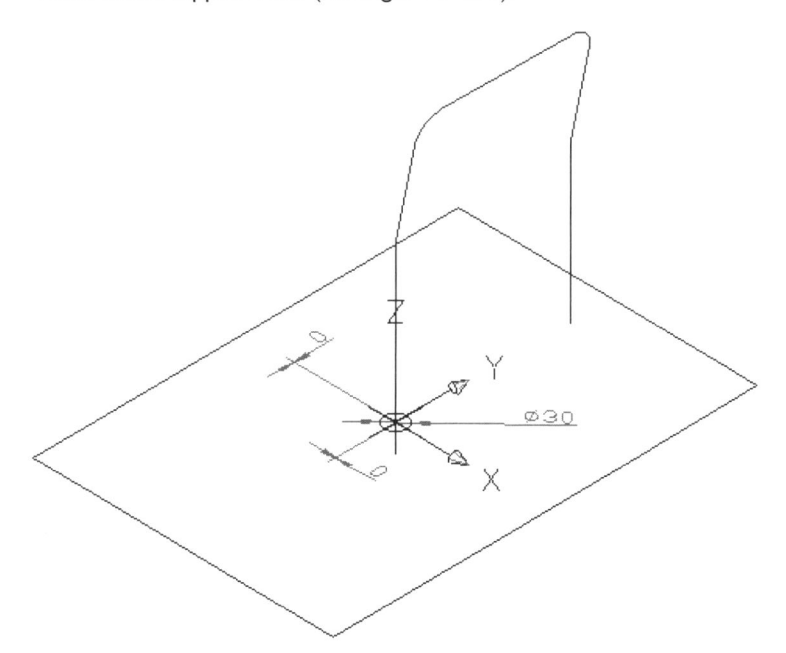

Figure 4–178 *Profile constructed*

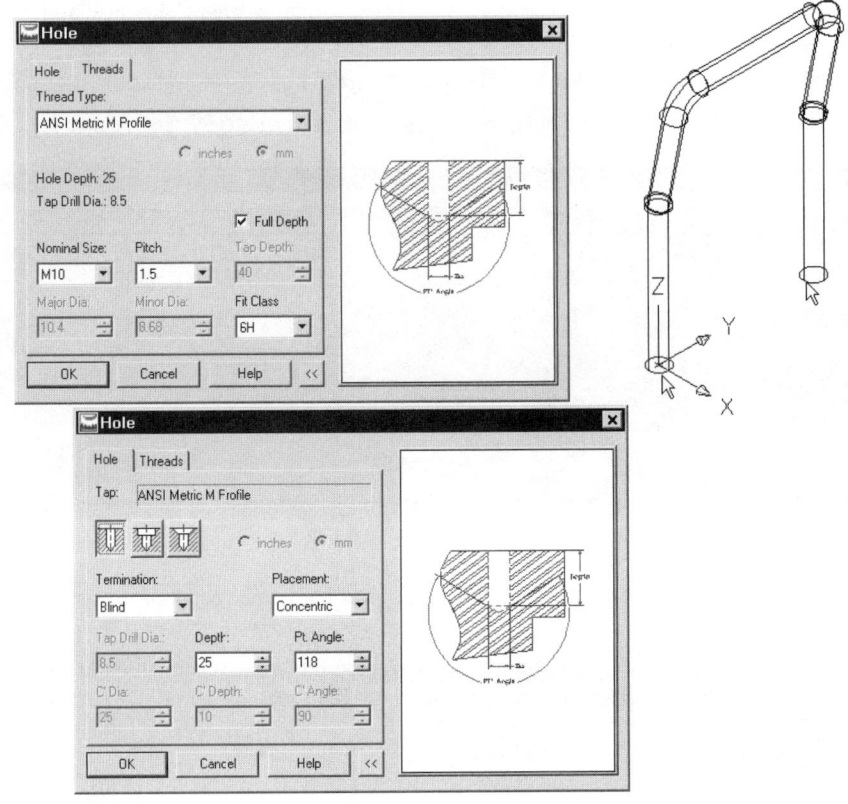

Figure 4–179 *Sweep solid constructed and holes being placed*

The solid part is complete (see Figure 4–180). Save and close your file (file name: *Handle.dwg*).

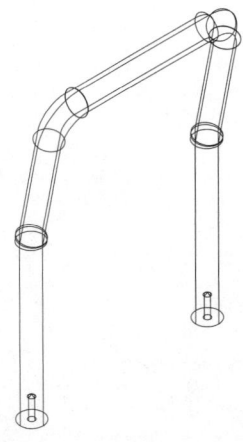

Figure 4–180 *Completed solid part*

STEERING WHEEL

Construct the steering wheel of the infant scooter. The main body of it is a loft solid.

1. Start a new part file. Use metric as the default.
2. Construct three basic work planes and construct a profile on work plane 1 and another profile on work plane 3 (see Figure 4–181).
3. Construct a work plane that offsets a distance of 60 mm from work plane 3 (see Figure 4–182).

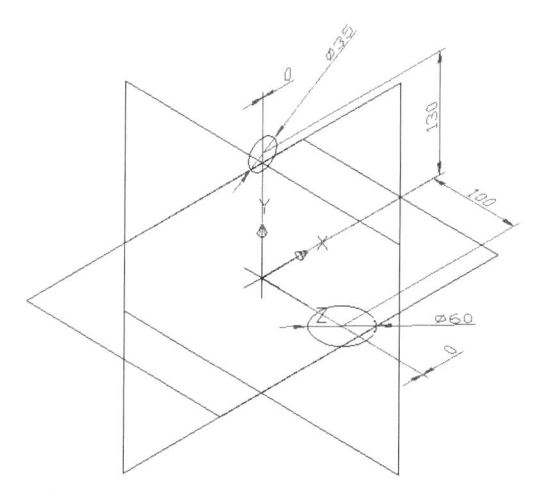

Figure 4–181 *Two profiles constructed*

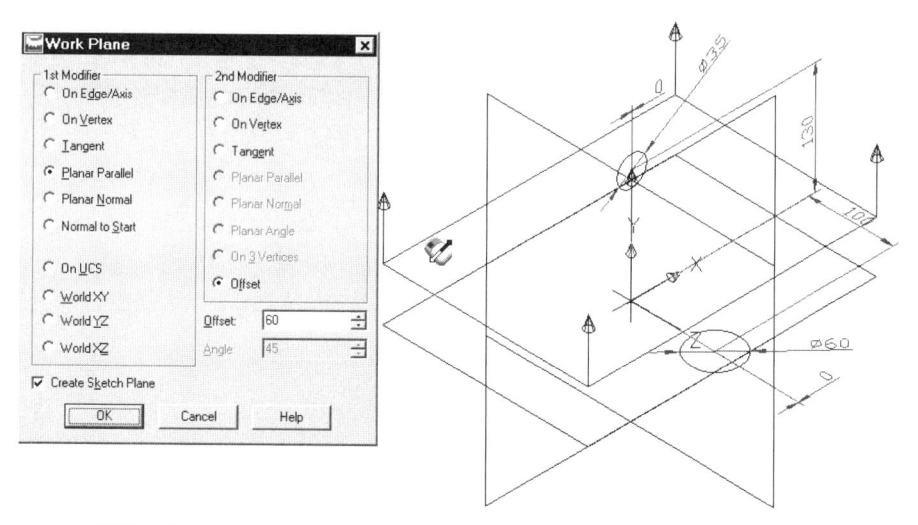

Figure 4–182 *Offset work plane constructed*

4. Construct a profile on the new work plane (see Figure 4–183).

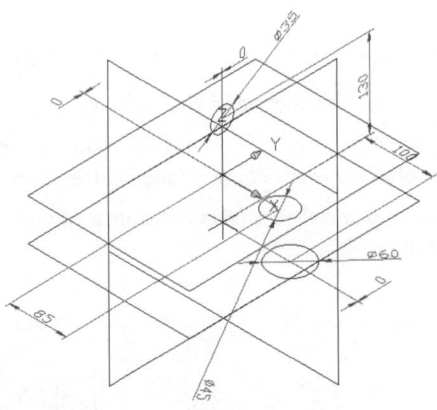

Figure 4-183 *Profile constructed*

5. Referring to Figure 4-184, construct a loft solid.

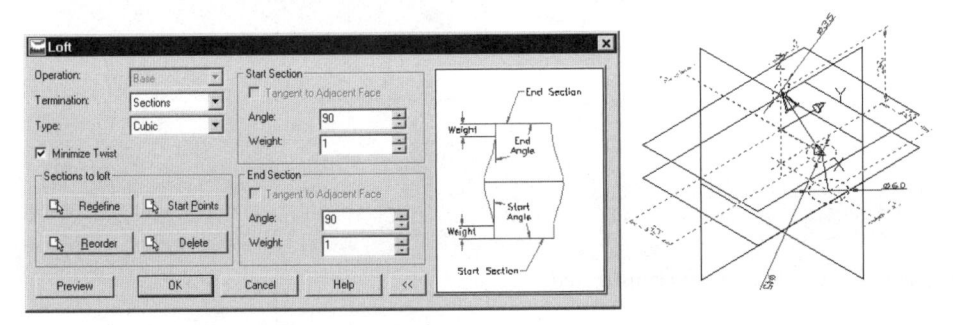

Figure 4-184 *Loft solid being constructed*

6. Referring to Figure 4-185, establish a sketch plane, construct a circle and a hidden line, resolve them to a profile, and add four parameter dimensions.

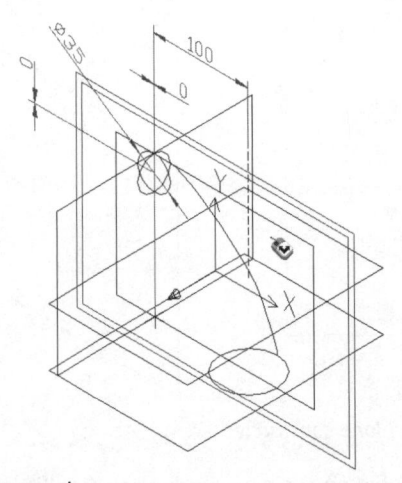

Figure 4-185 *Profile constructed*

7. Revolve the profile to form a revolved solid and join it to the solid part (see Figure 4–186).
8. Hide the work features.
9. Referring to Figure 4–187, establish a sketch plane and construct a profile.
10. Extrude the profile a distance of 45 mm to join the solid part (see Figure 4–188).

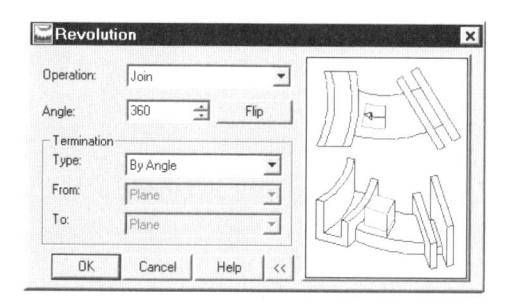

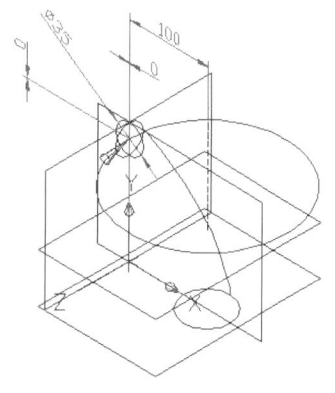

Figure 4–186 *Revolve solid being constructed*

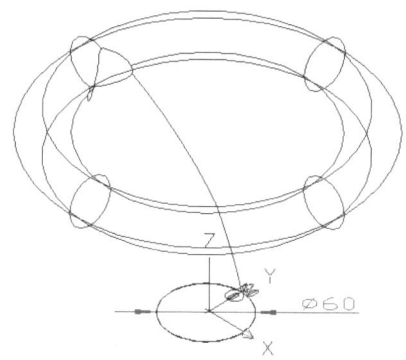

Figure 4–187 *Profile constructed*

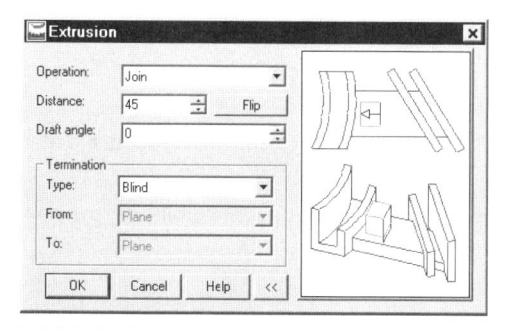

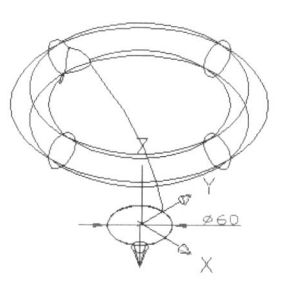

Figure 4–188 *Profile being extruded*

11. Construct a square, resolve it to a profile, and add parametric dimensions (see Figure 4–189).

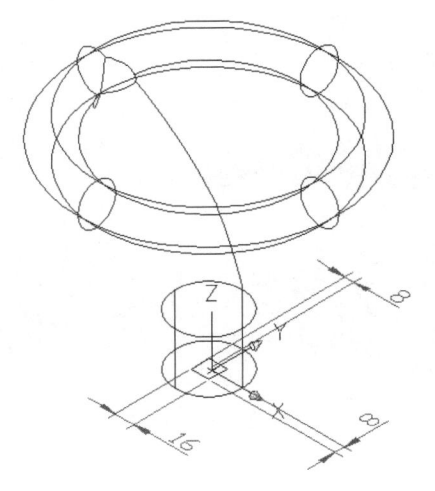

Figure 4–189 *Profile constructed*

12. Extrude the profile a distance of 70 mm to cut the solid part (see Figure 4–190).

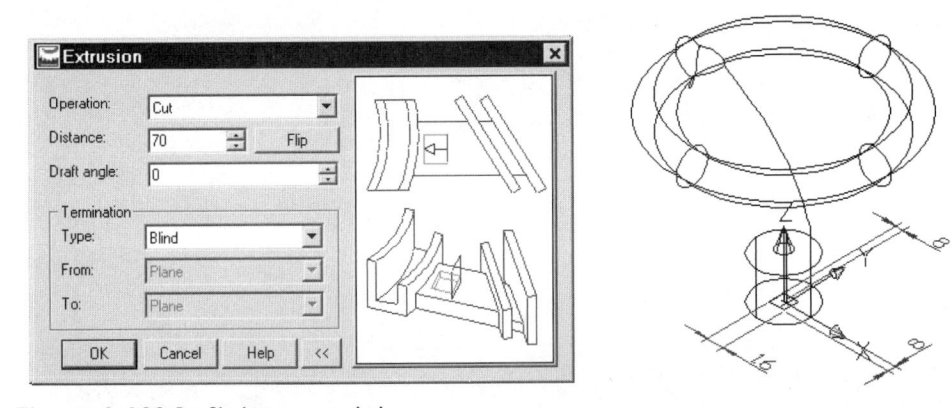

Figure 4–190 *Profile being extruded*

13. Referring to Figure 4–191, place a concentric blind-hole with 25 mm diameter and a depth of 50 mm.
14. Unhide work plane 3 and construct a work plane that offsets a distance of 75 mm from work plane 3 (see Figure 4–192).
15. Set the display to the top view and construct a work point on the new work plane (see Figure 4–193).

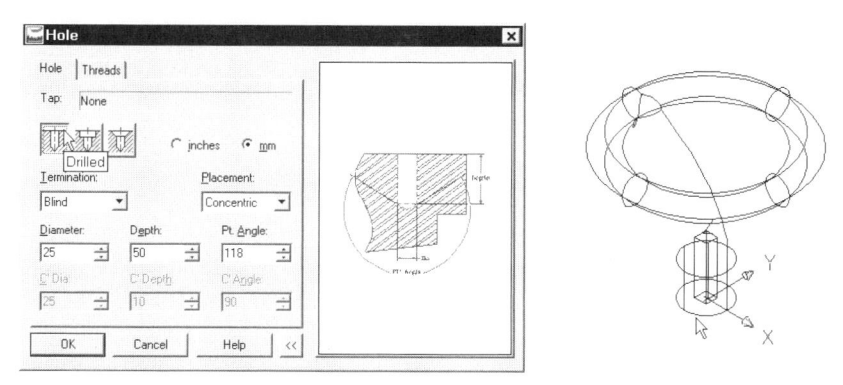

Figure 4–191 *Blind-hole being placed*

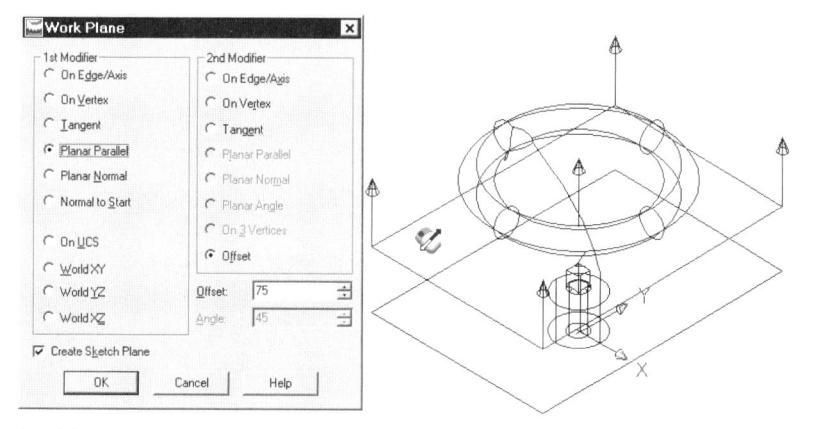

Figure 4–192 *Work plane constructed*

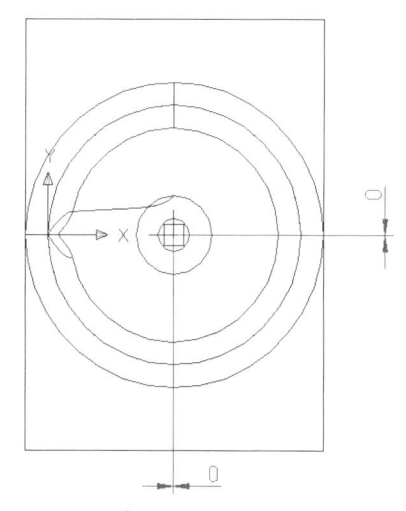

Figure 4–193 *Work point constructed*

16. Set the display to an isometric view and place a counter-bored hole on the work point (see Figure 4–194).
17. Hide the work features.

The solid part is complete (see Figure 4–195). Save and close your file (file name: *SteeringWheel.dwg*).

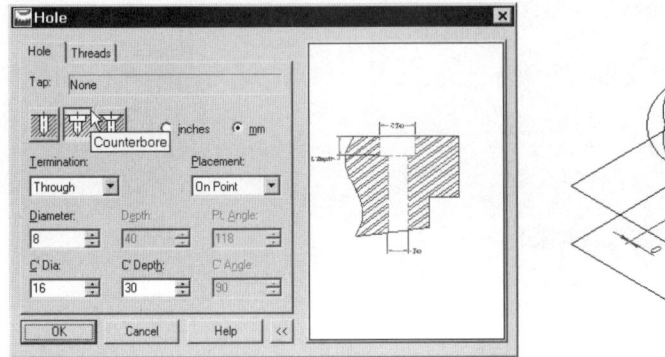

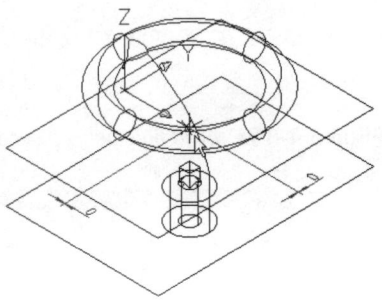

Figure 4–194 *Counter-bored hole being placed*

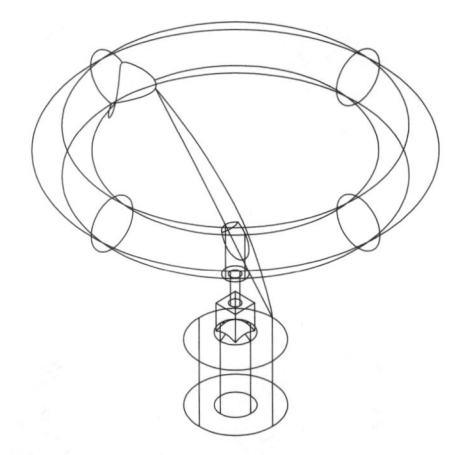

Figure 4–195 *Steering wheel*

COMPONENT PARTS FOR THE INFANT SCOOTER

Now you have completed ten component parts of the infant scooter: the frontal frame, central frame, rear frame, seat, steering shaft, steering wheel, thread, handle, wheel, and tire (see Figure 4–196). In the next chapter you will assemble them and the component parts that you completed in Chapter 3 to form an infant scooter.

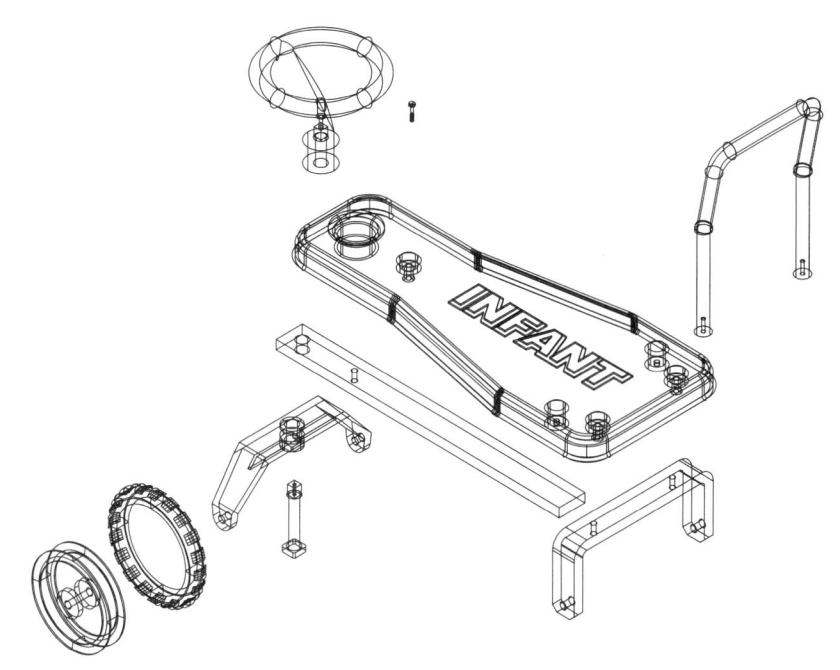

Figure 4–196 *Component parts completed in this chapter*

REVIEW QUESTIONS

1. Explain the ways to construct work planes, a work axis, and a work point.

2. Give examples to illustrate how you will use work features in solid modeling.

3. Delineate the processes of constructing rib, sweep, and loft solid features.

4. State the major difference between sketched solid features and placed solid features.

5. Explain the ways to construct hole, fillet, chamfer, shell, and array features in a solid part.

6. What is the difference between a linear and a cubic fillet?

7. Name two reasons for constructing a work axis.

8. A work axis can only be placed in the center of an arc or circular edge of a solid part. True or false?

CHAPTER 5

Assembly Modeling

OBJECTIVES

The aims of this chapter are to introduce the concepts of assembly modeling, to outline different design approaches to constructing a set of solid parts for an assembly, to show ways to assemble a set of component parts to form a virtual assembly, and to familiarize you with the use of various assembly modeling utilities. After studying this chapter, you should be able to do the following:

- Describe the key concepts of assembly modeling
- Use different design approaches to design a set of components
- Construct virtual assemblies of 3D component parts
- Use assembly utilities

OVERVIEW

Take a look around your desk. With the exception of very simple objects, such as a ruler, most objects have more than one component part. For example, a pencil has two component parts: the wooden body and the graphite core. When you design a set of component parts, the relative dimensions and positions of the component parts, and how they fit together, are crucial. You need to know whether there is any interference among them. If there is any interference, you need to find out where it occurs. Then you have to find a way to eliminate it. To shorten the design lead time, you can construct virtual assemblies in the computer to validate the integrity of a set of component parts.

ASSEMBLY MODELING CONCEPTS

An assembly is a collection of component parts that are put together properly to form a device that serves a purpose. In the computer, a virtual assembly is a set of data regarding how a set of solid parts is put together. For complex devices that have a lot of component parts, it is common practice to first organize the parts in a number of subassemblies and then put the subassemblies in the final assembly. (In constructing an engine, for example, you put the piston, piston ring, and connecting rods in a subassembly and then put the subassembly in the engine assembly.) If we collectively define solid parts and subassemblies as components, the definition of an assembly becomes a collection of components that can be solid parts or subassemblies. Figure 5–1 shows the hierarchy of the assembly of the infant scooter, including the solid parts and subassemblies.

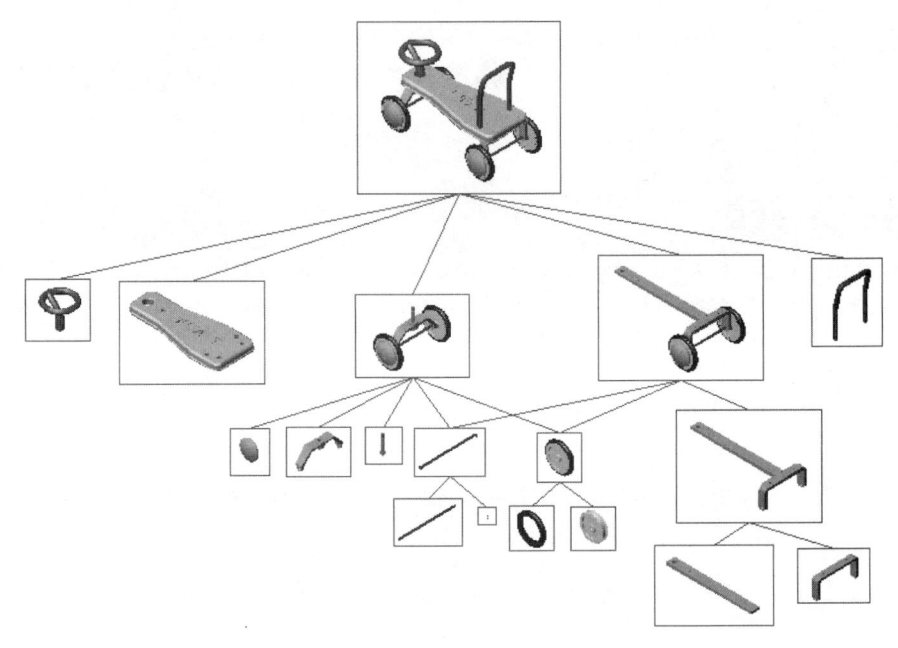

Figure 5–1 *Hierarchy of assembly file, subassembly files, and part files*

To construct a virtual assembly in a computer, the process goes far beyond gathering and translating a set of components together in 3D space. Most importantly, it involves the establishment of a relationship between the features of a component and the features of another component by using assembly constraints. In general, construction of a virtual assembly involves two major tasks: Linking a set of solid part files or subassembly files in an assembly file and applying assembly constraints on the parts and subassemblies in the assembly file.

LINKING TO SOLID PART FILES

The basic way to construct an assembly of components is to link a set of part or subassembly files to an assembly file. Information on the details of the components is kept in individual files and the assembly file concerns only the data regarding the location of the components and how the components are assembled together. Every time you open an assembly file, the latest information of the component parts is loaded. As a result, the changes you make in the part or subassembly files will automatically be incorporated in the assembly file.

ASSEMBLY CATALOG

To link a set of component part files to an assembly file, you use the assembly catalog to set up a search directory and select component files from the search directory. Now you will learn how to construct an assembly of two component parts of the infant scooter. Before you start work, move or copy all the component part files for the Infant Scooter that you constructed in previous chapters to the folder "C:\Projects\InfantScooter" of your computer.

1. Start a new file by selecting File>New. Use metric as the default.
2. Set the display to an isometric view.
3. Select Assembly>Catalog or the Catalog button from the browser.
4. Select the External tab, if it is not already selected.

Initially, the Part and Subassembly Definitions box of the Assembly Catalog dialog box is empty because the directories in which you saved the solid part files are not included in the Directories box. Now you will add a search directory.

5. Right-click inside the Directories box, and select Add Directory to add a search directory (see Figure 5–2).

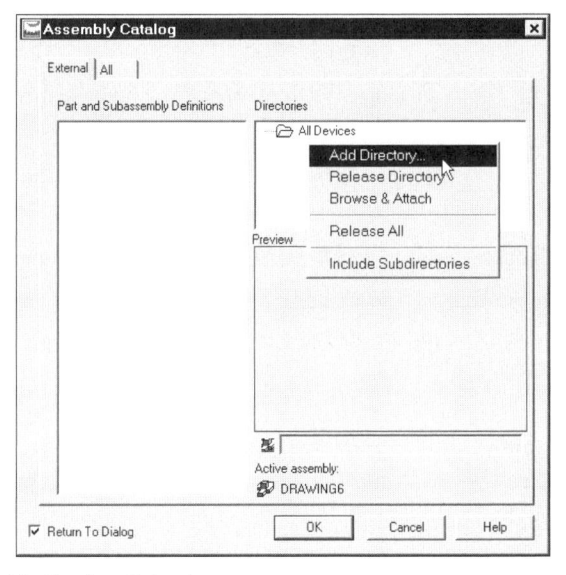

Figure 5–2 *Assembly Catalog dialog box*

Let us assume that you already put the solid part files in the directory C:\Projects\InfantScooter.

6. In the Browse for Folder dialog box, select the C:\Projects\InfantScooter directory.

When you return to the Assembly Catalog dialog box, the directory "C:\Projects\InfantScooter" is added and the files saved in that directory are displayed in the Part and Subassembly Definitions box. In the Preview box, you will see a preview of a selected file (see Figure 5–3).

Figure 5–3 *Component parts in the Part and Subassembly Definitions box*

Now you will link the files (Tire and Wheel) from the search directory and place an instance of them in the assembly file.

7. Select Tire in the Part and Subassembly Definition box and double-click.
8. Select a location on the screen and press the ENTER key.
9. Select Wheel in the Part and Subassembly Definition box and double-click.
10. Select a location in the screen and press the ENTER key.
11. After returning to the Assembly Catalog dialog box, select the All tab (see Figure 5–4).

In the All tab of the Assembly Catalog dialog box, you will find that the files (Tire and Wheel) are included in the External Assembly Definitions box, showing that they are now linked to the assembly file as external assembly definitions.

12. Select the OK button.

INSTANCE

Double-clicking the part files (Tire and Wheel) in the browser and selecting a location in the screen will link the files to the assembly and place an instance in the assembly. Now you will find an instance of the component parts in the model tab of the desktop browser (see Figure 5–5).

Sometimes several copies of a component are used in an assembly. For example, you use four identical wheels and tires in a car assembly. To construct multiple copies of a component in an assembly, you need only one set of definitions of the component and can repeat the other copies as instances that are referenced to the definition.

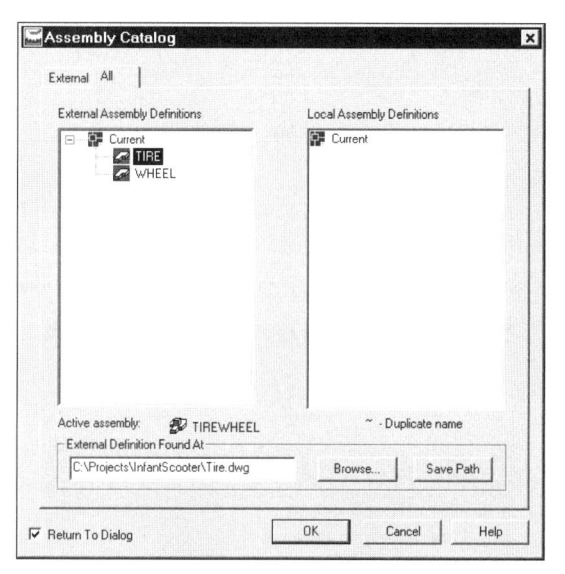

Figure 5–4 *All tab of the Assembly Catalog dialog box*

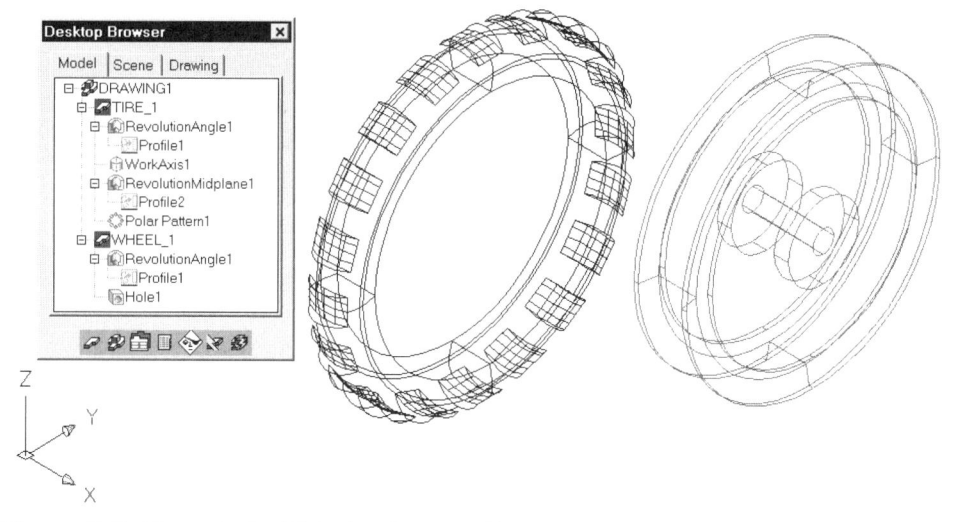

Figure 5–5 *Tire and wheel linked to the assembly*

ASSEMBLING THE COMPONENT PARTS

Initially component parts in the virtual assembly are free to translate. To put them together, you translate them to an appropriate location and apply assembly constraints.

Degree of Freedom

In 3D space in the virtual assembly, each component part is free to translate in three linear directions and rotate about three axes. These free translations are called degrees of freedom (DOF) and are depicted by a DOF symbol. To display the DOF symbol, you use the Desktop Visibility dialog box.

13. Select the Visibility button from the browser.
14. Select the Assembly tab.
15. Select the Unhide, Centers of Geometry, and Degrees of Freedom buttons, and select the OK button (see Figure 5–6).

The degrees-of-freedom symbols of the component parts are displayed (see Figure 5–7).

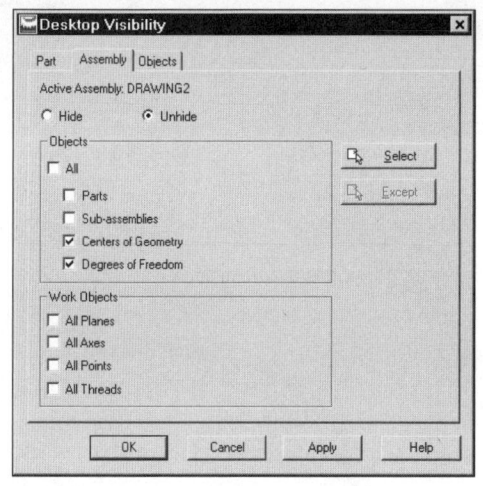

Figure 5–6 *Assembly tab of the Desktop Visibility dialog box*

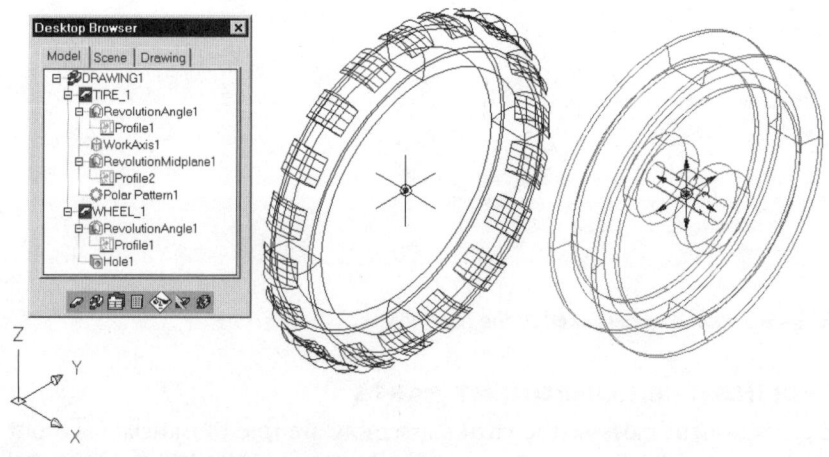

Figure 5–7 *Degrees-of-freedom symbols displayed*

As can be seen, the degrees-of-freedom symbols for the two components in the assembly are not the same. The DOF symbol of the wheel has three linear arrows, three rotational arrows, a circle, and a number. The linear arrows depict the three degrees of linear translation freedom. The rotational arrows show the three degrees of rotational freedom. The circle indicates the center of the geometry of the solid part. The number shows the order of the solid part in the hierarchy. The first solid part of the drawing is number 1. With two or more components in an assembly file, a hierarchy forms. The first component is said to be grounded. It has no degrees of freedom. Hence, the DOF symbol of the tire shown in Figure 5–7 has only a circle and number but not any arrows.

Power Manipulator

Initially components in an assembly file are located at their insertion points. To translate a component in 3D, you use the 3D manipulator.

16. Select Assembly>Power Manipulator

    Command: AMMANIPULATE

17. Select the wheel and press the ENTER key (see Figure 5–8).

The power manipulator displays. It has three axes and seven grip handles (in the shape of a ball) with one at the center and six at the ends of the three axes.

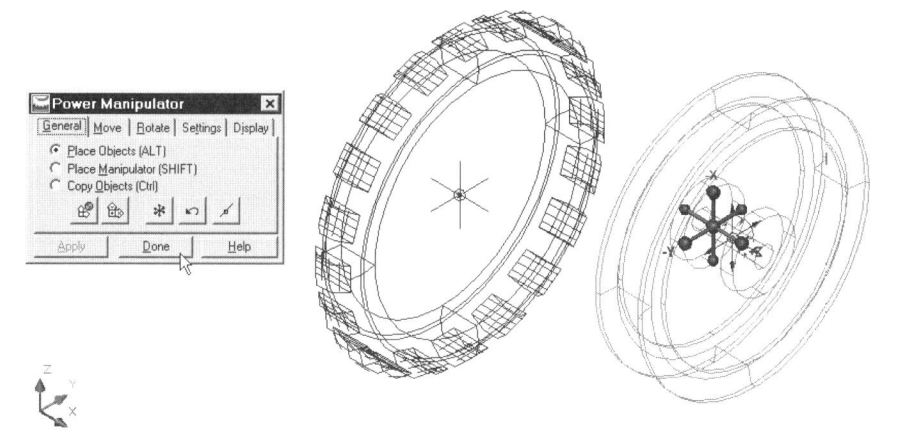

Figure 5–8 *3D manipulator*

18. To move the selected object freely in 3D space, select the central handle and drag the mouse.
19. To translate the selected object along the X, Y, or Z direction, select one of the handles at the end of an axis and drag the mouse.
20. To rotate the selected object about the X, Y, or Z axis, click one of the handles at the end of an axis to change to rotate mode, select the handle again, and drag the mouse.

Along with the power manipulator, there is a Power Manipulator dialog box. The dialog box has five tabs: General, Move, Rotate, Settings, and Display.

The General tab has three check boxes and five buttons:

Place Objects	enables you to reposition the selected object
Place Manipulator	enables you to reposition the location of the power manipulator icon
Copy Objects	enables you to make a copy of the selected object each time you translate the object
Aligned to WCS	enables you to align the power manipulator icon and the selected component with the WCS
Aligned to UCS	enables you to align the power manipulator icon and the selected component with the current UCS
Center View on Manipulator	enables you to align the power manipulator icon to the center of the screen
Undo	enables you to undo the last operation
Make a new selection	enables you to select another component

The Move tab has three entry boxes:

Exact Distance	enables you to specify the exact distance to move
Number of Copies	enables you to specify the number of copies to be made when the selected object is translated and copied
Use Distance snap	enables you to specify whether to use the distance snap and specify the snap distance

The Rotate tab also has three entry boxes:

Exact Angle	enables you to specify the exact angle of rotation
Number of Copies	enables you to specify the number of copies to be made when the selected object is rotated and copied
Use Angle snap	enables you to specify whether to use the angle snap and specify the snap angle

The Settings tab has two entry boxes and a check box:

Manipulator Size	enables you to specify the size of the power manipulator icon
Handle and Text Size	enables you to specify the size of the handles and axes
Double-click for Rotate operations	enables you to click twice to use the rotate operation

The Display tab enables you to specify the color of various components of the power manipulator and to specify the number of grips.

21. Use the power manipulator to translate the components in the assembly.

Assembly Constraints

It must be emphasized that using the power manipulator or other translate commands such as move or rotate only changes the position of the components in the assembly. They do not affect the component's degrees of freedom.

To impose restriction to the degrees of freedom and to properly align a component with another component in the assembly, you apply assembly constraints. (While the geometric constraints we studied in Chapter 2 restrict the forms and shapes of the parametric sketches of a solid, the assembly constraints discussed here concern the way the features of one component are associated with the features of another component.)

By using Mechanical Desktop, you apply four kinds of assembly constraints: mate, flush, angle, and insert.

Constraint	Functions
Mate constraint (AMMATE)	causes the points, lines/axes, or planes of a pair of solid parts to align with each other. You align a vertex to a vertex (see Figure 5–9), an edge to an edge (see Figure 5–10), a face to a face (see Figure 5–11), vertex to an edge, vertex to a face, or edge to a face.
Flush constraint (AMFLUSH)	causes the planes of a pair of solid parts to align in the same direction. Figure 5–12 shows two planes aligned in the same direction.
Angle constraint (AMANGLE)	causes the lines/axes or planes of a pair of solid parts to align at an angle to each other. Figure 5–13 shows two planes aligned at an angle.
Insert constraint (AMINSERT)	causes the circular edges of a pair of solid parts to be concentric and the planes defined by the circular edges to be coplanar (see Figure 5–14).

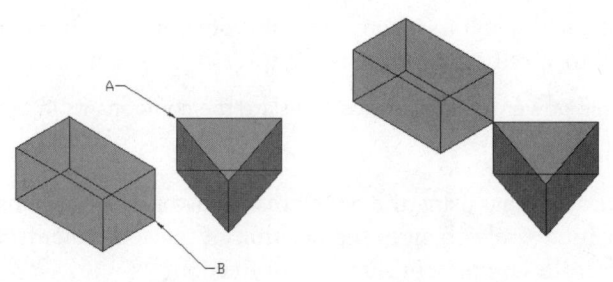

Figure 5–9 *Mate constraint – Vertex B mated to vertex A*

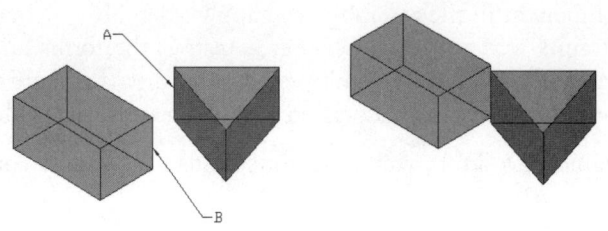

Figure 5–10 *Mate constraint – Edge B is mated to edge A*

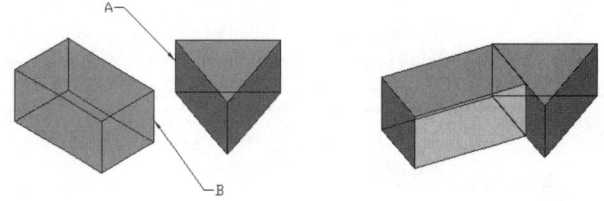

Figure 5–11 *Mate constraint – Vertical plane B is mated to vertical plane A*

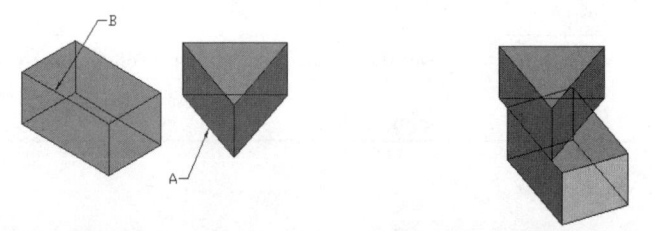

Figure 5–12 *Flush constraint – Vertical plane B is flush with vertical plane A*

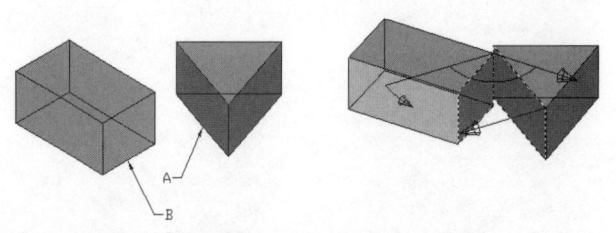

Figure 5–13 *Angle constraint – Plane B is set at an angle with plane A*

Assembly Modeling 217

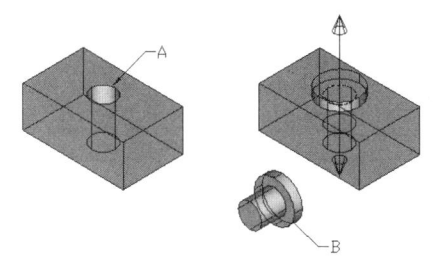

Figure 5–14 *Insert constraint – Circular edge B is inserted into circular edge A*

Assembly Constraint and Degrees of Freedom

When you apply an assembly constraint to selected features from a pair of components, the number of degrees of freedom of the components decreases and the second component in the assembly hierarchy translates toward the first component of the assembly. If there is a third component in the assembly, it translates toward the second component or the first component.

Now you will apply a mate constraint to align a face of the wheel to the tire.

22. Select Assembly>3D Constraints>Mate

    Command: AMMATE

23. Referring to Figure 5–15, select the highlighted face (if the face is not highlighted, left-click to cycle), and right-click to accept.

24. Select the face highlighted in Figure 5–16 and right-click to accept.

25. Press the ENTER key to accept the default amount of offset (zero) between the two selected faces.

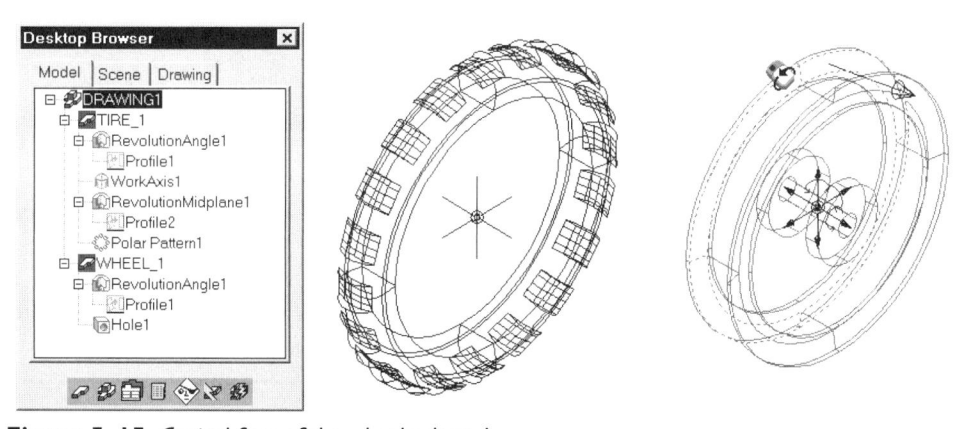

Figure 5–15 *Conical face of the wheel selected*

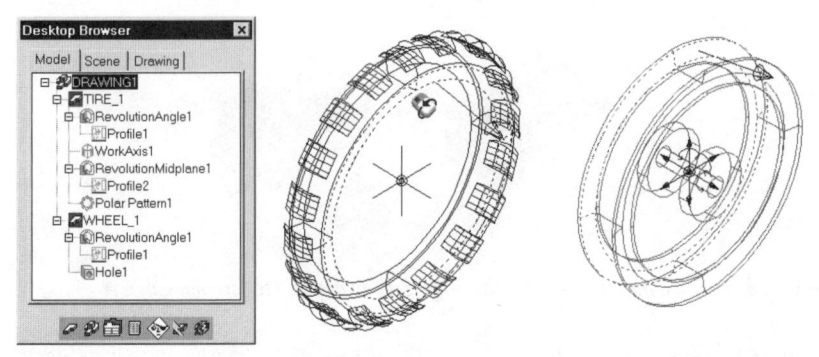

Figure 5-16 *Conical face of the tire selected*

Now the conical face of the wheel is mated to the conical face of the tire (see Figure 5–17). In the browser, you will find two Tan cn/cn symbols in the hierarchy, depicting a mate constraint between the faces of two components. Now the wheel can no longer translate along the X, Y, and Z axes and rotate about the Y and Z axes. However, it can still rotate freely about X. Thus, it has one degree of freedom left. In reality, a tire and wheel assembled this way can still have relative rotational motion about the wheel axis. Therefore, the assembly of the tire and wheel is complete. Save your file (file name: *TireWheel.dwg*).

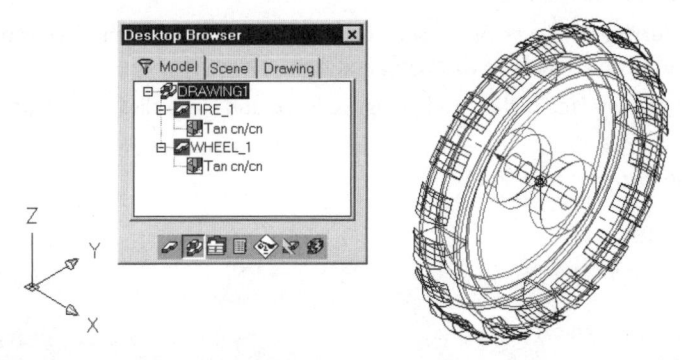

Figure 5-17 *Wheel assembled to the tire*

Editing/Deleting Assembly Constraints

To edit or delete an assembly constraint, you can select the constraint from the browser and double-click, or select the constraint from the browser, right-click, and select Edit or Delete. In the right-click menu, there are three menu items: Edit, Select Other End, and Delete.

Edit	displays the Edit 3D Constraints dialog box
Select Other End	enables you to select the other constraint symbol (Because a constraint is applied to a pair of components, the constraint appears in both components in the browser.)
Delete	deletes the selected constraint

Now you will learn how to edit the constraint.

26. Select the constraint from the browser, right-click, and select Edit or select the constraint from the browser and double-click.

In the Edit 3D Constraints dialog box, you change the value in the expression box to modify the offset distance between the mating faces and select the Update Constraint button to see the change (see Figure 5–18).

27. Now select the Cancel button because we are not going to make any changes.

Save and close your file.

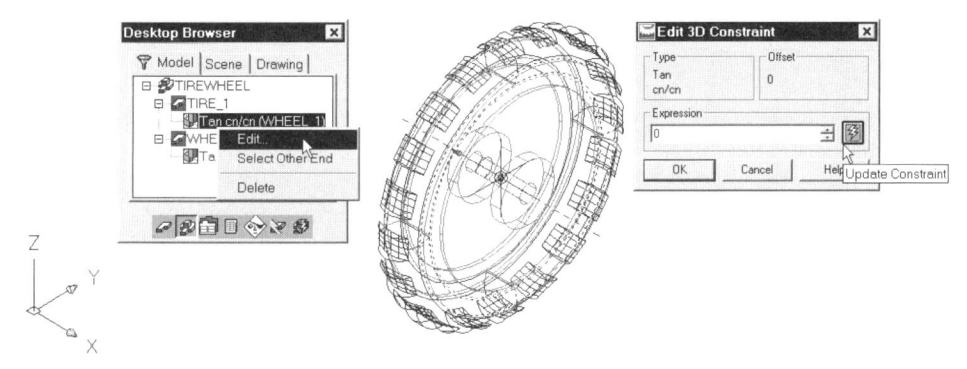

Figure 5–18 *Editing an assembly constraint*

DESIGN APPROACHES

There are three design approaches to construct an assembly of components: the bottom-up approach, top-down approach, and hybrid approach.

BOTTOM-UP APPROACH

The first approach to construct an assembly is to construct all the solid parts in separate part files, start a new assembly file, link the part files to the assembly file, and apply assembly constraints to the instances of the parts. To the assembly file, the part files are external definitions and are called external solid parts. Here you start from the bottom of the assembly hierarchy to construct individual part files and then move up to the assembly file. This approach is called the bottom-up approach. In making the TireWheel assembly, you constructed the tire and wheel, put them in an assembly file, and applied assembly constraints. Hence, you used the bottom-up approach (see Figure 5–19).

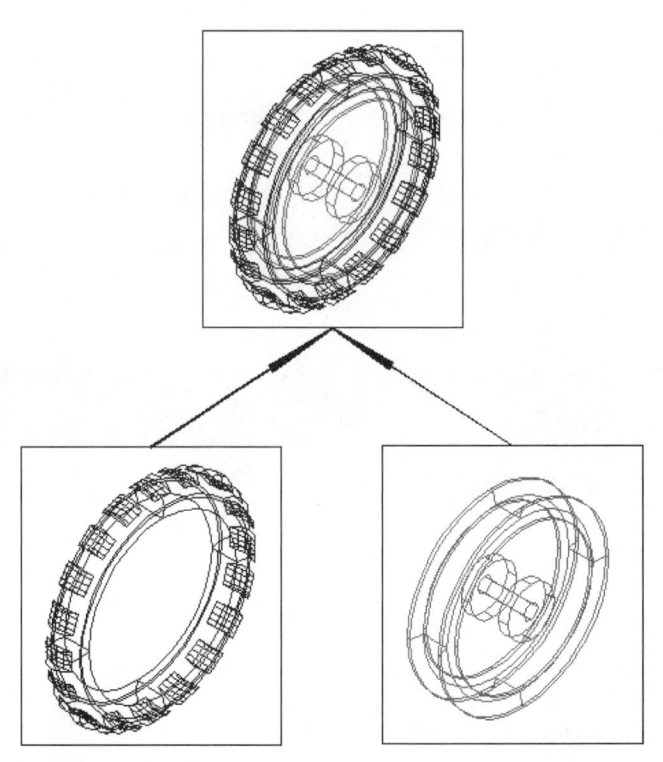

Figure 5–19 *The bottom-up approach to construct the assembly of a tire and wheel*

TOP-DOWN APPROACH

The second approach to construct an assembly is to start an assembly file, construct the solid parts while seeing the other component parts, and apply assembly constraints to the solid parts. By constructing all the solid parts in a single assembly file, you have a better perception of their relative shapes and sizes. The solid parts, being resident in the assembly file, are called local solid parts. In terms of manufacturing, we need individual part files to produce the components of a product. Practically, you should derive individual part files from the local solid parts. Deriving individual linked part files from local solid parts is called externalizing. This approach of starting from the top level of the assembly hierarchy to make all the solid parts in the assembly file and working downward to externalize the solid parts is called a top-down approach.

You will use the top-down approach to construct an assembly of two components (see Figure 5–20). You will construct the components in the assembly file, assemble the components, and externalize the components.

1. Start a new file. Use metric as the default.
2. Select Part>Part>New Part
3. Type **TOYBASE** at the command line area to specify a new part name.

An instance of the solid part definition is listed in the browser and the solid part is activated. Now you will construct features for the solid part in the assembly.

4. Referring to Figure 5–21, construct a rectangle, resolve it to a profile, and add parametric dimensions.

5. Set the display to an isometric view and extrude the profile a distance of 1 mm (see Figure 5–22).

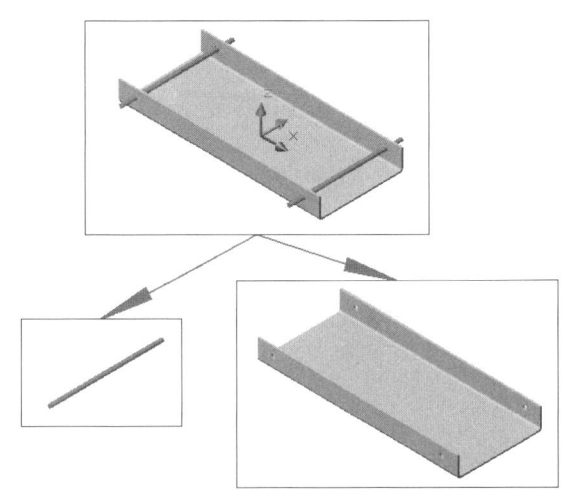

Figure 5–20 *Top-down approach - Construct components in the assembly and externalize the components*

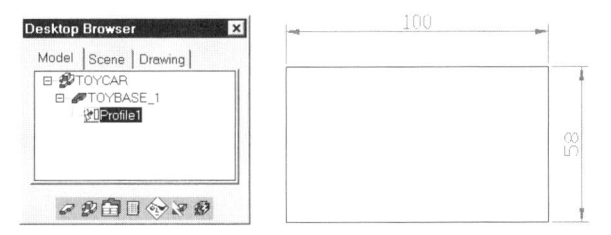

Figure 5–21 *Rectangle constructed in the solid definition in the assembly*

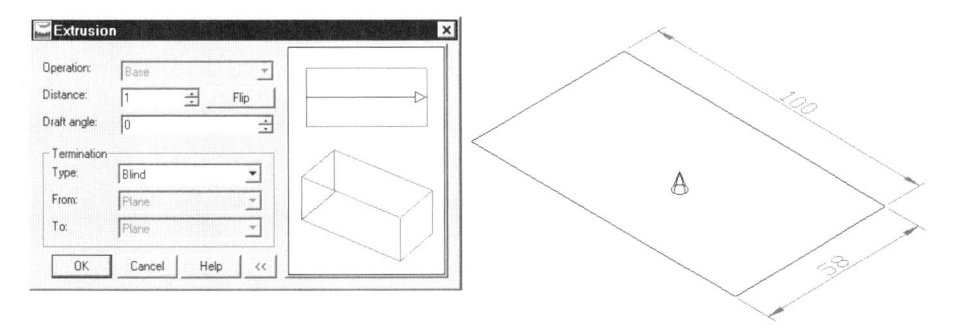

Figure 5–22 *Profile of the solid definition being extruded*

6. Establish a sketch plane on the top face of the extruded solid.
7. Construct two horizontal lines on the sketch plane and resolve them one by one into two open profiles.
8. Add parametric dimensions to the profiles (see Figure 5–23).

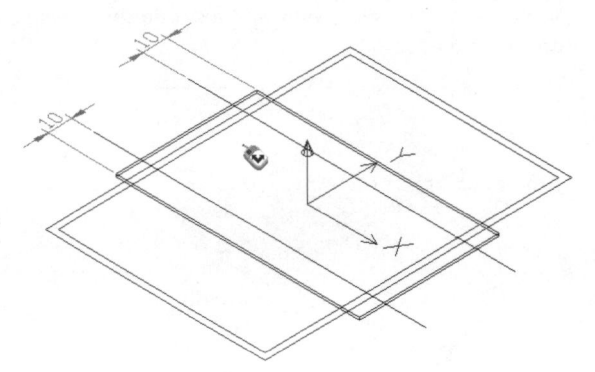

Figure 5–23 *Two open profiles constructed*

9. Construct two bend features one by one to bend the solid part about the open profiles (see Figures 5–24 and 5–25).
10. Referring to Figure 5–26, place two drilled through-holes.

A solid part definition is complete.

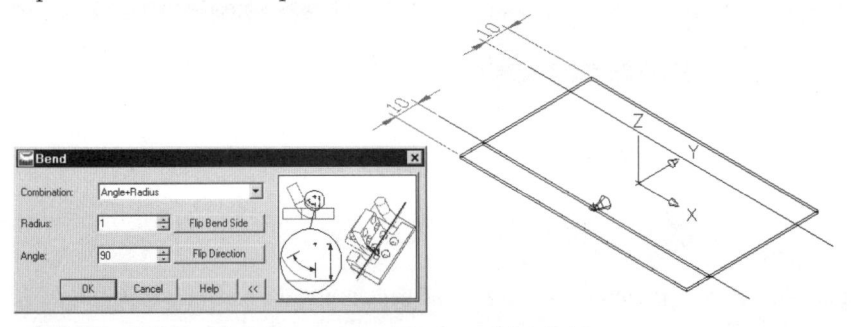

Figure 5–24 *Bend feature being constructed on the solid definition*

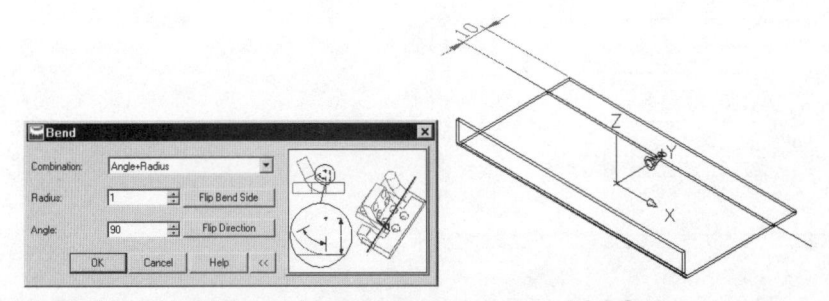

Figure 5–25 *Second bend feature being constructed on the solid definition*

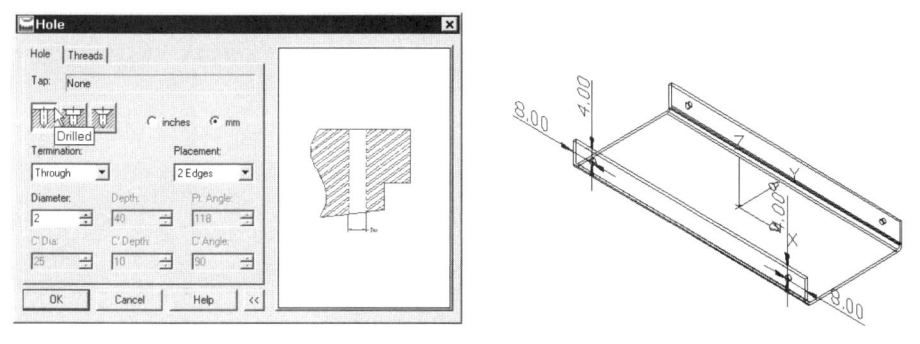

Figure 5-26 *Hole features placed on the solid definition*

Now construct the second solid part definition in the assembly.

11. Select Part>Part>New Part
12. Type TOYSHAFT at the command line area to specify the name of the part definition.
13. Referring to Figure 5-27, construct a circle of 2 mm diameter, resolve it to a profile, add parametric dimensions, and extrude it a length of 56 mm.

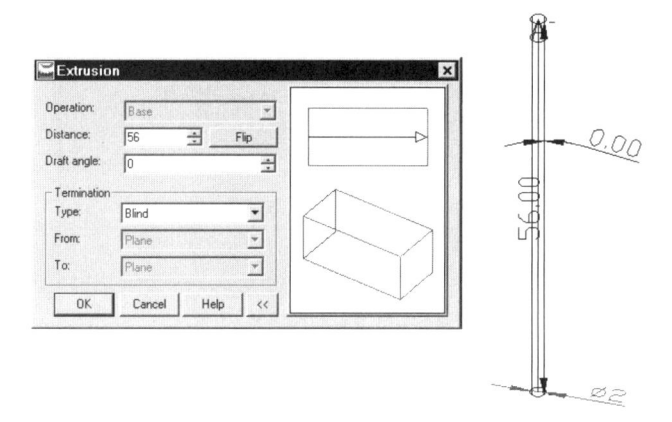

Figure 5-27 *Second solid definition being constructed*

In the browser shown in Figure 5-28, you will find two solid definitions. These two definitions are called local definitions because they are stored in the assembly file. Before you apply assembly constraints to the components, it may be necessary for you to find out the degrees of freedom of the components and translate the components. If so, display the DOF symbol and use the power manipulator.

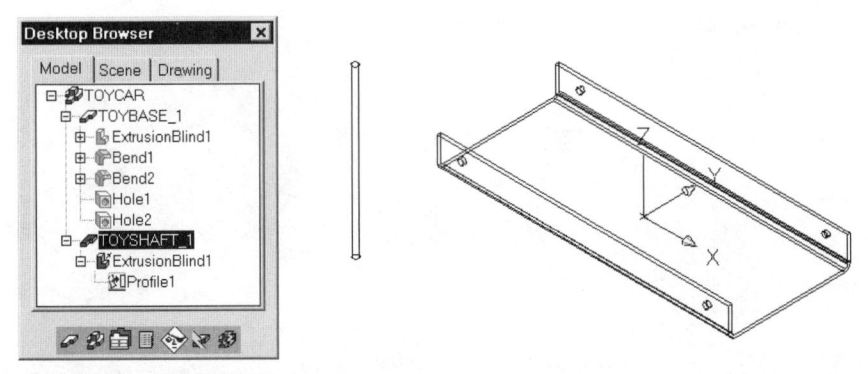

Figure 5–28 *Two solid definitions*

Now you will place one more instance of the solid definition (ToyShaft) in the assembly.

14. Select the Assembly Catalog button from the browser.
15. In the Assembly Catalog dialog box, select the All tab (see Figure 5–29).

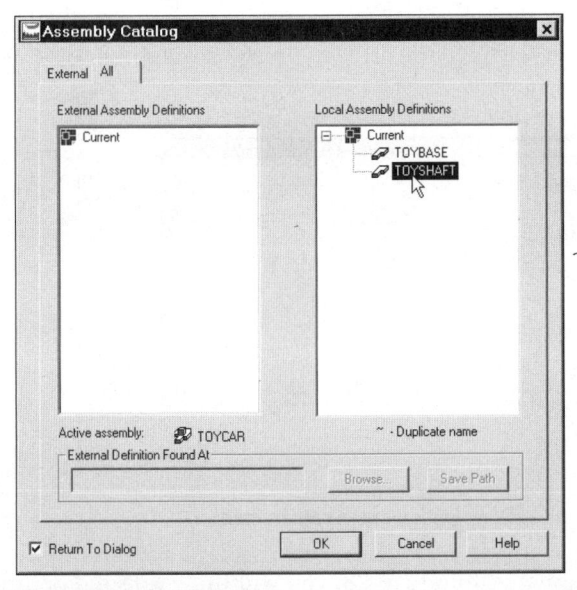

Figure 5–29 *Local assembly definitions*

16. Select ToyShaft from the Local Assembly Definitions box of the All tab of the Assembly Catalog dialog box, double-click, select an insertion location, and press the ENTER key.
17. On returning to the Assembly Catalog dialog box, select the OK button.

Now you have two instances of the solid definition (ToyShaft; see Figure 5–30).

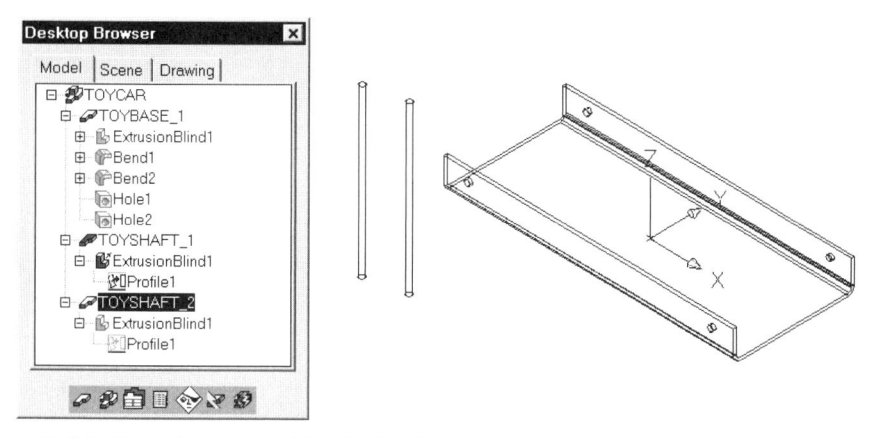

Figure 5–30 *Second instance of the ToyShaft inserted*

Now assemble the components.

18. Select Assembly>3D Constraints>Insert

    Command: AMINSERT

19. Referring to Figure 5–31, select a circular edge of the ToyShaft, left-click to flip the direction of insertion, and right-click to accept.

20. Referring to Figure 5–32, select a circular edge of the ToyBase and right-click to accept.

21. Type **7** at the command line area to specify an offset distance of 7 mm between the selected circular edges.

22. Repeat steps 18 through 21 above to assemble the other ToyShaft instance (see Figure 5–33).

Assembly of components is complete. Now you will externalize the local solid part definitions. By externalizing, you export the solid part definitions to separate solid part files while maintaining link and assembly information.

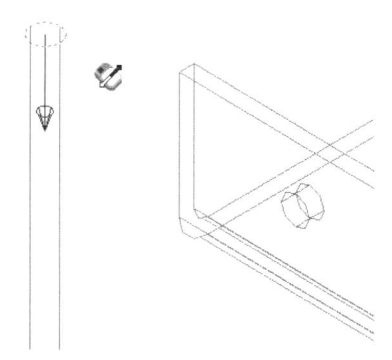

Figure 5–31 *Circular edge of a component selected for application of the insert constraint*

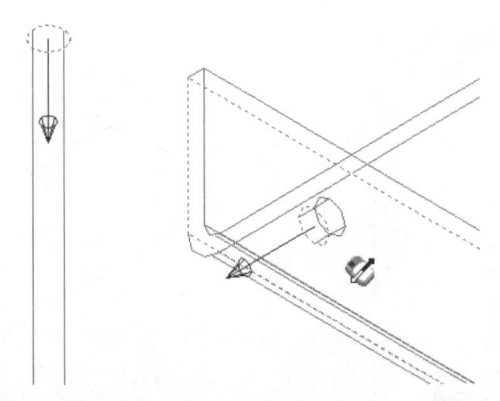

Figure 5-32 *Circular edge of the other component selected*

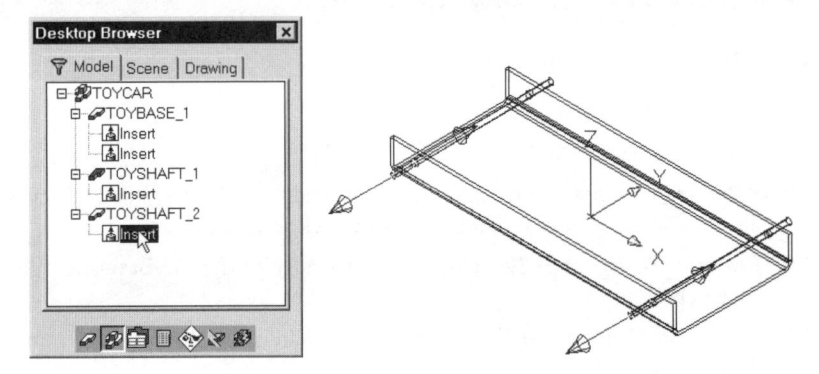

Figure 5-33 *Components assembled*

Let us assume that you have already constructed a folder "C:\Projects\ToyCar" in your computer and that you have decided to put the solid parts in this directory.

23. Select the Assembly Catalog button from the browser.
24. Select the All tab.
25. Select the definitions in the Local Assembly Definitions box one by one, right-click, and select Externalize (see Figure 5-34).

In the New External File dialog box shown in Figure 5-35, you will find a Use Template File button. Checking this button applies the selected template file to the external solid part. This way, appropriate drafting standards stored in the template file are applied to the externalized solid part. Thus you can merge solid part and standards in one step.

26. In the New External File dialog box, check the Use Template File button and select a template file if you wish to merge standards stored in the template with the externalized solid part.

27. Select the "C:\Projects\ToyCar" directory, specify a file name, and select the Save button.
28. On returning to the Assembly Catalog dialog box, select the OK button.

Now the solid parts are externalized (see Figure 5–36). Although they are now stored in individual external files, the assembly information is still retained and they are still linked to the assembly file. After externalizing, the final effect of the top-down approach is identical to the bottom-up approach. The set of assembly and part files is complete. Save your file (file name: *ToyCar.dwg*).

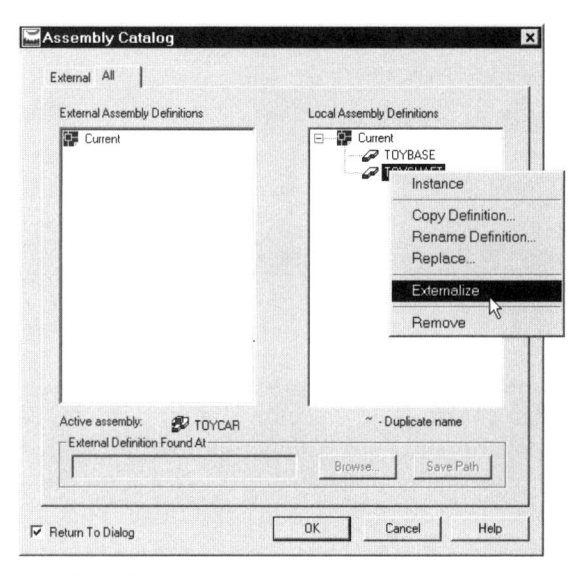

Figure 5–34 *Local definitions being externalized*

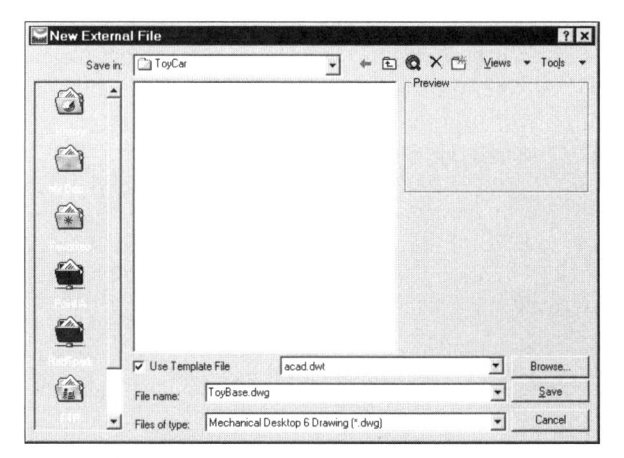

Figure 5–35 *New External File dialog box*

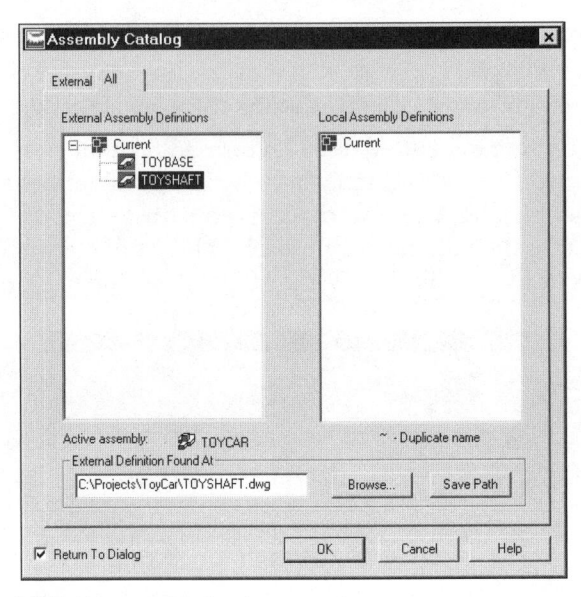

Figure 5-36 *Local definitions externalized*

HYBRID APPROACH

The third approach to construct an assembly is to use the bottom-up approach to construct some components and the top-down approach to construct the others. As a result, some solid parts reside in individual files external to the assembly and some solid parts reside within the assembly as local definitions. This approach is called the hybrid approach.

As in the top-down approach, it is practical to externalize the local solid parts after completing the assembly.

Now you will construct a component part in a separate part file and link it to the ToyCar assembly. Because you use the top-down approach to construct two components and use the bottom-up approach to construct the third component, you are effectively using the hybrid approach (see Figure 5-37).

1. Start a new part file. Use metric as the default.
2. Referring to Figure 5-38, construct a sketch, resolve it to a profile, add parametric dimensions, and revolve it to a revolved solid. (Use the lower edge indicated in the figure as the revolution axis.)
3. Set the display to an isometric view.
4. Place two fillet features with a radius of 2 mm (see Figure 5-39).

The solid part is complete. Save the file in the directory "C:\Projects\ToyCar" (file name: *ToyWheel.dwg*).

Assembly Modeling 229

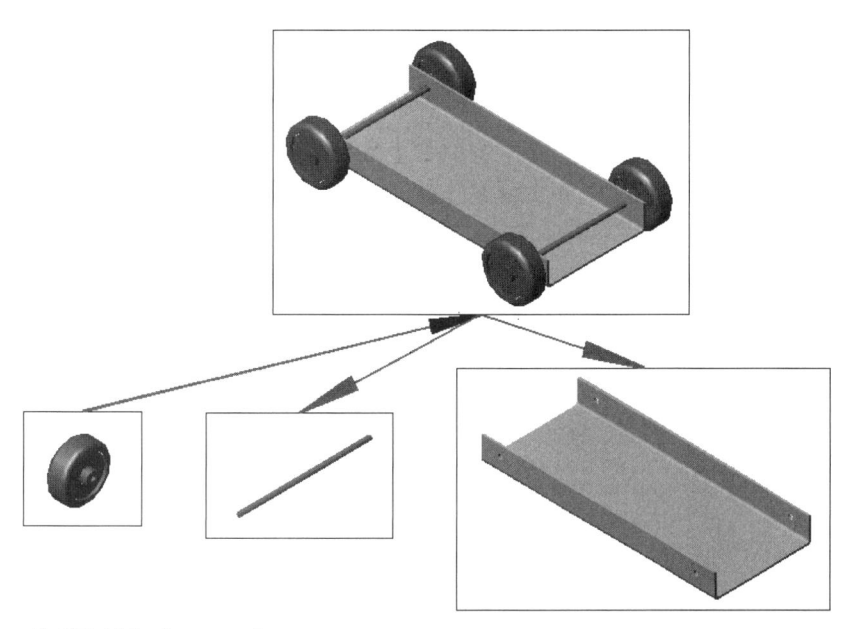

Figure 5–37 *Hybrid approach*

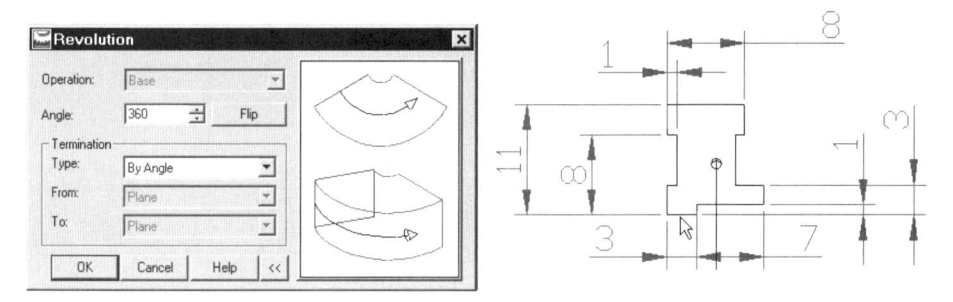

Figure 5–38 *Profile being revolved*

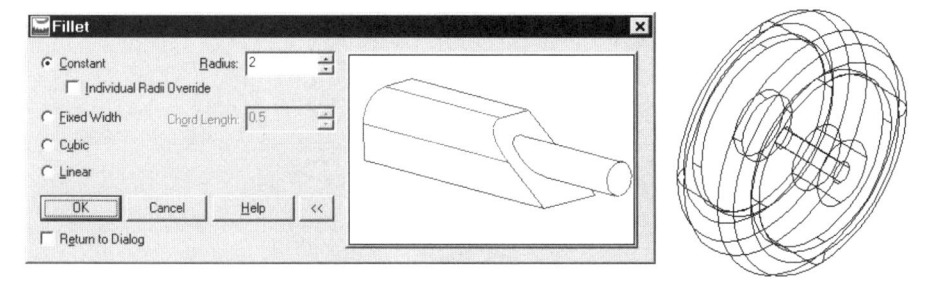

Figure 5–39 *Fillet features placed*

5. Open the assembly file ToyCar, if you have already closed the file.
6. Select the Assembly Catalog button from the browser.
7. In the External tab of the Assembly Catalog dialog box, add a directory "C:\Projects\ToyCar" (see Figure 5–40).
8. Select the part file (ToyWheel) from the Part and Subassembly Definitions box, and double-click.
9. Referring to Figure 5–41, select four locations in the assembly to place four instances of the solid part definition and press the ENTER key.
10. Use the power manipulator to translate the components.

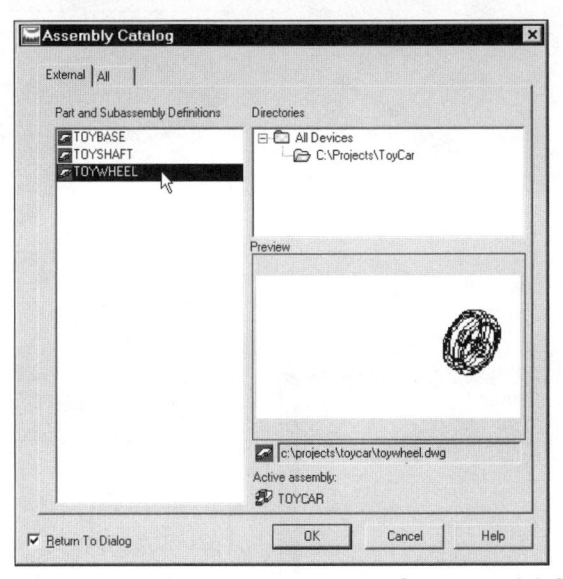

Figure 5–40 *Adding a directory and inserting an instance of an external definition*

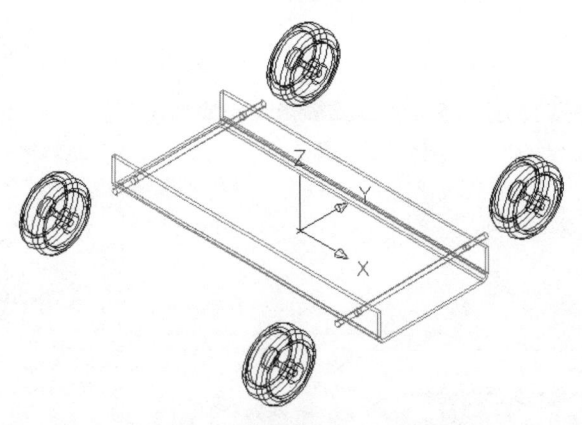

Figure 5–41 *Four instances of the external definition (ToyWheel) placed*

11. Select Assembly>3D Constraints>Insert
12. Referring to Figure 5–42, select a circular edge and right-click to accept.
13. Referring to Figure 5–43, select the second circular edge and right-click.
14. Type **0** at the command line area to specify a zero offset distance.

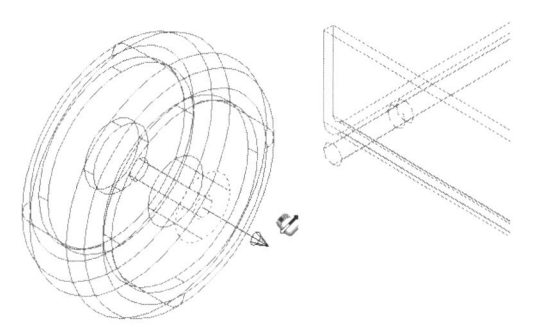

Figure 5–42 *Circular edge of the first component selected*

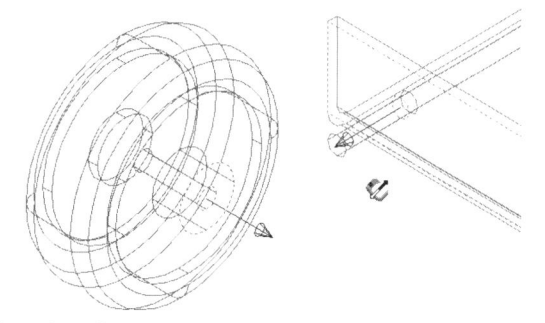

Figure 5–43 *Circular edge of the second component selected*

15. Repeat steps 11 through 14 above to apply insert constraints to the other three wheels of the toy car (see Figure 5–44).

The assembly is complete. Save your file.

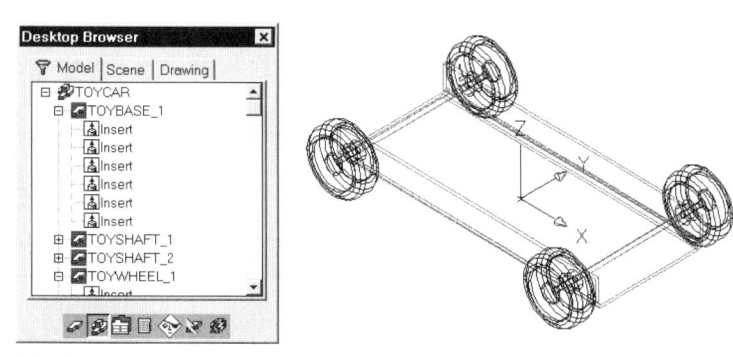

Figure 5–44 *Components assembled*

CHOICE OF APPROACH

The choice of the approach you will use to handle a project depends on a number of factors. Here are a few:

- The number of solid parts involved in the assembly
- The use of standard engineering parts in the assembly
- The number of engineering designers involved in the project

No matter which approach you use, it is common engineering practice to have an assembly drawing and a set of individual solid parts attached to the assembly drawing. Therefore, you should externalize all the local definitions.

MANIPULATING ASSEMBLY AND COMPONENT DEFINITIONS

There are two kinds of definitions in an assembly: external and local. You can localize an external definition and externalize a local definition. You can also copy in an external definition or copy out a local definition. To modify a part of the assembly, you can edit the part or the subassembly in the context of the assembly as well as opening part of the subassembly file for editing. Basic 3D work planes, when activated in an assembly, start a new part definition. Components in an assembly can be removed or replaced. After replacement, assembly constraints applied to the original components will be removed. Apart from modifying the assembly, you can find out the location of a component by specifying its name, retrieve information about a component, check whether the external files are up-to-date, and reload the external files if they are not up-to-date.

EXTERNAL AND LOCAL DEFINITIONS

Stemming directly from the three kinds of design approach, you manage an assembly of components in three ways: You keep all of them in separate part files (bottom-up approach), you store all of them in the assembly file (top-down approach), and you keep some in the assembly files and some in the part files (hybrid approach). Components stored in individual files are called external definitions and components residing within the assembly file are called local definitions.

Local and external component parts are swappable. You can externalize a local component part to make it an individual file that is attached to the current assembly as an external component part. On the other hand, you can localize an external solid part by copying the component part definition to the current drawing; this makes it a local solid part; and a link to the external definition is removed. To externalize or localize a definition, you select the component from the assembly catalog, right-click, and select Externalize or Localize (see Figure 5–45).

Practically, component definitions should be kept external to the assembly file. Therefore, you externalize all the local definitions, no matter which approach you will take to construct an assembly. However, you may occasionally localize some solid parts so that you can combine them into a single part. (You will learn how to combine solid parts in Chapter 9.)

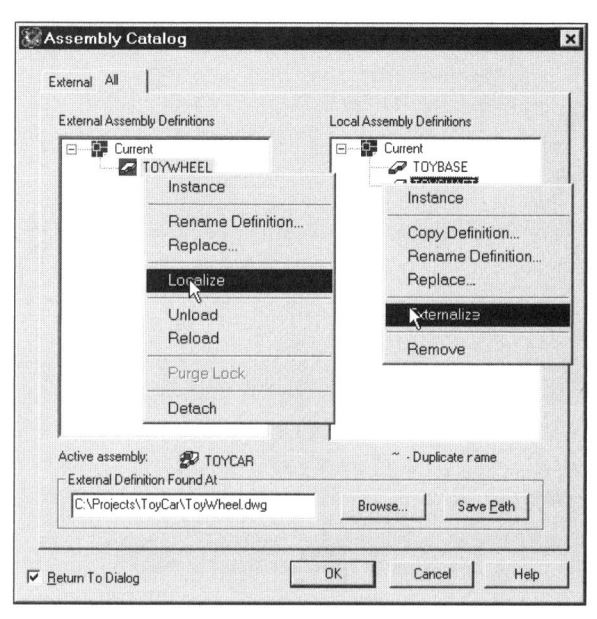

Figure 5–45 *Externalizing and localizing solid parts*

COPY IN AND COPY OUT

Similar to localizing and externalizing, you can copy in and copy out definitions.

Copy In

Copying in an external solid part constructs a local definition in the assembly catalog from an external file. To copy in an external component, select Assembly>Assembly>Copy In.

     Command: AMCOPYIN

In the File to Load dialog box, shown in Figure 5–46, select a file and select the Open button.

Copy Out

Copying out a local solid part constructs an external individual solid part file without externalizing the local solid part. To copy out a local definition, select Assembly>Assembly>Copy Out.

     Command: AMCOPYOUT

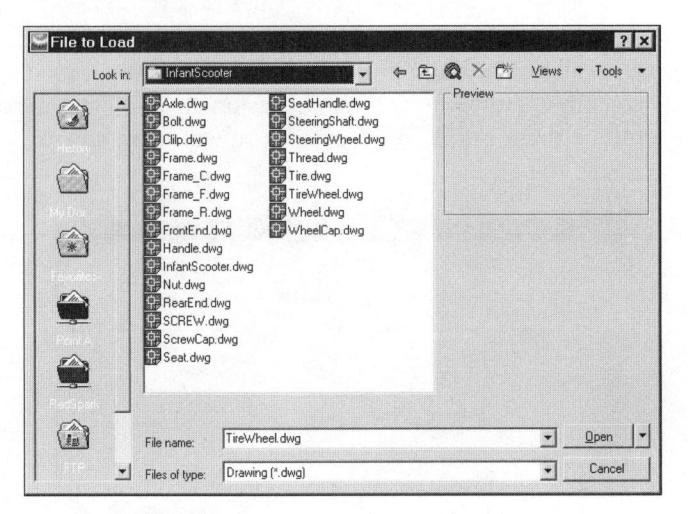

Figure 5–46 File to Load dialog box

In the Part/Assembly Out dialog box shown in Figure 5–47, select a local definition in the list, specify a file name, and select the OK button. If you do not find any definition in the list box, it means that your assembly file does not have any local definition for copying out.

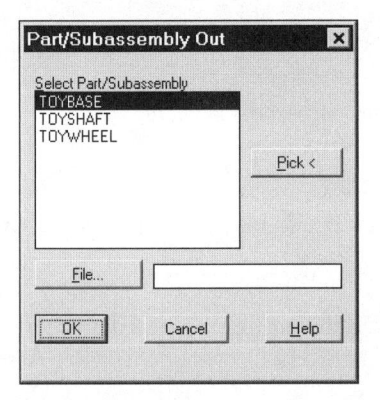

Figure 5–47 Part/Subassembly Out dialog box

IN-PLACE EDITING

There are two ways to edit a solid part:

- You open the solid part and make necessary change to it.
- You edit it in the context of the assembly.

The second way is called in-place editing. Now you will edit a component in the context of an assembly.

1. Open the assembly file *ToyCar.dwg*, if you have already closed it.
2. Select the ToyBase from the browser and double-click (see Figure 5–48).

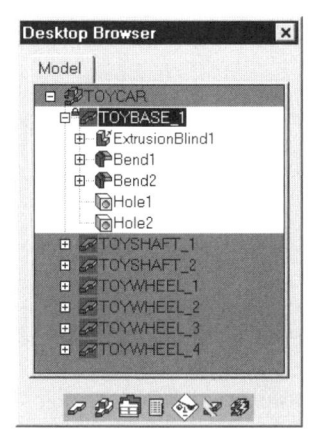

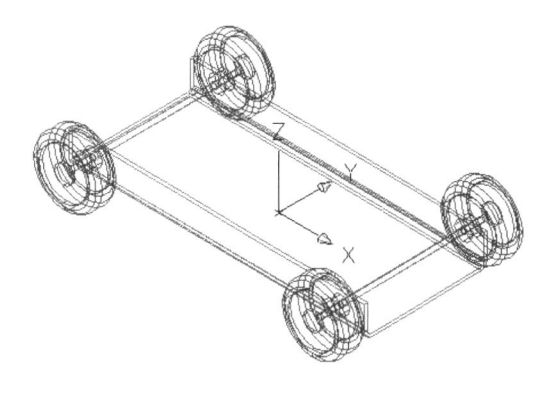

Figure 5–48 *Component selected and activated*

In the browser, there is only one tab (Model tab) available; the other components in the model tab are grayed out. There is also a small lock symbol along with the selected component, depicting that the component is locked for editing and no one else can edit the file until it is unlocked.

3. Select the profile highlighted in Figure 5–49 and double-click.
4. Select the dimension indicated in Figure 5–50 and double-click.
5. In the Power Dimensioning dialog box, change the dimension value to 60 and select the OK button.

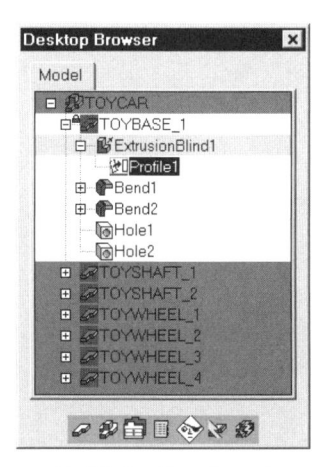

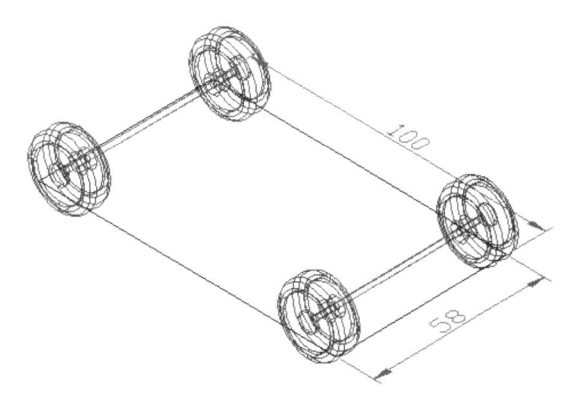

Figure 5–49 *Profile selected*

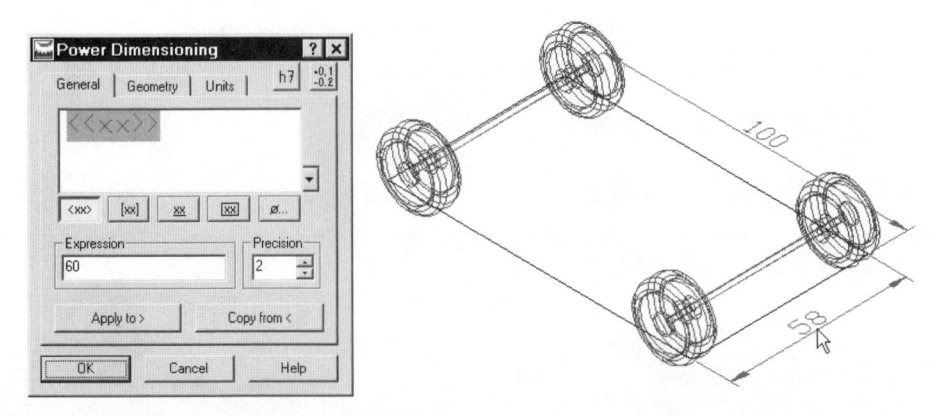

Figure 5–50 *Parametric dimension selected*

6. Select the Update button from the browser to update the change.
7. Select the assembly from in the browser and double-click to return to assembly mode.
8. Select the update assembly button from the browser.

The solid part is modified, and the solid part and the assembly are updated (see Figure 5–51). Note that there is still a lock symbol visible along with the component that you modified. If you close the file now and re-open it, the lock symbol will be removed. Now save the file. Because an external definition is modified, you need to confirm the change.

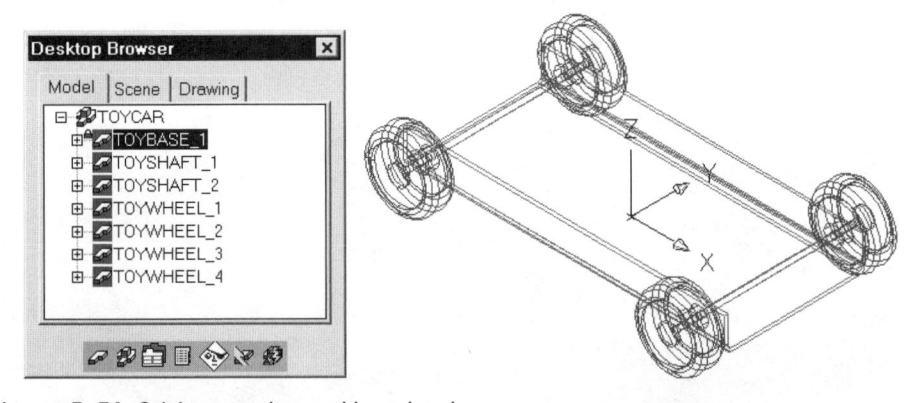

Figure 5–51 *Solid part and assembly updated*

9. Save the file.
10. In the External File Save dialog box, select the OK button (see Figure 5–52).

Figure 5–52 *External File Save dialog box*

BASIC 3D WORK PLANES

While working in an assembly file, setting up 3D work planes automatically starts a new local definition. To illustrate how new component part definitions are created, you will start a new file and use the AMBASICPLANES command twice.

1. Start a new file. Use metric default.
2. Select Part>Work Features>Basic 3D Work Planes
3. Repeat the AMBASICPLANES command.

Two component parts are created (see Figure 5–53).

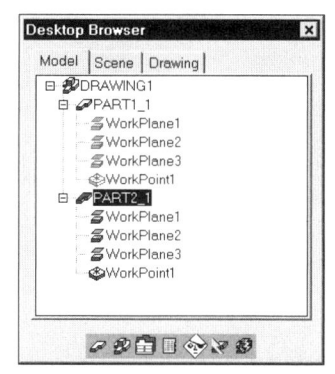

Figure 5–53 *Browser showing two component parts in the hierarchy*

REPLACE

To cope with design changes, you may wish to replace a component part with another component part. Replacement removes an existing component and attaches another component in one operation. Because the references with the removed component are lost, the assembly constraint you applied to the original component are removed. To replace a component, select Assembly>Assembly>Replace (see Figure 5–54).

Command: AMREPLACE

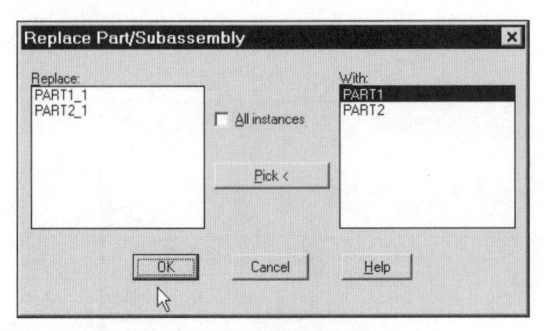

Figure 5-54 *Replace Part/Subassembly dialog box*

In the Replace Part/Subassembly dialog box, select a component to be replaced and a component to replace.

WHERE USED

Sometimes you have a component name but you do not know where it is placed in the assembly. To find out the location of component in a complex assembly, select Assembly>Assembly>Where Used (see Figure 5-55).

Command: AMWHEREUSED

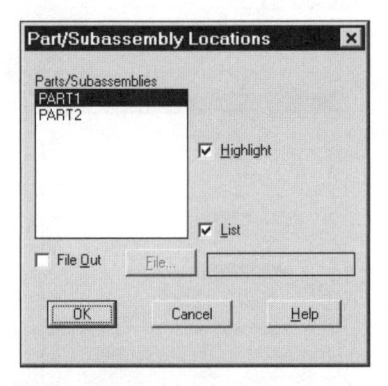

Figure 5-55 *Part/Subassembly Locations dialog box*

In the Part/Subassembly Locations dialog box, select a component and select the OK button. The selected component will be highlighted.

QUERY

To retrieve information regarding the component's name, degrees of freedom, and attributes, select Assembly>Assembly>Query.

Command: AMLISTASSM

After you press the ENTER key and select a component, details regarding the selected component will be displayed in the text window (see Figure 5-56).

```
AutoCAD Text Window - Drawing2.dwg
Edit
Command:
Command:
Command: _amlistassm

Enter an option (parts or subassemblies) [Name/Select] <Select>:

Select part and subassembly instances: 1 found

Select part and subassembly instances:
Part/Subassembly name: PART1_1    Base Part
Definition name: PART1

Degrees of freedom
-----------------------------------------------
 0 Rotational degrees of freedom
 0 Translational degrees of freedom

Attributes on definition
-----------------------------------------------

Command:
```

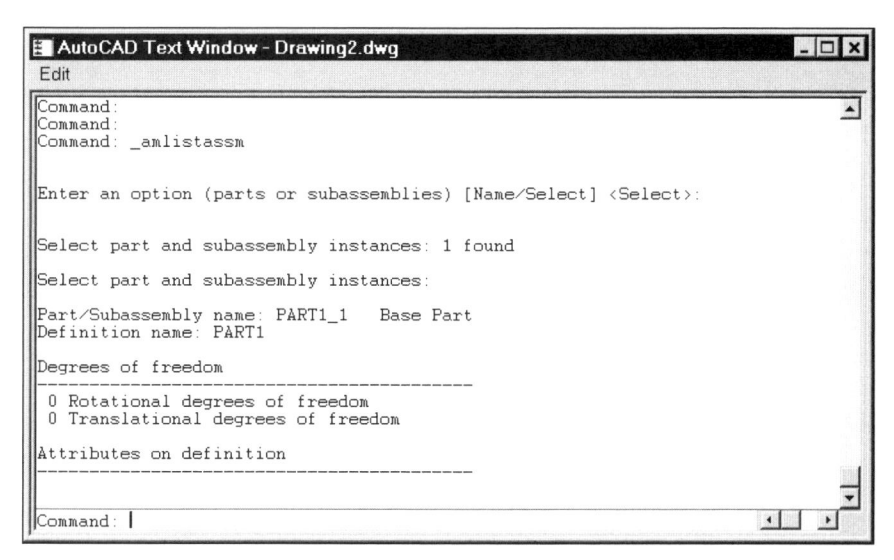

Figure 5–56 *Text window showing component name, degrees of freedom, and attributes*

AUDIT

While you are working on an assembly with external component parts, the external component parts may have been modified by some other person. To find out whether all the files in the assembly are up-to-date or not, you audit the assembly by selecting Assembly>Assembly>Audit. Figure 5–57 shows the message dialog box if all the files are up-to-date.

Command: AMAUDIT

Figure 5–57 *Message dialog box*

REFRESHING EXTERNAL COMPONENT PARTS

If the files in the assembly are not up-to-date, reload the external component parts to the assembly so that the latest updated versions of the component solids are reflected in the assembly. To refresh the component parts, select Assembly>Assembly>Refresh.

Command: AMREFRESH

ANALYSIS

With a virtual assembly, you evaluate mass properties of a set of components, check for interference between components, and find out the minimum distance between selected components.

MASS PROPERTIES

The way to evaluate mass properties of solid parts in a part file and an assembly file is the same: Select solid parts, assign materials, and evaluate various mass properties (mass, volume, surface area, centroid, mass moments of inertia, mass products of inertia, radii of gyration, principal mass moments, and principal axes).

1. Open the assembly file ToyCar, if you have already closed it.
2. Select Assembly>Analysis>Mass Properties

    Command: AMASSMPROP

3. Select all the components and press the ENTER key (see Figure 5–58).

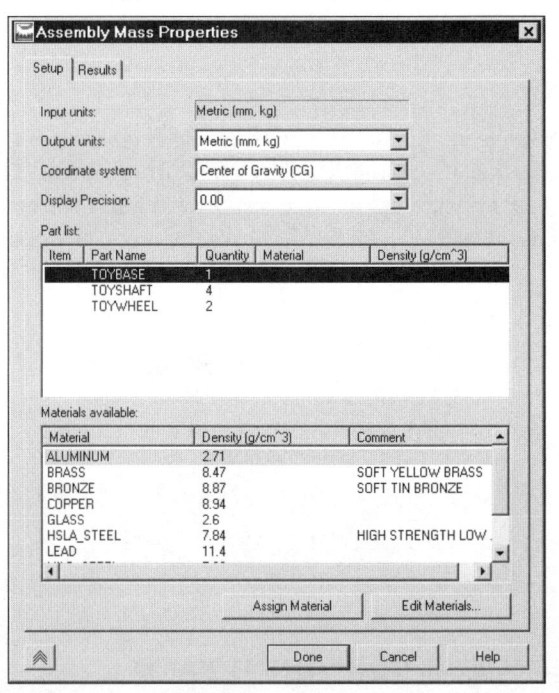

Figure 5–58 *Assembly Mass Properties dialog box*

4. Select the solid part one by one, pick a material, and select the Assign Material button.
5. Select the Results tab and select the Calculate button.

Material properties of the assembly are evaluated (see Figure 5–59).

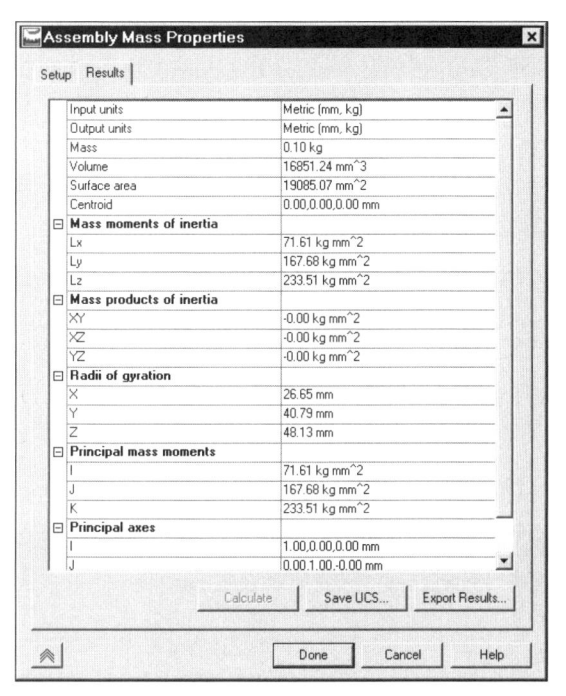

Figure 5–59 *Results tab*

Note that the result shown here may not be the same as yours because you may assign different kinds of materials to the components in the assembly.

INTERFERENCE

After you assemble a set of components in an assembly, it is sometimes necessary to find out whether there is any interference between the component parts and find out where the interference occurs, if any exists. Interference checking concerns interference between two sets of components. Select two sets of components and let the computer check for interference.

    6. Select Assembly>Analysis>Check Interference

       Command: AMINTERFERE

    7. Type **N** at the command line area to exclude nested components.

    8. Select the wheels one by one and press the ENTER key.

    9. Select the shafts one by one and press the ENTER key.

There should be no interference.

MINIMUM DISTANCE

In order for the components of some assemblies to function properly, it is sometimes necessary to leave a specific amount of clearance between the components. To check the size of these clearances, you measure minimum distance. Like interference checking, select two sets of components and let the computer tell you the minimum 3D distance between them.

10. Select Assembly>Analysis>Minimum 3D Distance

    Command: AMDIST

11. Select a wheel and press the ENTER key.
12. Select the base and press the ENTER key.

At the command line area, press the ENTER key again. The minimum distance is displayed.

ASSEMBLY OPTIONS

There are a number of options affecting the construction of an assembly.

13. Select the Options button from the browser (see Figure 5–60).

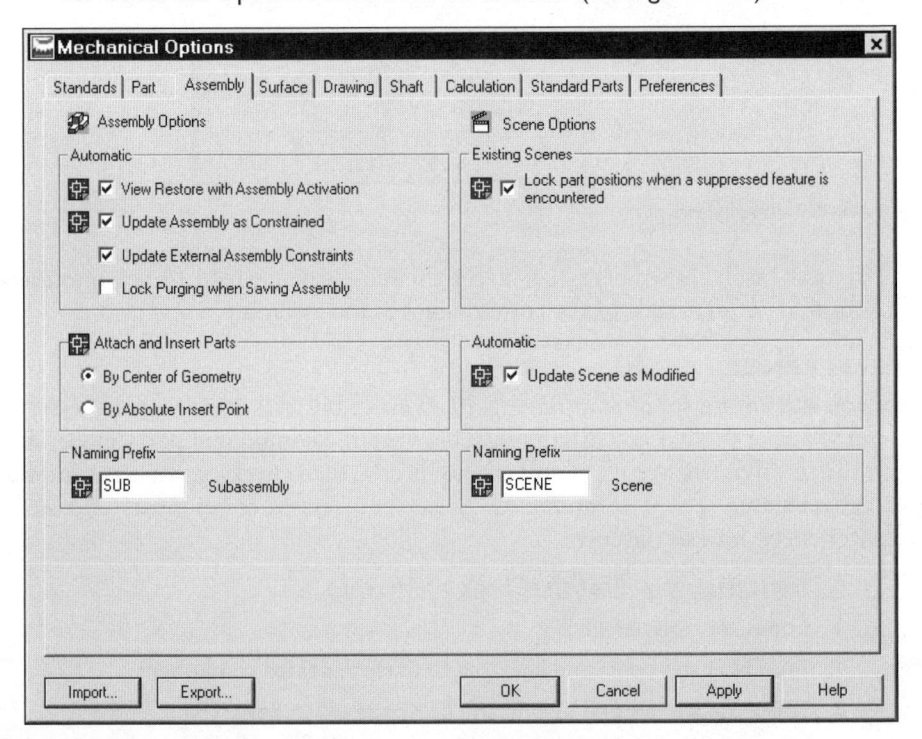

Figure 5–60 *Assembly tab of the Mechanical Options dialog box*

The Assembly tab of the Mechanical Options dialog box has six areas in two columns. In the left column, there are three areas concerning assembly: Automatic, Attach and Insert Parts, and Naming Prefix (see the table below).

Automatic	View Restore with Assembly Activation	restores the last display view when you change from one subassembly to another
	Update Assembly as Constrained	causes the parts to update as the assembly constraints are applied
	Update External Assembly Constraints	causes external assembly files to be updated if they are modified
	Lock Purging when Saving Assembly	causes lock files to be purged when you save an assembly
Attach and Insert Parts	By Center of Geometry	uses the center of geometry as the insertion point
	By Absolute Insert Point	uses the absolute insertion point
Naming Prefix	Subassembly	sets the prefix of the name of the subassembly

SUMMARY

An assembly is a collection of components that are put together to form a useful whole. To manage a large number of components, you first form subassemblies of small collections of components and then put the subassemblies into a single assembly. Hence, an assembly file consists of an assembly of components that can be solid parts or subassemblies of solid parts. Constructing an assembly with the computer concerns two major tasks: linking the component files (part or subassembly) to the assembly file and applying assembly constraints to relate the components to each other.

To link the components to an assembly, you use the assembly catalog. You include a search directory and select components from the search directory. Initially, components (except the first grounded component) in an assembly are free to translate linearly along three axes and rotate about three axes. These freedoms are called six degrees of freedom. To find out the degrees of freedom of a component, you display the DOF symbol. To translate the components in 3D space, you use the power manipulator. However, translating does not affect the degrees of freedom of the component in the assembly. To relate components to each other, you apply a set of assembly constraints. They are mate, flush, angle, and insert.

There are three design approaches to handling a set of components: bottom-up, top-down, and hybrid. In the bottom-up approach, you construct all the components in individual part files and link them to an assembly file. Because the component parts are external to the assembly file, they are called external definitions. In the top-down approach, you construct all the component parts in the assembly file. Because all the component parts reside in the assembly file, they are called local definitions. The third approach is a hybrid of the first and second approaches. You construct some components in the assembly file and construct the others in separate part files. To meet practical manufacturing requirements, it is common practice to have individual files for individual components. Hence, you externalize any local definitions to become external definitions. You can also copy out local definition or copy in an external definition, which is similar to but not the same as externalizing and localizing.

While working in the assembly, you may edit the component parts, even if they are external definitions (in-place editing). When you construct basic work planes in an assembly, a new part definition is created automatically. In an assembly, you can replace a component with another, find out the locations of the component in the assembly, discover information about a component, audit the integrity of the assembly, reload external definitions, evaluate mass properties, check interference, and find out minimum distances between components.

Assembly Modeling 245

PROJECTS

Now you will put your newly acquired knowledge to work with the following projects.

CONSTRAINT MANIPULATION

You will construct an assembly to familiarize yourself with the use of mate, flush, and angle constraints. You will use the top-down approach to construct two component parts and apply constraints to assemble them together.

1. Start a new file. Use metric as the default.
2. Select Part>Part>New Part to start a new part.
3. Referring to Figure 5–61, construct a sketch, resolve it to a profile, add parametric dimensions as shown, and extrude it a distance of 70 units.

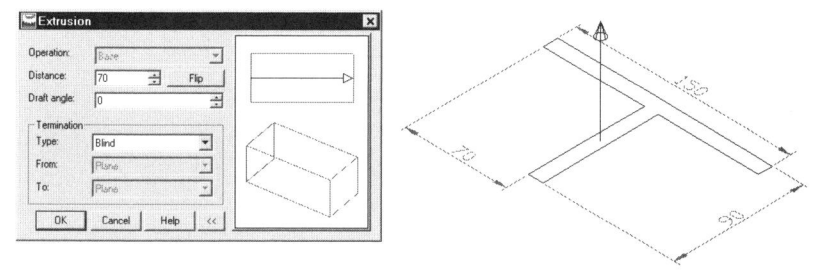

Figure 5–61 *Profile being extruded*

4. Construct a rectangle, resolve it to a profile, add four collinear constraints, and extrude it a distance of 10 units to join the first extruded solid feature (see Figure 5–62).
5. Select Part>Part>New Part to start another new part.
6. Referring to Figure 5–63, construct a sketch, resolve it to a profile, add equal length constraints to the edges, add a parametric dimension, and extrude it a distance of 10 units.

Two component definitions are constructed.

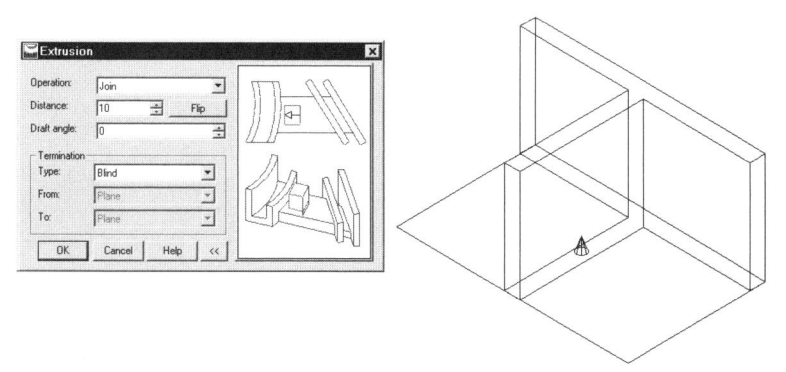

Figure 5–62 *Second profile being extruded*

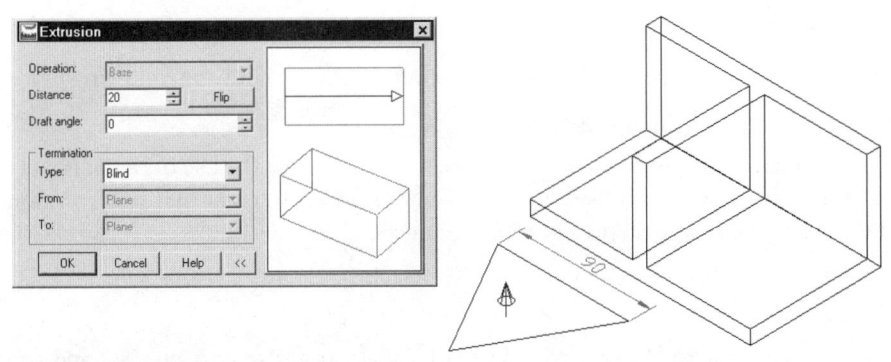

Figure 5-63 *Second part being constructed*

Now apply mate constraints to assemble the vertices of a component to the vertex, edge, and face of another component.

    7. Select Assembly>3D Constraints>Mate

    8. Select the vertices indicated in Figure 5-64.

    9. Type **0** at the command line area to specify an offset distance of zero.

Vertices of the components are mated.

    10. Repeat the AMMATE command.

    11. Select the vertex and the edge indicated in Figure 5-65.

    12. Press the ENTER key.

A vertex is mated to an edge.

    13. Repeat the AMMATE command.

    14. Select the vertex and the face indicated in Figure 5-66.

    15. Press the ENTER key.

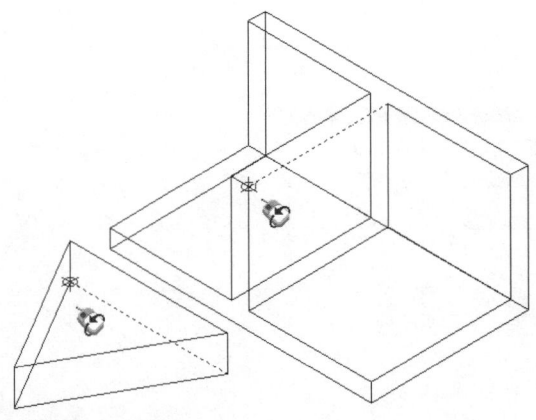

Figure 5-64 *Vertices being mated*

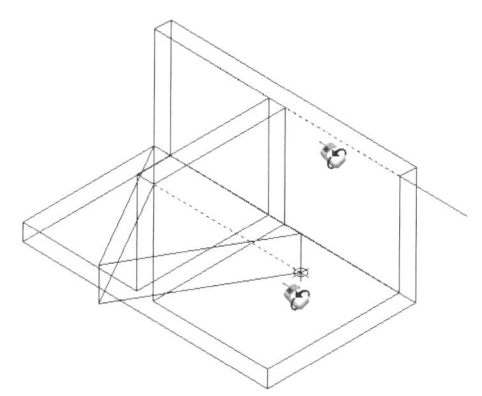

Figure 5–65 *Vertex and edge being mated*

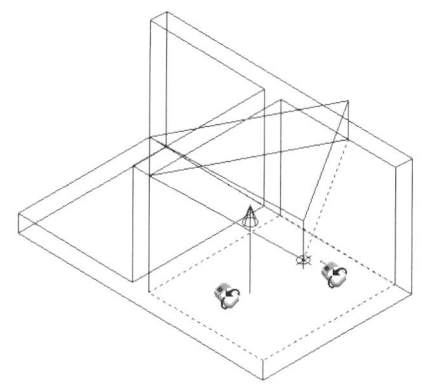

Figure 5–66 *Vertex and face being mated*

A vertex and a face are mated. Now copy an instance of a component.

16. Copy an instance of a component (see Figure 5–67).

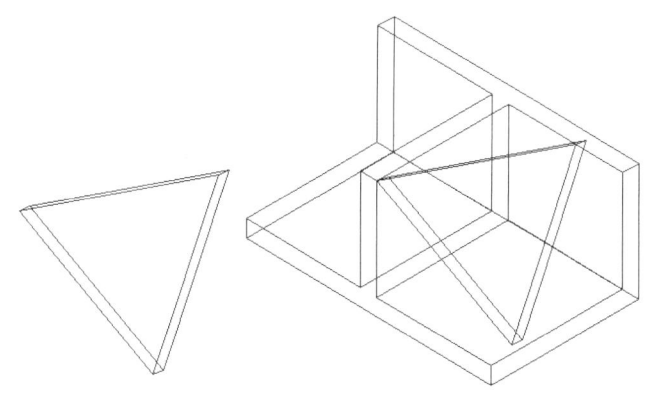

Figure 5–67 *Instance copied*

Use the flush constraint to align two faces.

17. Select Assembly>3D Constraints>Flush

    Command: AMFLUSH
18. Select the faces highlighted in Figure 5–68.
19. Press the ENTER key.

Two faces are flushed.

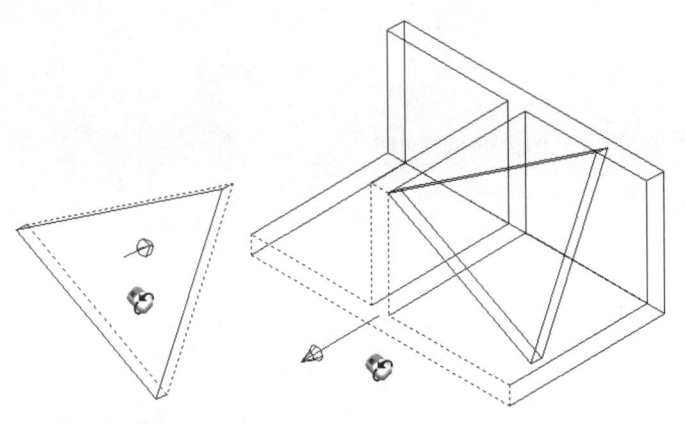

Figure 5–68 *Faces being flushed*

Now use the mate constraint to align two edges.

20. Select Assembly>3D Constraints>Mate
21. Select the edges highlighted in Figure 5–69.
22. Press the ENTER key.

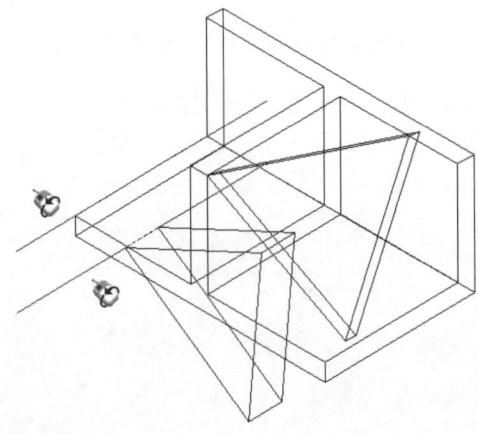

Figure 5–69 *Edges being mated*

Two edges are aligned. Now use the angle constraint to determine the angular position of a component.

23. Select Assembly>3D Constraints>Angle

    Command: AMANGLE

24. Select the edges highlighted in Figure 5–70.
25. Type **60** at the command line area to specify the angle between the two selected edges.
26. Press the ENTER key.

The assembly is complete (see Figure 5–71). Save and close your file (file name: *Constraints.dwg*).

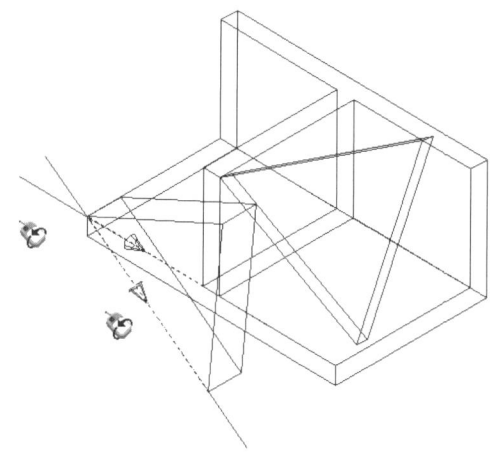

Figure 5–70 Angle constraint being applied

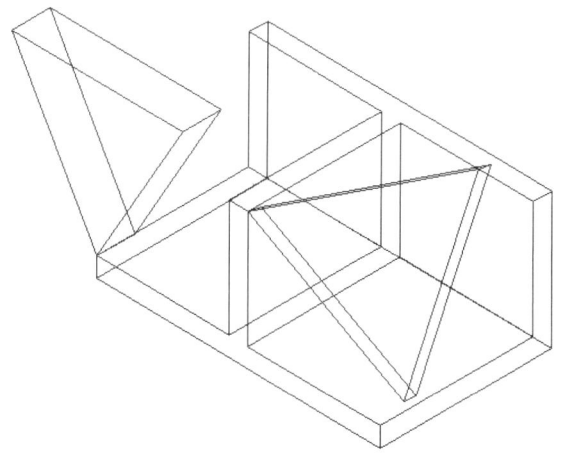

Figure 5–71 Assembly completed

FRONT END OF THE INFANT SCOOTER

Now you will construct the assembly of the front end of the infant scooter.

1. Start a new file. Use metric as the default.
2. Set the display to an isometric view.
3. Use the assembly catalog to include the search path "C:\Projects\InfantScooter." (We assume that you already put all the components of the infant scooter in this directory.)
4. Place one instance of the steering shaft, one instance of the Frame_F, one instance of Axle, two instances of Clip, two instances of the TireWheel (subassembly), and two instances of the WheelCap in the assembly file (see Figure 5–72).

Figure 5–72 *Instances placed in the assembly*

5. Referring to Figure 5–73, apply assembly constraints to assemble the components together.

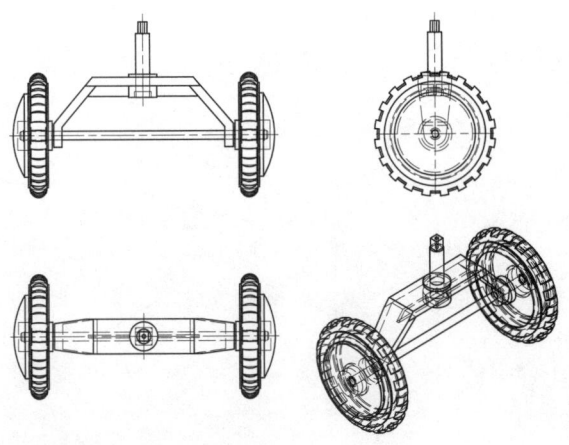

Figure 5–73 *Orthographic views of the assembly*

Save and close your file (file name: *FrontEnd.dwg*).

CENTRAL FRAME OF THE INFANT SCOOTER

Now you will construct the central frame assembly of the infant scooter.

1. Start a new file. Use metric as the default.
2. Add the search directory "C:\Projects\InfantScooter" in the assembly catalog.
3. Place an instance of Frame_C and Frame_R in the assembly (see Figure 5–74).
4. Referring to Figure 5–75, assemble the component parts. The dimension shown in the figure indicates the relative position between the components.

Save and close your file (file name: *Frame.dwg*).

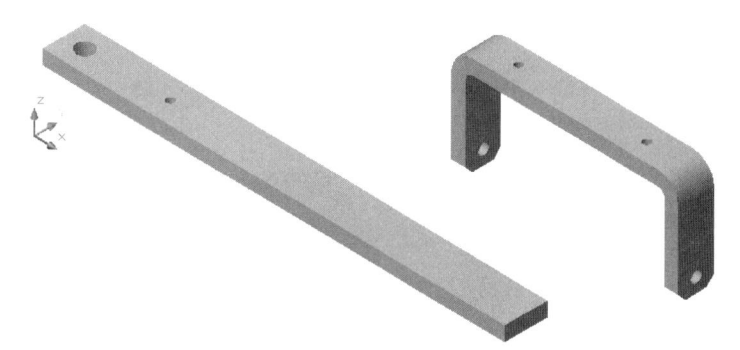

Figure 5–74 *Instances placed in the assembly*

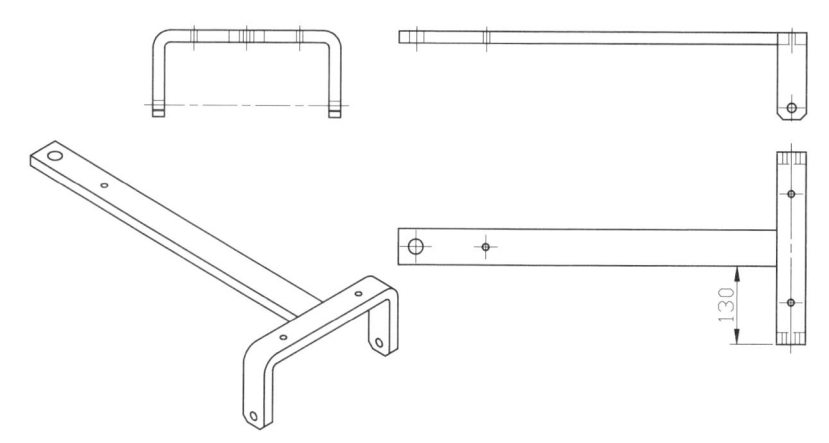

Figure 5–75 *Drawing showing the position between the two components*

REAR END OF THE INFANT SCOOTER

Now you will construct the rear end assembly of the infant scooter.

1. Start a new file. Use metric as the default.
2. Add the search directory "C:\Projects\InfantScooter" in the assembly catalog.

3. Place the following components in the assembly file: Frame, Axle, Clips, TireWheel, and WheelCap (see Figure 5–76).
4. Referring to orthographic views shown in Figure 5–77, assemble the components together.

Save and close your file (file name: *RearEnd.dwg*).

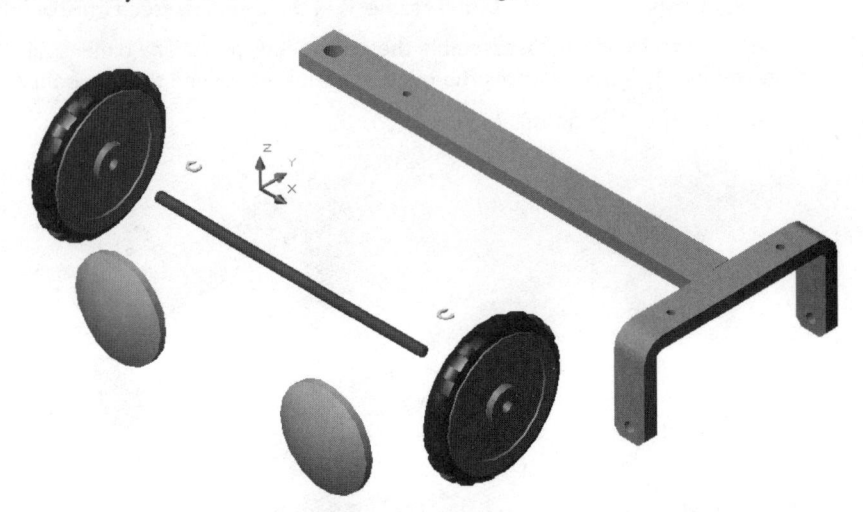

Figure 5–76 *Components of the assembly*

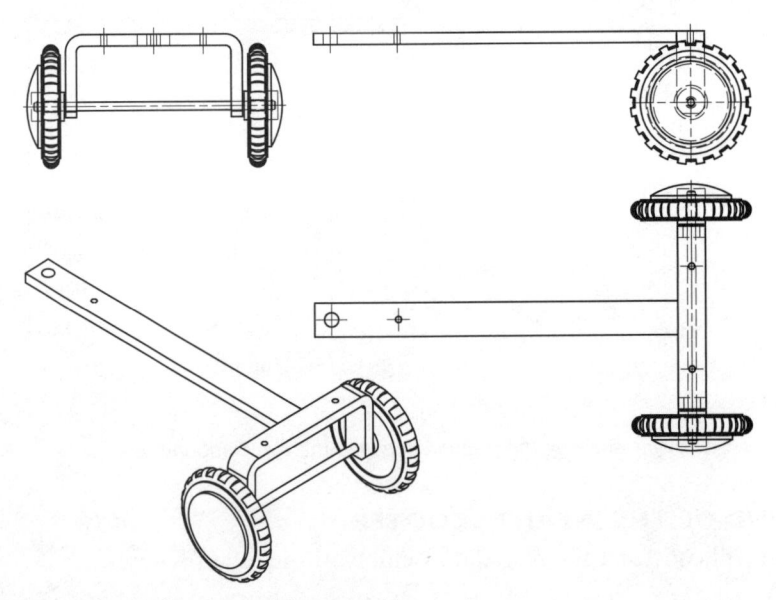

Figure 5–77 *Orthographic views of the assembly*

SEAT ASSEMBLY OF THE INFANT SCOOTER

Now you will construct the seat assembly of the infant scooter.

1. Start a new file. Use metric as the default.
2. Add the search directory "C:\Projects\InfantScooter" in the assembly catalog.
3. Place the following components in the assembly: Seat, Handle, and Bolt (see Figure 5–78).

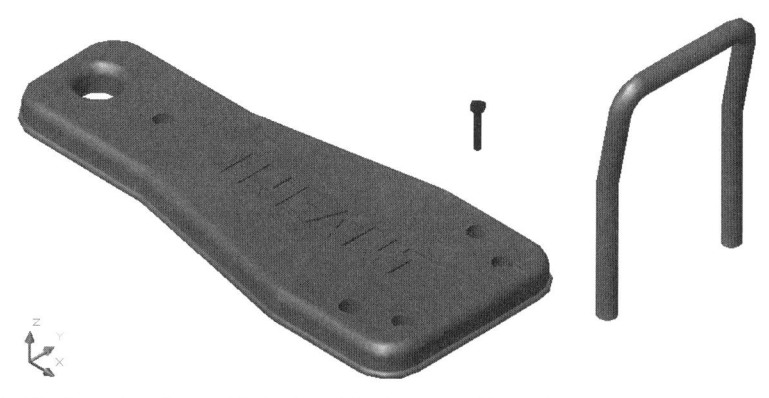

Figure 5–78 *Seat, handle, and bolt placed in the assembly*

4. By using the assembly catalog, localize the component Bolt (see Figure 5–79).

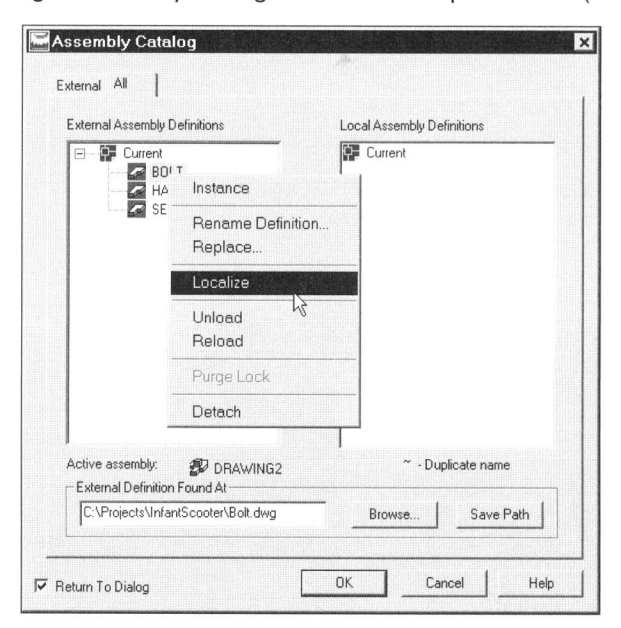

Figure 5–79 *Component being localized*

5. Rename the component Bolt to Screw (Check the All Instances box; see Figure 5–80).

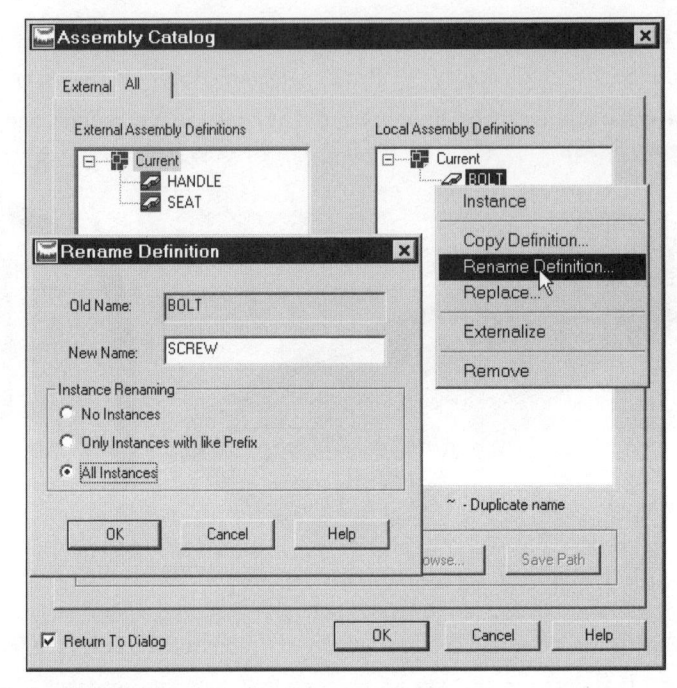

Figure 5–80 *Component renamed*

6. Select Screw from the Browser and double-click.
7. Referring to Figures 5–81 and 5–82, modify the extruded solid features.
8. Externalize the component Screw, and save it in the "C:\Projects\InfantScooter" directory (see Figure 5–83).

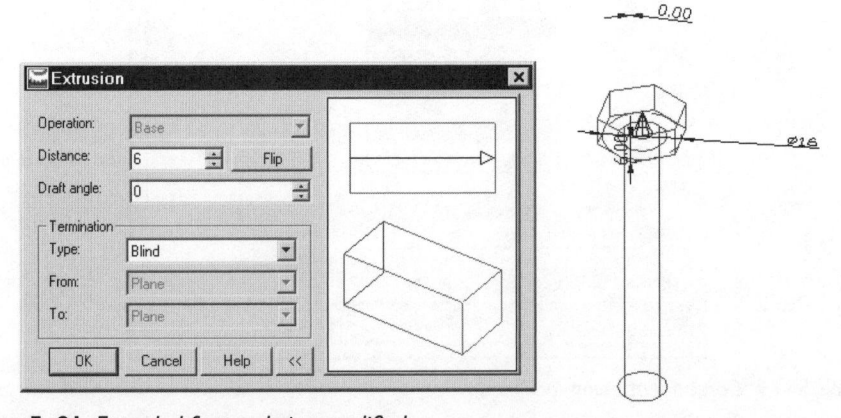

Figure 5–81 *Extruded feature being modified*

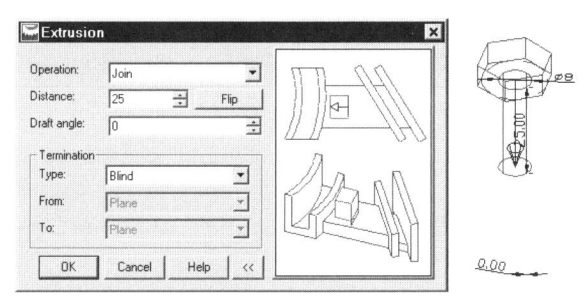

Figure 5–82 *Second extruded feature being modified*

Figure 5–83 *Component being externalized*

9. Place one more instance of Screw and assemble the components in accordance with the orthographic views shown in Figure 5–84.

Save and close the file (file name: *SeatHandle.dwg*).

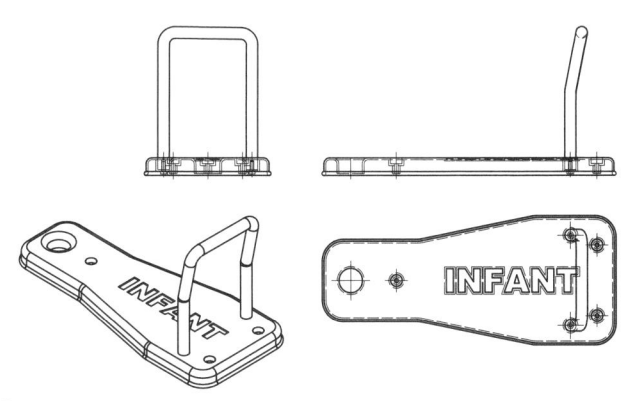

Figure 5–84 *Components assembled*

INFANT SCOOTER

Now you will construct the final assembly of the infant scooter.

1. Start a new file. Use metric as the default.
2. Add the search directory "C:\Projects\InfantScooter" in the assembly catalog.
3. Place the following components in the assembly file: RearEnd, FrontEnd, SteeringWheel, Thread, SeatHandle, Bolt, Nut, and ScrewCap (see Figure 5–85).
4. Referring to Figures 5–86 and 5–87, assemble the components.

Save and close your file (file name: *InfantScooter.dwg*).

Figure 5–85 *Components of the infant scooter*

Figure 5–86 *Components assembled*

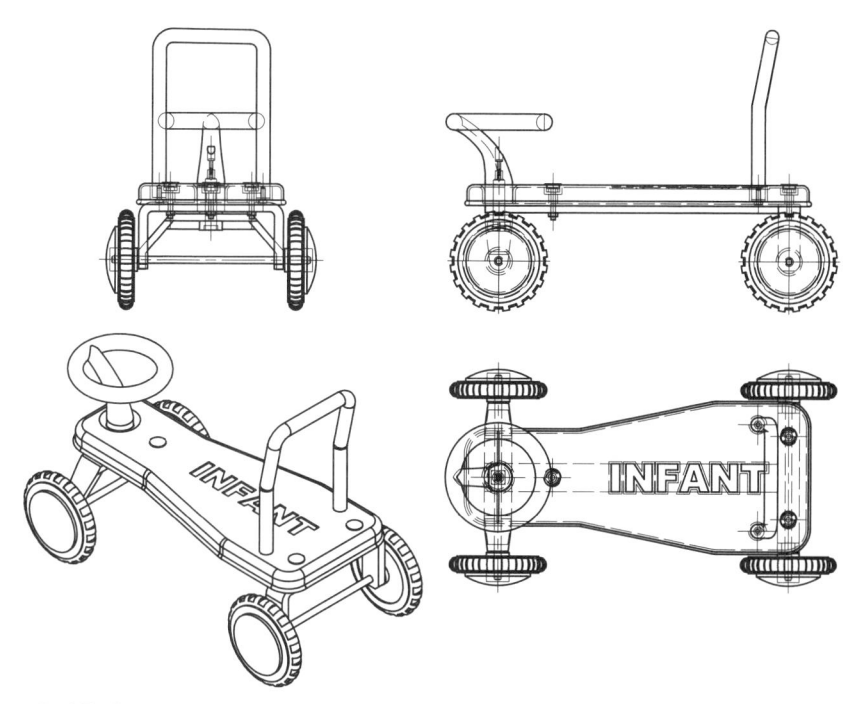

Figure 5–87 *Orthographic views of the final assembly*

REVIEW QUESTIONS

1. Explain the definition of an assembly.

2. Explain the six degrees of freedom.

3. Illustrate, with the aid of sketches, the application of different kinds of assembly constraints.

4. Describe the three approaches to designing an assembly.

5. Explain the meaning of internal definition and external definition.

6. State the difference between externalize and copy out.

7. State the difference between localize and copy in.

8. Explain the way to edit a component in the context of an assembly.

9. A bottom-up assembly has all the parts in the same file. True or false?

10. You can in-place edit a local or external part. True or false?

11. An assembly does not have to be fully constrained. True or false?

12. The solid part that is created from the CHECK INTERFERENCE command is parametric. True or false?

CHAPTER 6

Assembly Scene

OBJECTIVES

The goals of this chapter are to introduce the concept of an assembly scene and to illustrate the techniques needed to construct exploded assembly scenes. After studying this chapter, you should be able to do the following:

- Explain the concept of assembly scene
- Set up assembly scenes
- Construct exploded views in assembly scenes

OVERVIEW

It is a common engineering practice to explode apart a set of assembled components to depict how they are put together in the assembly. To illustrate the relative position of the components in the assembly, you add trail lines to link the exploded components. Figure 6–1 shows a scene of an assembly in which the components are displaced apart.

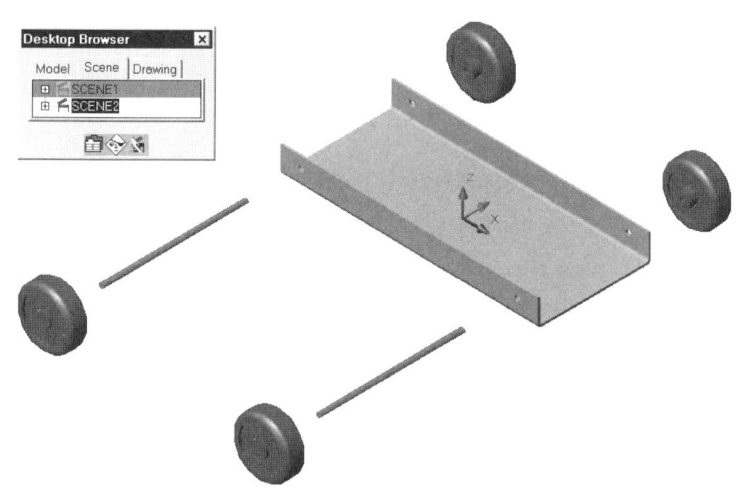

Figure 6–1 *Exploded scene of an assembly*

ASSEMBLY SCENE CONCEPTS

As you learned in Chapter 5, components in an assembly are put together by using a set of assembly constraints. To maintain integrity of the assembled components as a whole in the assembly and still obtain an exploded view, set up assembly scenes which are data sets in the assembly file to illustrate how components of an assembly are related to each other. In the assembly scene, displace all the components apart by specifying an overall explosion factor or translate the components apart individually by tweaking.

By setting up a number of scenes in an assembly, you can have a scene with no explosion or tweaking to show how the components are assembled, a scene in which some components are hidden, and one or more scenes to demonstrate how the components are related to each other by exploding or tweaking. Figure 6–2 shows two scenes of an assembly.

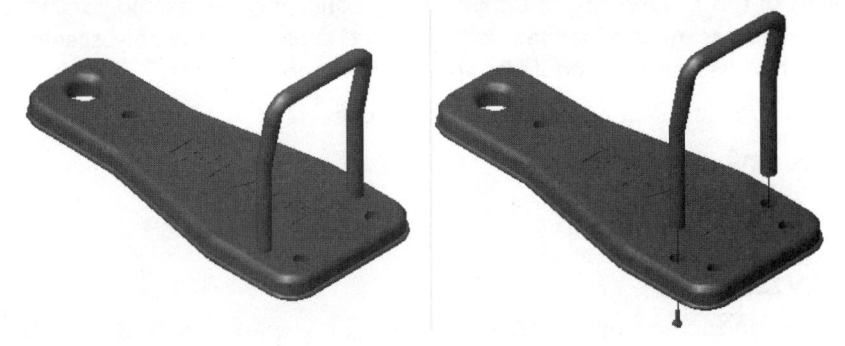

Figure 6–2 *Scenes of an assembly*

SCENE

A scene is a unique data set in the assembly file. You can have more than one scene in an assembly file. You use scenes to depict how component parts are related to each other by displacing them apart.

SCENE CONSTRUCTION

To construct scenes in an assembly, select the Scene tab from the browser.

1. Open the file *SeatHandle.dwg* that you constructed in Chapter 5.
2. Select the Scene tab from the browser.

At the bottom of the Scene tab browser, there are three buttons: options, visibility, and update. They enable you to set scene options, control scene visibility, and update the scene.

Now you will construct a new scene.

3. Move the mouse cursor over the background of the browser, right-click, and select New Scene (see Figure 6–3).

Assembly Scene 261

4. Press the ENTER key to accept the default scene name.

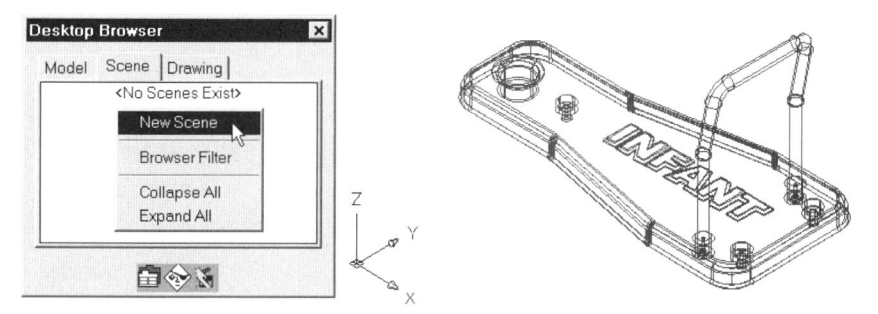

Figure 6–3 *New scene being constructed*

After you specify the scene name, the Create Scene dialog box (shown in Figure 6–4) displays. In the dialog box, there are several areas to choose from:

Target Assembly	enables you to select a target assembly, if there are subassemblies in the current assembly
Scene Name	enables you to specify the scene name
Scene Explosion Factor	enables you to explode all the components in the scene apart by a specified distance
Synchronize visibility with target assembly	enables you to synchronize the visibility with the selected target assembly

Figure 6–4 *Create Scene dialog box*

Because there is no subassembly in this assembly, there is only one object in the target assembly pull-down list box. By default, the explosion factor is zero. A zero explosion factor means that the components are not exploded apart and the position of components are the same as those in the assembly.

5. In the Create Scene dialog box, select the OK button.

A scene is constructed.

6. Repeat steps 3 through 5 three more times to construct three more scenes in the assembly file.

A total of four scenes are constructed (see Figure 6–5).

ACTIVATE SCENE

With more than one scene in the assembly, you activate one of them to make it current. There are two ways to activate a scene:

- You select the scene from the browser and double-click.
- You use the right-click menu.

Now you will use the right-click menu.

7. Select Scene1 from the browser, right-click, and select Activate Scene (see Figure 6–5).

Scene1 is activated.

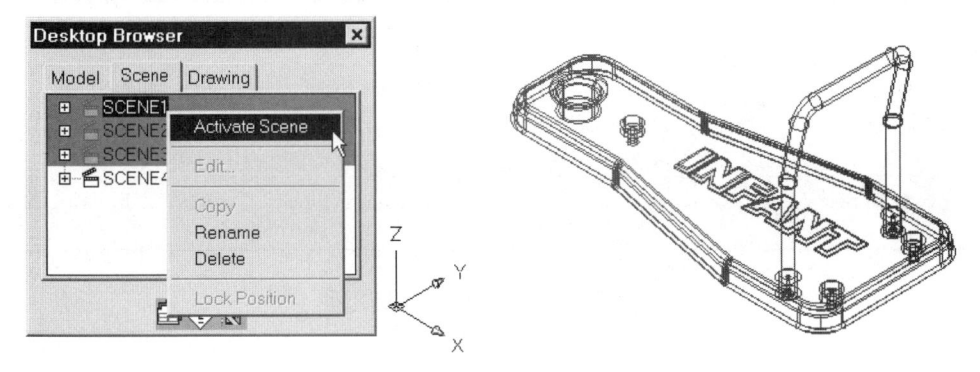

Figure 6–5 *Scene1 being activated*

SCENE VISIBILITY

Visibility of objects in a scene can be controlled independently from the other scenes.

8. Select the Visibility button from the browser (see Figure 6–6).

By using the Scene tab of the Desktop Visibility dialog box, you control the visibility of parts, subassemblies, and trails in a scene. Trails are lines linking displaced components in a scene. You will learn how to construct trail lines later.

9. In the Scene tab of the Desktop Visibility dialog box, select the Select button.
10. Select the seat and press the ENTER key.
11. On returning to the Desktop Visibility dialog box, select the OK button.

The seat in Scene1 is hidden (see Figure 6–7). Note that the component hidden in Scene1 is not hidden in all other scenes.

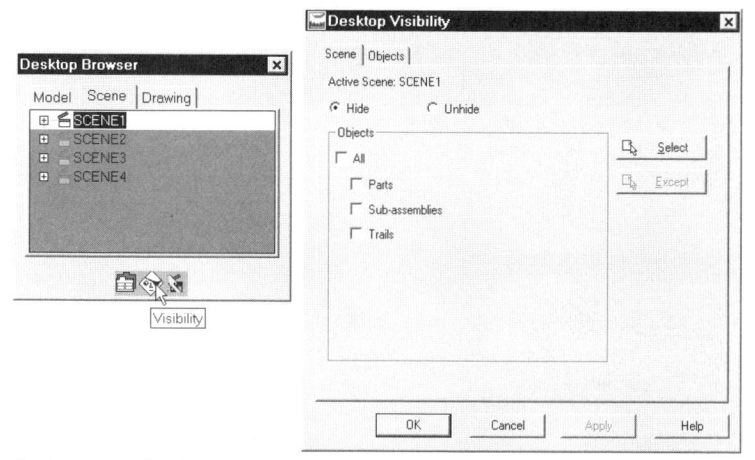

Figure 6–6 *Activating the Scene tab of the Desktop Visibility dialog box*

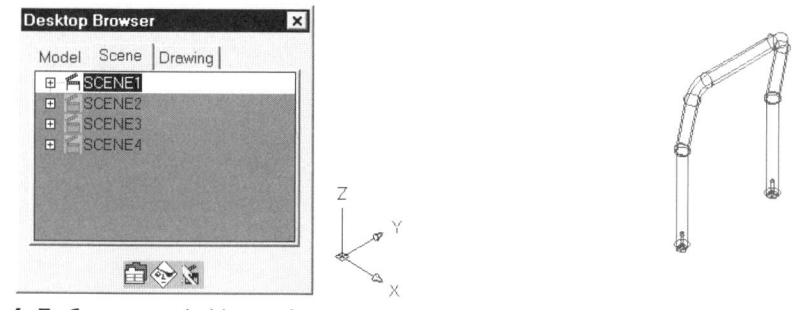

Figure 6–7 *Component hidden in the scene*

EXPLODED VIEWS

There are two ways to displace the components apart in an assembly scene:

- Specify an explosion factor while or after the scene is constructed
- Tweak individual components apart manually

For simple assemblies, using the explosion factor is a convenient way to translate all components apart. For more complex assemblies, it would be better to control the translation of each component by tweaking. Remember that exploding or tweaking the components in a scene will not affect the assembly or other scenes because each scene is an individual illustration of the assembly.

EXPLOSION FACTOR

An explosion factor specifies a distance for all the components to displace apart. You may specify an explosion factor while you construct the scene or specify the explosion factor after you have constructed the scene. Now you will set an explosion factor to Scene2.

12. Activate Scene2 by selecting it from the browser and double-clicking.
13. Select Scene2 from the browser, right-click, and select Edit.
14. In the Explode Factor dialog box, set the explosion factor to 100 and select the OK button (see Figure 6–8).

All the components in the scene are exploded apart at a distance of 100 mm (see Figure 6–9).

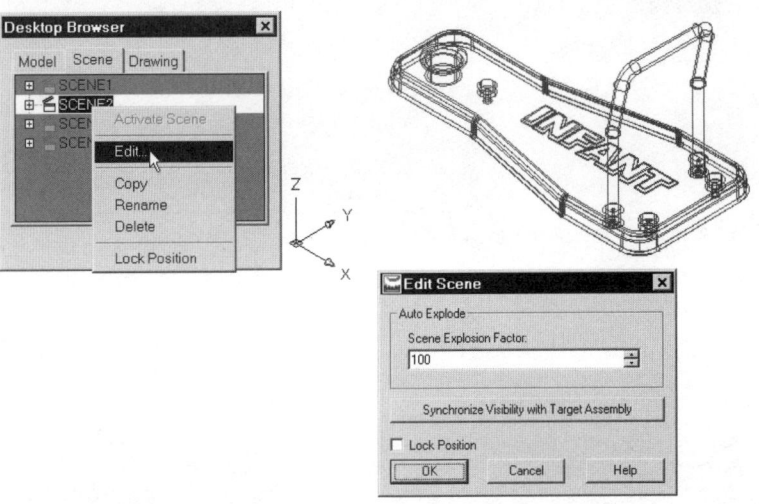

Figure 6–8 *Explosion factor specified*

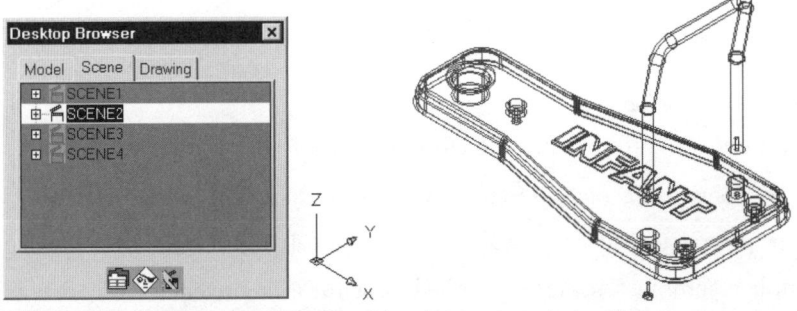

Figure 6–9 *Components in Scene2 displaced by setting an explosion factor*

To reiterate, the explosion factor set in Scene2 will not affect the positions of the components in all other scenes.

TWEAK

To displace the components apart individually, you can tweak them by translating and rotating them. Now you will activate Scene3 and tweak the components apart individually.

15. Double-click Scene3 in the browser to activate it.

16. Select an instance of the Screw from the browser, right-click, and select New Tweak.

    Command: AMTWEAK

The power manipulator displays.

17. Select the handle of the power manipulator indicated in Figure 6–10.

18. Referring to Figure 6–11, drag the handle to a new position, right-click, and select Enter.

The selected component is tweaked. Now you will edit the tweak distance.

19. Select Move Tweak from the browser, right-click, and select Edit (see Figure 6–12).

20. In the Edit Tweak dialog box, enter **150** in the Expression box and select the OK button.

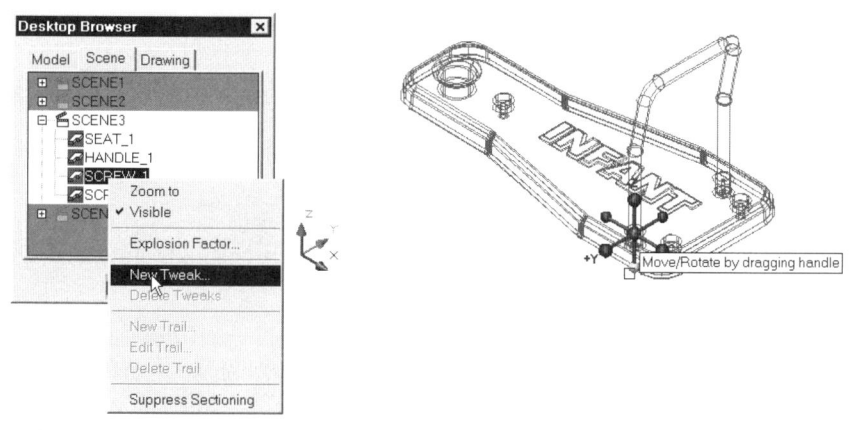

Figure 6–10 *Selected component being tweaked*

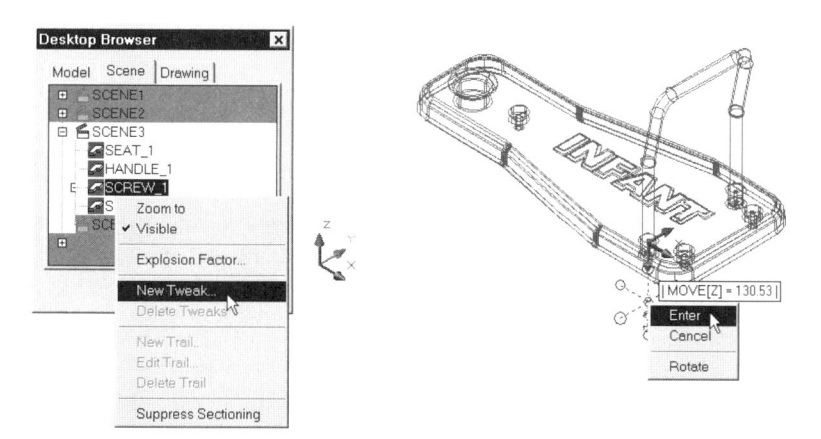

Figure 6–11 *Selected component tweaked*

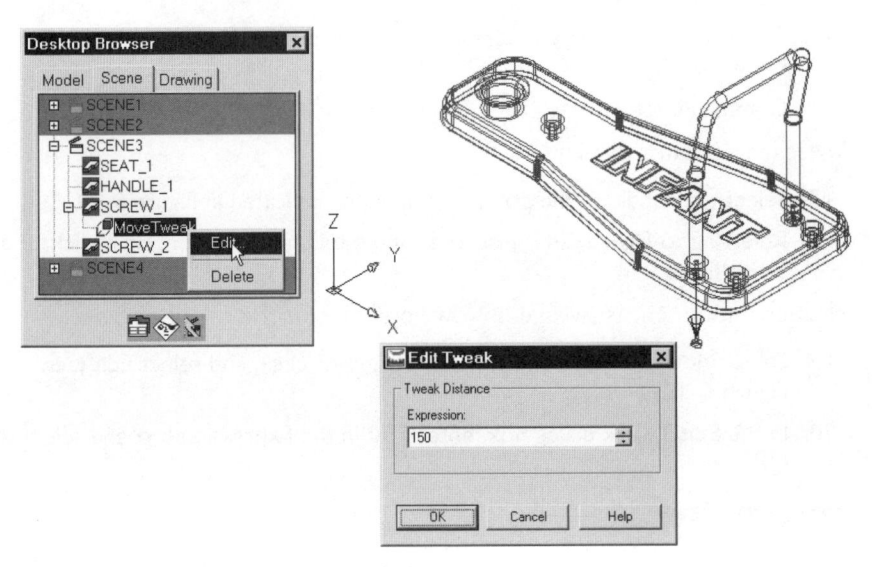

Figure 6-12 *Tweak distance being modified*

21. Repeat steps 16 through 20 to tweak the other screws and the handle in accordance with Figure 6–13.

The components in Scene3 are tweaked. Note that trail lines are added automatically between the tweaked components. These trail lines can be made invisible by using the Desktop Visibility dialog box shown in Figure 6–6.

Now the components in the four scenes in the assembly are different. In Scene1, a component is hidden. In Scene2, all the components are displaced apart by an explosion factor. In Scene3, the components are tweaked. In Scene4, the positions of the components are the same as those in the assembly.

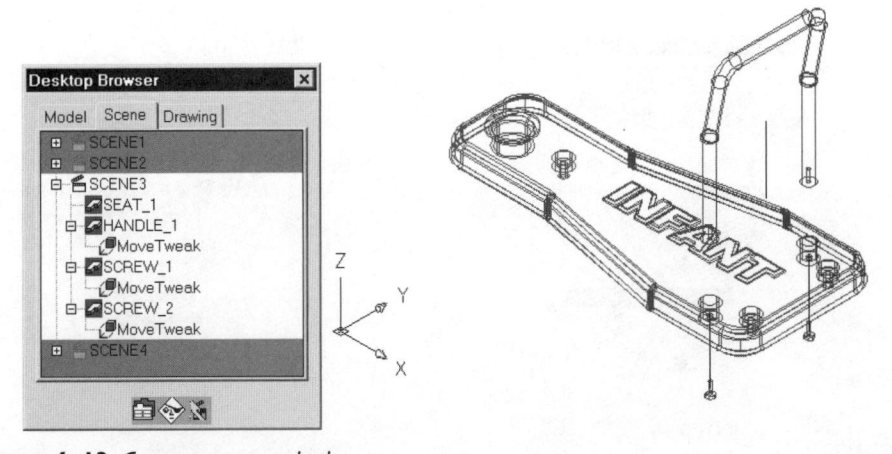

Figure 6-13 *Components tweaked*

TRAIL LINES

After a component is tweaked apart, a trail line is added automatically. For a scene with an explosion factor, you add the trail lines manually.

Now you will add trail lines to the displaced components in Scene2.

22. Select Scene2 from the browser and double-click to activate it.
23. Select Screw2 of Scene2 from the browser, right-click, and select **New Trail** (see Figure 6–14).

    Command: AMTRAIL
24. Select the upper part of the screw indicated in Figure 6–14.

The Trail Offsets dialog box displays. It enables you to determine the start and end positions of the trail lines in relation to the selected component. A zero offset means that the trail starts or ends at the selected component.

25. In the Trail Offsets dialog box, select the OK button.

A trail line is added (see Figure 6–15).

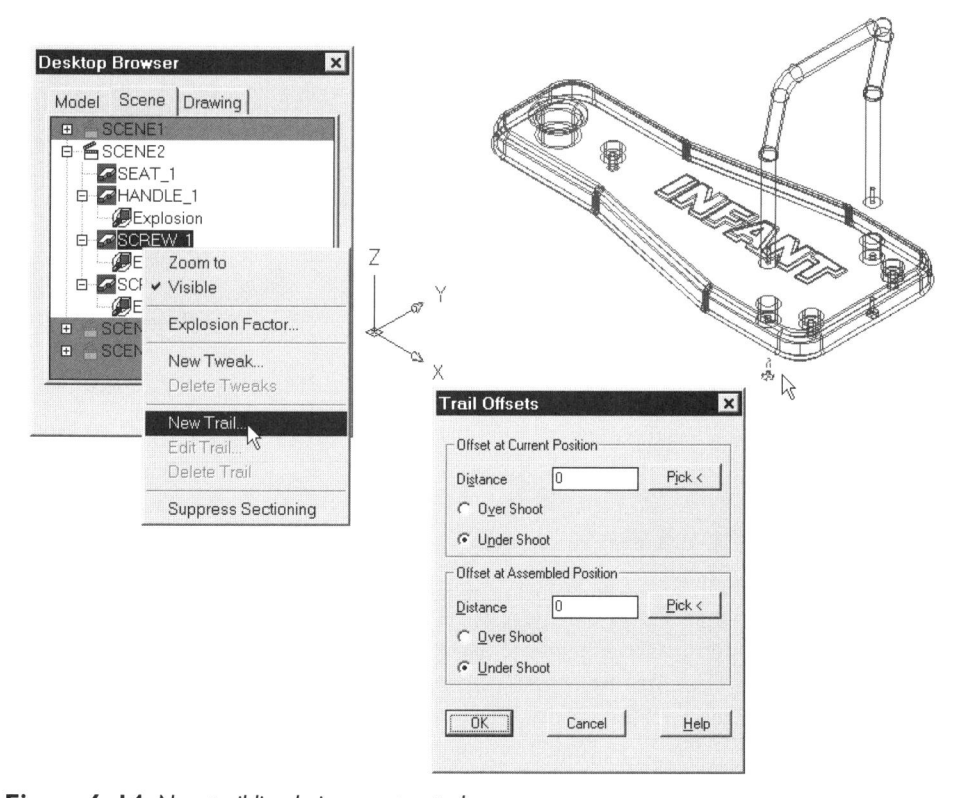

Figure 6–14 *New trail line being constructed*

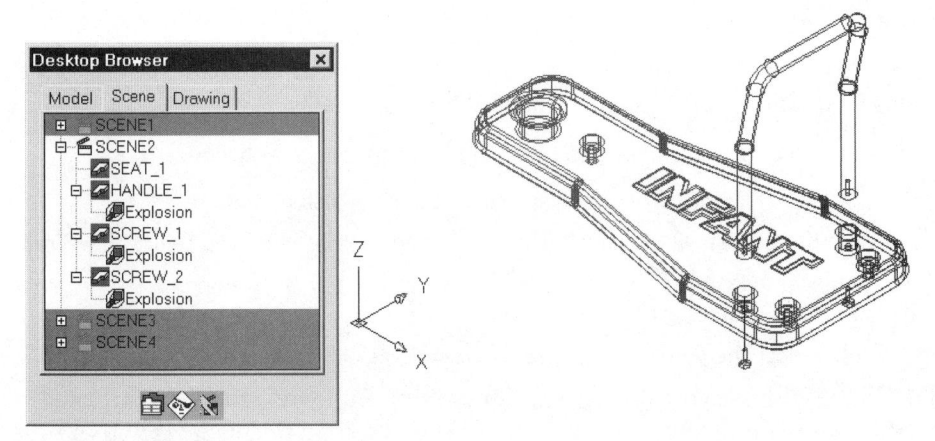

Figure 6-15 *Trail line constructed*

26. Repeat steps 23 through 25 to add trail lines in accordance with Figure 6-16. The scenes are complete. Save your file.

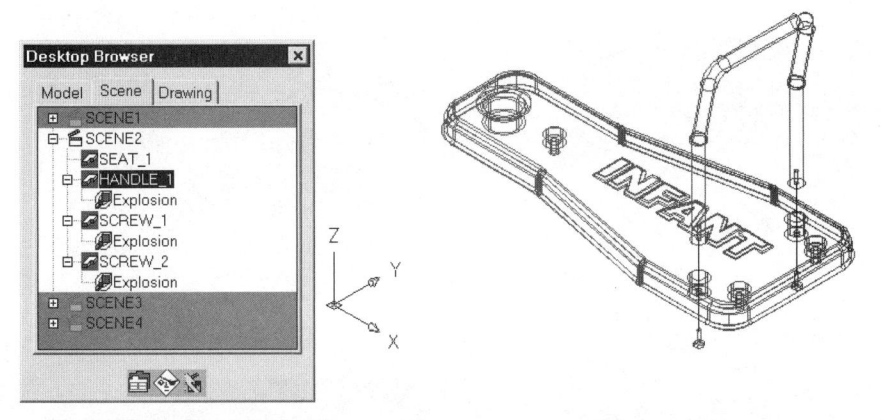

Figure 6-16 *Trail lines constructed*

SCENE MANIPULATION

Apart from constructing new scenes, you can copy an existing scene and improvise further. You can rename a scene as necessary. You lock a scene to prevent the positions of the components in the scene from accidentally being translated. When an assembly or the components of the assembly are modified, you update the scene. To construct sectional views across a set of components in an assembly, some component may need to be suppressed from sectioning to meet standard engineering drafting requirements. (For example, a shaft is normally not sectioned in an assembly if the section plane goes through the axis of the shaft.)

Assembly Scene   269

COPY SCENE

To save time in constructing a new scene, you can copy an existing scene and modify. Now you will copy Scene2.

27. Double-click Scene2 in the browser to activate it.
28. Select Scene2 from the browser, right-click, and select Copy (see Figure 6–17).
29. Press the ENTER key to accept the default name (Scene5).
30. Type **120** at the command line area to change the explosion factor to 120.
31. Press the ENTER key to activate the new scene.

A new scene is copied from Scene2. Note that the trail lines in Scene2 are inherited in Scene5.

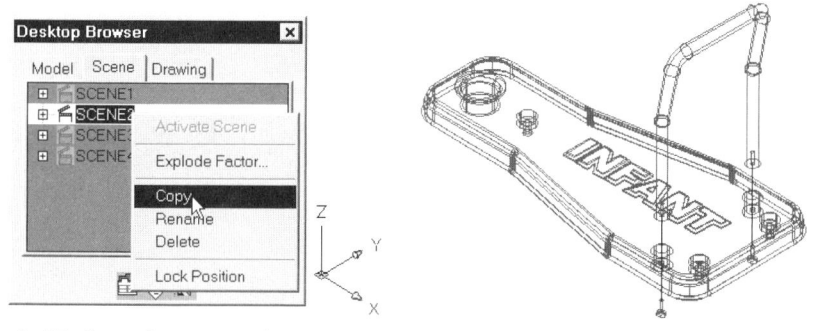

Figure 6–17 *Scene being copied*

RENAME SCENE

Suppose you want to change the name of a scene.

32. Select Scene5 from the browser, right-click, and select Rename (see Figure 6–18).
33. Type a new name (**SEATHANDLE**) in the browser.

The selected scene is renamed.

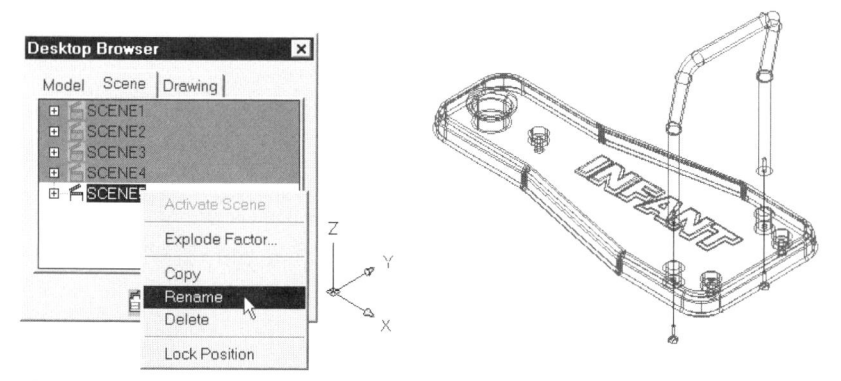

Figure 6–18 *Scene name being changed*

LOCK SCENE

To prevent the positions of the components from accidentally becoming displaced by exploding or tweaking, you can lock the scene. Now you will lock Scene4.

34. Select Scene4 from the browser and double-click to activate it.
35. Select Scene4 from the browser, right-click, and select Lock Position (see Figure 6–19).

The positions of the components in Scene4 are locked. To unlock the scene, select it from the browser, right-click, and unselect Lock Position.

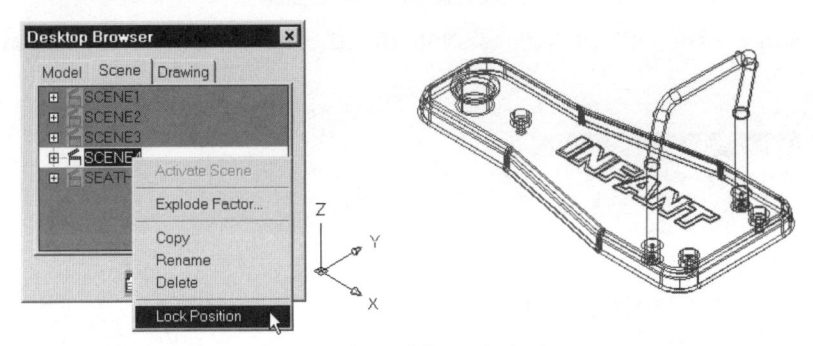

Figure 6–19 *Positions of components in Scene4 being locked*

UPDATE SCENE

To update a scene after the assembly or the components of the assembly are modified, you can select the Update Scenes button from the browser (see Figure 6–20).

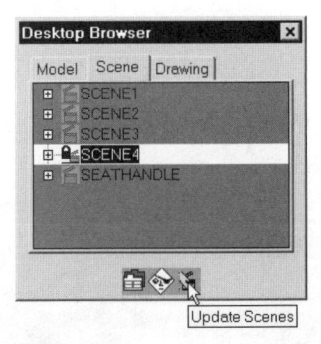

Figure 6–20 *Update Scenes button*

SUPPRESS SECTIONS

To prepare for a sectional engineering drawing of an assembly in which some components are not to be sectioned, you can construct a scene, activate the scene, select the components to be suppressed, and use the scene in engineering drawing construction. (You will learn how to construct an engineering drawing for an assembly in Chapter 11.)

To suppress a component from being sectioned, select the component from the browser, right-click, and select Suppress Sections (see Figure 6–21).

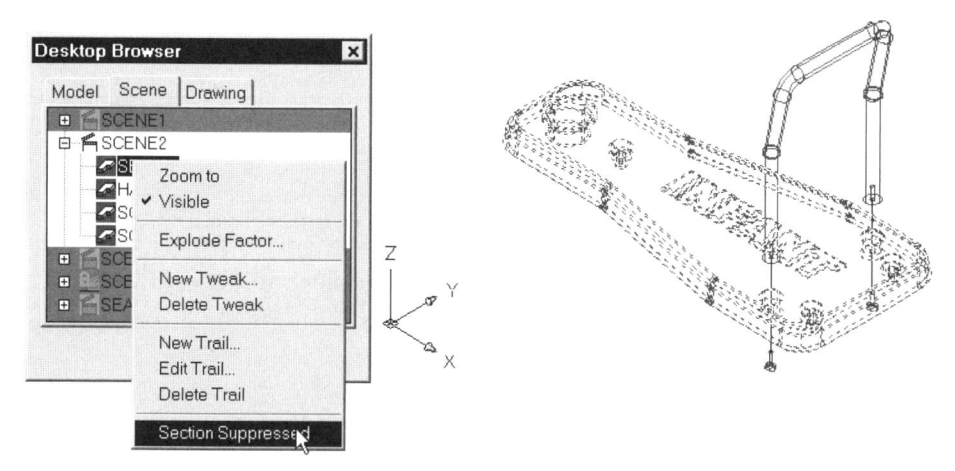

Figure 6–21 *Component being suppressed in a sectional view of an engineering drawing*

SCENE OPTIONS

Now you will take some time to study the scene options. Select the Options button from the browser (see Figure 6–22).

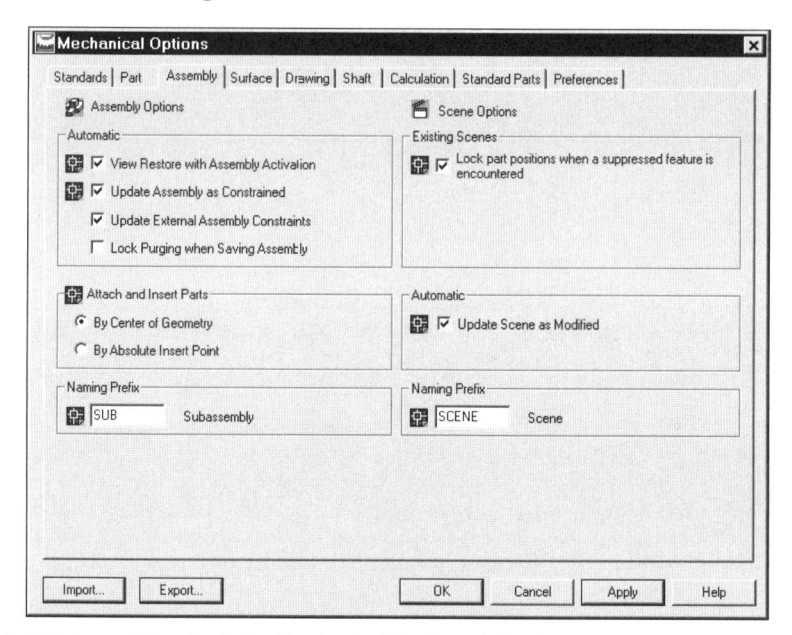

Figure 6–22 *Assembly tab of the Mechanical Options dialog box*

Scene options reside in the Assembly tab of the Mechanical Options dialog box. In the right column, there are three areas (Naming Prefix, Existing Scenes, and Automatic).

Existing Scenes	Lock part positions when a suppressed feature is encountered	determines whether the parts are locked at their current positions if a suppressed feature is encountered
Automatic	Update Scene as Modified	determines whether the exploded solid parts are moved apart by applying the explosion factor or tweaking
Naming Prefix	Scene	sets the prefix of the name of the scenes

The scenes of the SeatHandle assembly are complete. Save and close your file.

SUMMARY

After you have assembled a set of components together properly in an assembly file, you can set up a number of scenes. In the scenes, you hide some components, and explode or tweak the components apart in various ways to demonstrate how components are assembled. Because an assembly scene is a separate set of data in the assembly file, hiding components and exploding or tweaking the components in the assembly scene does not affect the integrity of the components in the assembly. Using assembly scenes, you can construct exploded engineering drawing views and suppress sectioning of components in sectional views.

PROJECTS

You will work on the following projects to enhance your knowledge.

FRONT END OF THE INFANT SCOOTER

Construct three scenes in the assembly of the front end of the infant scooter.

1. Open the file *FrontEnd.dwg* that you constructed in Chapter 5.
2. Right-click in the Scene tab and select New Scene to construct a new scene.
3. In the Create Scene dialog box, select TIREWHEEL from the Target Assembly pull-down box, set the explosion factor to 50, and select the OK button (see Figure 6–23).

An exploded scene of the subassembly is constructed (see Figure 6–24).

4. Construct two scenes for the FRONTEND assembly without a zero explosion factor.
5. Tweak one of the scenes in accordance to Figure 6–25.

Save and close your file.

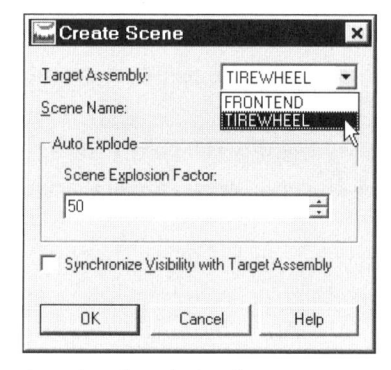

Figure 6–23 *Subassembly selected and explosion factor set*

Figure 6–24 *Exploded scene of the subassembly constructed*

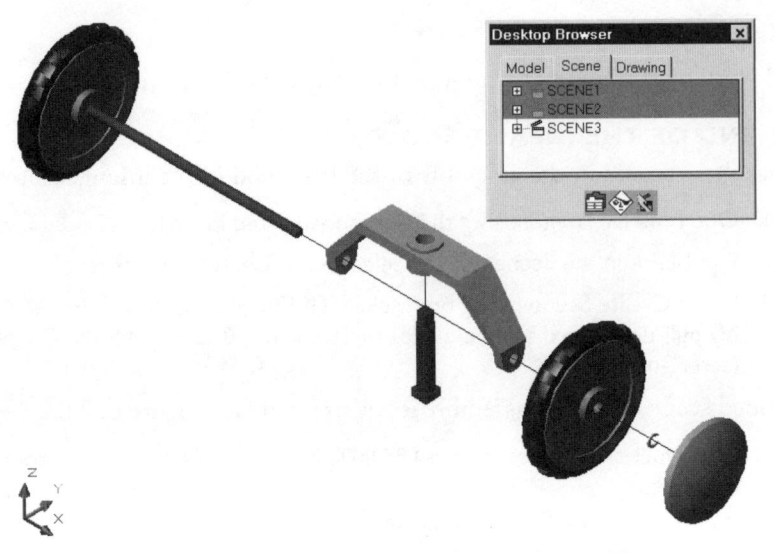

Figure 6–25 *Components tweaked in a scene of the front end assembly*

REAR END OF THE INFANT SCOOTER

Now you will construct two scenes for the rear end assembly of the infant scooter and tweak the components in a scene.

    1. Open the file *RearEnd.dwg* that you constructed in Chapter 5.

    2. Referring to Figure 6–26, construct two scenes and tweak the components in a scene.

Save and close your file.

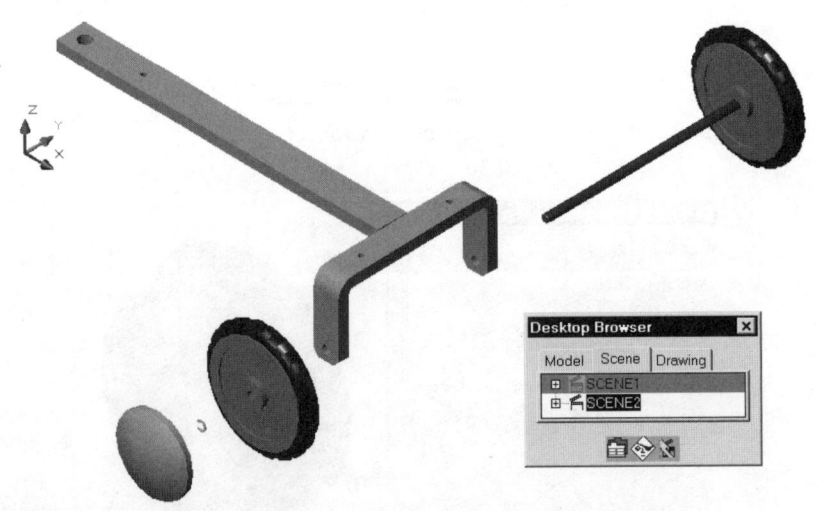

Figure 6–26 *Components tweaked in the scene of the rear end assembly*

INFANT SCOOTER

Now construct scenes for the infant scooter.

1. Open the file *InfantScooter.dwg* that you constructed in Chapter 5.
2. Construct two scenes, tweak the components in a scene, and add trail lines (see Figure 6–27).

Save and close your file.

Figure 6–27 *Tweaked components in the infant scooter*

REVIEW QUESTIONS

1. Explain what is meant by an assembly scene.

2. Describe how you will construct an assembly scene in an assembly file.

3. Explain the ways to displace the components of an assembly apart.

4. How will you add trail lines in the scenes of an assembly?

5. Components hidden in a scene will be invisible in all other scenes. True or false?

6. Trail lines are added manually in an exploded scene. True or false?

CHAPTER 7

Surface Modeling I

OBJECTIVES

The aims of this chapter are to delineate the key concepts of surface modeling, guide you towards mastering the techniques of constructing 3D wires, and familiarize you with the use of 3D wires in making 3D free-form surfaces and 3D free-form solids. After studying this chapter, you should be able to do the following:

- Describe the key concepts of surface modeling
- Construct and modify 3D wires
- Construct and edit 3D surfaces
- Construct 3D free-form solids from surfaces

OVERVIEW

In our daily lives, you encounter many objects with free-form body shapes. Examples include the handle of a razor, the casing of a computer pointing device, the casing of a mobile phone, and the body panels of an automobile. To construct these free-form objects as 3D models in the computer, you use the surface modeling tool.

The way to construct a surface in the computer is entirely different from the solid modeling method that you learned about in previous chapters. The fundamental way to construct a free-form surface is to construct a set of wires that depict the profile and silhouette of the surface and let the computer generate a surface on the framework of wires. In the projects in this chapter, you will see that the most tedious job in surface modeling is making the wires, and that making the surfaces from the wires is quite simple. You need only use the appropriate surface construction commands.

In general, surfaces can be categorized into three major types according to how they are constructed: primitive surfaces, free-form surfaces, and derived surfaces. After you construct a set of contiguous surfaces, you can stitch them together to form a quilt or a solid, thicken a surface or a quilt to become a solid, and use surfaces or quilts to cut an AutoCAD native solid or a Mechanical Desktop solid. Working in the opposite direction, you can convert a set of stitched surfaces, an AutoCAD native solid, and a Mechanical Desktop solid to a set of surfaces.

SURFACE MODELING CONCEPTS

A surface in a computer is a mathematical expression that represents a 3D shape with no thickness. There are two basic ways to represent a surface in the computer: using polygon meshes to approximate a surface and using complex mathematics to obtain an exact representation of the surface.

POLYGON MESH

Polygon mesh is an approximate method to represent a surface. It simplifies and reduces a smooth free-form surface to a set of planar polygonal faces and wire edges and silhouettes of the surface, to sets of straight line segments. Accuracy of representation is inversely proportional to the size of the polygon faces and line segments. A mesh with smaller polygon size represents a surface better but the memory required to store the mesh is larger. Figure 7–1 shows the polygon mesh of a scale model car body.

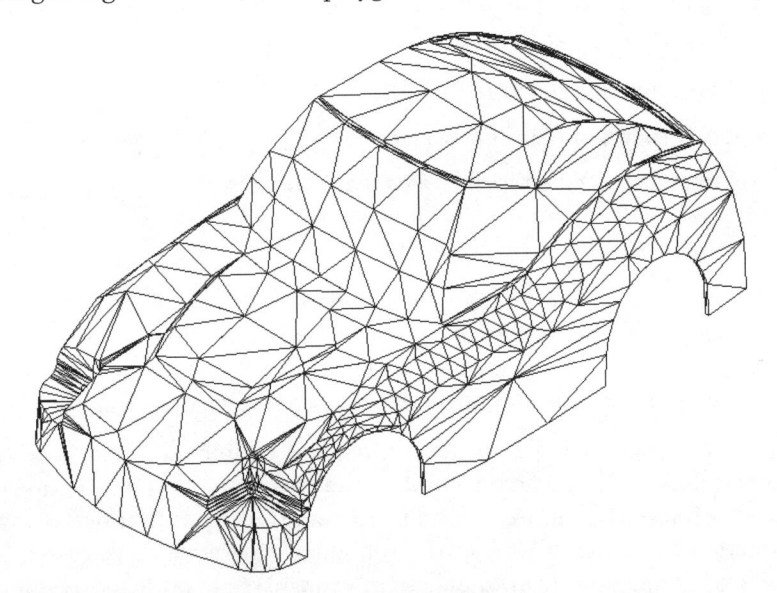

Figure 7–1 *Using polygon meshes to approximate surfaces*

A severe drawback of this method is that, despite using a very small polygon, it can never represent a surface accurately because the surface is always faceted. As a result, this method can only be used for visualization of the real object and cannot be used in most downstream computerized manufacturing systems. The AutoCAD surface tools construct surfaces by using polygon meshes. In this chapter, you will focus only on the second, more precise method: the NURBS surface.

NURBS SURFACE

To accurately represent free-form smooth surfaces in 3D design applications and computerized manufacturing systems, you use a higher order spline surface, called the Non-Uniform Rational B-Spline (NURBS) surface. NURBS mathematics is an advanced tool in surface modeling. A NURBS surface uses NURBS mathematics to define a set of control vertices and a set of parameters (knots). The distribution of control vertices together with the values of the parameters control the shape of the surface. The use of NURBS mathematics allows the implementation of multi-patch surfaces with cubic surface mathematics and maintains full continuity control even with trimmed surfaces. Figure 7–2 shows a NURBS surface model.

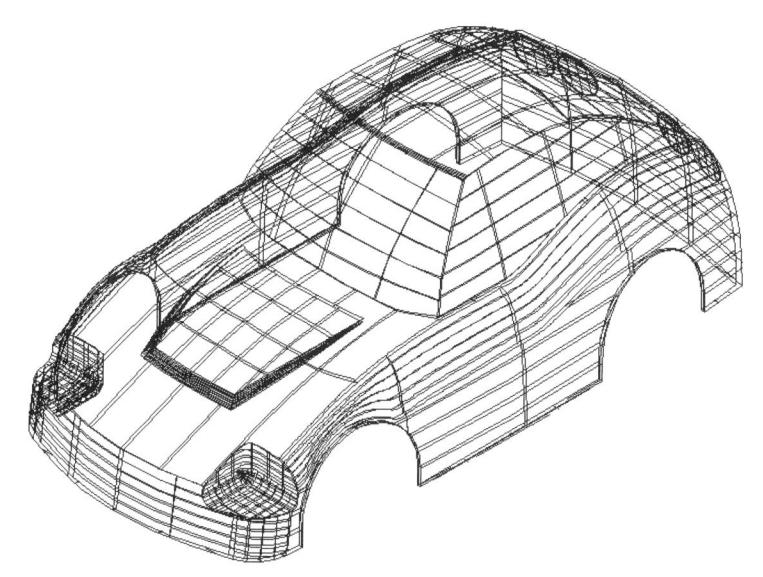

Figure 7–2 *NURBS surface model*

The surface modeling tool of Mechanical Desktop uses NURBS mathematics for the construction of spline wires and surfaces, and offers additional sophisticated tools for creating and editing wires for subsequent generations of NURBS surface models.

MODELING APPROACH

To construct a surface model, think holistically about the entire 3D object to be represented, analyze, and decompose the model into discrete elements of surfaces. Once you know the shape of the surfaces to construct, build the surfaces one by one. Figure 7–3 shows the surface model of a car body and Figure 7–4 shows the surfaces exploded apart. The fundamental way to build a free-form surface is to construct a framework of 3D wires and let the computer construct the surface. Because the work of framework construction is tedious, the process of surface modeling is quite time-consuming and requires considerable skills.

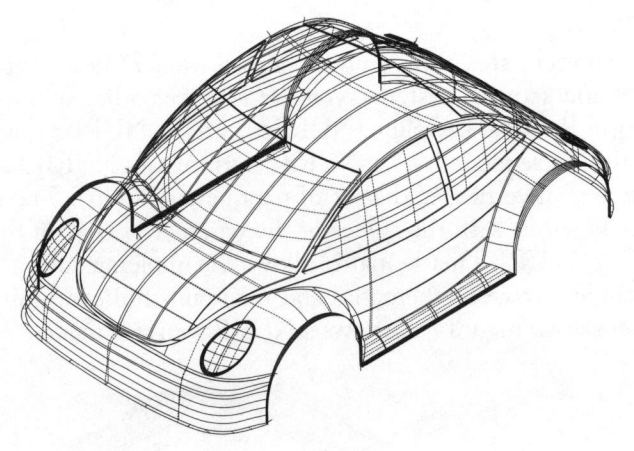

Figure 7-3 *Surface model of a car body*

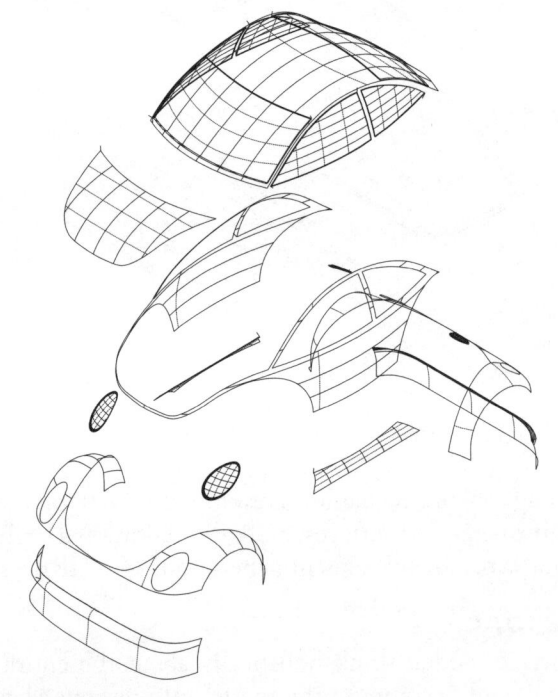

Figure 7-4 *Surfaces exploded apart*

SIGNIFICANCE OF SURFACE MODEL

In contrast to a solid part which is an integrated mathematical data for a 3D object, a surface model is a set of surfaces put together to resemble the boundary faces of the object. Each surface is an individual entity in the computer having edge, vertex, and surface data. The only drawback of a surface model is that there is no volume data. As compared to a solid model, a surface model can better represent complex 3D shapes because you can construct individual surfaces one by one and put them together. Figure 7–5 shows a rendered image of the free-form shape of the car body model. It is a set of individual surfaces.

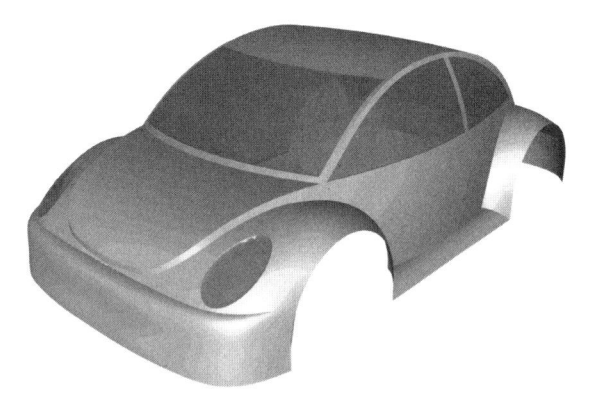

Figure 7–5 *Rendered image of the computer model of a car body*

Because you can retrieve the coordinates of any point on the surface of a model, you can use the model in most downstream computerized manufacturing operations such as CNC milling. Figure 7–6 shows the car model being machined on a computerized numerical control milling machine.

Figure 7–6 *A foam model being machined in a CNC milling machine*

VOLUME REPRESENTATION

Because a surface has no thickness, a surface model, being a set of surfaces, does not have any explicit volume data. To construct a solid with free-form body shapes, there are three ways:

- Thicken a surface
- Use a surface to cut a solid
- Stitch a set of surfaces enclosing an "air-tight" volume into a solid

The simplest way to incorporate volume data is to assign a thickness to a surface to obtain a solid with uniform thickness (see Figure 7–7). The solid constructed this way is a base solid feature. (You will learn how to edit base solid features in Chapter 9.)

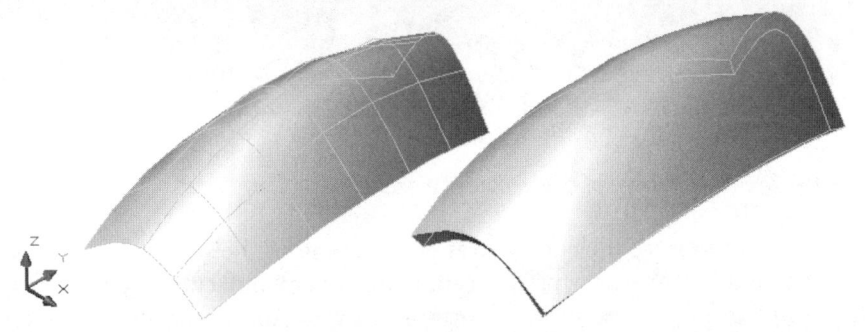

Figure 7–7 *Surface (left) assigned a thickness to become a solid of uniform thickness (right)*

The second way to construct a solid with a free-form shape is to use a surface to cut a solid (AutoCAD native solids or Mechanical Desktop parametric solids). Figure 7–8 shows an AutoCAD native solid cut by a surface. (You will learn how to use a surface to cut a Mechanical Desktop solid in Chapter 9.)

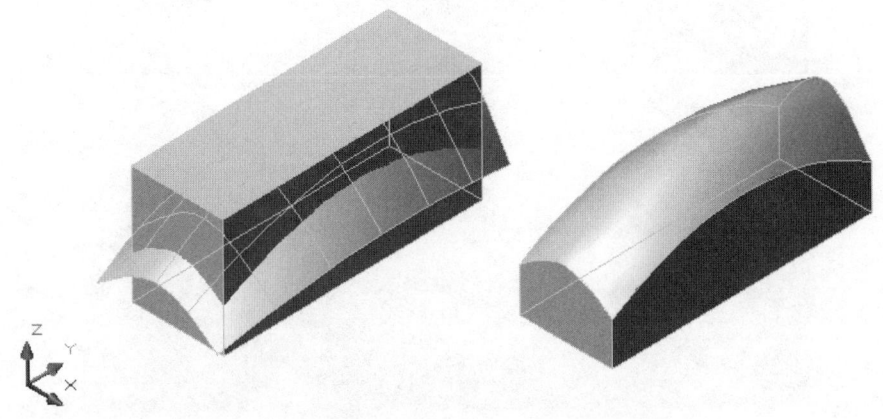

Figure 7–8 *Surface and solid (left), and solid cut by the surface (right)*

The third way to construct a free-form solid is to construct a set of surfaces to represent the outer boundary faces of a 3D object and stitch the surfaces to a solid.

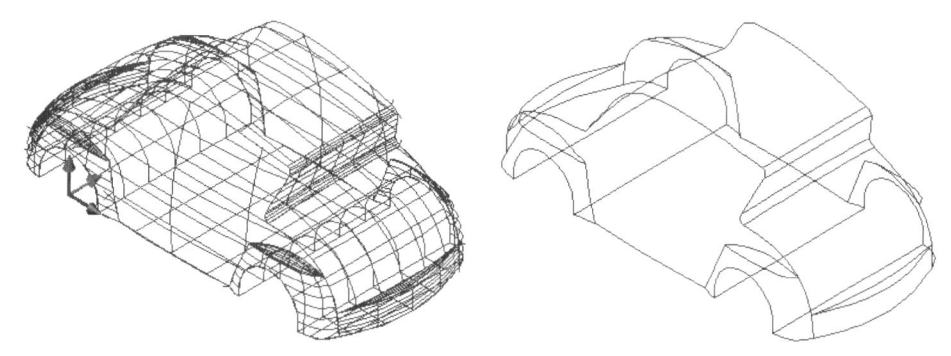

Figure 7–9 *A set of surfaces (left) and surfaces stitched to a solid (right)*

NURBS SURFACE CONSTRUCTION

Using Mechanical Desktop's surface modeling tool, you can construct three major kinds of NURBS surfaces: primitive surfaces, free-form surfaces, and derived surfaces. In addition, you can convert AutoCAD objects and Mechanical Desktop solid objects into sets of surfaces. To access Mechanical Desktop surface modeling tools, use the Surface toolbar or the Surface modeling toolbar. To display the toolbar, select Surface>Launch Toolbar (see Figure 7–10).

Figure 7–10 *Surface Modeling toolbar*

PRIMITIVE SURFACES

As the name implies, primitive surfaces are basic geometric surface shapes. To produce a primitive surface, you specify a geometric shape and state its dimensions and location. For example, if you want to produce a cylindrical surface, you need only specify the location of the center of the base, the diameter, and the height. These primitive surfaces, although easy to construct, have very limited use. Besides, equivalent parametric solid shapes are also available. If your design concerns only these kinds of shapes, you should use solid parts to represent your model.

Basic Shapes

Basic primitive surface shapes consist of conical surfaces, cylindrical surfaces, spherical surfaces, toroidal surfaces, and planar surfaces (see Figures 7–11 and 7–12).

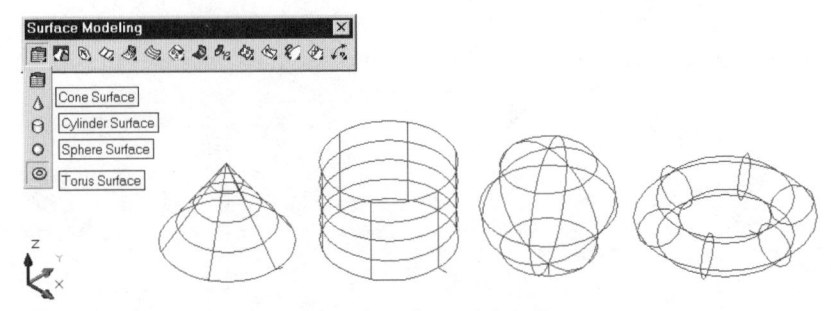

Figure 7–11 *Conical, cylindrical, spherical, and toroidal surfaces*

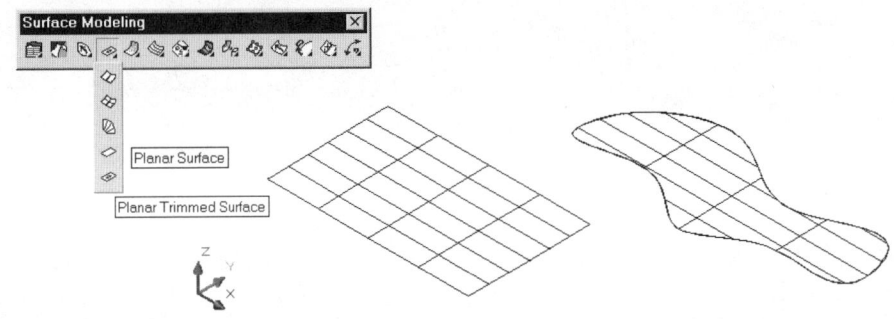

Figure 7–12 *Planar surfaces*

In essence, a planar surface is rectangular in shape. You select the command and specify two diagonal corners. To construct a trimmed planar surface, you construct a close loop curve and let the computer use the curve to trim the surface. (You will learn more about surface trimming later in this chapter.)

Connected Shapes

In addition to making individual primitive surfaces, you can make a tubular surface with straight sections and circular elbows from a series of connected cylindrical and toroidal surfaces. To make a tubular surface, construct a 3D polyline with a number of straight line segments. After that, you specify the diameter of the tube and the radii of the bends (see Figure 7–13).

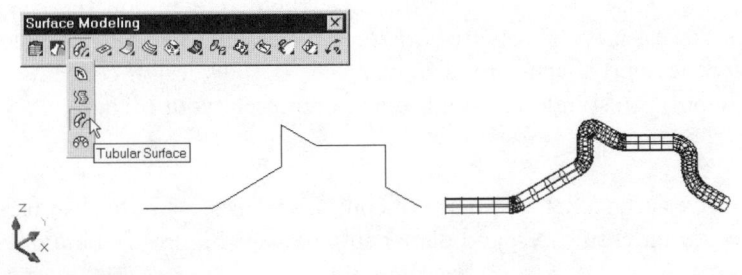

Figure 7–13 *Tubular surface*

FREE-FORM SURFACES

The most significant type of surface in free-form model making is the free-form surface. There are six kinds of free-form surfaces: the revolved surface, the extruded surface, the rule surface, the Loft U surface, the swept surface, and the Loft UV surface. To make a free-form surface, the first step is to construct a set of wires that define the profiles and silhouettes of the surface. Based on the wires, the computer then computes and constructs a set of surface data.

In general, a free-form surface needs two sets of wires in two orthogonal directions to define its profile and silhouette. To distinguish these two directions from the X axis and the Y axis, they are called the U direction and the V direction. Wires in these directions are called U-wires and V-wires, respectively.

Revolved and Extruded Surfaces

The revolved surface and the extruded surface are the simplest types of free-form surface because each needs only a single wire to define its shape. For a revolve surface, you construct a wire that defines the cross section along the axis of revolution. To make the surface, you specify a section wire, an axis of rotation, and an angle of rotation. Figure 7–14 shows a wire and a revolved surface constructed by rotating the wire 90° around an axis.

Similar to a revolved surface, an extruded surface also needs a single wire to define its cross section. To make the extruded surface, you specify a section wire, a direction of extrusion, and a taper angle. Figure 7–15 shows a wire and an extruded surface constructed by extruding the wire.

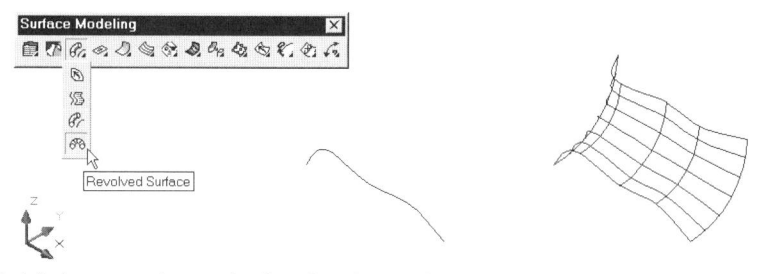

Figure 7–14 *A wire and a revolved surface by revolving the wire about an axis*

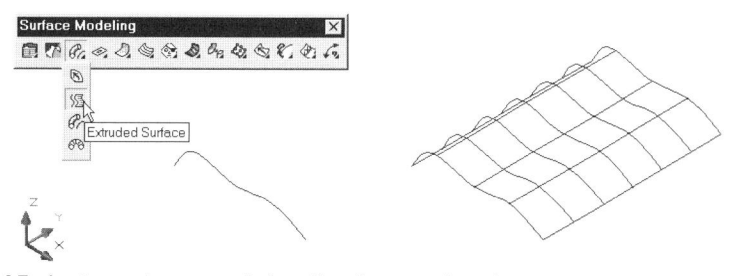

Figure 7–15 *A wire and an extruded surface by extruding the wire*

Rule Surface

To construct a surface that varies linearly in cross section from one edge to the other, you supply two U-wires. The resulting surface is called a rule surface. It changes in cross section uniformly from the first wire to the second wire in one direction. In the other direction, the cross sections are straight lines. Figure 7–16 shows two U-wires and a rule surface constructed from the wires.

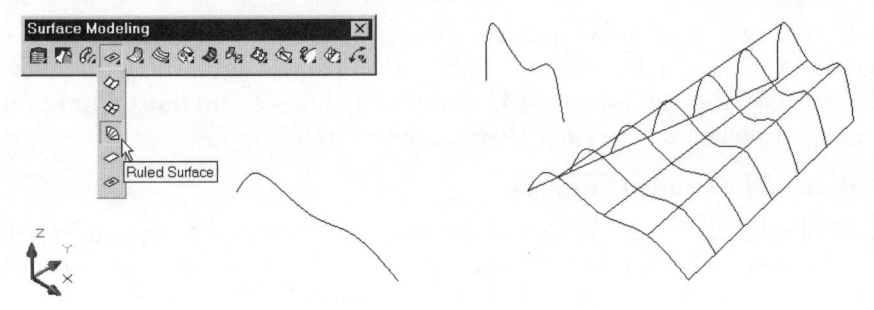

Figure 7–16 *Two U-wires and a rule surface by interpolating the wires*

Loft U Surface

To make a surface that has a variable cross section, you use three or more U-wires as input wires. The resulting surface is called a Loft U surface. Its cross section in the U direction will change smoothly from the first wire to the second, and then from the second wire to the third. If you have a fourth wire, the cross section changes from the third wire to the fourth as well. As a result, the cross section in the V direction also changes smoothly from one edge to the other. Figure 7–17 shows three U-wires and a Loft U surface constructed from the wires.

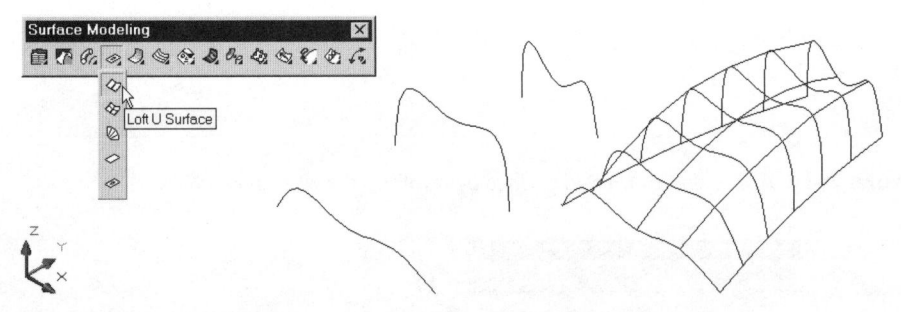

Figure 7–17 *Three U-wires and a Loft U surface by interpolating the wires*

Swept Surface (Single Rail)

In Figure 7–17, the cross sections of a Loft U surface in the V direction are smooth spline wires that pass through the U-wires. Their shapes are determined by the ways the U-wires change from one section to another. To exercise more control over the contours of the surface in the V direction, you can specify one or two rails for the U-wires to transit. The surface is called a swept surface. Figure 7–18 shows four wires (three U-wires as the cross sections and a V-wire as the single rail) and a swept surface constructed from the wires.

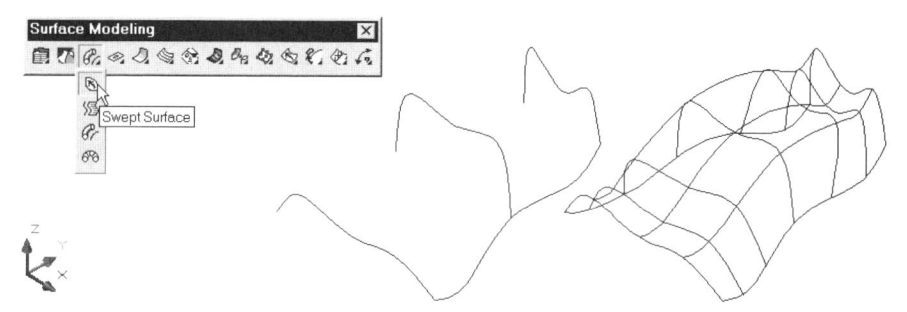

Figure 7–18 *Three U-wires and a V-wire, and a swept surface*

By comparing Figure 7–17 with Figure 7–18, you will see how the rail controls the transition of the U-wires as they change from one section to another.

Swept Surface (2-Rail)

In Figure 7–18, a single rail is used to control one end of each U-wire. To control both ends of the U-wires, you use two rails. When you use two rails, the U-wires' cross sections change in two ways. First, they change from one section to another. Second, they change shape according to the distance between the rails. Figure 7–17 shows five wires (three U-wires as cross sections and two V-wires as rails) and a swept surface. Compare Figure 7–19 with Figure 7–18 to see the difference.

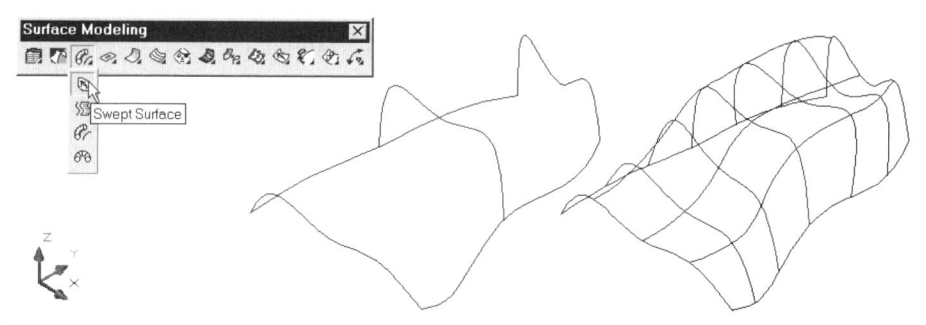

Figure 7–19 *Three section and two rail wires, and a swept surface*

Loft UV Surface

To control the V direction cross sections of a surface fully, you specify two sets of wires, one set of U-wires and one set of V-wires. The resulting surface is called a Loft UV surface. Figure 7–20 shows a Loft UV surface constructed from three U-wires and three V-wires.

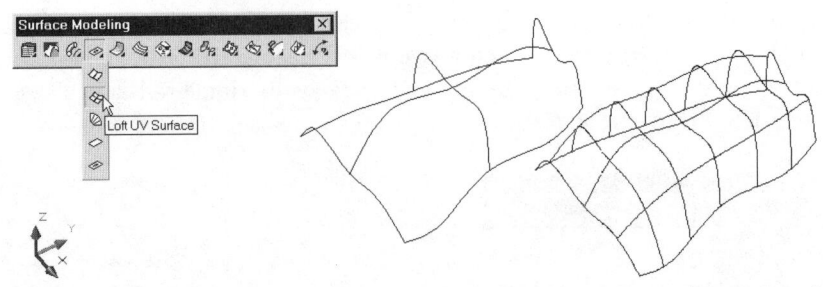

Figure 7-20 *Wires in two orthogonal directions and a Loft UV surface by interpolating in two directions*

DERIVED SURFACES

Derived surfaces include the fillet surface, corner surface, blend surface, and offset surface.

Fillet Surface

A fillet surface treats the edges of two intersecting surfaces by providing a connecting surface with a circular cross section. Figure 7–21 shows two surfaces and a fillet surface of constant radius formed between them.

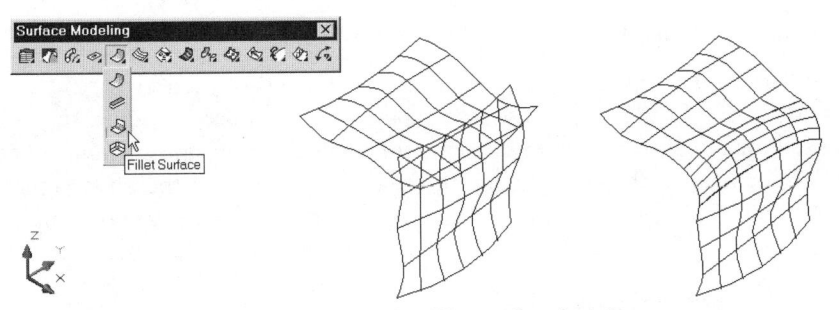

Figure 7-21 *Surfaces (left) and constant-radius fillet surface (right)*

When you make a fillet surface, you can set the fillet radius to vary linearly or cubically. In a linear variable fillet surface, the fillet radius changes linearly from one set value to another set value. In a cubical variable fillet surface, the fillet radius changes cubically from one set value to another. Figure 7–22 shows a pair of surfaces, a linear variable fillet, and a cubical variable fillet.

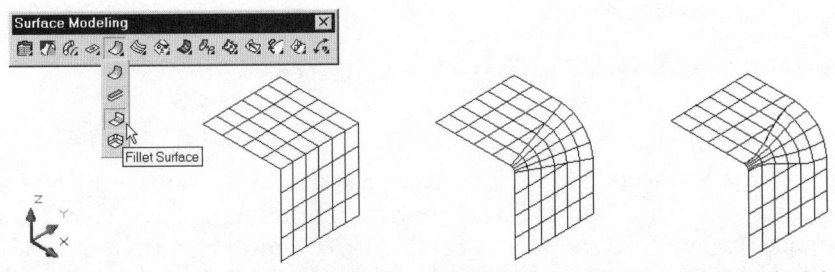

Figure 7-22 *Surfaces (left), linear variable fillet (middle), and cubical variable fillet (right)*

Corner Surface

A fillet surface is used to treat two intersecting surfaces. If you have three intersecting surfaces to treat, you form fillets in pairs, then treat the intersecting fillets with a corner fillet. Figure 7–23 shows three surfaces, fillet surfaces formed among the surfaces, and a corner fillet surface formed at the intersection of three intersecting fillet surfaces.

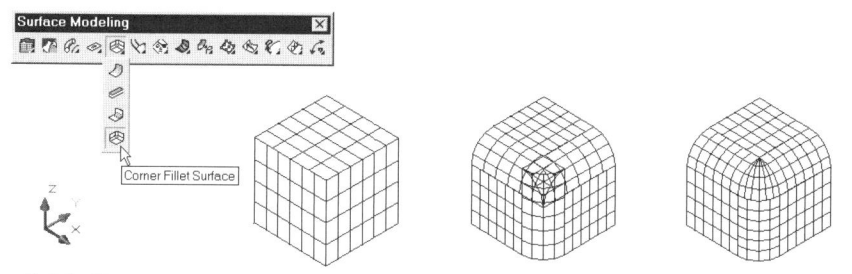

Figure 7–23 *Three surfaces (left), intersecting fillet surfaces (middle), and a corner fillet surface (right)*

Blend Surface

When you make a surface model, two adjacent surfaces do not always have to intersect. To treat the joint between two non-intersecting surfaces, fill in the gap between them by blending. Figure 7–24 shows a blend surface formed between two non-intersecting surfaces.

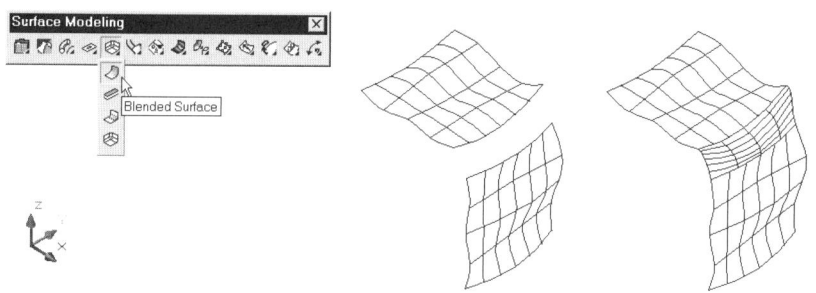

Figure 7–24 *Non-intersecting surfaces (left) and the blend surface formed between them (right)*

You can also blend three or four surfaces (see Figures 7–25 and 7–26).

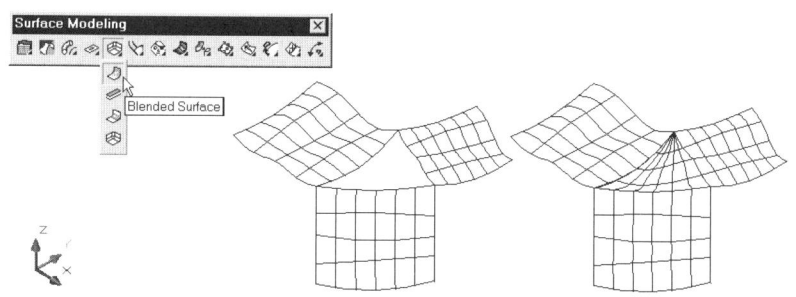

Figure 7–25 *Three surfaces (left) and a blend surface among edges of three surfaces (right)*

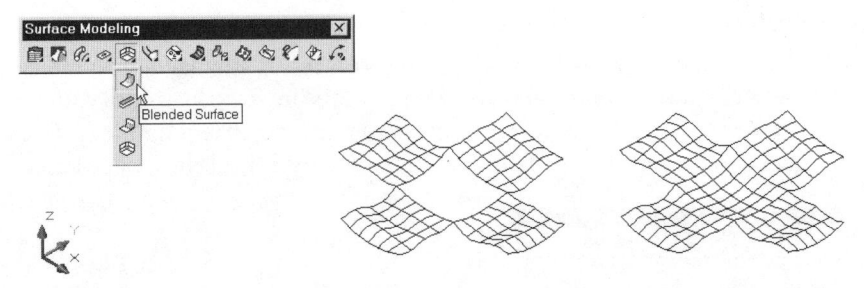

Figure 7–26 *Four surfaces (left) and a blend surface among edges of four surfaces (right)*

Offset Surface

Sometimes you may have to construct a surface to run at a constant distance from another surface. To make such a second surface, derive an offset surface from an existing surface. Figure 7–27 shows a Loft U surface and an offset surface derived from the Loft U surface.

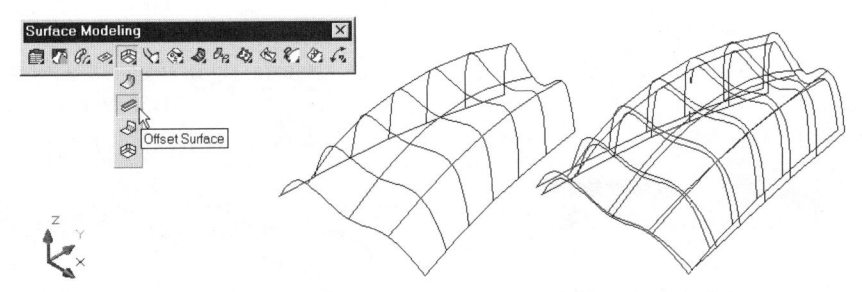

Figure 7–27 *Loft U surface (left) and offset surface and Loft U surface (right)*

CONVERSION

Apart from making the three basic kinds of surfaces (primitive surfaces, free-form surfaces, and derived surfaces), you can construct surfaces by conversion. You can convert a solid and AutoCAD arcs, circles, lines, and polylines with thickness to surfaces.

NURBS SURFACE EDITING

Surface editing tools can be categories into three groups. The first group concerns the change of the boundaries of a surface, the second group concerns the change of the surface's profile and silhouettes, and the third group concerns the surface's normal direction and solid operations.

BOUNDARY EDITING

A smooth free-form surface needs to be constructed from smooth defining wires. Smooth surfaces with irregular edges occur in many designs. If you use the irregular edges to construct the surface directly, you will get a surface with many sudden changes in curvature. The surface will not be smooth at all. To obtain a smooth surface with irregular edges, build a larger surface from smooth wires. Then you trim the smooth surface with the irregular edges. The resulting surface is a trimmed surface.

Contrary to trimming, you untrim a trimmed surface to revert the surface to its untrimmed state. The reason why it is possible to revert a trimmed surface to its untrimmed state is because the original untrimmed surface is also stored in the database.

To reduce the memory requirement to store a trimmed surface, you reduce the original untrimmed surface to its minimum size.

A surface can be broken into two surfaces and you can join two untrimmed surfaces into a single surface. If you have a set of contiguous surfaces, you stitch them to a quilt.

Trimmed Surfaces

To produce a smooth surface, it is necessary to use smooth defining wires and smooth boundary lines. However, most of the surfaces that we use to compose a design do not necessarily have smooth boundaries, although they have smooth profiles. Figure 7–28 shows the surface model of an automobile body panel. This is a smooth surface, but its boundary is irregular.

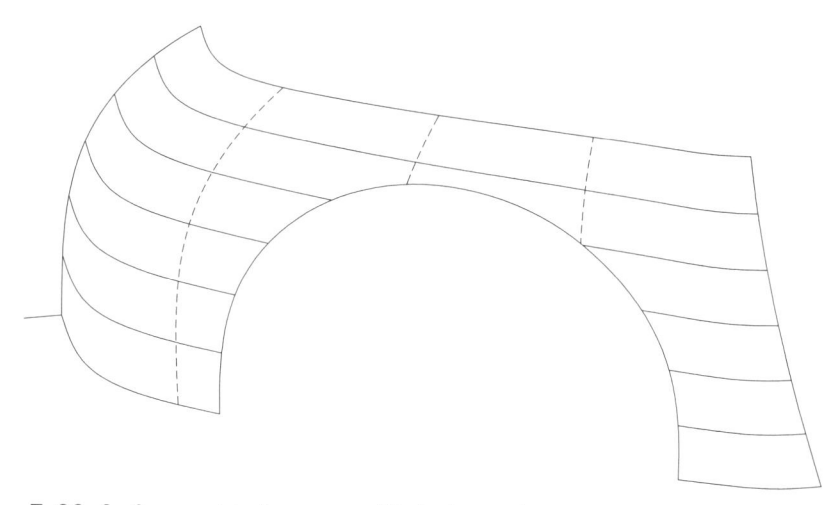

Figure 7–28 *Surface model of an automobile body panel*

Given this problem, you might intuitively use the boundary wires that you see as the defining wires to construct the surface model. If you did, you would probably get an irregular surface like the one shown in Figure 7–29.

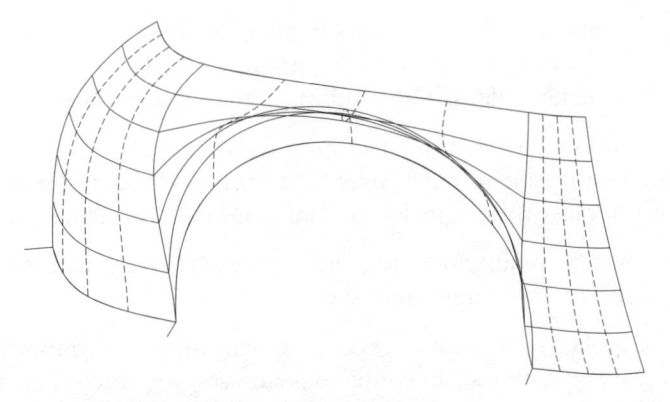

Figure 7–29 *Irregular surface defined by irregular boundaries*

Obviously, the surface shown in Figure 7–29 is not the one we want (which is shown in Figure 7–28). What has gone wrong? The answer is that a set of irregular wires will generate an irregular surface. Unless the boundary lines are smooth wires, they cannot be used as defining wires for the surface.

To obtain a smooth surface with an irregular boundary, you have to perform two steps. You use a set of smooth wires to produce a smooth surface that is much larger than the required surface. This is called the base surface. Then you use the irregular boundary wire to trim the smooth surface.

In the computer, a trimmed resulting surface consists of the original untrimmed surface (base surface) and the trim boundaries (trimmed edges). Although both of these are saved in the database, only the boundary and the remaining part of the trimmed surface are displayed. As a result, we obtain a smooth free-form surface with irregular boundaries. This is called a trimmed surface.

To produce a free-form surface that is large enough for subsequent trimming, define a set of wires that encompass the required surface. To make such wires, you need to be able to visualize the defining wires that are outside the required surface. In Figure 7–30, the construction of the smooth automobile body panel starts from a set of smooth wires. From the smooth wires, a smooth surface that is much larger than the required surface is made. To obtain the required surface, an irregular boundary is used to trim the large smooth surface.

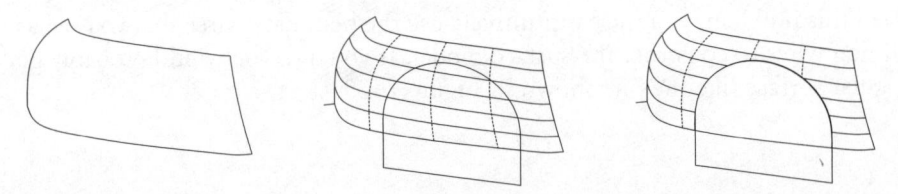

Figure 7–30 *Defining wires (left), the untrimmed surface (center), and the irregular boundary (right)*

There are two ways to trim a surface:

- Project and trim
- Intersect and trim

The first way to trim a surface is to construct a wire, and project the wire to trim the surface. Figure 7–31 shows a Loft U surface trimmed by a circle.

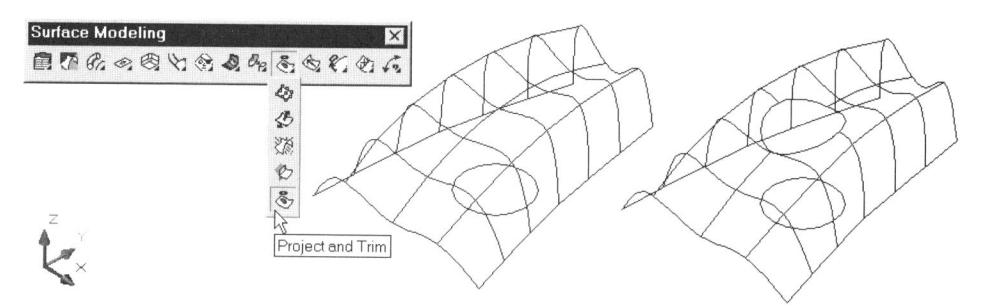

Figure 7–31 *Surface and wire (left) and surface trimmed by projecting the wires (right)*

Another way to trim a surface is to construct two intersecting surfaces and trim the unwanted portions of the surfaces away to form a sharp edge at the intersection. Figure 7–32 shows two intersecting surfaces trimmed.

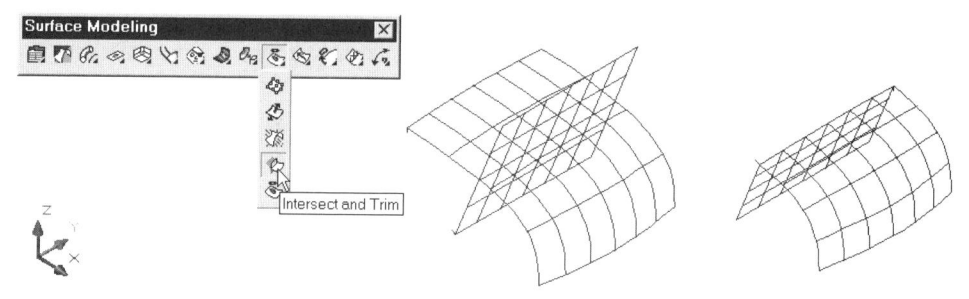

Figure 7–32 *Two intersecting surfaces (left) and trimmed surfaces (right)*

Untrimming a Surface

We all make mistakes and sometimes change our minds. You may need to change a trimmed surface back to its original untrimmed state. Because a trimmed surface in the computer consists of the original base surface and the trimmed edge, you remove its trimmed boundary to change it back to its untrimmed state. In Figure 7–33, the trimmed surface (left) has a number of trimmed boundaries. Untrimming it removes all the trim boundaries and returns it to its untrimmed state (right). Sometimes you may only want to remove one of the trim boundaries instead of removing all the boundaries. To remove only selected trimmed boundaries, you extract them. You will learn how to extract trim boundaries later in this chapter while working on 3D wire construction.

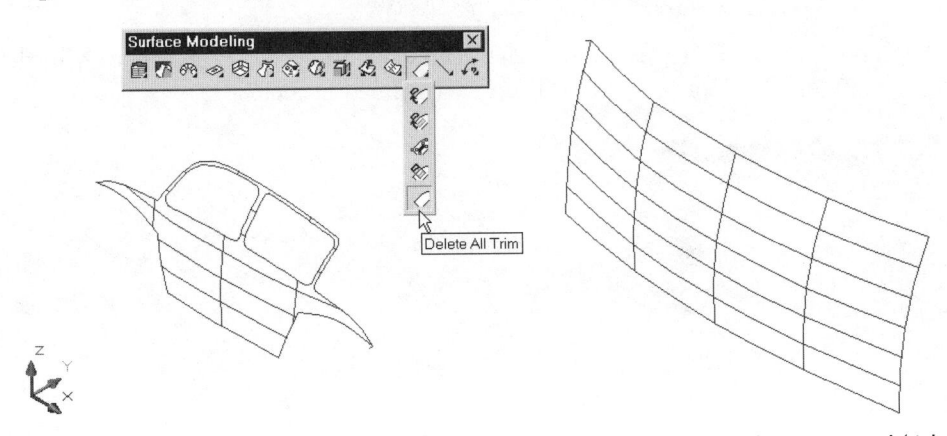

Figure 7–33 *A trimmed surface (left) and the surface with its trimmed boundaries removed (right)*

Truncating a Trimmed Surface

A trimmed surface retains its original smooth defining boundaries in the database while possessing a new trimmed boundary. The original surface is called the base surface, and the trimmed boundary is called the trim edge. Sometimes you may use a base surface that is much larger than required. If so, unnecessary memory space is wasted to store the unwanted part of the base surface. To reduce the memory used, you truncate the base surface of a trimmed surface. Figure 7–34 shows the original base surface and the truncated base surface.

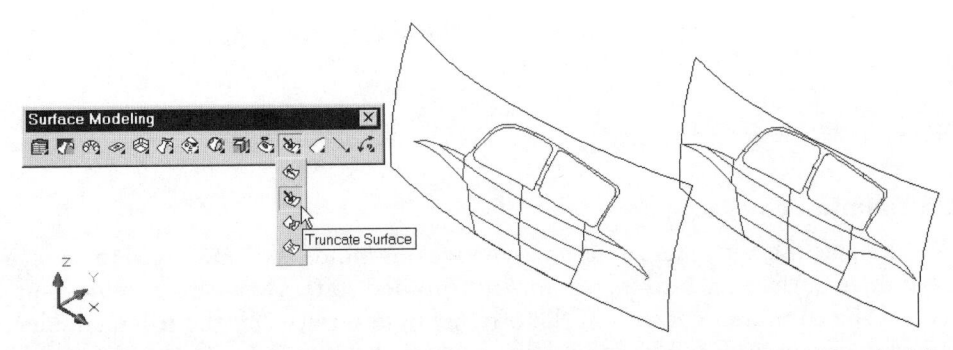

Figure 7–34 *Trimmed surface with a large base surface (left) and a truncated surface (right)*

Breaking a Surface

A surface can be broken into two surfaces along a selected U or V line. After breaking, the surfaces still maintain the original continuity. In Figure 7–35, the profiles

Surface Modeling I **295**

and silhouettes of the broken surfaces (right side) will be the same as those of the original surface (left side).

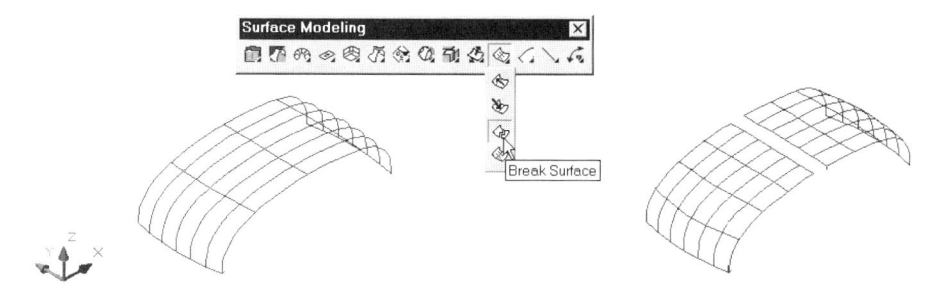

Figure 7–35 *A single surface (left) broken into two surfaces and moved apart (right)*

Joining Two or More Surfaces

If you have two or more contiguous surfaces with shared untrimmed edges, you can join them together to form a single surface (see Figure 7–36).

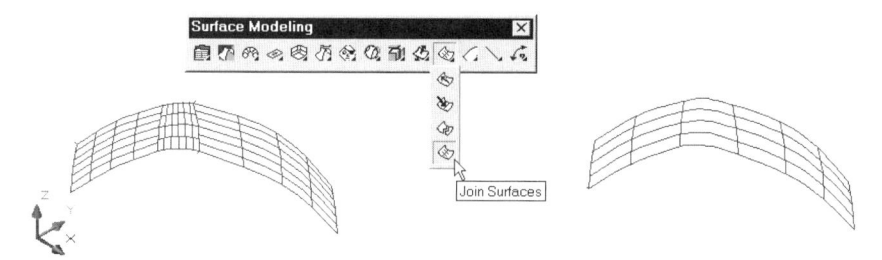

Figure 7–36 *Three surfaces (left) joined together to form one single surface (right)*

Stitching to a Quilt

Surfaces can only be joined if their contiguous boundaries are untrimmed surface boundaries. If the contiguous surfaces are trimmed surfaces, you may stitch them together to form a quilt. A quilt is a set of connected surfaces. You will not find any visual difference after a set of surfaces is stitched to a quilt (see Figure 7–37).

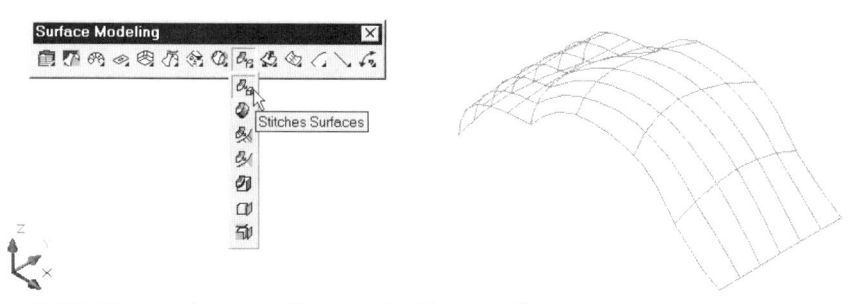

Figure 7–37 *Two contiguous surfaces stitched into a quilt*

Using a quilt, you can intersect a set of connected surfaces with another surface or another quilt. Figure 7–38 shows a quilt intersected and trimmed by a surface.

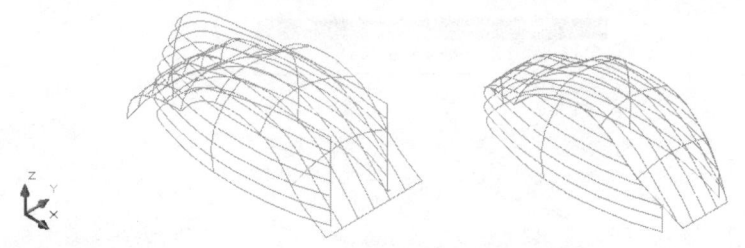

Figure 7–38 *Quilt and a surface intersected (left) and trimmed (right)*

PROFILE EDITING

The profile and silhouettes of a surface are determined by the profiles of the curves and the method you use to construct the surface from the curves. Because the surface modeling tool is not parametric, there is no relationship between the original curves and the surface constructed from the curves. One way to modify the shape of the surfaces is to delete the surface, modify or reconstruct the curves, and build the surfaces again. However, this can be very time-consuming. To make minor modifications to the profile of the surface, you may lengthen an untrimmed edge to enlarge it, scale the surface either to enlarge or reduce its size, replace an edge of the surface with another edge, refine the surface patch, change the grip points, and adjust the span of the grip points.

Lengthening a Surface

You can lengthen a surface along the untrimmed edge. Note that a trimmed surface cannot be lengthened because the trimmed edge is a boundary within the base surface. Figure 7–39 shows a lengthened surface.

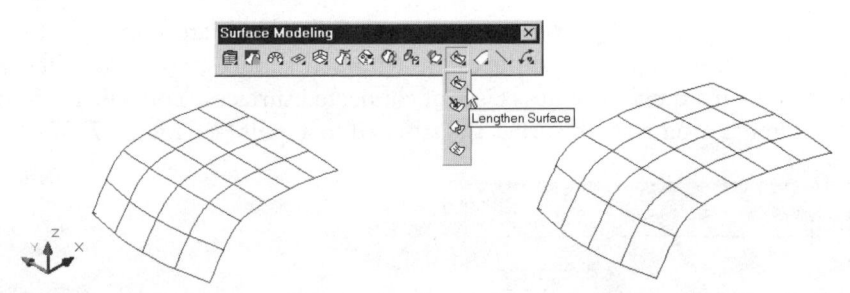

Figure 7–39 *Original surface (left) and untrimmed edges of the lengthened surface (right)*

Scaling a Surface

To change the size of a surface and yet maintain the overall proportions of the profiles and contours, you scale it in 3D. The X, Y, and Z scale can be different. Figure 7–40 shows a set of surfaces scaled.

Surface Modeling I 297

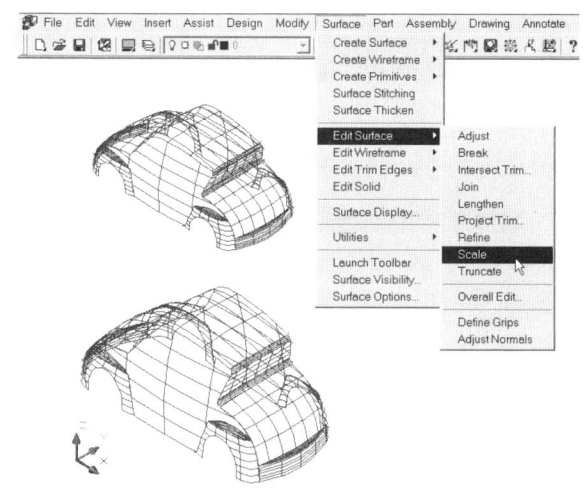

Figure 7–40 *Original set of surfaces (top) and scaled surfaces (bottom)*

Adjusting the Edges of Two Surfaces

The gap between two untrimmed surfaces can be closed by adjusting the edges of the surfaces. The effect of edge adjustment is very similar to making a blended surface. Figure 7–41 shows a portion of the adjacent surfaces modified.

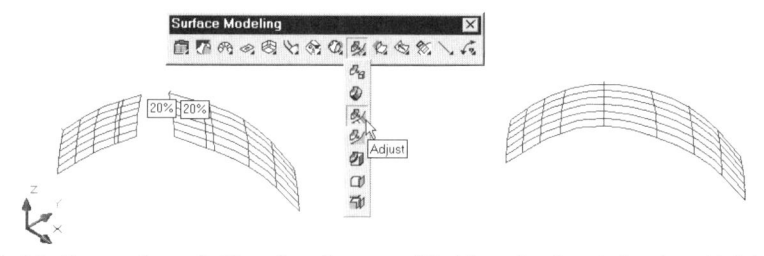

Figure 7–41 *Two surfaces (left) and surfaces modified by adjusting their edges (right)*

Replacing Surface Edge

Similar to edge adjustment, edge replacement also fills the gap between two surfaces. Unlike with edge adjustment, only one surface is modified (see Figure 7–42).

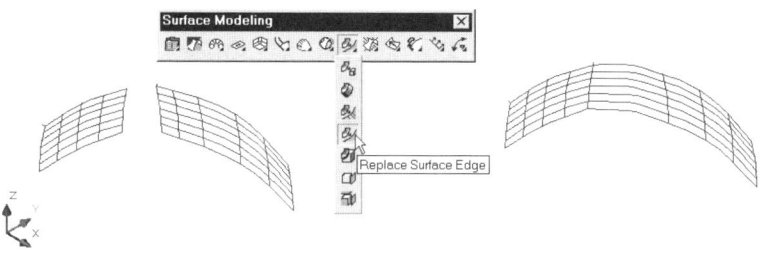

Figure 7–42 *Two surfaces (left) and a surface's edge replaced by the edge of another surface*

Refining the Patches of a Surface

The accuracy of a surface in relation to the input wires is determined by the number of UV patches used to construct the surface. You refine a surface by changing its UV patches. Reducing the number of patches decreases the accuracy of a surface. Figure 7–43 shows the effect of reducing the UV patches of a surface.

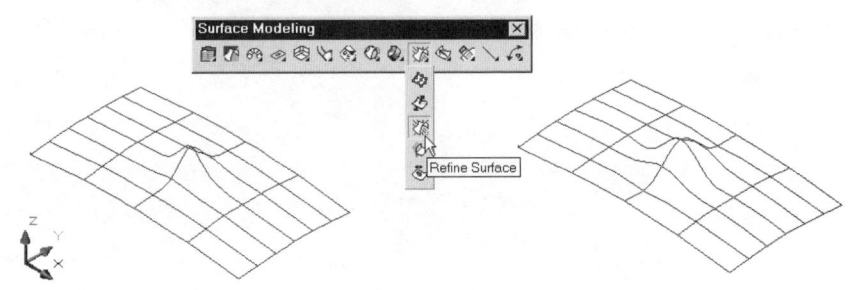

Figure 7–43 *Original surface (left) and patch number changed (right)*

Changing the Grip Points of a Surface

Grip points on a surface are locations where you can grip and move to a new position in order to change the surface's profiles and silhouettes. Figure 7–44 shows two different grip point settings.

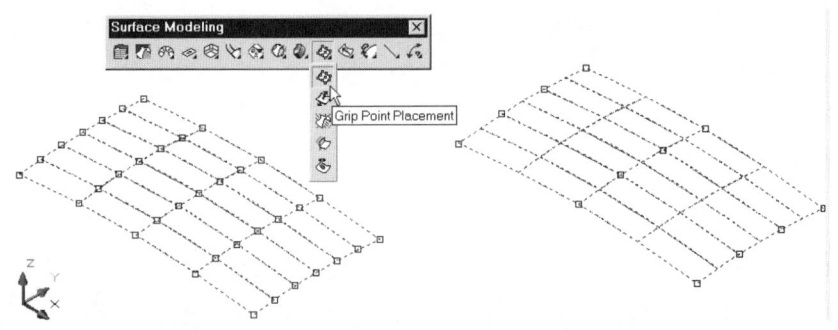

Figure 7–44 *Grip point settings changed*

Modifying the Span of a Surface

When you move a grip point, a circular area of the surface will deform. The radius of this circular area is called span. Changing the span affects the way the surface is deformed when a grip point is manipulated. Figure 7–45 shows the effect of grip deformation on two surfaces with different span size.

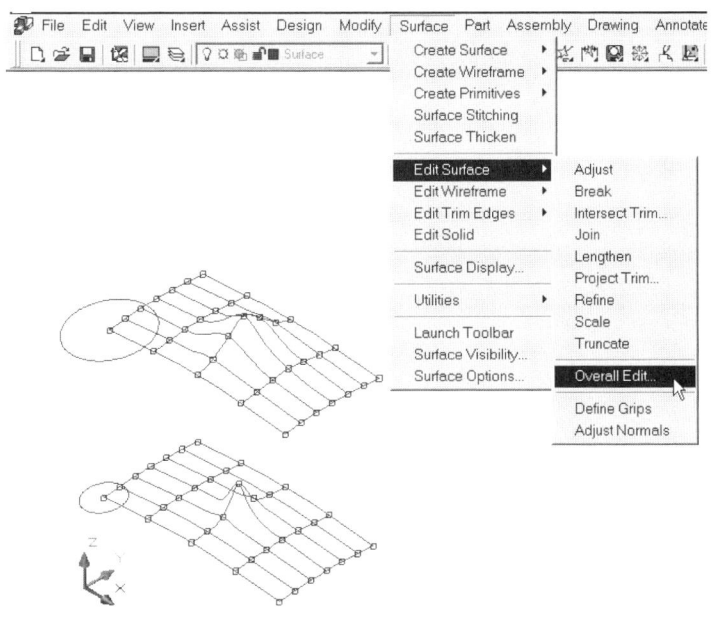

Figure 7–45 *Small span size (top) and large span size (bottom)*

NORMAL DIRECTION

A surface has no thickness. To represent a 3D object in a computer, you need a number of surfaces. For the computerized downstream manufacturing operations to recognize which side of the surface represents a void and which side represents a volume, a normal vector is used. The direction of normal is determined by the direction of the curves, the curve patterns, and also the sequence of selection when you construct a surface from the curves. This sounds too complicated for us to memorize. Hence, you may simply disregard the normal direction when you first construct the surface and flip the normal in a later stage. The normal direction of a surface is depicted by a line normal to and at the corner of the surface. Figure 7–46 shows two surfaces with different normal directions.

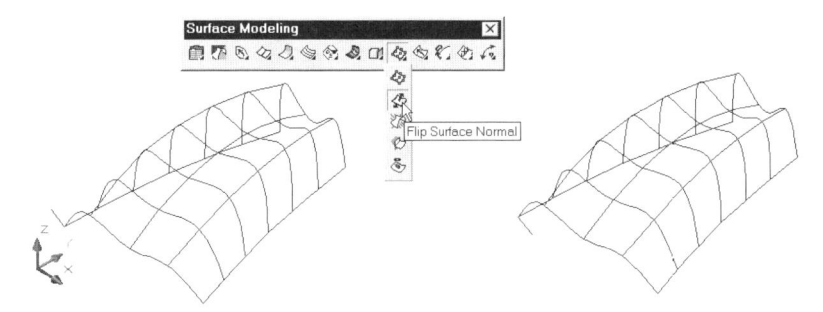

Figure 7–46 *Two surfaces with different normal directions*

SOLID OPERATION

There are three ways to use surfaces in construction of solid parts: thicken a surface by assigning a thickness (see Figure 7–47); use a surface to cut an AutoCAD native solid or a Mechanical Desktop solid (Figure 7–48 shows a surface cutting an AutoCAD native solid); or stitch a set of surfaces that encloses an "air tight" volume into a solid (see Figure 7–49).

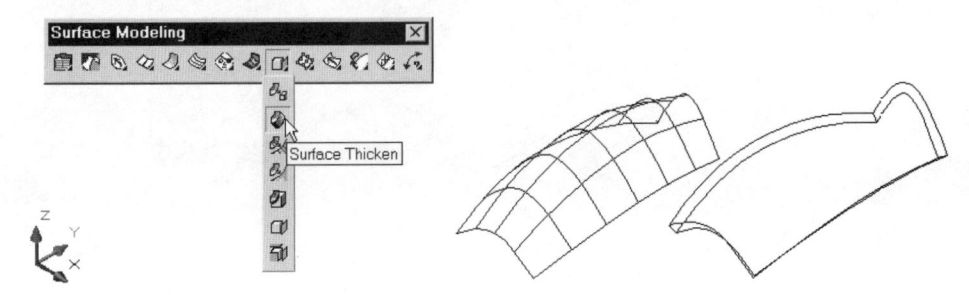

Figure 7–47 *Thickening a surface*

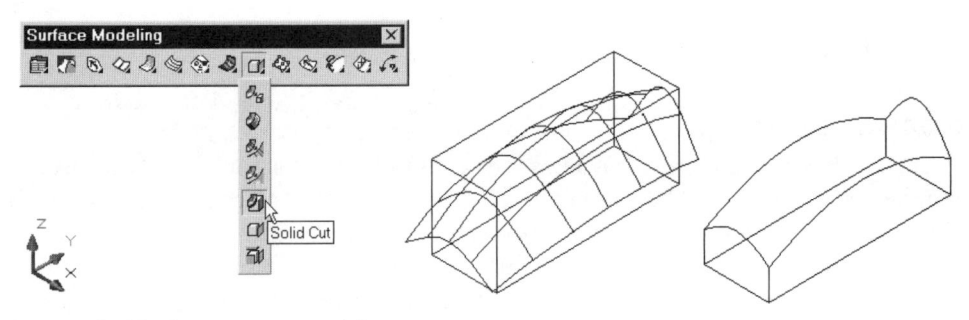

Figure 7–48 *Cutting a native solid*

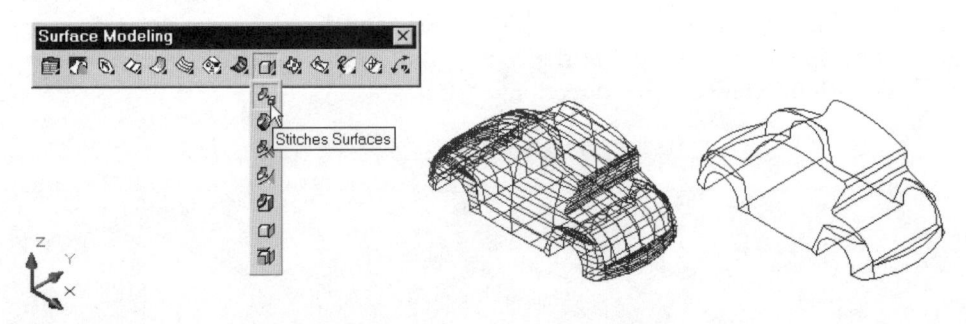

Figure 7–49 *Stitching a set of surfaces*

CONVERSION

You can convert the following objects into individual NURBS surfaces.

- Arcs, circles, lines, and polylines with thickness
- 3D polygon meshes
- AutoCAD solids and bodies
- Mechanical Desktop solid parts
- Surface quilt

Figure 7–50 shows a solid converted to a set of surfaces.

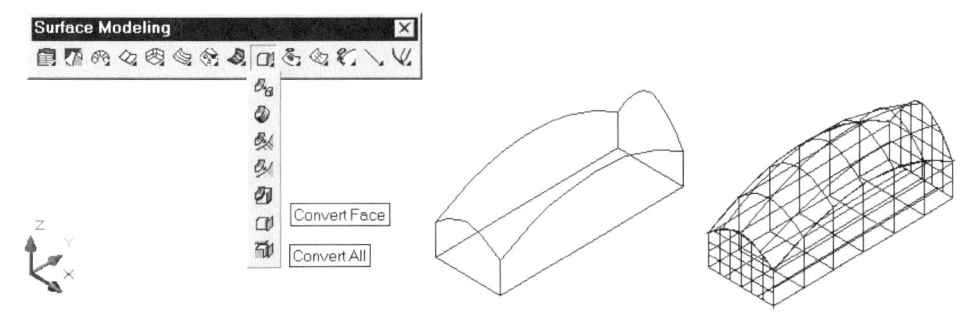

Figure 7–50 *Solid (left) converted to a set of surfaces (right)*

3D WIRES CONSTRUCTION

To make free-form surfaces, derived surfaces, and trimmed surfaces, 3D wires are needed. Basically, you use AutoCAD tools to construct splines, lines, arcs, and ellipses as wires. In addition, you use Mechanical Desktop tools to perform the following tasks:

- Construct augmented lines
- Construct a wire by joining existing wires
- Construct splines from existing wires
- Construct a spline tangent to an existing spline
- Construct a 3D wire by offsetting a 3D wire
- Construct 3D fillet wire
- Construct a wire at the intersection of two surfaces
- Construct a wire by projecting a wire onto a surface
- Copy the edges of a surface
- Extract the edges of trimmed boundaries

- Generate a set of section lines across a surface
- Produce a set of flow lines along the U and V directions of the surface
- Construct a parting line along a surface

AUGMENTED LINES

An augmented line is a special kind of wire along which normal vectors are placed at regular intervals. You make use of the normal vectors of the augmented lines to control machines that operate in 4- or 5-axes. You also take an augmented line as a rail and use its normal vector to guide the transition of the section wires for making a swept surface. Figure 7–51 shows augmented lines generated along UV surface lines of a free-form surface.

> **Note:** The length of the augmented line vector is unimportant. They only indicate a direction.

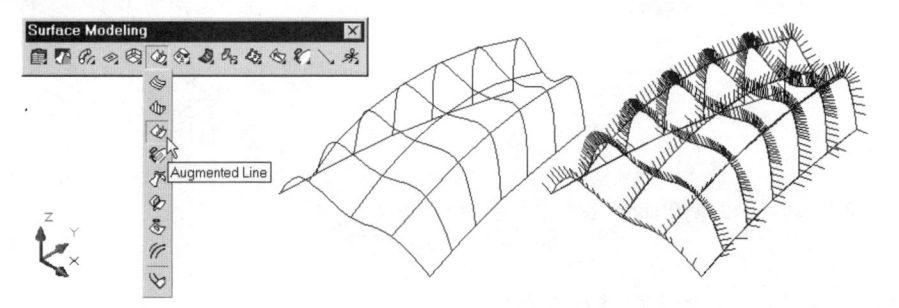

Figure 7–51 *Surface (left) and augmented lines constructed along the UV lines of the surface (right)*

JOINING WIRES

To construct a wire in accordance with a specific shape you have in mind, you may find it easier to construct lines, arcs, and circles than to construct splines. To use connected wires of this kind, you join them together to form a 3D polyline, spline, or augmented line. Joining can be set to manual or automatic. Figure 7–52 shows three curves manually joined into a single spline (note the gap between the source wires).

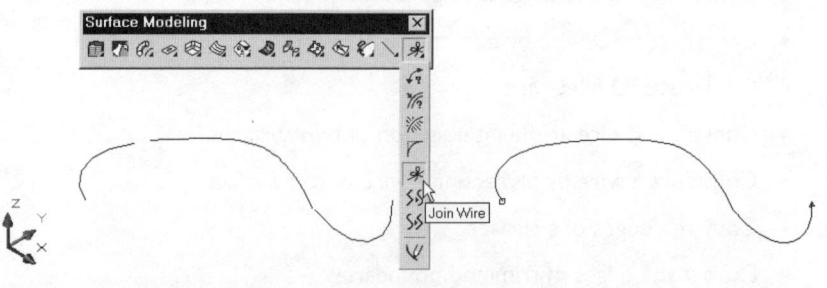

Figure 7–52 *Separate wires (left) joined into a single wire (right)*

SPLINE FITTING

Instead of joining wires together, you fit a polyline/line/arc into a spline. Both joining and fitting create splines from existing wires. The difference between them is that joining constructs a single spline from connected wires and fitting constructs a spline for each selected wire. There are two fit modes: tolerance and control points. Using tolerance mode, construct a spline using the maximum chordal deviation between the resulting spline and the vertices of the input polyline. Using control point mode, construct a spline using a specified number of control points. Figure 7–53 shows a polyline fitted into a spline by using control point mode.

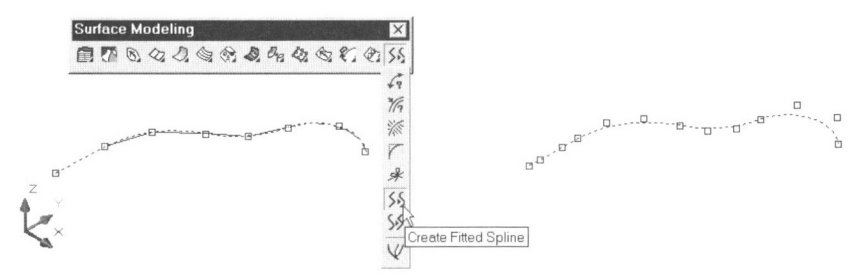

Figure 7–53 *Polyline (left) fitted to a spline (right)*

After an object is used in an operation, such as fit spline operation here, you can keep or delete the original object by using the DELOBJ variable. To toggle between deleting or retaining the original object in an operation, use the Toggles Keep Original button of the Surface Modeling toolbar before using the operation (see Figure 7–54).

Figure 7–54 *Toggles Keep Original button*

TANGENT SPLINE

If you have already constructed a spline, you can construct a spline with its starting point tangent to the existing spline (see Figure 7–55).

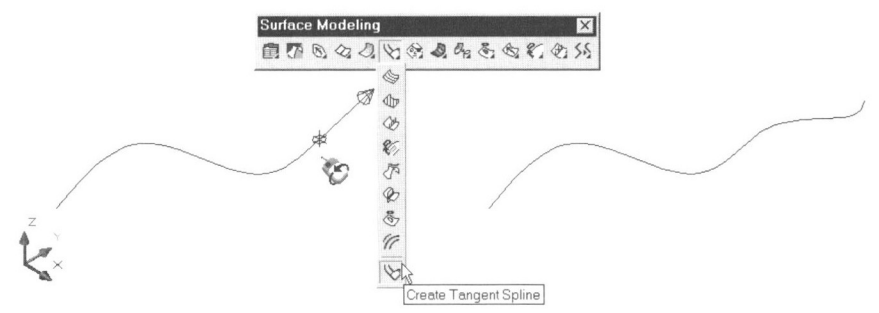

Figure 7–55 *Existing spline (left) and tangent spline (right)*

OFFSETTING 3D WIRES

You offset a 3D wire from a polyline wire. The new wire is constructed at an offset distance that is normal to the selected wire, and the offset is relative to the current display view (see Figure 7–56).

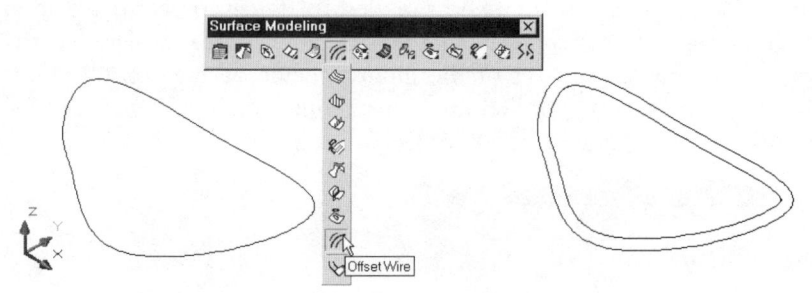

Figure 7–56 *3D wire (left) and wire offset in the current viewing plane*

FILLETING 3D WIRES

Regardless of the UCS location, you can construct a rounded corner between two coplanar wires. While filleting, you can choose to trim or not to trim the selected wires (see Figure 7–57).

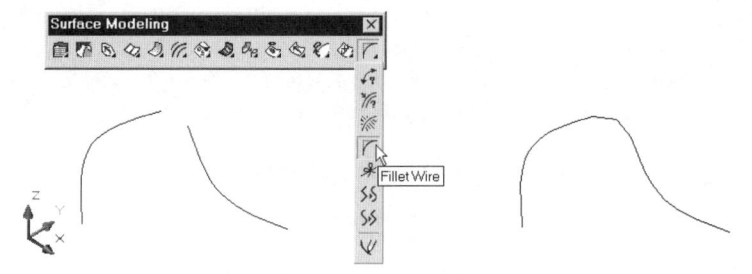

Figure 7–57 *Coplanar wires (left) and fillet constructed (right)*

INTERSECTION OF TWO SURFACES

At the intersection of two intersecting surfaces, you can construct a wire. Figure 7–58 shows a wire constructed at the intersection of two surfaces.

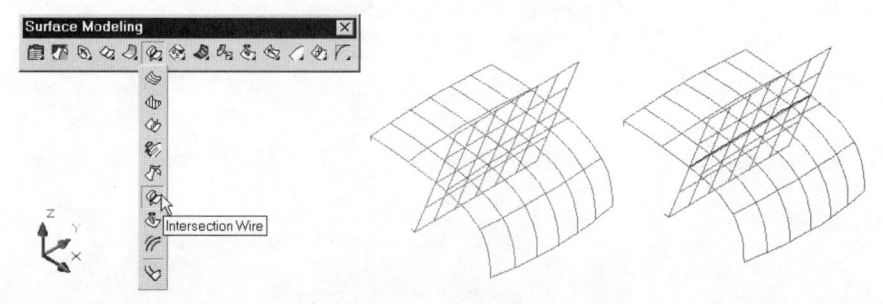

Figure 7–58 *Two intersecting surfaces (left) and wires constructed at the intersection (right)*

PROJECTING A WIRE

Constructing 3D wires on a free-form surface is more difficult and time-consuming than making 2D wires on any specific construction plane (User Coordinate Systems). To make 3D wires on a surface, you construct 2D wires and then project them onto the surface (see Figure 7–59). The setting of the DELOBJ variable determines whether the source wire is deleted after it is projected.

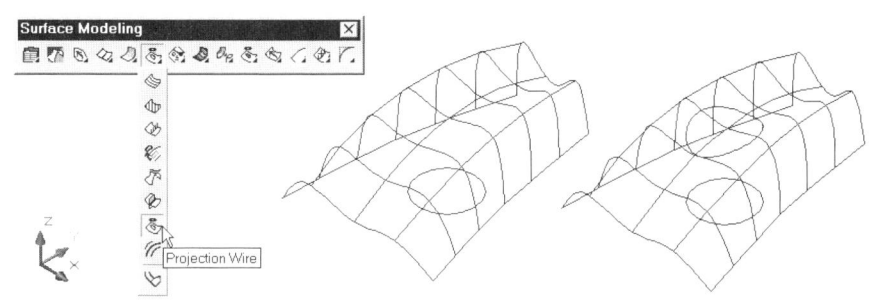

Figure 7–59 Wire and a surface (left) and wire projected on the surface (right)

COPYING SURFACE EDGES

Boundary edges of existing surfaces can be useful in making other surfaces. You can copy them so that they become wires. Figure 7–60 shows the edges copied from a trimmed surface.

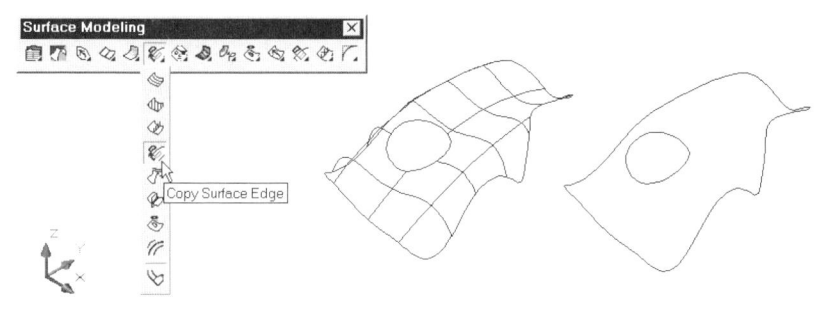

Figure 7–60 Surface (left) and edges copied (moved apart) from the surface (right)

EXTRACT TRIM BOUNDARIES

Untrimming a surface removes all the trimmed boundaries and reverts the trimmed surface to its untrimmed state. By extracting trim boundaries, you remove selected trim boundaries and extract them as wires. Figure 7–61 shows a surface loop extracted from a trimmed surface.

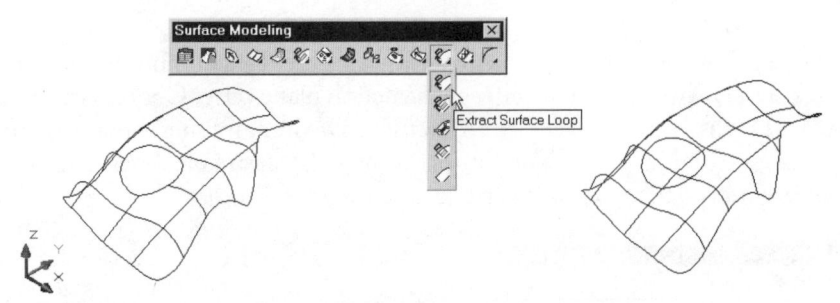

Figure 7-61 *Trimmed surface (left) and a surface loop extracted (right)*

CUTTING A SERIES OF SECTION LINES

To visualize and inspect a 3D surface model, you generate section lines from it. Apart from visualization and inspection, you use the section lines as tool paths for machining. Figure 7-62 shows a set of parallel section lines generated on a surface.

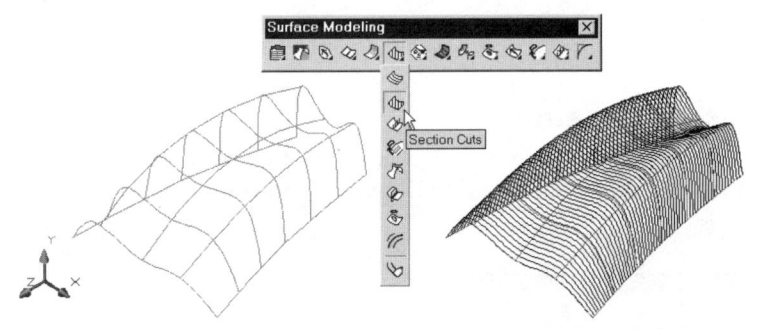

Figure 7-62 *Surfaces (left) and section lines generated on the surfaces (right)*

GENERATING FLOW LINES

On the computer display, lines along the U and V directions are used to depict the profile and silhouette of a surface. They are called flow lines. To use them as tool paths for manufacturing or other purposes, you construct 3D wires from them. Figure 7-63 shows a set of flow lines constructed from a surface.

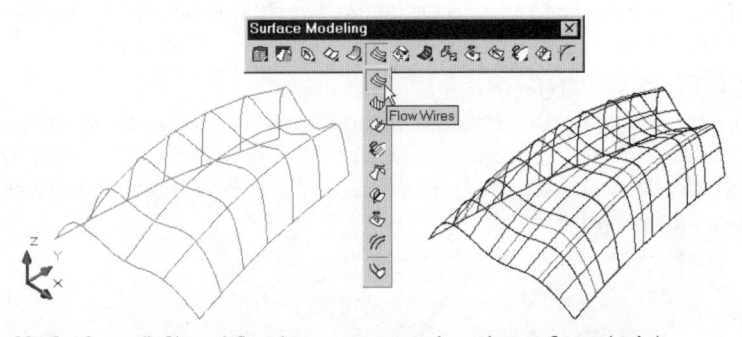

Figure 7-63 *Surfaces (left) and flow lines constructed on the surfaces (right)*

GENERATING A PARTING LINE

To make a mold from a surface model, you need a parting line. Figure 7–64 shows a parting line generated on a surface, separating the surface in halves in the Y direction.

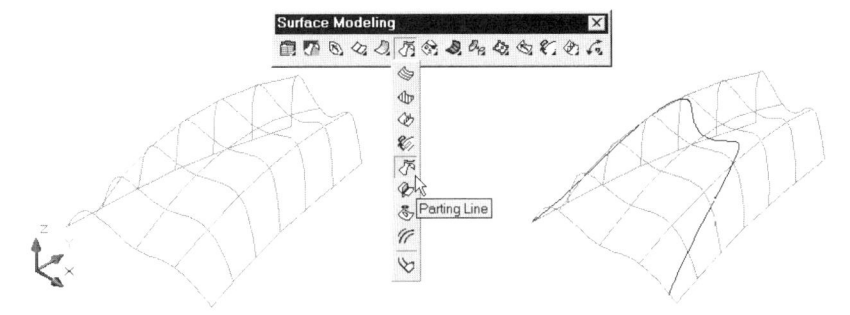

Figure 7–64 *Surfaces and parting line generated on the surfaces (right)*

3D WIRES EDITING

To modify curves, you use AutoCAD tools together with Mechanical Desktop tools. By using Mechanical Desktop, you can:

- Edit a spline
- Revert a spline to a polyline by unsplining
- Change the number of control points of a wire to refine it
- Change the direction of a wire
- Edit augmented line

SPLINE EDIT

Because the surface patches' accuracy depends on the splines that are used to construct the surface, fine-tuning the splines enables you to better control the surface's profile and silhouette. By using the AMSPLINEDIT command, a spline's parameter can be edited through a dialog box.

To illustrate how a spline can be edited this way, construct a spline.

1. Start a new file. Use metric as the template.
2. Select Design>Spline

    Command: AMSPLINEDIT
3. Type **10,10** to specify the first point.
4. Type **30,20** to specify the second point.
5. Type **40,50** to specify the third point.
6. Type **60,10** to specify the fourth point.

7. Press the ENTER key to end the point selection.
8. Press the ENTER key to use the first point as the start tangent point.
9. Press the ENTER key to use the end point as the end tangent point.
10. Select the Spline Edit button from the Surface Modeling toolbar.
11. Select the spline.

The Edit Spline Dialog box displays (see Figure 7–65).

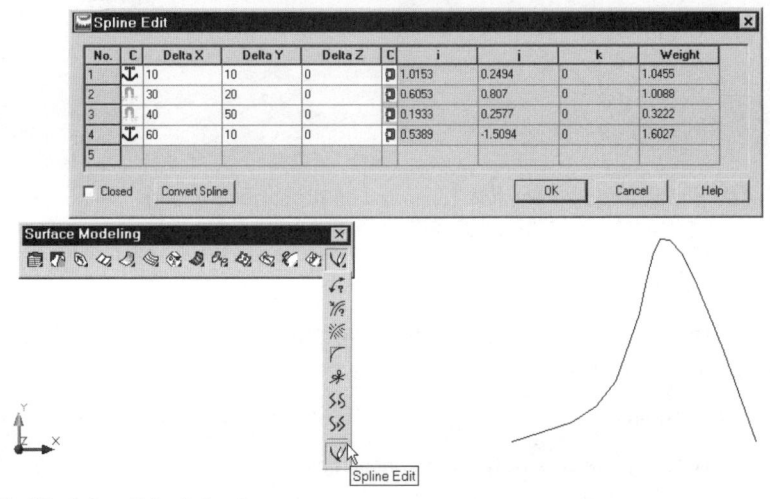

Figure 7–65 *Spline Edit dialog box*

In the Spline Edit dialog box, the spline's fit points or controls are listed in ten columns:

No	Lists the spline's fit or control points
C (constraint)	Indicates whether the fit point or control point is anchored or floating
Delta X	Indicates the fit point or control point's distance from the origin along the X axis
Delta Y	Indicates the fit point or control point's distance from the origin along the Y axis
Delta Z	Indicates the fit point or control point's distance from the origin along the Z axis
C (clamp)	Toggles on and off to make the tangency vectors (i, j, and k) editable
i	Indicates the tangency vector (along the X axis) into and out of a spline
j	Indicates the tangency vector (along the Y axis) into and out of a spline
k	Indicates the tangency vector (along the Z axis) into and out of a spline
Weight	Determines the distance a spline segment maintains tangency before it transitions to the next fit or control point

To change the location of the fit point or control point, you modify the values in the Delta X, Y, and Z columns.

12. To change a fit point or control point's constraint status, select the row listing of the point, right-click, and select Anchor or Float (see Figure 7–66).

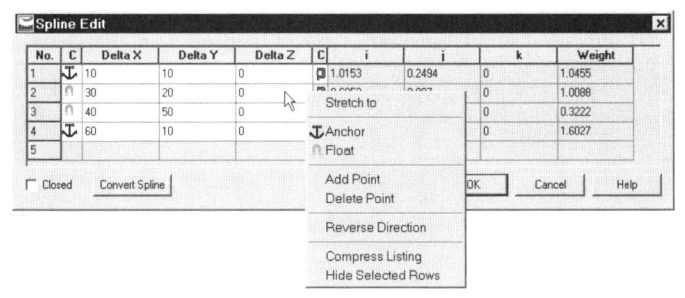

Figure 7–66 *Right-click menu*

In the right-click menu shown in Figure 7–66, there are 8 buttons:

Stretch To	Enables you to stretch the point of the spline to another object
Anchor	Enables you to fix the fit point or control point so that it is independent of the other points of the spline and will not move when the spline is recomputed
Float	Enables you to remove the anchor of the fit point or control point so that it is dependent on all other points of the spline and will move when other points are edited
Add Point	Enables you to add a new point to the spline and opens a new row in the dialog box
Delete Point	Enables you to delete the selected point of the spline
Reverse Direction	Enables you to reverse the direction of the spline
Compress Listing	Compresses the listing and lists only the first three and the last threepoints of the spline
Hide Selected Rows	Enables you to hide the selected rows

If you select the first or the last point of the spline and right-click, the right-click menu has a few more buttons (see Figure 7–67).

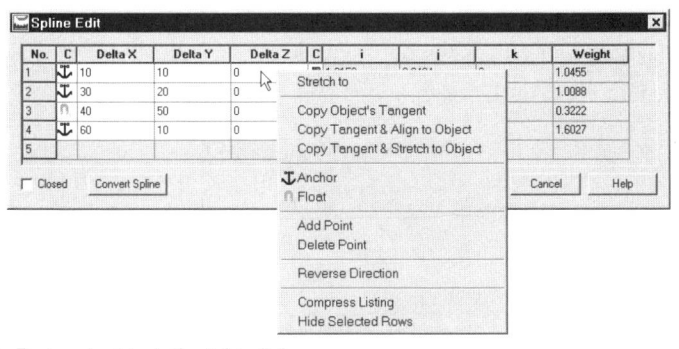

Figure 7–67 *First or last point's right-click menu*

Copy Object's Tangent	Enables you to edit the spline by copying the slope direction of another object
Copy Tangent & Align to Object	Enables you to edit the spline by copying the slope direction of another object and aligning the spline, along that slope, to that object
Copy Tangent & Stretch to Object	Enables you to edit spline by copying the slope direction of another object and stretching the spline, along that slope, to that object

At the lower left of the Spline Edit dialog box, there are a check box and a button:

| Closed | Enables you to convert the spline into a close loop spline |
| Convert Spline | Enables you to convert a fit point spline to a control point spline, and vice versa |

Initially, the spline is a fit point spline because the spline passes through all the input points.

13. Select the Convert Spline button.
14. In the alert message box, select the OK button.

The spline is converted to a control point spline (see Figure 7–68).

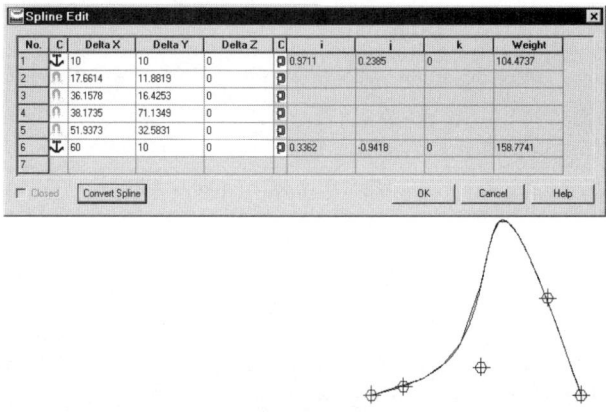

Figure 7–68 *Spline converted to a control point spline*

15. Select the Convert Spline button again.

The Fit Point Conversion dialog box displays (see Figure 7–69).

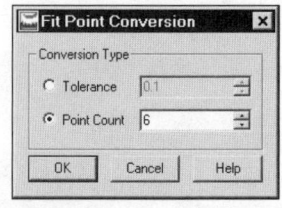

Figure 7–69 *Fit Point Conversion dialog box*

16. In the Fit Point Conversion dialog box, check the Point Count button, set point count to 6, and select the OK button.
17. Select the OK button in the alert dialog box.

The spline is converted back to a point fit spline (see Figure 7–70).

Figure 7–70 *Spline converted back to point fit spline*

Save and close your file (file name: *SplineEdit.dwg*).

UNSPLINING

Sometimes, you need to change a spline back to a polyline. For example, to construct a wire offset from a spline, you change the wire to a polyline because a spline cannot be offset. After offsetting, you can fit it to a spline again (see Figure 7–71).

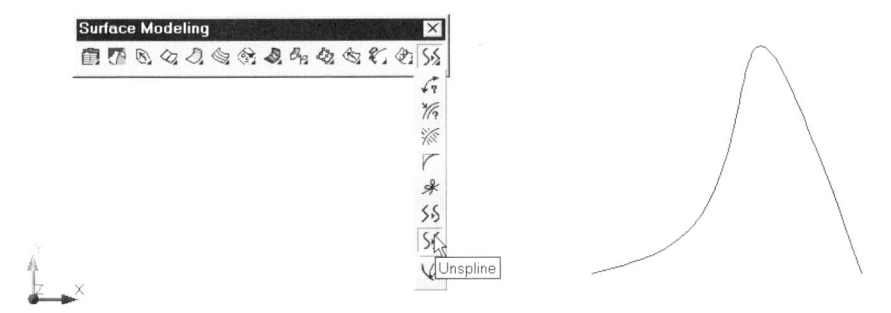

Figure 7–71 *Unsplining a spline*

REFINING A WIRE

Wires have control points. You edit a wire by manipulating its control points. To define a more accurate surface, you may have to refine the control points of wires before using them in surface construction (see Figure 7–72). Note that this command does not apply to splines.

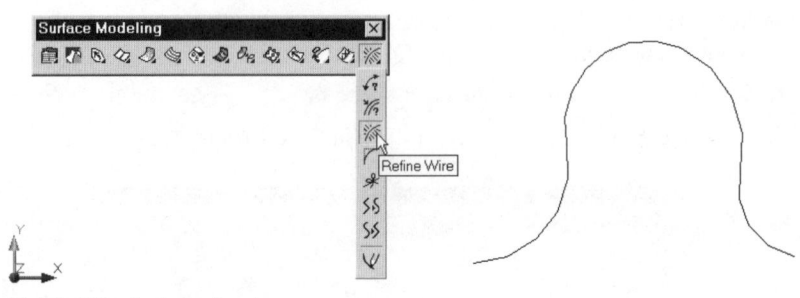

Figure 7-72 *Wire being refined*

CHANGING THE DIRECTION OF A WIRE

Wires have direction. A wire has a starting point and an ending point. The profiles and silhouettes of a surface constructed from a given set of wires are affected by the directions of the individual wires. To control the surface as it is constructed from a set of wires, you may have to change the directions of the wires (see Figure 7-73).

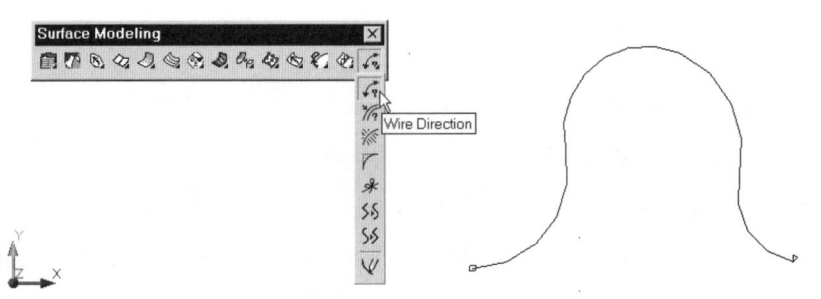

Figure 7-73 *Direction of a wire being reversed*

Figure 7-74 shows how the wire directions can affect the profile of the surface constructed.

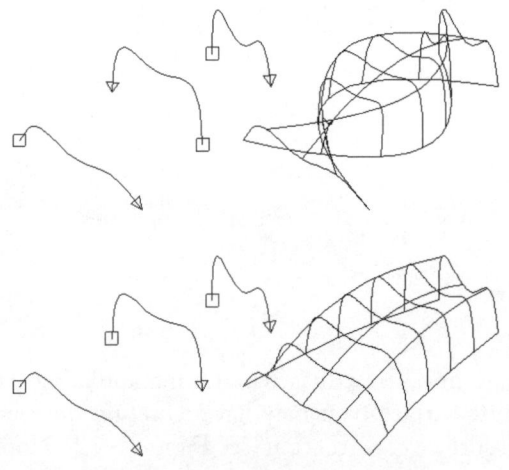

Figure 7-74 *Wire directions and the Loft U surfaces constructed*

EDITING AUGMENTED LINES

The vectors along the augmented lines can be modified in six ways: add vectors, blend, copy, normal length, rotate, and twist. Figure 7–75 shows the toolbar buttons.

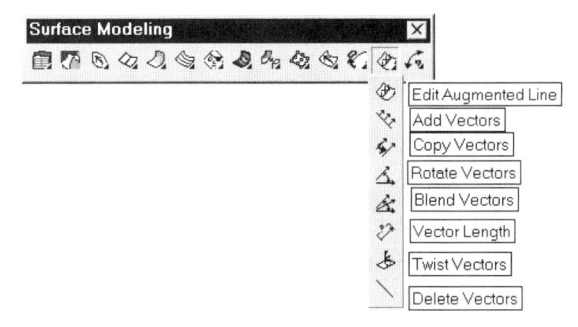

Figure 7–75 *Augmented line editing tools*

UTILITIES

Now you will learn various surface modeling utilities.

SURFACE ANALYSIS

By using the AMANALYZE command, you can check a surface's curvature, draft angle accuracy, and curvature (see Figure 7–76).

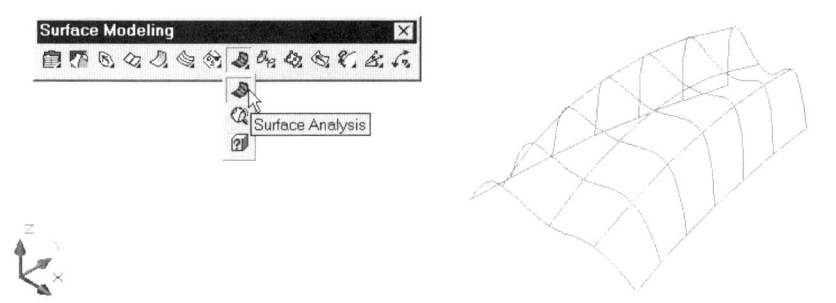

Figure 7–76 *Surface analysis*

In the Surface Analysis dialog box shown in Figure 7–77, the selected surface's minimum, maximum, mean, and Gaussian curvature are represented by a color-shaded display.

By selecting Draft from the Analysis pull-down box of the Surface Analysis dialog box, the surface's draft angles are illustrated by a color-shaded display (see Figure 7–78).

To examine the fit and finish of a surface, and assess the regularity of tangency and curvature conditions between mating surfaces, and the regularity of curvature and curvature change within a surface boundary, you select Reflection Lines from the Analysis pull-down box (see Figure 7–79).

Figure 7–77 Curvature analysis

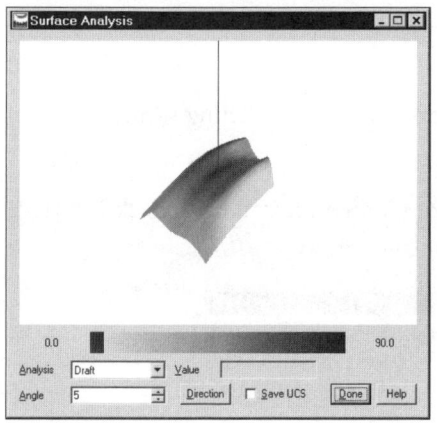

Figure 7–78 Draft angle analysis

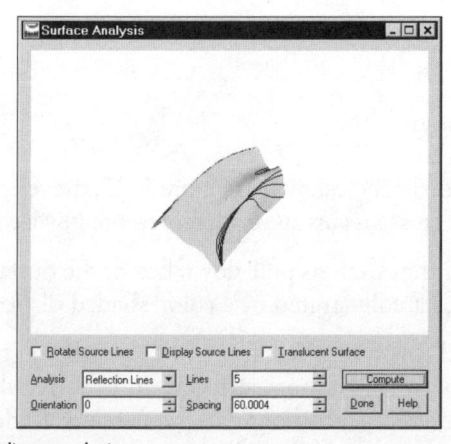

Figure 7–79 Reflection line analysis

SURFACE DISPLAY

A surface's display consists of two parts: persistent and temporary. The U and V lines that depict the profile and silhouette and the normal vector that indicates the normal direction are displayed persistently. The patch normal, base surface, and control points are temporary displays (see Figure 7–80).

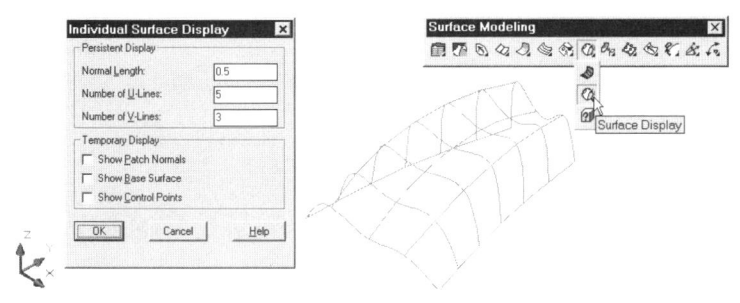

Figure 7–80 *Display of a surface*

SHOW EDGE NODES

Beginning and end of each of the individual edge lines of a surface can be displayed (see Figure 7–81).

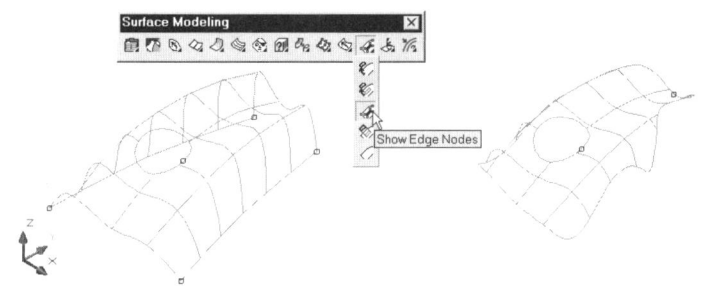

Figure 7–81 *Edge nodes of two surfaces displayed*

SURFACE MASS PROPERTIES

A surface has no thickness. Therefore, the mass of a surface is zero. By assigning a thickness to a surface, you evaluate its mass properties (see Figure 7–82).

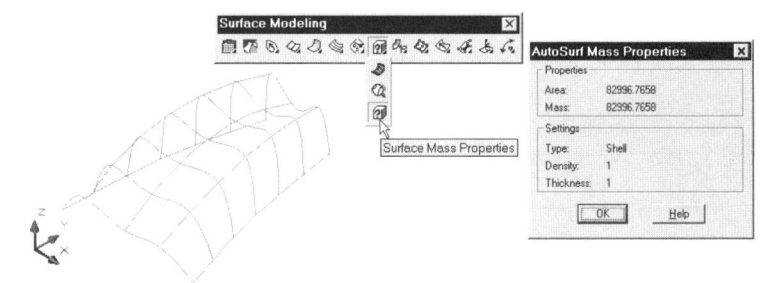

Figure 7–82 *Surface's mass properties evaluation*

CHECK FIT

Fitness between two wires or a wire and a surface can be found and displayed graphically in the screen (see Figure 7–83).

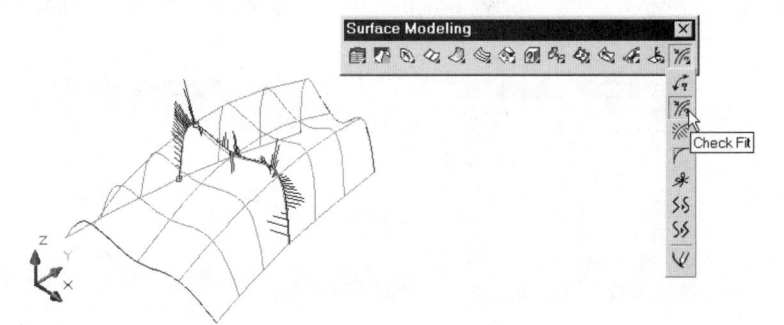

Figure 7–83 *Checking fitness between a wire and a surface*

SURFACE MODELING OPTIONS

System options affect the outcome of the 3D surfaces that you construct. From the Surface pull-down menu, select Surface Options (see Figure 7–84).

1. Select Surface>Surface Options

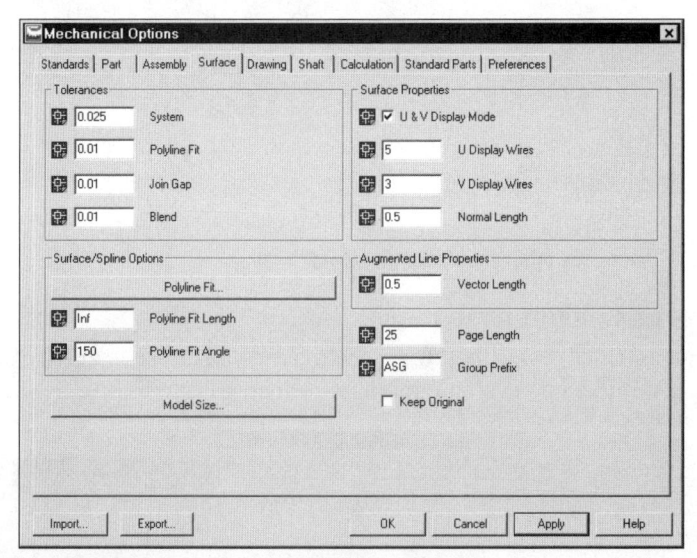

Figure 7–84 *Surface tab of the Mechanical Options dialog box*

If you are a surface modeling novice, you may find it difficult to set the Surface tab of the Desktop Options dialog box. To avoid the trouble of setting each of the options individually, you set them collectively by selecting the Model Size button.

2. In the Surface tab of the Mechanical Options dialog box, select the Model Size button (see Figure 7–85).

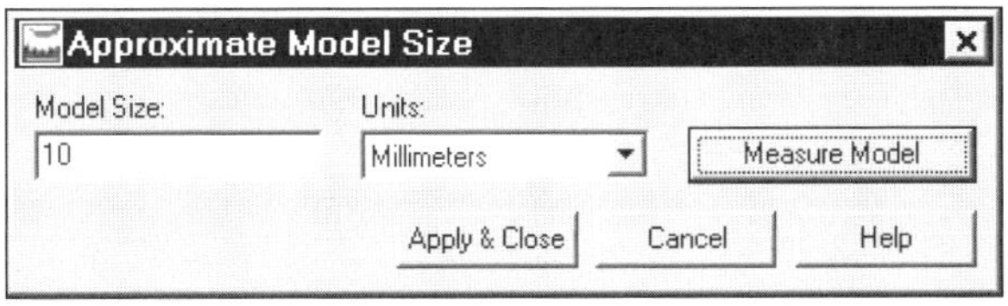

Figure 7–85 *Approximate Model Size dialog box*

Using this dialog box, you simply select the model size and the unit of measurement. If you have already constructed some objects on your screen, you select the [Measure Model] button to let the computer measure the model for you.

3. Type a value at the model size box or select the Measure Model button, and select the Apply & Close button.

In the Tolerances area of the Surface Tab of the Desktop Options dialog box, you set four kinds of tolerances: System Tolerance, Polyline Fit Tolerance, Join Gap Tolerance, and Blend Tolerance.

System	controls the system tolerance for constructing flow lines on the surfaces.
Polyline Fit	controls the tolerance for fitting a polyline into a spline.
Join Gap	controls the tolerance of the gap (if any) between two endpoints.
Blend	determines whether to create C0 breaks in the blend joint.

In the Surface/Spline Options area, you determine approximately how the spline is fitted. As we have said, Mechanical Desktop employs NURBS mathematics for defining wires and surfaces. Basically, it uses 3D NURBS splines to construct 3D NURBS surfaces. However, it also accepts lines, arcs, 2D polylines, and 3D polylines. While using these objects as wires, Mechanical Desktop fits them into splines according to a set of rules. There are two options: Polyline Fit Length and Polyline Fit Angle.

Polyline Fit Length	sets the polyline length for holding splines and surfaces flat.
Polyline Fit Angle	sets the polyline angle to break splines and surfaces at corners.

To set these options, you enter values in the appropriate boxes or select the Polyline Fit button to bring out the Polyline Fit dialog box.

4. In the Surface tab of the Mechanical Options dialog box, select the Polyline Fit button (see Figure 7–86).

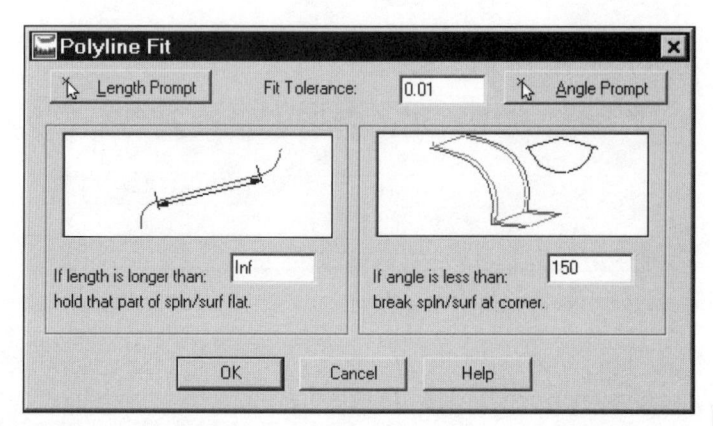

Figure 7–86 *Polyline Fit dialog box*

In the Polyline Fit dialog box, set the length and angle graphically by selecting the Length Prompt and the Angle Prompt buttons.

    5. Select the [OK] button to return to the Desktop Options dialog box.

In the Surface Properties area, you set the display components of the surfaces. On your screen, a surface has three display components: boundary edge, U- and V- wires, and normal vector. The boundary edge is the edge of the surface, the U- and V- wires are wires in two orthogonal directions to depict the profile of the surface, and the normal vector is used to indicate the normal direction of the surface.

U & V Display Mode	determines the kind of linetypes to be used to display the U- and V- wires.
U Display Wires	sets the number of U display wires.
V Display Wires	sets the number of V display wires.
Normal Length	sets the length of the normal vector.

In the Augmented Line Properties area, you set the vector length.

Vector Length	controls the length of the augmented lines. Augmented lines are series of normal vectors along a spline, a line, or a polyline.

In the remaining area of the dialog box, there are a few boxes.

Page Length	sets the text page length of the text window.
Group Prefix	sets the prefix name.
Keep Original	determines whether the original object used to make another object is retained.

    6. Select the OK button.

The surface options are set.

VISIBILITY

3D surfaces require 3D wires as framework. As you continue to construct more wires and surfaces, selection of individual objects might become difficult. Naturally, you can put them into different layers. However, you might end up with a lot of layers, which can lead to confusion. To solve the problem of object manipulation, you control their visibility individually rather than changing the visibility setting of the layer on which they reside.

The visibility tool lets you hide and unhide objects. Select Surface Visibility from the Surface pull-down menu (see Figure 7–87).

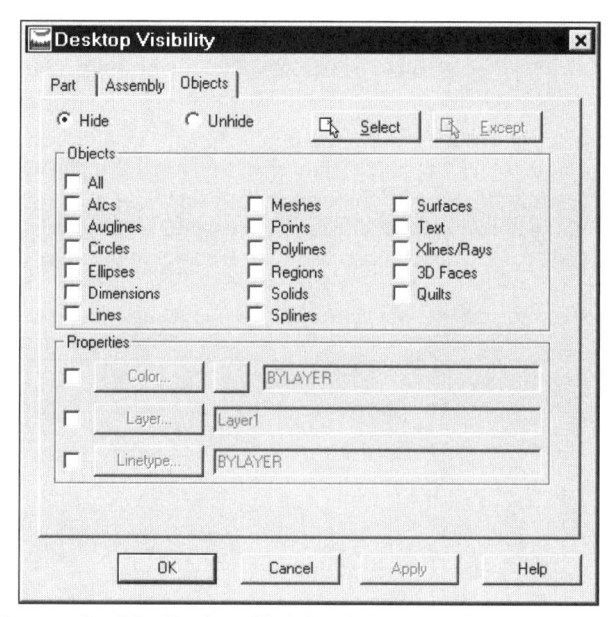

Figure 7–87 *Objects tab of the Desktop Visibility dialog box*

Refer to the Surface tab of the Desktop Visibility dialog box. You can hide or unhide objects by selection or in accordance with the object types or object properties.

SUMMARY

To construct a 3D free-form object in the computer, you use a set of free-form surfaces. Before making the surfaces, you first study and analyze the 3D object to determine what kinds of surfaces are needed. You can consider using primitive surfaces, free-form surfaces, derived surfaces, or trimmed surfaces. Among the various kinds of surfaces, free-form surfaces are the ones most commonly used. All free-form surfaces have one thing in common: They all need to be constructed from smooth wires.

It is intuitive to start thinking about the surfaces but not the wires while you are designing and making a surface model. Because the computer constructs surfaces from defining wires, the first task that you need to tackle in surface modeling is to think about what kinds of wires are needed and how they can be constructed. After making the wires, you can then let the computer generate the required surfaces.

In the surface modeling projects of this chapter, you will learn to construct 3D wires as well as 3D surfaces. Here the particulars of all the wires are given to you. While working on these projects, you should try to relate the 3D wires to the 3D surfaces. It is hoped that you will reverse the process, seeing the wires when a surface is given. Every time you start to construct a surface model, remember to set the related system variables, because these settings affect the way the surfaces are constructed and the accuracy of the surfaces in relation to the inputting wires.

PROJECTS

Now you will work on the following projects to enhance your surface modeling knowledge.

OIL COOLER

Figure 7–88 shows the surface model of an oil cooler. Obviously, this oil cooler is a series of cylindrical and toroidal surfaces joined together. To make this surface model, you construct a 3D polyline as the control wire and use the AMTUBE command to construct the surfaces.

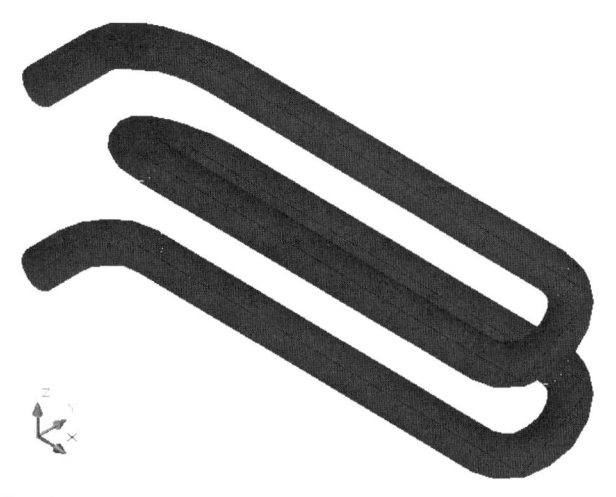

Figure 7–88 *Oil cooler*

1. Start a new drawing. Use metric as the default.
2. Set up two layers: Wire and Surface.
3. Set layer Wire as the current layer.
4. Select Design>3D Polyline
   Command: 3DPOLY
5. Type the following coordinates at the command line area and press the ENTER key.

0,0	@320<180	@320<180
@55<90	@0,-55,30	@0,-55,30
@320<0	@320<0	@55<270
@0,55,30	@0,55,30	

A 3D polyline is constructed.

6. Set the current layer to Surface.

7. Select Surface>Create Surface>Tubular or select Tubular Surface from the Surface Modeling toolbar (see Figure 7–89).

   Command: AMTUBE

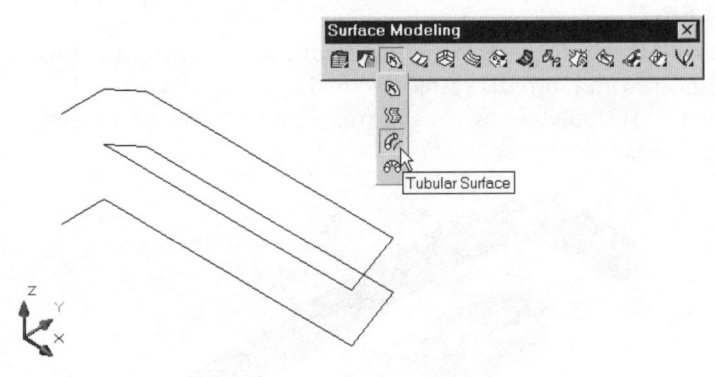

Figure 7–89 *3D polyline constructed*

8. Select the 3D polyline near the lower left end.
9. Type **25** at the command line area to specify the tube diameter.
10. Type **A** to use the automatic option.
11. Type **30** to specify the bend radius.
12. Select continue until surfacing is complete.
13. Turn off layer Wire.

The model is complete. Save your drawing (file name: *OilCooler.dwg*).

HELICAL SPRING

The purpose of this project is to illustrate how you will use an augmented line in constructing a swept surface. Figure 7–90 shows the surface model of a helical coil spring. The free length of the coil spring is 40 units, its pitch is 8 units, its wire diameter is 2 units, and its mean diameter is 12 units. You will construct an augmented line and a circle. Using the augmented line as the sweeping rail, you will construct a swept surface.

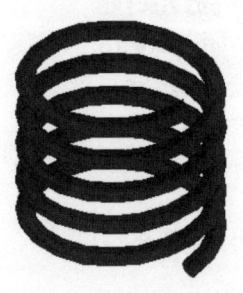

Figure 7–90 *Helical spring*

1. Start a new drawing. Use metric as the default.
2. Construct two layers: Wire and surface, and set current layer to Wire.
3. Select Surface>Surface Options
4. In the Surface tab of the Mechanical Options dialog box, set Vector length in the Augmented Line Properties box to 6 and select the OK button.
5. Set the display to an isometric view.

Now you will use the line (see Figure 7–91).

6. Construct a line from **0,0** to **0,0,40**.

Figure 7–91 *Line constructed and display set*

Now you will refine the line and change it to an augmented line. (For a free length of 40 units, there will be 5 coils in the spring. Suppose you want to place 12 control points for each coil; you will need 60 control points.)

7. Select Surface>Edit Wireframe>Refine or select Refine Line from the Surface Modeling toolbar.

   Command: AMREFINE3D

8. Select the line and press the ENTER key.
9. Type **P** to use the control point option.
10. Type **60** to change the number of control points to **60**.

The line is refined to have 60 control points.

11. Select Surface>Edit Wireframe>Augment Polyline or select Add Vectors from the Surface Modeling toolbar.

    Command: AMEDITAUG

12. Select the line and press the ENTER key.
13. Press the ENTER key to exit.

The line is converted into an augmented line with 60 vectors (see Figure 7–92).

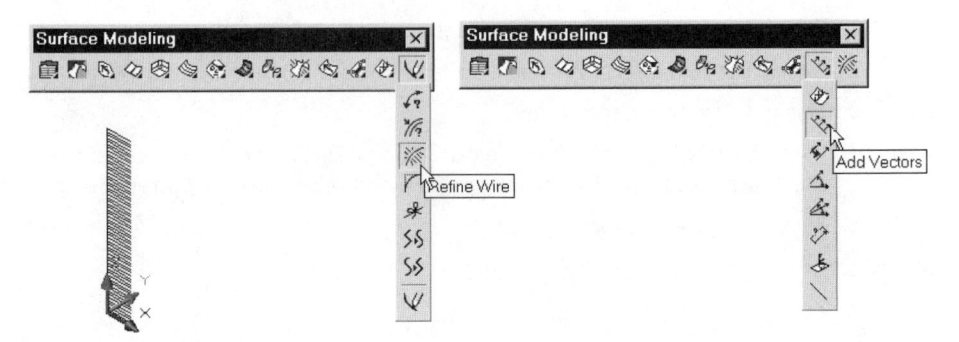

Figure 7-92 *Line converted to an augmented line*

14. Select Surface>Edit Wireframe>Twist Vectors or select Twist Vectors from the Surface Modeling toolbar.
15. Type **1800** at the command line area to twist the vectors 1800 degrees.
16. Type **A** to twist all the vectors.
17. Select the augmented line and press the ENTER key.

The vectors are twisted (see Figure 7–93). Now you will rotate the UCS 90° around the X-axis and construct a circle.

18. Select Assist>New UCS>X
19. Type **90** at the command line area.
20. Select Design>Circle>Center, Radius
21. Type **20,0** to specify the center.
22. Type **2** to specify the radius.

A circle is constructed (see Figure 7–94).

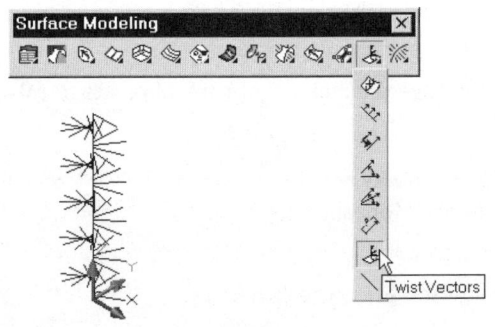

Figure 7-93 *Vectors of the augmented line twisted*

Figure 7–94 *UCS rotated and a circle constructed*

Now you will construct a swept surface.

23. Set the current layer to Surface.
24. Select Surface>Create Surface>Sweep

    Command: AMSWEEPSF

25. Select the circle to specify the section and press the ENTER key.
26. Select the augmented line to use it as the rail and press the ENTER key.
27. In the Sweep Surface dialog box (Figure 7–95), select the OK button.

The model is complete (see Figure 7–96). Save and close your file (file name: *SurfaceHelicalSpring.dwg*).

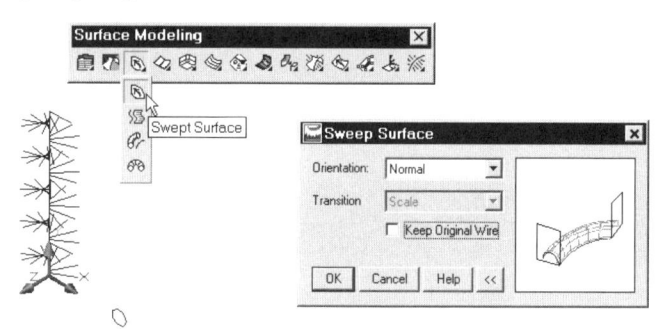

Figure 7–95 *Sweep Surface dialog box*

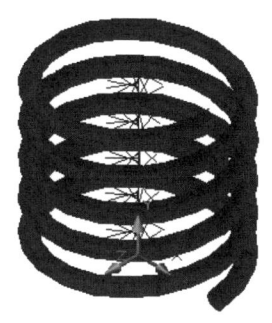

Figure 7–96 *Sweep surface constructed*

REMOTE CONTROL SURFACE MODEL

Now you will construct the surface model of a remote control shown in Figure 7–97. To make this model, you need the wires shown in Figure 7–98.

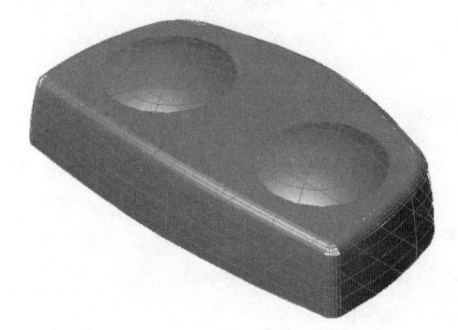

Figure 7–97 *Remote control*

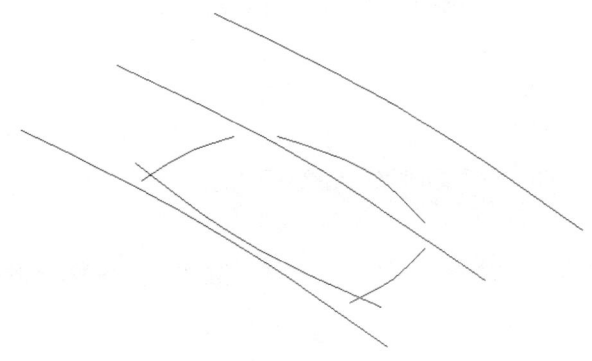

Figure 7–98 *Wires required for the remote control*

1. Start a new file. Use metric as the default.
2. Construct two layers: Wire and surface, and set current layer to Wire.
3. Set the display to an isometric view.
4. Select Design>Spline.

    Command: SPLINE

5. Select any three points on the screen to specify the first, second, and third points of the spline.
6. Press the ENTER key to end the point selection.
7. Press the ENTER key to use the first point as the start tangent point.
8. Press the ENTER key to use the end point as the end tangent point.
9. Select Surface>Edit Wireframe>Spline Edit
10. Select the spline you constructed. In the Spline Edit dialog box (see Figure 7–99), set the data of the spline as follows:

Delta X	Delta Y	Delta Z
−150	−35	20
0	−35	30
150	−35	20

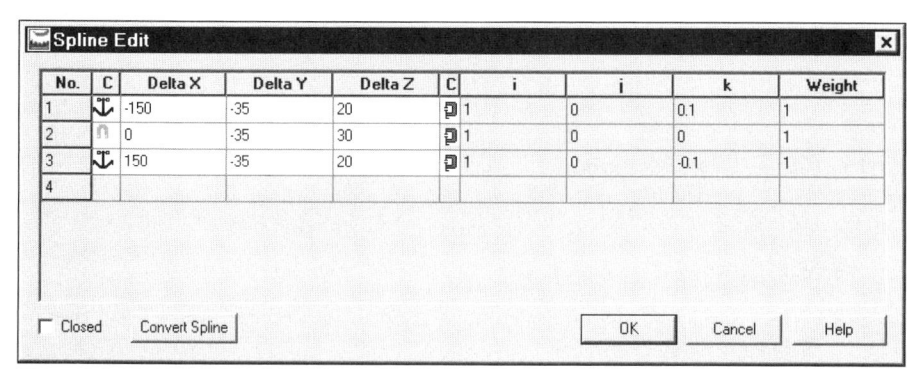

Figure 7–99 *Spline Edit dialog box*

11. Repeat steps 4 through 10 above twice to construct two more splines that pass through the following points:

    Second spline:

    First point: **−150,45,25**

    Next point: **0,45,35**

    Next point: **150,45,25**

    Next point: ENTER

    Start tangent: ENTER

    End tangent: ENTER

    Third spline:

    First point: **−150,125,20**

    Next point: **0,125,30**

    Next point: **150,125,20**

    Next point: ENTER

    Start tangent: ENTER

    End tangent: ENTER

Three splines are constructed (see Figure 7–100).

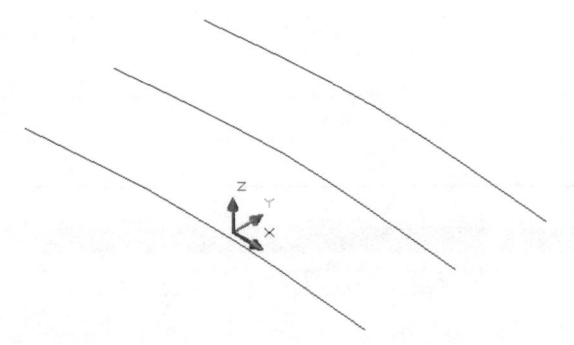

Figure 7–100 *Three splines constructed*

Using the splines as wires, you will construct a Loft U surface.

12. Set the current layer to Surface.
13. Select Surface>Create Surface>Loft U or select Loft U Surface from the Surface Modeling toolbar.

    Command: AMLOFTU

14. Select the splines one by one (from left to right) and press the ENTER key (see Figure 7–101).

In the Loft Surface dialog box, the Align button enables you to adjust the direction of selected wires to make them unidirectional, the Smooth button enables you to fit the input wires to reduce the complexity of the surface created, and the Respace button enables you to adjust any poorly proportioned wire ends.

15. Select the OK button.

Figure 7–101 *Loft Surface dialog box*

A Loft U surface is constructed. Now you will learn how to modify the grip points of the surface.

16. Select the surface to show up the grip points (see Figure 7–102).

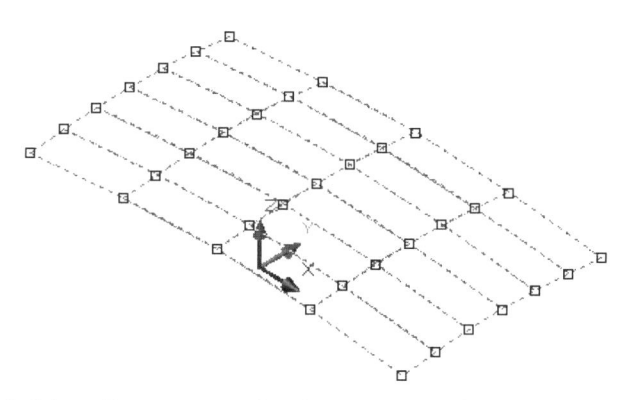

Figure 7–102 Loft U surface constructed and grip points displayed

The grip points are highlighted. Now you will change the number of grip points.

17. Select Surface>Edit Surface>Define Grips or select Grip Point Placement from the Surface Modeling toolbar.

    Command: AMEDITSF

18. Type **8** at the command line area to specify the grip point numbers along the U direction.

19. Type **6** at the command line area to specify the grip point numbers along the V direction.

The number of grip points is changed (see Figure 7–103).

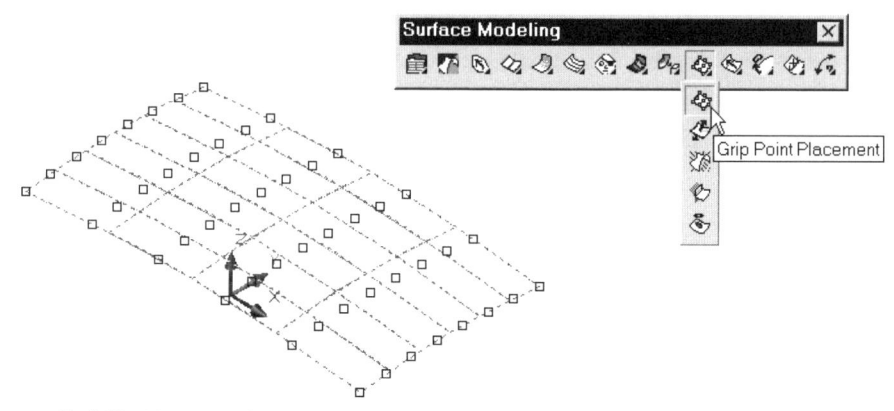

Figure 7–103 Number of grip points changed

Now you will discover by yourself how a surface can be edited by manipulating its grip points. You will copy the Loft U surface, edit the copied surface, and compare the surfaces to find out the effect of editing.

20. Select Modify>Copy

Command: COPY

21. Select the surface and press the ENTER key.
22. Type **200,200** at the command line area and press the ENTER key.

The surface is copied to a distance of 200,200 from the original surface (see Figure 7–104).

Now you will edit the span size of the surface. When you select a grip point of a surface and move it to a new location, a circular area of the surface around the selected grip point is deformed. The size of this circular area is called the span.

23. Select Surface>Edit Surface>Overall Edit

Command: AMEDIT

24. Select the surface (right) and press the ENTER key.
25. In the Surface Edit dialog box, set span size to 20 (see Figure 7–105).
26. Select the Preview button.
27. Press the ENTER key to return to the Surface Edit dialog box.
28. Select the OK button.

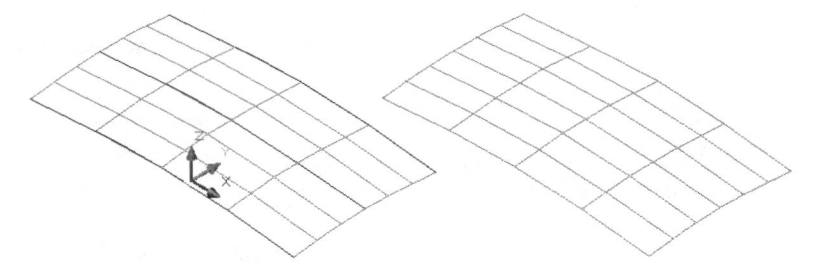

Figure 7–104 *Surface copied*

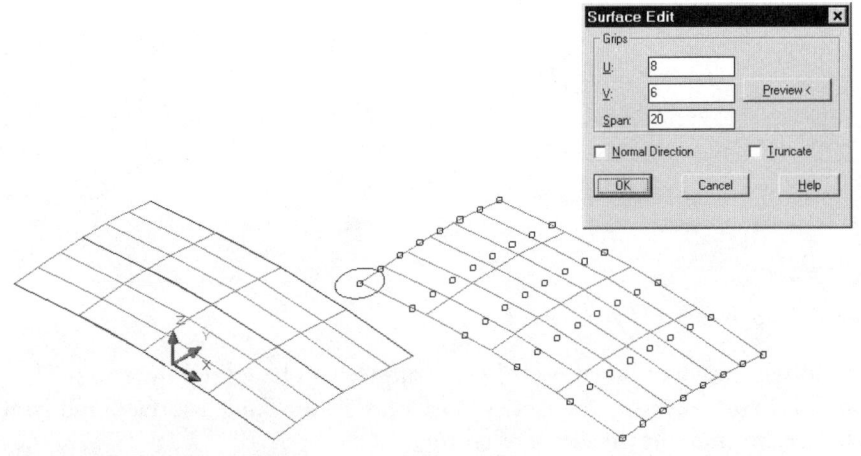

Figure 7–105 *Span size modified*

The span is modified. Now you will move a grip point to deform the surface.

29. Referring to Figure 7–106, select the surface to display the grip point and then select a grip point.
30. Type **@0,0,15** at the command line area.

    ** STRETCH **

    Specify stretch point or [Base point/Copy/Undo/eXit]: **@0,0,15**

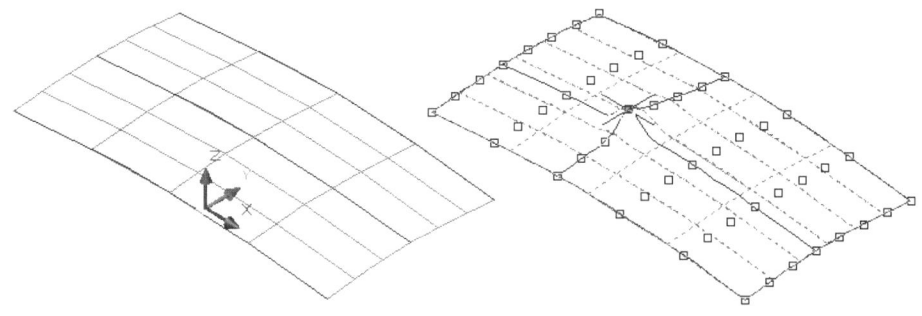

Figure 7–106 *Surface modified by moving a grip point*

A surface has U and V patches that control the shape and silhouette. The U and V patches are not U and V display lines. To rebuild a surface, you specify either patch size or tolerance. Now you will refine the deformed surface by specifying the patch size.

31. Select Surface>Edit Surface>Refine or select Refine Surface from the Surface Modeling toolbar.

    Command: AMREFINESF
32. Select the surface (right) and press the ENTER key.
33. Type **10** at the command line area to set the tolerance value to 10.

The surface is refined (see Figure 7–107).

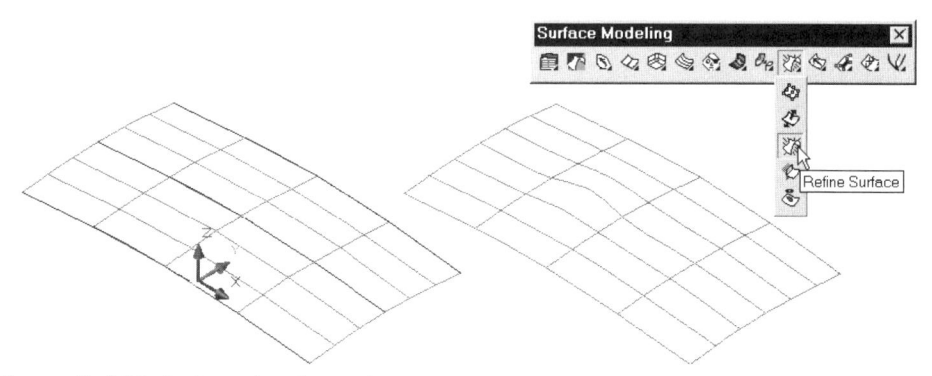

Figure 7–107 *Deformed surface refined*

Now you will learn how to construct flow lines and augmented flow lines.

34. Select Surface>Create Wireframe>Flow

    Command: AMFLOW

35. Select the surface (right) and press the ENTER key.
36. In the Surface Flow Lines dialog box (see Figure 7–80), set U and V wires both to 10.
37. Select the Preview button.

Flow lines are displayed (see Figure 7–108).

38. Press the ENTER key to return to the Surface Flow Lines dialog box.
39. In the Surface Flow Lines dialog box, check the Augmented box and select the Preview button.

Augmented flow line preview is displayed (see Figure 7–109). Note that the length of the augmented line vector is determined by the setting in the Surface tab of the Mechanical Options dialog box.

40. Press the ENTER key to return to the Surface Flow Lines dialog box and select the OK button.

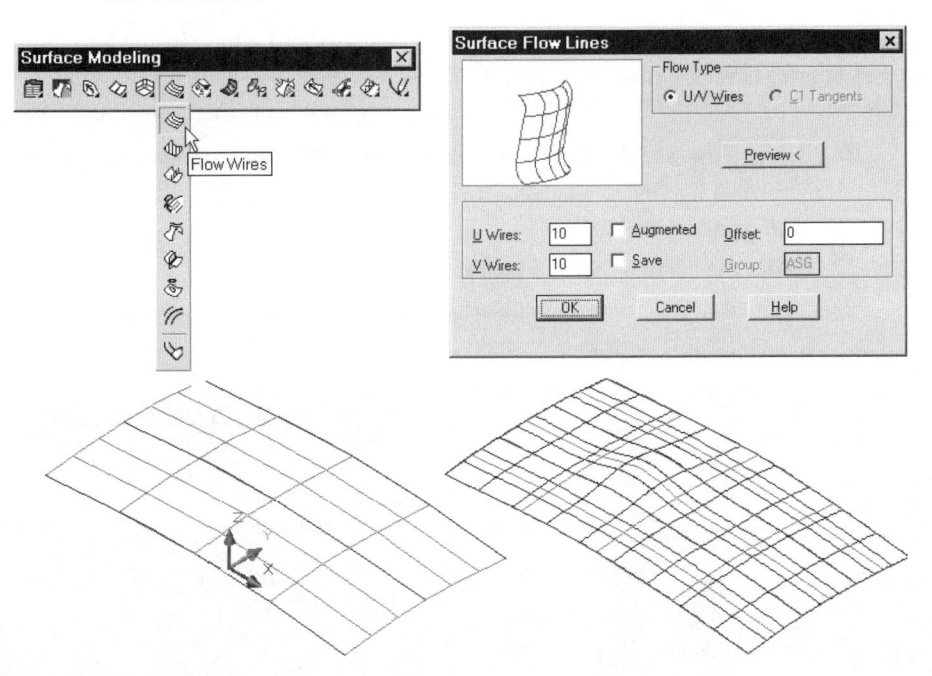

Figure 7–108 *Surface flow lines*

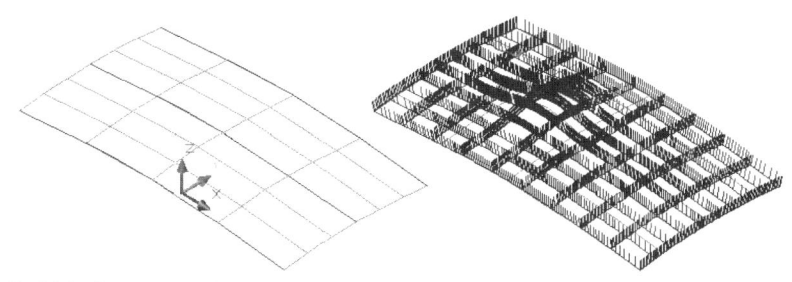

Figure 7–109 *Augmented flow lines*

Because the deformed surface (right) is not required in the remote control, you will now erase it.

41. Select the surface (right) and press the DELETE key.

Now you will construct two primitive spherical surfaces.

42. Select Surface>Create Primitives>Sphere or select Sphere Surface from the Surface Modeling toolbar.

    Command: AMPRIMSF

    SPHERE

43. Type **–40,45,–20** at the command line area to specify the center.
44. Type **60** to specify the radius.
45. Press the ENTER key to accept the default starting angle.
46. Press the ENTER key to construct a full circle.
47. Repeat the command to construct another spherical surface of the same radius at **40,45,–20**.

Two spherical surfaces are constructed (see Figure 7–110).

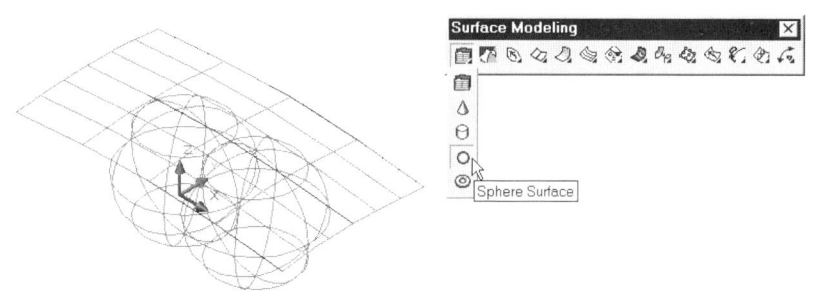

Figure 7–110 *Two primitive spheres constructed*

Now you will trim the surfaces.

48. Select Surface>Edit Surface>Intersect Trim or select Intersect and Trim from the Surface Modeling toolbar.

49. Select the portion of the surface that you want to retain after trimming. Here, select the upper part of the right sphere and the outer edge of the Loft U surface.

In the Surface Intersection dialog box (see Figure 7–111), there are several areas. The Type area enables you to either trim or break the intersecting surfaces. The Trim area enables you to trim the first or second selected surfaces. The Output Polyline box enables you to construct a 3D polyline at the intersection of the surfaces.

50. Select the OK button.
51. Repeat steps 48 through 50 to trim the other spherical surface and the Loft U surface.

The surfaces are trimmed (see Figure 7–112).

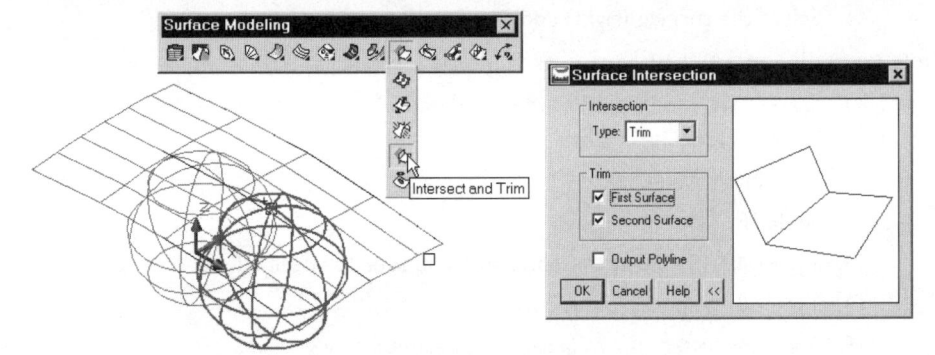

Figure 7–111 *Surface Intersection dialog box*

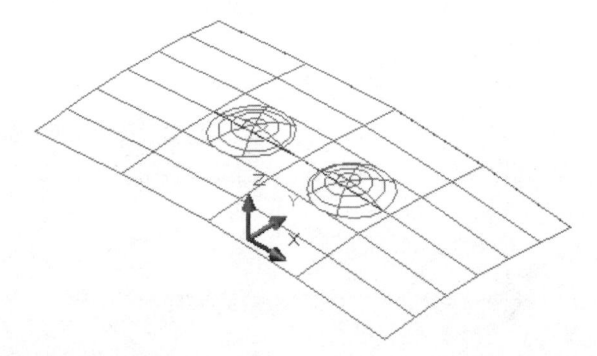

Figure 7–112 *Surfaces trimmed*

Now you will hide the surfaces and construct four splines.

52. Select Surface>Surface Visibility
53. In the Objects tab of the Surface Visibility dialog box, check the Hide box and the Surfaces box, and select the OK button.

54. Set the current layer to Wire.
55. Construct four splines (see Figure 7–113) in accordance with the following data:

Spline	A	B	C	D
First Point	−85,0	−60,86	78,68	−100,10
Next Point	−85,40	0,98	85,40	0,0
Next Point	−78,68	60,86	85,0	100,10
Next Point	ENTER	ENTER	ENTER	ENTER
Start Tangent	ENTER	ENTER	ENTER	ENTER
End Tangent	ENTER	ENTER	ENTER	ENTER

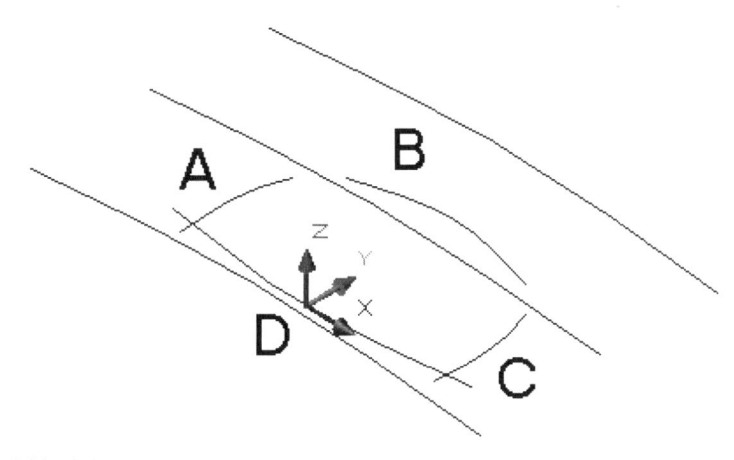

Figure 7–113 *Splines constructed*

56. Set the current layer to Surface.
57. Select Surface>Create Surface>Revolve or select Revolved Surface from the Surface Modeling toolbar.

    Command: AMREVOLVESF
58. Select spline A (shown in Figure 7–113) and press the ENTER key.
59. Type **60,0** at the command line area to specify the axis start point.
60. Type **@1<90** to specify the axis end point.
61. Type **0** to specify the start angle.
62. Type **15** to specify the included angle.

The points (60,0) and (@1<90) above specify an axis that starts from the point (60,0) and points at a direction of 90°.

63. Now repeat the AMREVOLVESF command to construct three more revolve surfaces:

Path wire	B (Figure 7–113)	C (Figure 7–113)	D (Figure 7–113)
Axis start point	0,–60	–60,0	0,160
Axis end point	@1<0	@1<270	@1<180
Start angle	0	0	0
Included angle	15	15	15

Four revolve surfaces are constructed (see Figure 7–114).

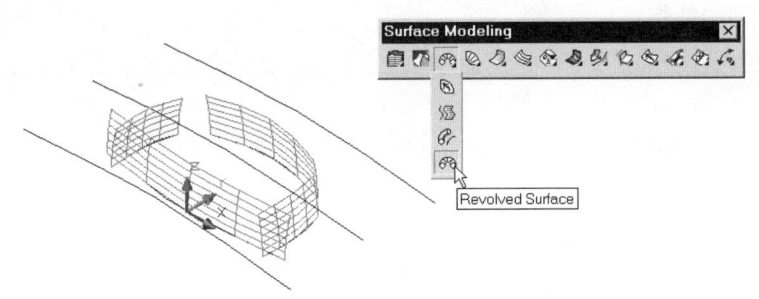

Figure 7–114 *Revolve surfaces constructed*

Now you will hide all the splines and construct a blended surface.

64. Select Surface>Surface Visibility
65. In the Objects tab of the Surface Visibility dialog box, check the Hide box and the Splines box, and select the OK button.
66. Select Surface>Create Surface>Blend or Blended Surface from the Surface Modeling toolbar.

    Command: AMBLEND

67. Select the edges indicated in Figure 7–115 and press the ENTER key.
68. Type **W** to use the weight option.
69. Type **1** to set the first weight.
70. Type **1** to set the second weight.

A blended surface is constructed (see Figure 7–116).

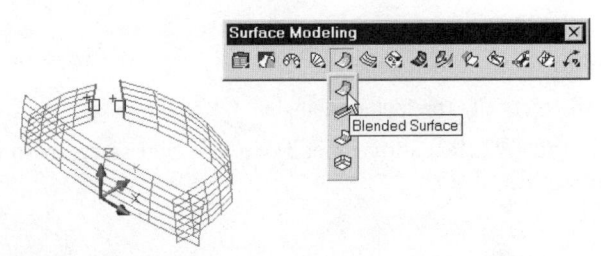

Figure 7–115 *Blended surface being constructed*

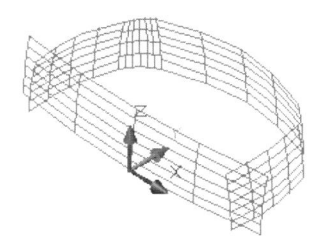

Figure 7–116 *Blended surface constructed*

Now you will learn how to adjust the edges of a pair of surfaces.

71. Rotate the display in accordance with Figure 7–117.
72. Select Surface>Edit Surface>Adjust or select Adjust from the Surface Modeling toolbar.

    Command: AMADJUSTSF

73. Select the edges indicated in Figure 7–117.
74. Press the ENTER key.

The edges of the surfaces are adjusted (see Figure 7–118).

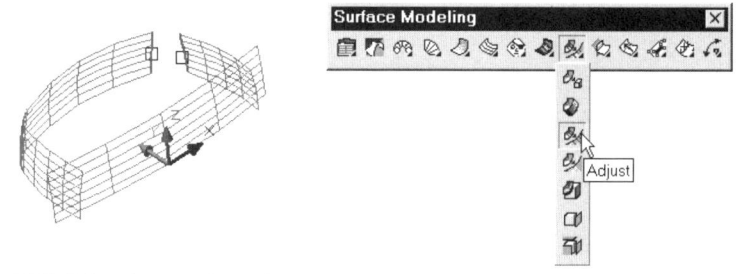

Figure 7–117 *Edges being adjusted*

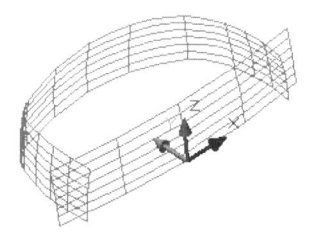

Figure 7–118 *Edges adjusted*

For the sake of constructing a symmetrical model about the center of the model, you will undo the AMADJUSTSF command and construct a blended surface.

75. Undo the AMADJUSTSF command. (Do not undo more than once. Otherwise, you will undo too many commands.)

76. Construct a blended surface in accordance with Figure 7–119.

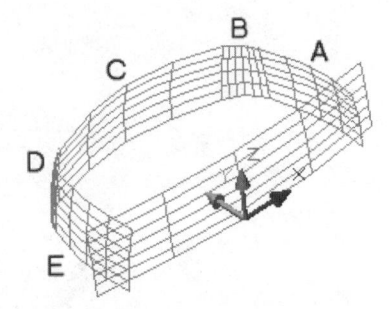

Figure 7–119 *Edge adjustment undone and blended surface constructed*

Now you will join five surfaces into a single surface.

77. Select Surface>Edit Surface>Join or select Join Surfaces from the Surface Modeling toolbar.

    Command: AMJOINSF

78. Select surfaces A, B, C, D, and E (shown in Figure 7–119) and press the ENTER key.

Three surfaces are joined (see Figure 7–120).

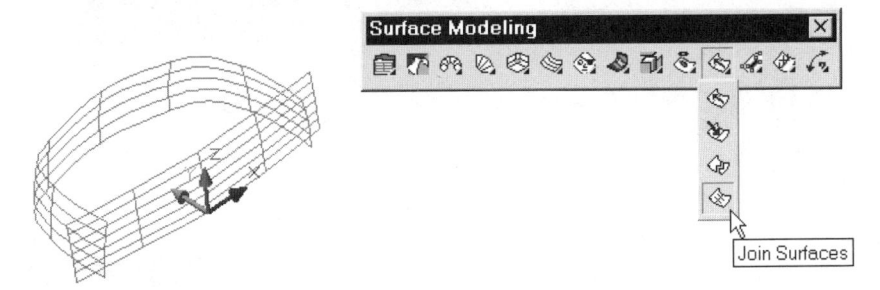

Figure 7–120 *Surfaces joined*

Now you will unhide the hidden surface and construct fillet and corner surfaces.

79. Select Surface>Surface Visibility or select Object Visibility from the Surface Modeling toolbar.
80. In the Objects tab of the Desktop Visibility dialog box, check the Unhide and Surfaces boxes, and select the OK button.

The hidden surface is unhidden (see Figure 7–121).

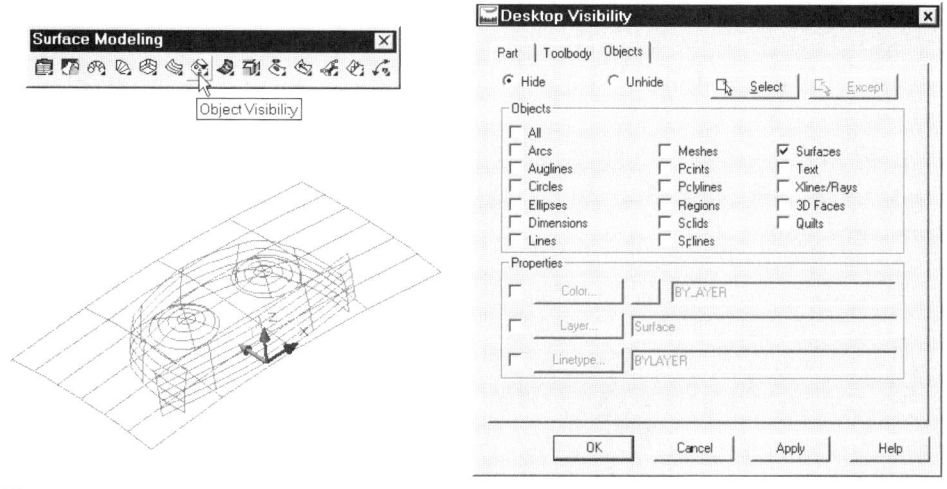

Figure 7–121 *Surfaces unhidden*

81. Select Surface>Create Surface>Fillet or select Fillet Surface from the Surface Modeling toolbar.

    Command: AMFILLETSF

82. Referring to Figure 7–122, select the central part of the Loft U surface and the lower edge of the joined surface.

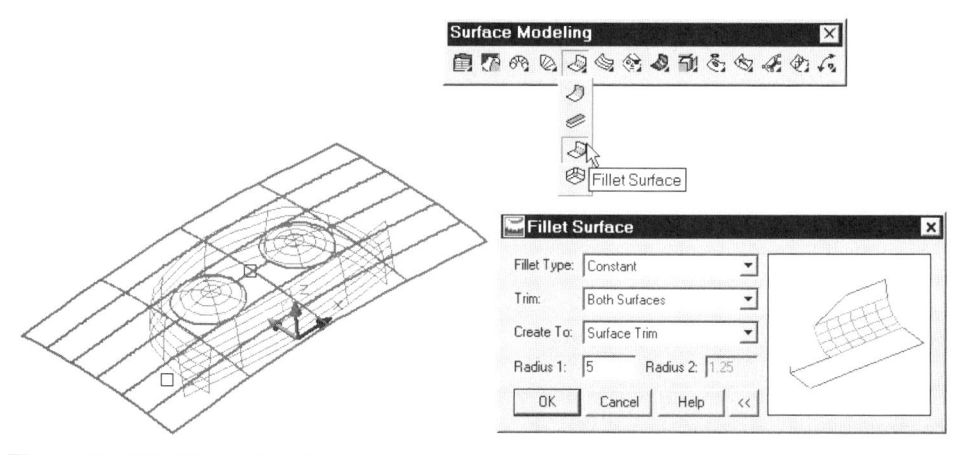

Figure 7–122 *Fillet surface being constructed*

In the Fillet Surface dialog box, the Fillet Type area enables you to decide which type of fillet surface to construct. If you select the Variable button, you can select either the Linear or Cubic button. The fillet radii of a linear variable fillet change linearly, and the fillet radii of a cubic variable fillet change cubically from one end to the other. Because the surfaces selected for making a fillet may not have the same edge length, you can choose to extend the fillet surface to the longer edge by selecting the Extended button. In some situations, surfaces selected for filleting may have been trimmed. To construct the fillet surface in accordance with the original untrimmed surface, select the Base Surface button.

83. Select Constant type, set radius to 5 mm, and select the OK button.

The selected surfaces are trimmed and a constant radius fillet is constructed between them (see Figure 7–123).

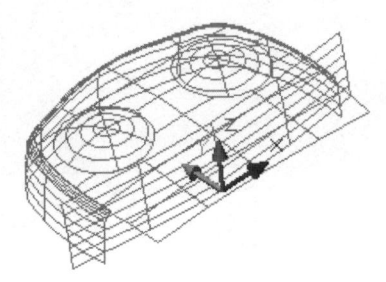

Figure 7–123 *Fillet surface constructed*

84. Repeat the AMFILLETSF command.
85. Select the surfaces indicated in Figure 7–124.
86. In the Fillet Surface dialog box, select Constant Extended End(s) in the Fillet Type box, select Base Surface in the Create To box, set radius to 5 mm, and select the OK button.

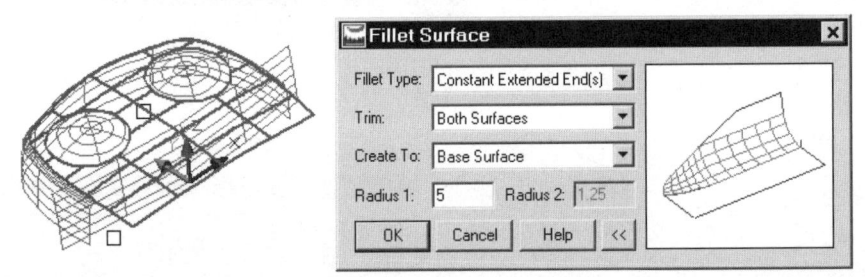

Figure 7–124 *Fillet surfaces constructed*

The surfaces are trimmed and an extended fillet surface of 5 mm constant radius is constructed to the base surfaces (see Figure 7–125).

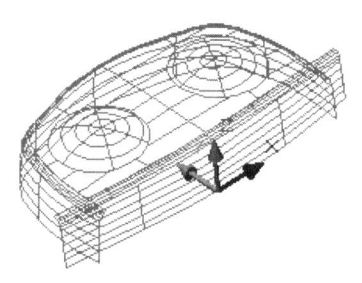

Figure 7–125 *Extended fillet surface constructed to the base surfaces*

A constant radius fillet extended to the base surface is created between the selected surfaces.

87. Referring to Figure 7–126, construct two more constant extended fillet surfaces (created to the base surfaces) of radius 5 mm.

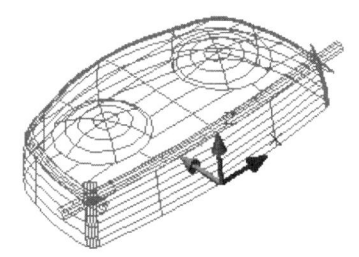

Figure 7–126 *Fillet surfaces constructed*

Now you have four fillet surfaces. Among them two surfaces intersect at two locations. In order to construct two corner fillet surfaces, you will break a fillet surface into two.

88. Hide all the surfaces except the fillet surface shown Figure 7–127.

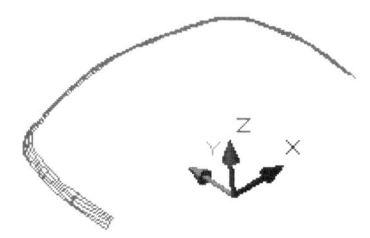

Figure 7–127 *Surfaces hidden*

89. Use the shortcut key 9 to set the display to the top view (see Figure 7–128).
90. Select Surface>Edit Surface>Break or select Break Surface from the Surface Modeling toolbar.
91. Select the surface indicated in Figure 7–128 and press the ENTER key.

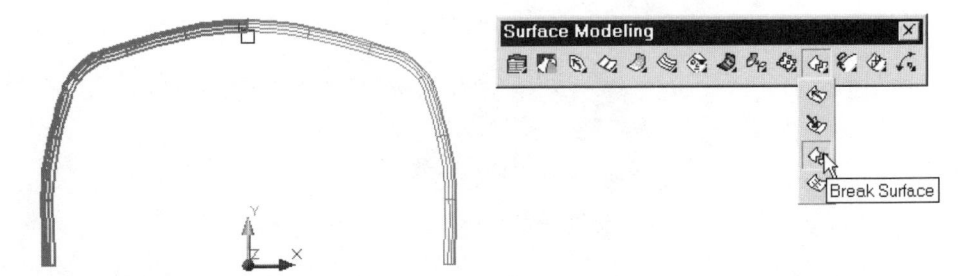

Figure 7–128 *Fillet surface broken into two surfaces*

Now you will construct a corner surface.

   92. Unhide all the hidden surfaces.
   93. Set the display to an isometric view by using the shortcut key **8**.
   94. Select Surface>Create Surface>Corner Fillet or select Corner Fillet Surface from the Surface Modeling toolbar.

      Command: AMCORNER

   95. Select the fillet surfaces indicated in Figure 7–129.

A corner fillet is constructed (see Figure 7–130).

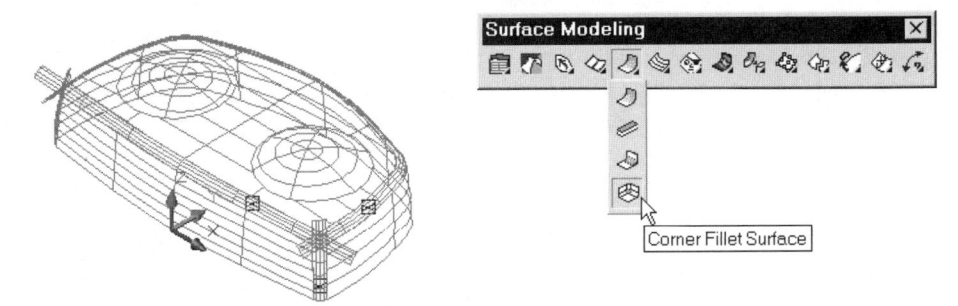

Figure 7–129 *Corner fillet being constructed*

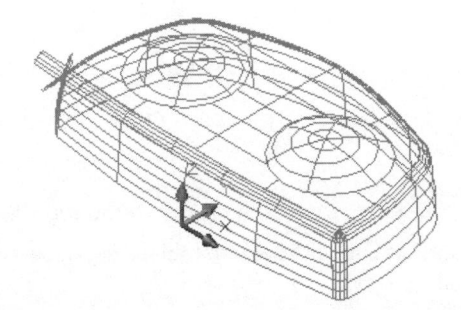

Figure 7–130 *Corner fillet constructed*

Surface Modeling I 343

96. Referring to Figure 7–131, construct another corner fillet surface.

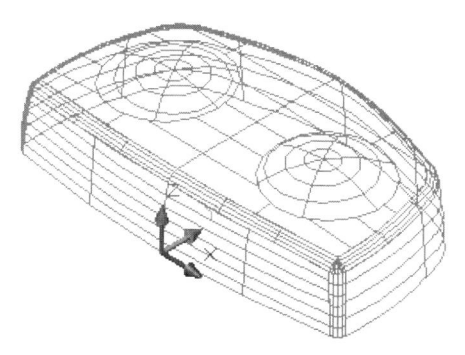

Figure 7–131 *Corner fillet constructed*

The upper part of the remote control is complete. Now you will construct the bottom surface of the remote control. You will copy the lower edges of the surfaces and construct a trimmed planar surface.

97. Set the current layer to Wire.
98. Use the shortcut key 9 to set the display to top view.
99. Select Surface>Create Wireframe>Copy Edge or select Copy Surface Edge from the Surface Modeling toolbar.
100. Type **S** to use the Spline option, or press the ENTER key if the default output option is already set to spline.
101. Select the lower edges indicated in Figure 7–132 and press the ENTER key.

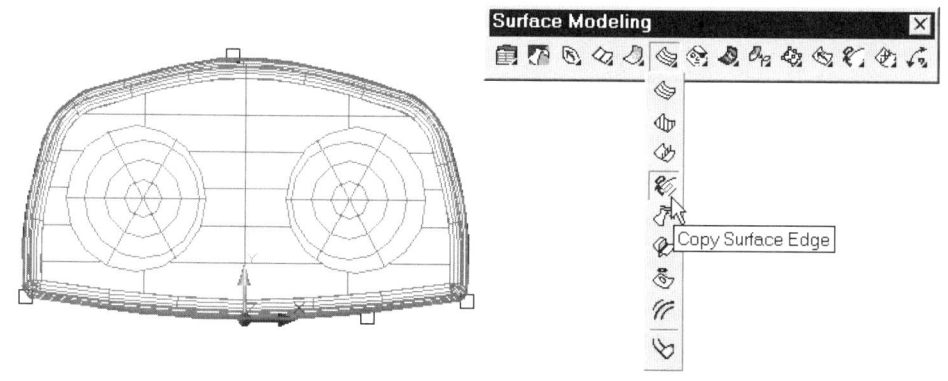

Figure 7–132 *Edges being copied*

102. Hide all the surfaces (see Figure 7–133).

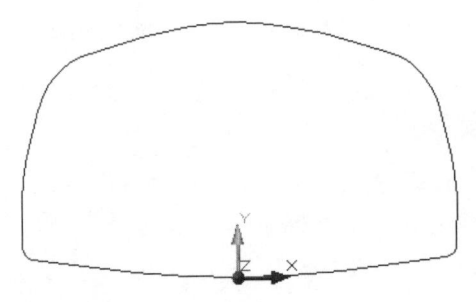

Figure 7-133 *Edges copied and surfaces hidden*

103. Set the current layer to Surface.
104. Select Surface>Create Surface>Planar Trim or select Planar Trimmed Surface from the Surface Modeling toolbar.
105. Select all the copied edges and press the ENTER key.

A trimmed planar surface is constructed (see Figure 7–134).

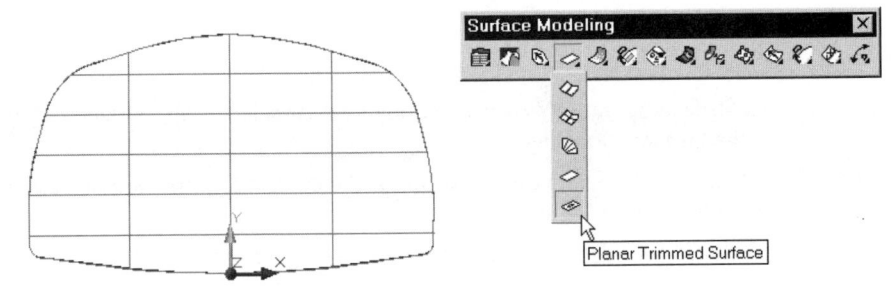

Figure 7-134 *Trimmed planar surface constructed*

106. Unhide all the surfaces.
107. Set the display to an isometric view (see Figure 7–135).

The model is complete. Save your file (file name: *RemoteControl1.dwg*).

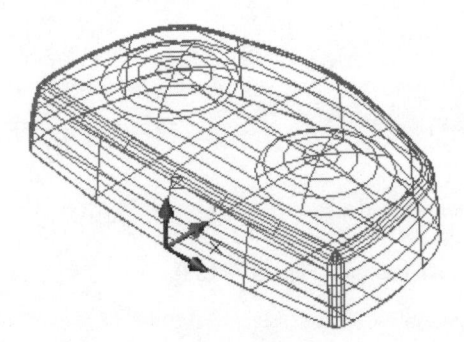

Figure 7-135 *Completed model*

REMOTE CONTROL SOLID MODEL 1

Now construct a solid model from the remote control surface model by using a surface quilt to cut an AutoCAD native solid.

1. Open the file *RemoteControl1.dwg*, if you have already closed it.
2. Select File>Save As and specify a new file name (*RemoteControl2.dwg*).
3. Select Surface>Surface Stitching

   Command: AMSTITCH
4. Select all the surfaces except the bottom surface and press the ENTER key.
5. In the Surface Stitching dialog box (shown in Figure 7–136), select Optimal Stitching and Quilt, and select the OK button.
6. Hide all the surfaces (see Figure 7–137).

Note that there will not be any visual difference after a set of surfaces is stitched to a quilt.

Now construct an AutoCAD native solid box and use the quilt to cut the box. (To cut a solid, you use either a single surface or a quilt. All the boundary edges of the surface or the quilt must lie on or outside the boundary of the solid that is to be cut.)

7. Select Design>Solids>Box

   Command: BOX
8. Type **–100,–10** to specify the first corner.
9. Type **100,110** to specify the opposite corner.
10. Type **70** to specify the height.

A solid box is constructed (see Figure 7–138).

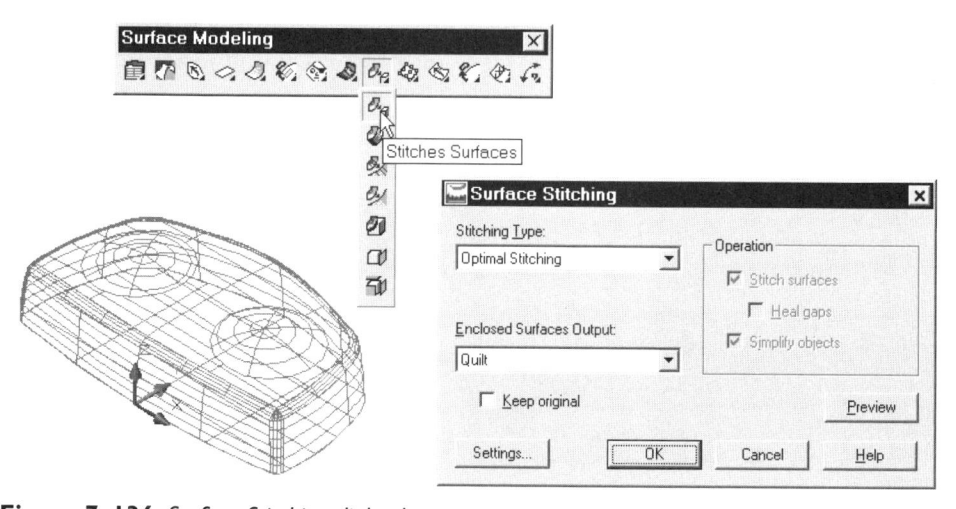

Figure 7–136 *Surface Stitching dialog box*

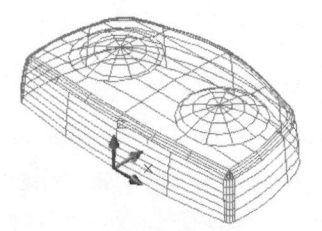

Figure 7-137 *Quilt constructed and surface hidden*

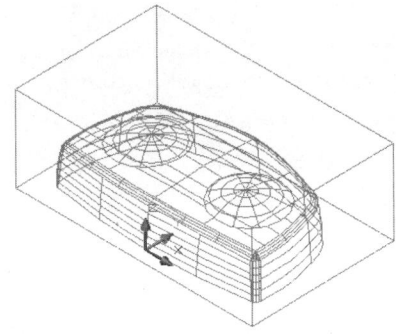

Figure 7-138 *Solid box constructed*

11. Select Surface>Edit Solid or select Solid Cut from the Surface Modeling toolbar (see Figure 7-139).

    Command: AMSOLCUT

12. Select the AutoCAD solid.
13. Select the quilt.
14. Left-click if necessary to flip the direction of the portion to remove so that the upper part of the solid is removed.
15. Right-click to accept.

The solid is cut by the quilt (see Figure 7-140). Save and close the file.

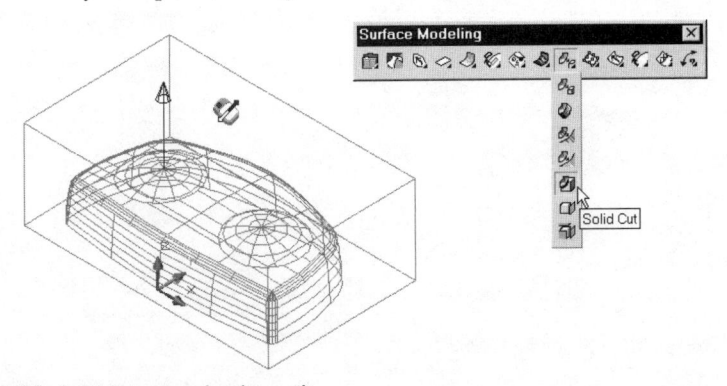

Figure 7-139 *Solid being cut by the quilt*

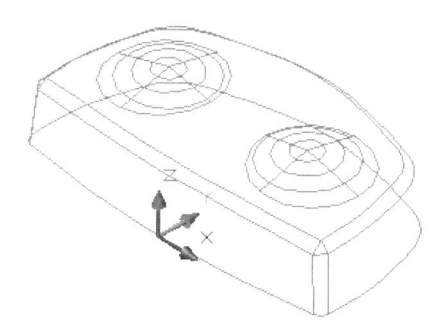

Figure 7-140 *Solid cut by the quilt*

REMOTE CONTROL SOLID MODEL 2

Another way of making a solid is to stitch all the surfaces together.

1. Open the file *RemoteControl1.dwg*.
2. Select File>Save As and specify a new file name (*RemoteControl3.dwg*).
3. Select Surface>Surface Stitching or select Stitch Surfaces from the Surface Modeling toolbar.
4. Select all the surfaces and press the ENTER key.
5. In the Surface Stitching dialog box, select Optimal Stitching and Part, and select the OK button (see Figure 7-141).
6. Save and close your file.

The surfaces are stitched to a static base solid part (see Figure 7-142). The static solid part, although non-parametric, enables you to add parametric solid features to further elaborate the design. The added solid features, however, are parametric.

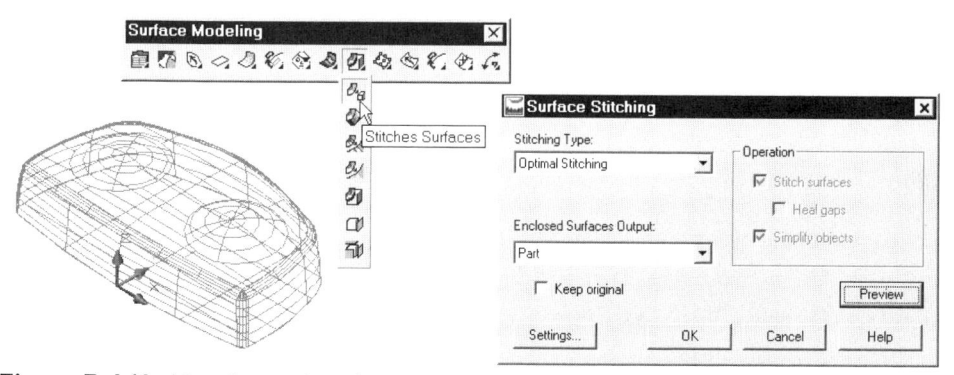

Figure 7-141 *All surfaces selected*

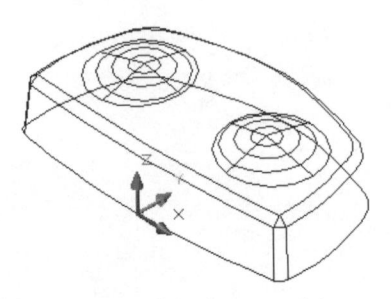

Figure 7–142 *Surfaces stitched into a solid*

The different between surface cutting and surface stitching are:

- An AutoCAD native solid cut by a surface is still an AutoCAD solid. The boundary of the surface of quilt that is used to cut an AutoCAD solid must be lying on or outside the boundary of the AutoCAD solid.
- A Mechanical Desktop base solid is constructed when a set of surfaces are stitched. The set of surfaces to be stitched to a solid must enclose an "airtight" volume.

MOUSE

Now you will construct a surface model of a computer mouse and thicken the surfaces to become a solid model. Figure 7–143 shows the completed solid model.

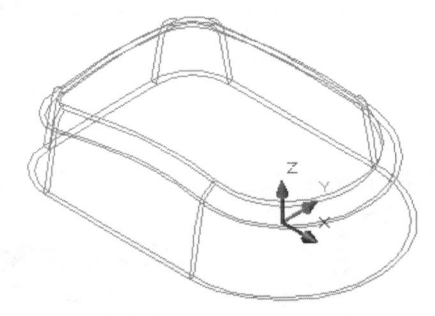

Figure 7–143 *Mouse*

1. Start a new part file. Use metric as the default.
2. Set up two layers: Wire and Surface.
3. Set layer Wire the current layer.
4. Construct a line segment running from **–32,20** to **0,20**.
5. Construct another line segment running from **–32,–20** to **0,–20**.
6. Construct the third line segment running from **–40,12** to **–40,12**.

Three line segments constructed (see Figure 7–144).

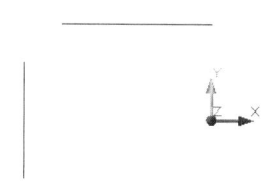

Figure 7–144 *Three line segments constructed*

7. Referring to Figure 7–145, construct two fillets of radius 8 mm.
8. Select Design>Circle>2 Points
9. Press the SHIFT key and right-click.
10. Select Endpoint and select an endpoint indicated in Figure 7–146.
11. Press the SHIFT key and right-click.
12. Select Endpoint and select the other endpoint indicated in Figure 7–146.
13. Referring to Figure 7–147, trim the circle.

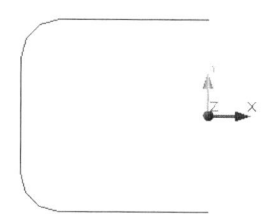

Figure 7–145 *Fillets constructed*

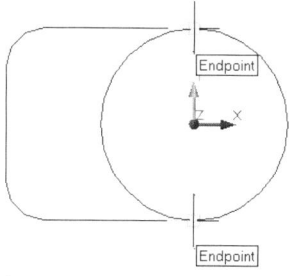

Figure 7–146 *Circle constructed*

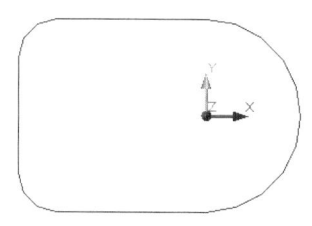

Figure 7–147 *Circle trimmed*

14. Set the display to an isometric view.
15. Construct three splines as follows:

First point	−45,0,12	25,−30,7	−40,12,0
Next point	−10,0,17	25,0,12	−38,12,11
Next point	25,0,12	25,30,7	−35,12,20
Next point	ENTER	ENTER	ENTER
Start tangent	ENTER	ENTER	ENTER
End tangent	ENTER	ENTER	ENTER

Three splines are constructed (see Figure 7–148).

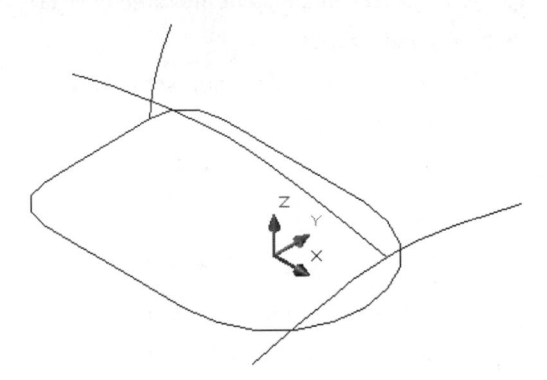

Figure 7–148 *Splines constructed*

Now you will construct a number of swept surfaces.

16. Set the current layer to Surface.
17. Select Surface>Create Surface>Sweep or select Swept Surface from the Surface Modeling toolbar.

    Command: AMSWEEPSF

18. Referring to Figure 7–149, select the spline as the cross section and press the ENTER key.
19. Select the line as the rail and press the ENTER key.
20. In the Sweep Surface dialog box, select the OK button.
21. Using the swept surface's edge as the cross section and an arc as the rail, construct a swept surface as shown in Figure 7–150.
22. Referring to Figure 7–151, construct four more swept surfaces one by one, each using its adjacent surface's edge as a cross section and a wire on the XY plane as a rail.
23. Now construct one more swept surface (see Figure 7–152).

Surface Modeling I 351

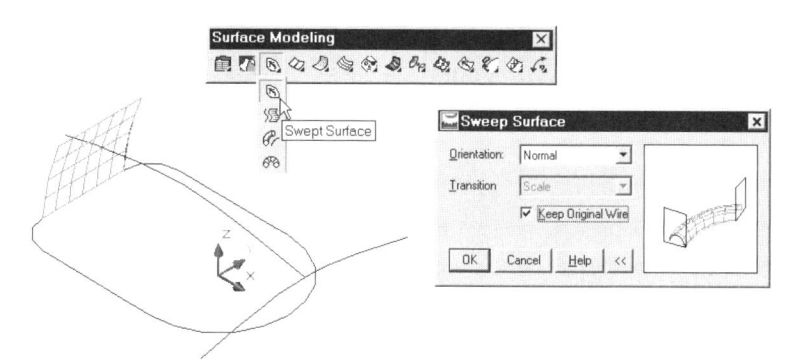

Figure 7-149 Swept surface constructed

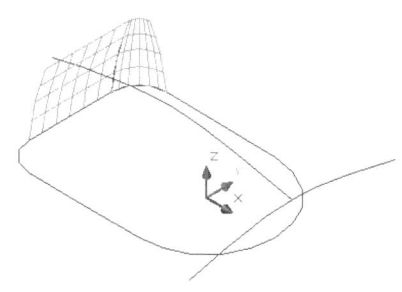

Figure 7-150 Swept surface constructed

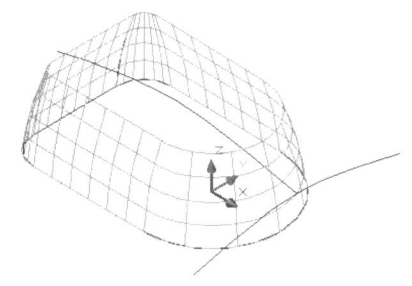

Figure 7-151 Four more swept surfaces constructed

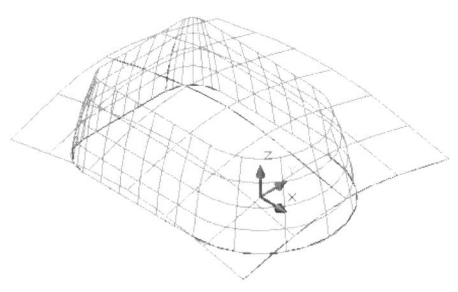

Figure 7-152 One more swept surface constructed

24. Turn off layer Wire.
25. Referring to Figures 7–153 through 7–155, construct six constant fillet surfaces of 3 mm radius.

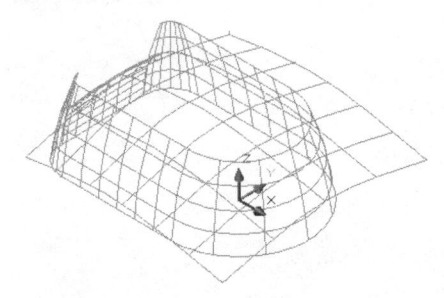

Figure 7–153 *Constant radius fillet surface constructed*

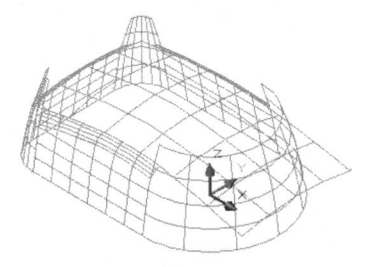

Figure 7–154 *Two more constant fillet surfaces constructed*

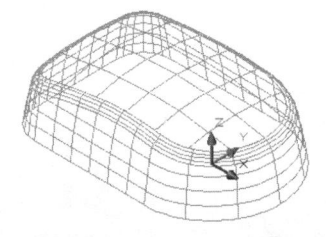

Figure 7–155 *Three more constant fillet surfaces constructed*

Now you will check the normal directions of the surfaces.

26. Select Gouraud Shaded, Edges On from the Mechanical View toolbar (see Figure 7–156).
27. Select 3D Orbit from the Mechanical View toolbar (see Figure 7–157).

As can be seen, some surfaces are not shaded because their normals are in the wrong direction. (Note that the display may not be exactly the same as yours.)

28. Select a point on the screen and drag to rotate the display.
29. Visually examine the surfaces for wrong direction of normals.

30. Select Surface>Edit Surface>Adjust Normals or select Flip Surface Normal.
31. Select those surfaces with a wrong direction of normals so that all the surfaces' normals are the same (see Figure 5–158).

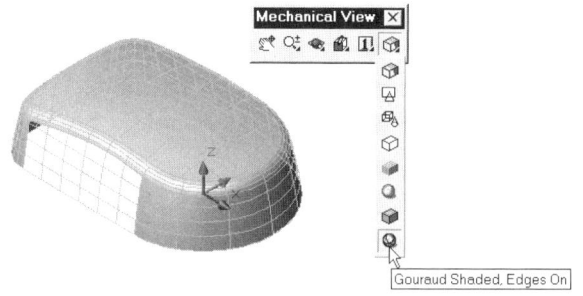

Figure 7–156 *Display shaded*

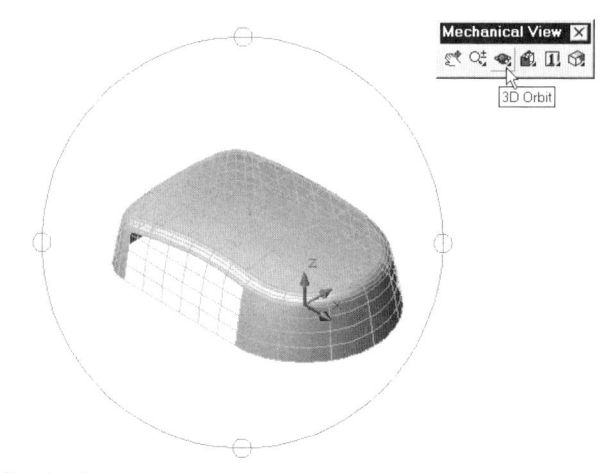

Figure 7–157 *Display being rotated*

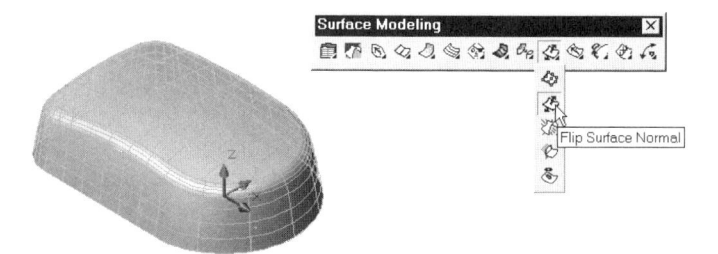

Figure 7–158 *Normals flipped*

Now you will stitch the surfaces into a single quilt (see Figure 7–159).

32. Select Surface>Surface Stitching or select Stitches Surfaces from the Surface Modeling toolbar.
33. Select all the surfaces and press the ENTER key.
34. In the Surface Stitching dialog box, set output to a quilt and select the OK button.

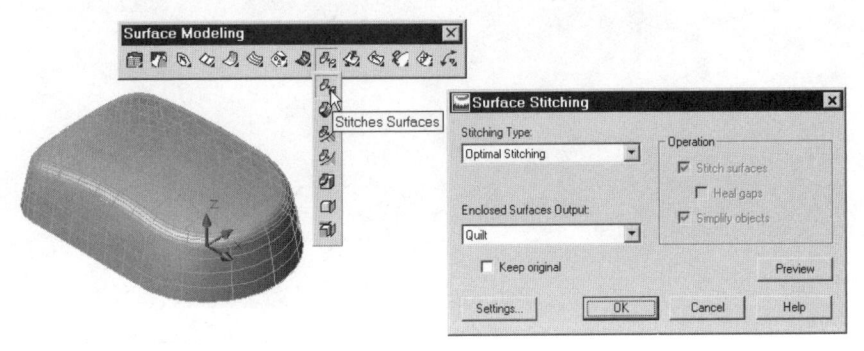

Figure 7–159 *Surfaces being stitched to a quilt*

Now you will thicken the quilt.

35. Select Surface>Surface Thicken or select Surface Thicken from the Surface Modeling toolbar.
36. Select the quilt and press the ENTER key.
37. Right-click to accept the direction of thickening, if it is the same as that shown in Figure 7–160.
38. Type **1** to specify the thickening distance.
39. Select 3D Wireframe from the Mechanical View toolbar.

A solid part is constructed (see Figure 7–161). Save and close your file (file name: *Mouse.dwg*).

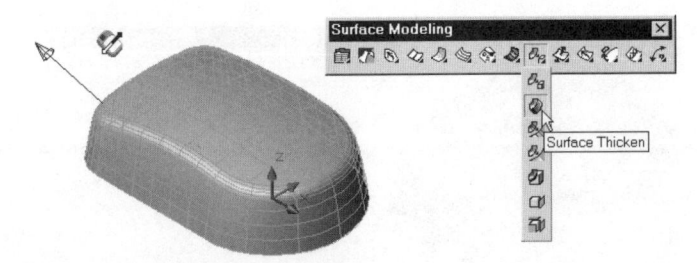

Figure 7–160 *Quilt being thickened*

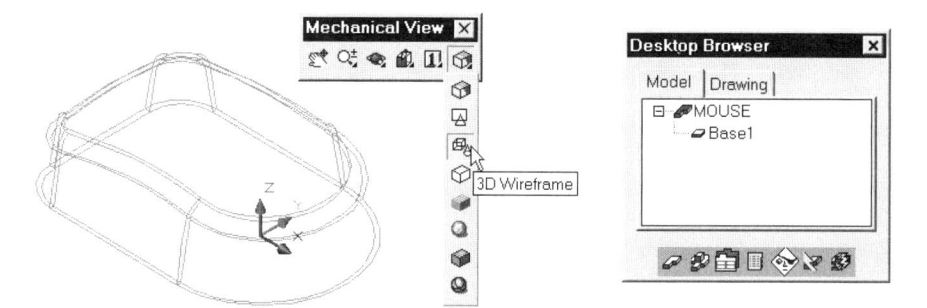

Figure 7–161 *Display set*

Similar to constructing a shell feature in a solid part, thickening concerns offsetting the surfaces. If the radius of curvature of the surface is smaller than or equal to the offset distance, thickening will fail to complete. Therefore, thickening a quilt may not always be successful if any of the individual surface members cannot be thickened. If a problem is encountered during thickening a quilt, convert the quilt back to a set of individual surfaces by using the AM2SF command and thicken each surface one by one to find out where the problem is occurring.

REVIEW QUESTIONS

1. Compare surface modeling with solid modeling with regard to the representation of 3D objects in the computer.

2. Give a brief account of the types of surfaces that you construct using Mechanical Desktop.

3. What kind of AutoCAD and Mechanical Desktop objects can you convert to surfaces?

4. What are the editing processes that you apply on a surface?

5. List the Mechanical Desktop wire construction and editing tools.

6. What is the major difference between joining wires to form a spline and fitting wires to form splines?

7. How does the direction of a wire affect a surface constructed from it?

8. Explain how you use augmented lines in constructing a swept surface.

9. Explain the ways to construct free-form solid parts.

CHAPTER 8

Surface Modeling II

OBJECTIVES

The goals of this chapter are to familiarize you with using the surface modeling module in free-form surface design. After studying this chapter, you should be able to do the following:

- Thoroughly understand the NURBS surface modeling tools of Mechanical Desktop
- Apply Mechanical Desktop surface modeling tool in designing free-form objects

OVERVIEW

Car bodies involve many surfaces of different kinds. Constructing model car bodies will help you learn more about surface modeling. While constructing the wires and surfaces for the car bodies in the projects in this chapter, you will learn various surface and wire construction methods and work on various utility tools.

SURFACE MODELING TOOLS

There are two basic ways to represent a surface in the computer: polygon mesh and NURBS mathematics. To accurately represent free-form surfaces in the computer, you use NURBS surfaces. A three-dimensional surface model usually consists of a number of surfaces put together. Depending on the shapes of each surface, you use various surface construction tools to construct primitive surfaces, free-form surfaces, derived surfaces, trimmed surfaces, and converted surfaces.

SURFACE CONSTRUCTION

There are many ways to construct a NURBS surface. Using Mechanical Desktop, you can construct four categories of surfaces: primitive surfaces, free-form surfaces, derived surfaces, trimmed surfaces, and converted surfaces.

Primitive Surfaces

Primitive surfaces are surfaces of regular geometric shapes that you construct by specifying their types, dimensions, and locations.

Primitive surfaces consist of: cylindrical surfaces, conical surfaces, spherical surfaces, toroidal surfaces, and planar surfaces.

To make cylindrical and toroidal surfaces together, you construct a tubular surface.

Free-Form Surfaces

Free-form surfaces are the kind that you use most in making 3D surface models.

Free-form surfaces include: revolved surfaces, extruded surfaces, rule surfaces, Loft U surfaces, swept surfaces, and Loft UV surfaces.

To make free-form surfaces, you have to define a set of wires. Making free-form surfaces is simple once the wires have been constructed. However, making the wires is a tedious job.

Derived Surfaces

Derived surfaces are surfaces that are derived from any existing surfaces. They may consist of: fillet surfaces, corner surfaces, blend surfaces, and offset surfaces.

Trimmed Surfaces

Smooth surfaces are constructed from smooth wires. However, most smooth surfaces that you can use to form a 3D model have irregular boundary edges. There are two ways to make a smooth surface with irregular edges:

- Build surfaces and use wires to trim away the unwanted portions.
- Construct two intersecting surfaces and trim them.

Converted Surfaces

Some AutoCAD objects, such as 2D objects with thickness and native solids, and Mechanical Desktop solids, can be converted to sets of surfaces.

SURFACE MANIPULATION

There are many ways to manipulate a surface. They include:

- Trimming a surface
- Untrimming a trimmed surface
- Truncating a trimmed surface
- Breaking a surface into two surfaces
- Joining surfaces into a single surface
- Stitching surfaces to a quilt

- Lengthening a surface
- Scaling a surface
- Adjusting the edge of a surface
- Replacing the edge of a surface
- Refining the patch of a surface
- Changing the grip points of a surface
- Modifying the span of a surface
- Flipping the normal of a surface

3D WIRE CONSTRUCTION

Before you can construct 3D surfaces, you must construct 3D wires. Using Mechanical Desktop, you can construct wires in many ways:

- Construct augmented lines
- Join wires to form polyline or spline
- Fit splines from existing wires
- Construct a tangent spline
- Construct an offset wire
- Construct a fillet wire
- Construct an intersection wire
- Project wires onto a surface to construct another set of wires
- Copy edges
- Extract edges
- Generate section cuts
- Generate flow wires
- Construct parting lines

WIRE MANIPULATION

Mechanical Desktop contains tools to manipulate 3D wires. These include:

- Editing a spline
- Unsplining a spline
- Changing the number of control points
- Changing wire direction
- Editing an augmented line

SOLID OPERATION

There are three ways to use a surface in solid modeling:

- Use a surface or a quilt to cut an AutoCAD solid.
- Use a surface or a quilt to cut a Mechanical Desktop solid.
- Stitch a set of surfaces that enclose an "air-tight" volume into a base solid part.

SUMMARY

In making a surface model, you must construct wires before making the surfaces. The process of surface modeling starts from making the wires. However, the first thing that comes to your mind when you want to make a surface model are the shapes, profiles, and silhouettes of the surfaces, not the wires. Because you need to construct the wires first, you have to think about how they look. Before you can do that, you have to identify what kinds of free-form surfaces are to be constructed. Once the kinds of surfaces are identified, you can make the wires. From the wires, you will construct the surfaces.

PROJECTS

Now you will work on the following car bodies to enhance your surface modeling knowledge.

MODEL CAR A

Figure 8–1 shows the three-dimensional surface model of a model car. Take some time to study the model and think about what surfaces are required and what wires are needed. To help you see more clearly what surfaces are required, the model is exploded in Figure 8–2. The wires for making the surfaces are shown in Figure 8–3. You will construct these wires and then construct surfaces from these wires.

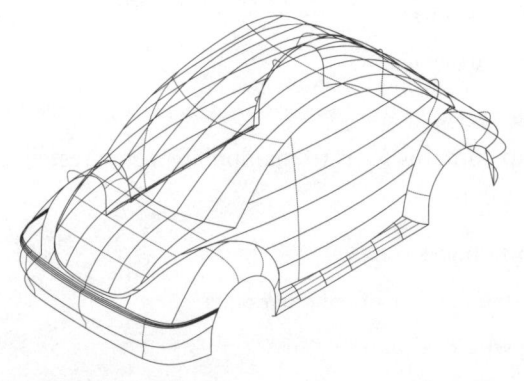

Figure 8–1 *Surface model of model car A*

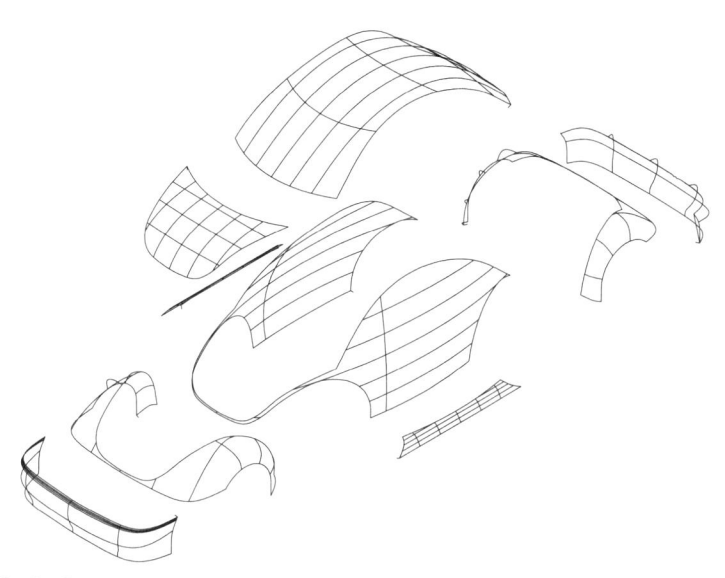

Figure 8–2 *Surfaces exploded apart*

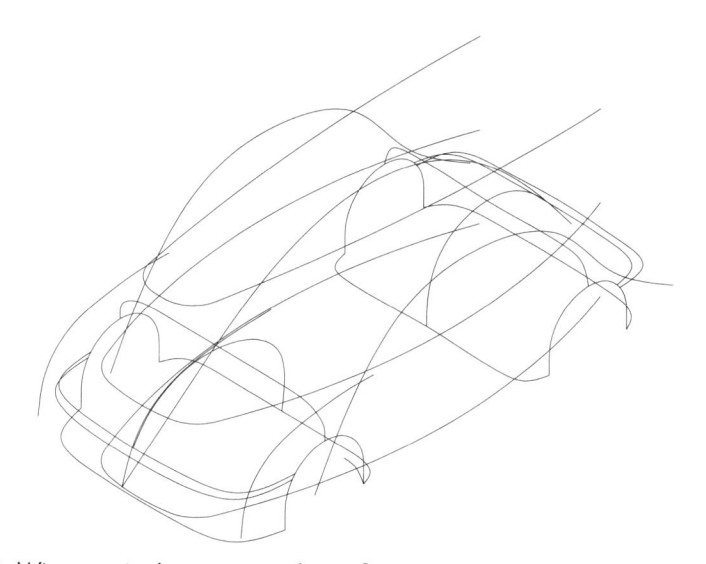

Figure 8–3 *Wires required to construct the surfaces*

1. Start a new file. Use metric as the default.
2. Construct two layers: Wire and Surface.
3. Set the layer Wire as the current layer.

Now you will construct six arcs (A, B, C, D, E, and F; see Figure 8–4).

4. Use the ARC command six times by selecting Design>Arc>3 Points.

5. Each time, type the coordinates of the start point, second point, and end point in accordance with the table below:

Arc	A	B	C	D	E	F
Start point	130,–25	–32,–20	–32,20	130,–25	–24,–12	–24,12
Second point	49,–38	–35,0	49,38	53,–23	–26,0	53,23
End point	–32,–20	–32,20	130,25	–24,–12	–24,12	130,25

Now you will construct four fillets.

6. Select Surface>Edit Wireframe>Fillet or select Fillet Wire from the Surface Modeling toolbar.

   Command: AMFILLET3D

7. Type **R** to use the radius option.
8. Type **14** to set the fillet radius.
9. Select A and B (shown in Figure 8–4).
10. Repeat the AMFILLET3D command.
11. Select B and C (shown in Figure 8–4).
12. Repeat the AMFILLET3D command.
13. Type **R** and then **12** to set the fillet radius.
14. Select D and E (see Figure 8–4).
15. Repeat the AMFILLET3D command.
16. Select E and F (see Figure 8–4).

The fillets are constructed (see Figure 8–5).

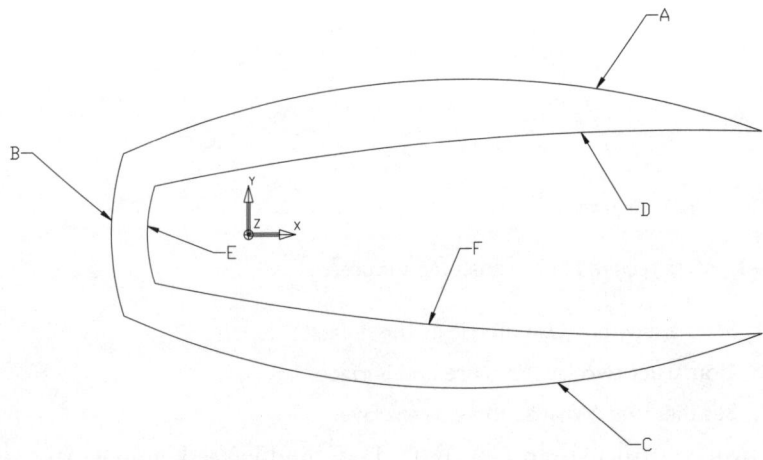

Figure 8–4 *Six arcs constructed*

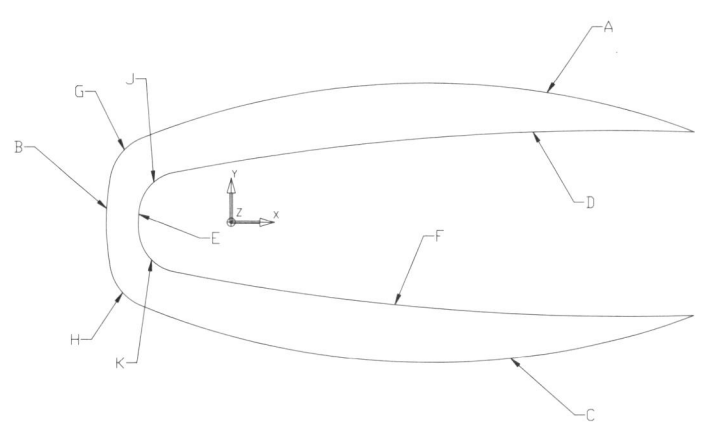

Figure 8–5 *Fillets constructed*

Now you will join the wires into two splines.

17. Select Surface>Surface Options or select AutoSurf Options from the Surface Modeling toolbar.
18. In the Surface tab of the Mechanical Options dialog box, uncheck the Keep Original box and select the OK button.

This way, the original wires will be deleted after the joined wire is constructed.

19. Select Surface>Edit Wireframe>Join or select Join Wire from the Surface Modeling toolbar.

    Command: AMJOIN3D

20. In the Join 3D dialog box (see Figure 8–6), select automatic mode and polyline output, and select the OK button.
21. Select A (see Figure 8–5) as the start wire.
22. Select G, B, H, and C and press the ENTER key.
23. Repeat the AMJOIN3D command to join wires D, J, E, K, and F (as shown in Figure 8–6).

The wires are joined into two splines.

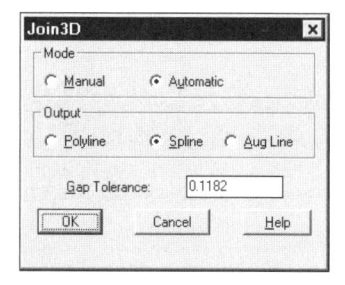

Figure 8–6 *Join 3D dialog box*

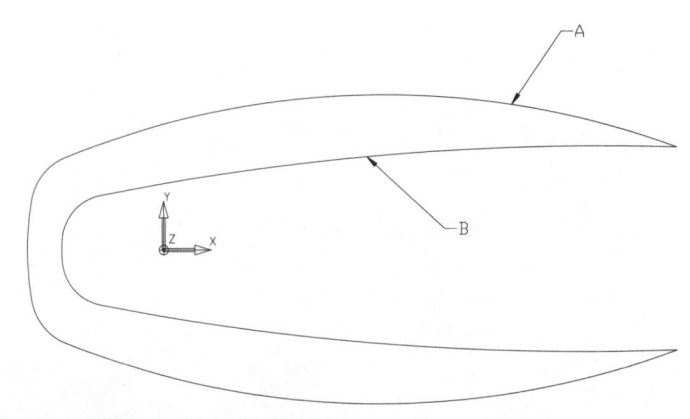

Figure 8–7 *Corners filleted and arcs joined*

Now you will move a spline and copy the other spline.

24. Set the display to an isometric view.
25. Select Modify>Copy.
26. Select A (see Figure 8–7) and press the ENTER key.
27. Type **0,0,32.5** at the command line area.
28. Press the ENTER key.

Spline A is copied a distance of 32.5 mm in the Z direction.

29. Select Modify>Move
30. Select B (see Figure 8–7) and press the ENTER key.
31. Type **0,0,65** at the command line area.
32. Press the ENTER key.

Spline B is moved a distance of 65 mm in the Z direction (see Figure 8–8). Now you will construct a Loft U surface.

33. Set the current layer to Surface.
34. Select Surface>Create Surface>Loft U or select Loft U Surface from the Surface Modeling toolbar.
35. Select splines A, B, and C (shown in Figure 8–8) one by one and press the ENTER key.
36. In the Loft Surface dialog box, select the OK button.

A Loft U surface is constructed (see Figure 8–9).

Now you will hide the surface and splines, set the current layer to Wire, and construct four splines.

37. Select Surface>Surface Visibility or select Object Visibility from the Surface Modeling toolbar.

38. In the Objects tab of the Mechanical Visibility dialog box, check the Hide and All boxes, and select the OK button.

The surface and splines are hidden.

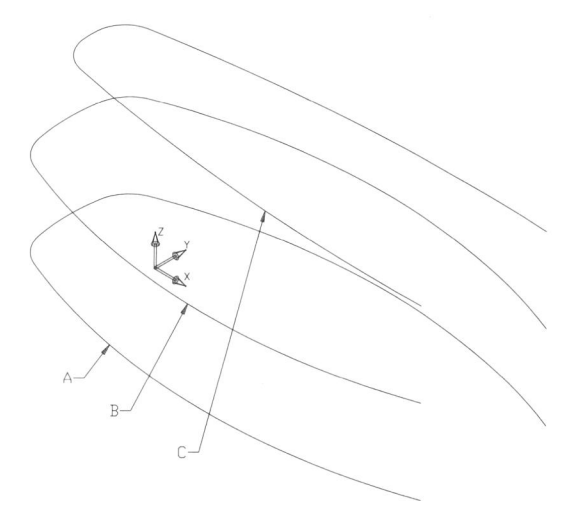

Figure 8–8 *Splines copied and moved*

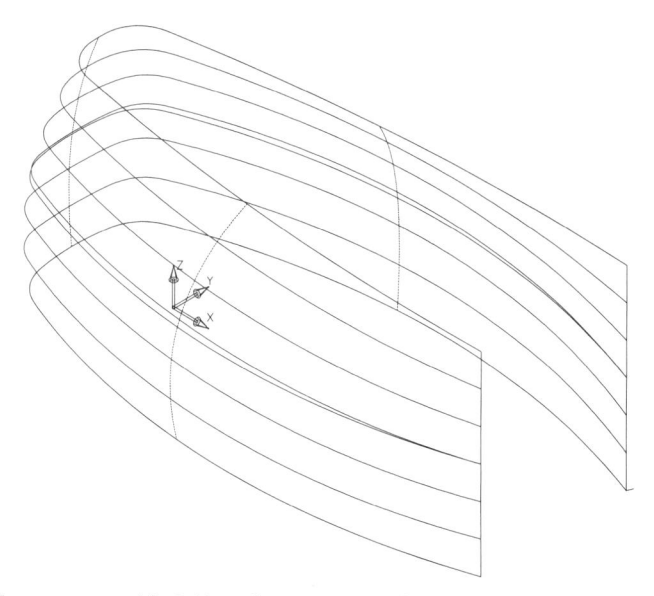

Figure 8–9 *Current set and Loft U surface constructed*

39. Set the current layer to Wire.
40. Referring to the following tables, construct four splines (see Figure 8–10).

Spline	1	2
First Point	−4,−42,6	−41,0,6
Next Point	12,−42,33	−12,0,29
Next Point	35,−42,54	18,0,50
Next Point	67,−42,58	53,0,61
Next Point	96,−42,47	89,0,55
Next Point	117,−42,24	119,0,34
Next Point	143,−42,6	143,0,6
Next Point	ENTER	ENTER
Start Tangent	ENTER	ENTER
End Tangent	ENTER	ENTER
Spline	**3**	**4**
First Point	−34,−42,6	−41,0,6
Next Point	−31,−42,20	−35,0,21
Next Point	−21,−42,30	−24,0,31
Next Point	−8,−42,34	−12,0,35
Next Point	9,−42,36	5,0,37
Next Point	20,−42,36	20,0,37.5
Next Point	ENTER	ENTER
Start Tangent	ENTER	ENTER
End Tangent	ENTER	ENTER

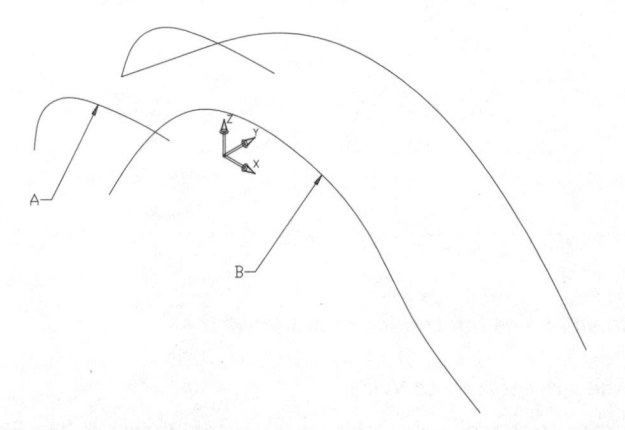

Figure 8–10 *Four splines constructed*

41. Select Modify>Copy
42. Select splines A and B (shown in Figure 8–10) and press the ENTER key.
43. Type **84<90** to specify a distance.
44. Press the ENTER key.

Two splines are copied (see Figure 8–11). Now you will construct two Loft U surfaces.

45. Set the current layer to Surface.
46. Select Surface>Creat Surface>Loft U or select Loft U surface from the Surface Modeling toolbar.
47. Select splines A, B, and C (shown in Figure 8–11) one by one and press the ENTER key.
48. In the Loft Surface dialog box, select the OK button.
49. Repeat the AMLOFTU command to construct another Loft U surface on splines E, F, and G (shown in Figure 8–11).

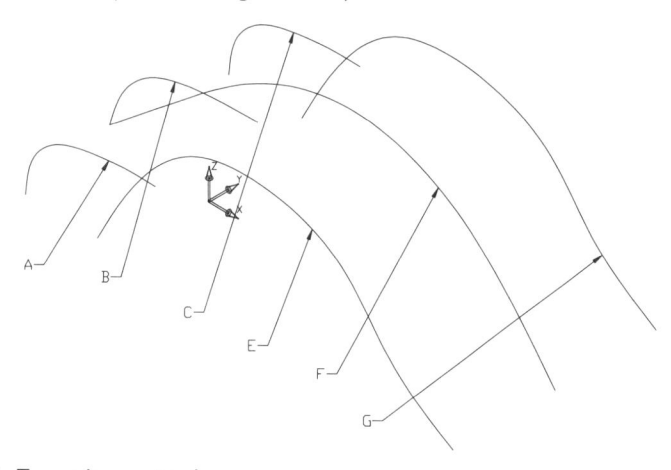

Figure 8–11 *Two splines copied*

Two Loft U surfaces are constructed (see Figure 8–12). Now you will trim the surfaces and stitch them into a single quilt.

50. Select Surface>Edit Surface>Intersect Trim or select Intersect and Trim from the Surface Modeling toolbar.
51. Select surfaces A and B (shown in Figure 8–12). You should select the portion of the surface that you want to retain.
52. In the Surface Intersection dialog box, select the OK button.
53. Select Surface>Surface Stitching or select Stitches Surfaces from the Surface Modeling toolbar.
54. Select the surfaces and press the ENTER key.

55. In the Surface Stitching dialog box, select optimal stitching type and quilt output, and select the OK button.

Now you will hide the splines and unhide the hidden surface.

56. Select Surface>Surface Visibility or select Object Visibility from the Surface Modeling toolbar.
57. In the Objects tab of the Mechanical Visibility dialog box, check the Hide and Spline boxes, and select the Apply button.
58. Check the Unhide and Surfaces box, and select the OK button.

The Loft U surfaces are trimmed, the splines are hidden, and the hidden surface is unhidden (see Figure 8–13).

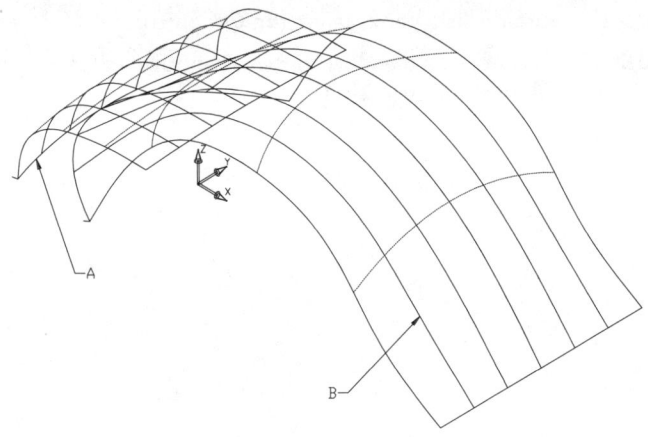

Figure 8–12 *Two Loft U surfaces constructed*

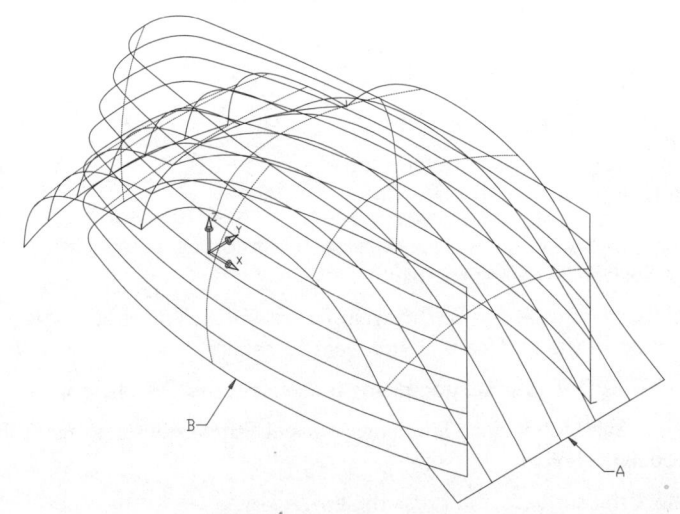

Figure 8–13 *Surfaces trimmed and stitched, splines hidden, and surface unhidden*

Now you will intersect and trim the surface and the quilt, and stitch the trimmed surface and quilt them into a quilt.

59. Select Surface>Edit Surface>Intersect Trim or select Intersect and Trim from the Surface Modeling toolbar.
60. Select surfaces A and B (shown in Figure 8–13).
61. In the Surface Intersection dialog box, select the OK button.
62. Select Surface>Surface Stitching or select Stitches Surfaces from the Surface Modeling toolbar.
63. Select the surface and the quilt, and press the ENTER key.
64. In the Surface Stitching dialog box, select optimal stitching type and quilt output, and select the OK button.

The surface and quilt are trimmed and stitched into a single quilt (see Figure 8–14).

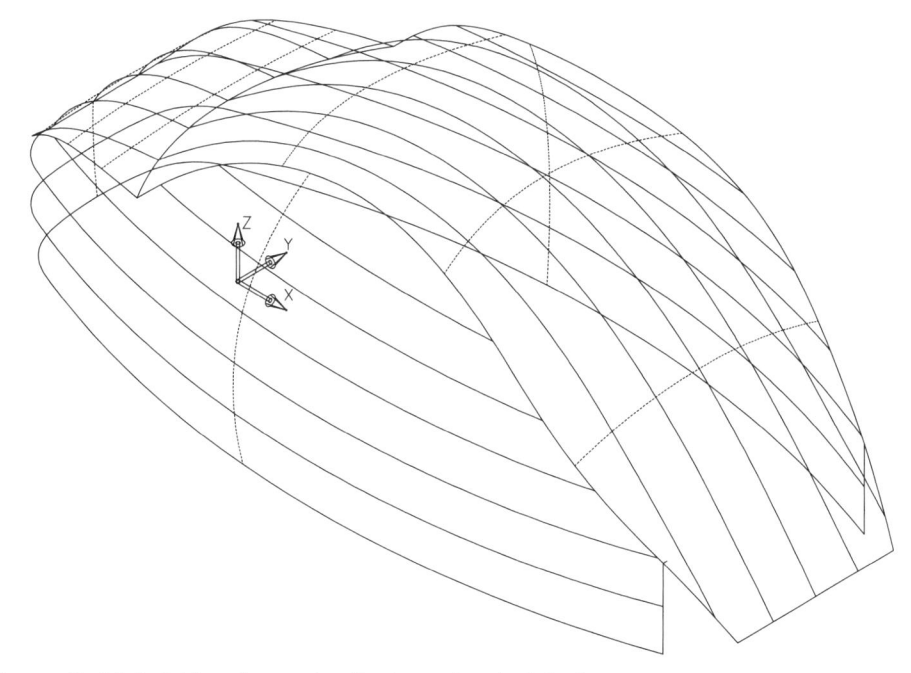

Figure 8–14 *Loft U surface and quilt trimmed and stitched*

Now you will construct six splines.

65. Set the current layer to Wire.
66. Referring to the information provided in the tables below, construct six splines.

Spline	1	2	3	4
First Point	16,−42	−12,−42,17.5	−13,−42,16	−16,−42
Next Point	20.5,−40	−30.5,−37,17.5	−31.5,−37.5,16	−30.5,−36
Next Point	24.5,−23.5	−35.5,−19,17.5	−38,−20,16	−35.5,−17
Next Point	24.5,0	−36.5,0,17.5	−39,0,16	−36,0
Next Point	24.5,23.5	−35.5,19,17.5	−38,20,16	−35.5,17
Next Point	20.5,40	−30.5,37,17.5	−31.5,37.5,16	−30.5,36
Next Point	16,42	−12,42,17.5	−13,42,16	−16,42
Next Point	ENTER	ENTER	ENTER	ENTER
Start Tangent	ENTER	ENTER	ENTER	ENTER
End Tangent	ENTER	ENTER	ENTER	ENTER

Spline	5		6
First Point	0,−42,23	First Point	24.5,0,0
Next Point	0,−41.5,28	Next Point	24,0,19.5
Next Point	0,−37.5,31.5	Next Point	−8,0,35
Next Point	0,−25,33.5	Next Point	−31.5,0,24.5
Next Point	0,25,33.5	Next Point	−36.5,0,17.5
Next Point	0,37.5,31.5	Next Point	ENTER
Next Point	0,41.5,28	Start Tangent	ENTER
Next Point	0,42,23	End Tangent	ENTER
Next Point	ENTER		
Start Tangent	ENTER		
End Tangent	ENTER		

Six splines are constructed (see Figure 8–15).

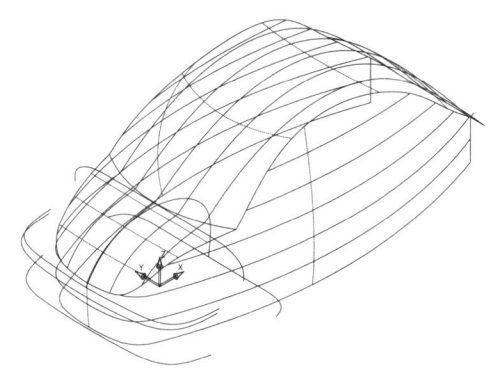

Figure 8-15 *Six splines constructed*

Now you will rotate the UCS and construct a polyline.

67. Select Assist>New UCS>X
68. Type **90** at the command line area to rotate the UCS 90 degrees about the X-axis.
69. Select Design>Polyline

    Command: PLINE
70. Type **16,0,42** to indicate the start point.
71. Type **@7<90** to indicate the next point.
72. Type **A** to use the arc option.
73. Type **@32<180** to indicate the end point of the arc segment.
74. Type **L** to use the line option.
75. Type **@7<270** to indicate the end point of the line.
76. Press the ENTER key.

A polyline is constructed (see Figure 8–16).

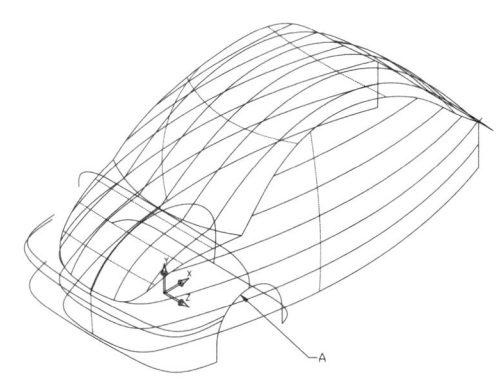

Figure 8-16 *UCS rotated and polyline constructed*

Now you will set the UCS to world and construct a rectangular array of the polyline.

77. Select Assist>New UCS>World
78. Select Modify>Array
79. Select A (shown in Figure 8–16) and press the ENTER key.
80. Type **R** to use the rectangular option.
81. Type **2** to specify the number of rows.
82. Type **2** to specify the number of columns.
83. Type **84** to specify the distance between the rows.
84. Type **108** to specify the distance between the columns.

The UCS is set to world and a rectangular array of the polyline is constructed (see Figure 8–17). Next you will construct a Loft UV surface.

85. Set the current layer to Surface.
86. Select Surface>Create Surface>Loft UV or select Loft UV Surface from the Surface Modeling toolbar.
87. Select A, B, and C (see Figure 8–17) one by one and press the ENTER key.
88. Select D, E, and F (see Figure 8–17) one by one and press the ENTER key.

A Loft UV surface is constructed (see Figure 8–18).

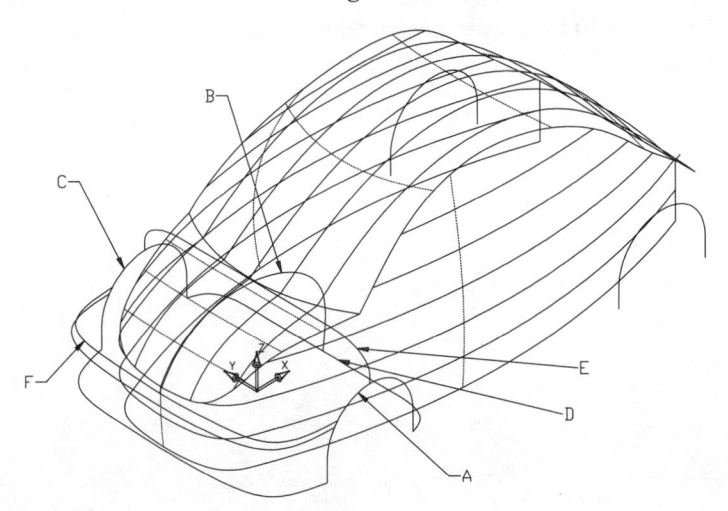

Figure 8–17 *UCS set and array constructed*

Surface Modeling II 373

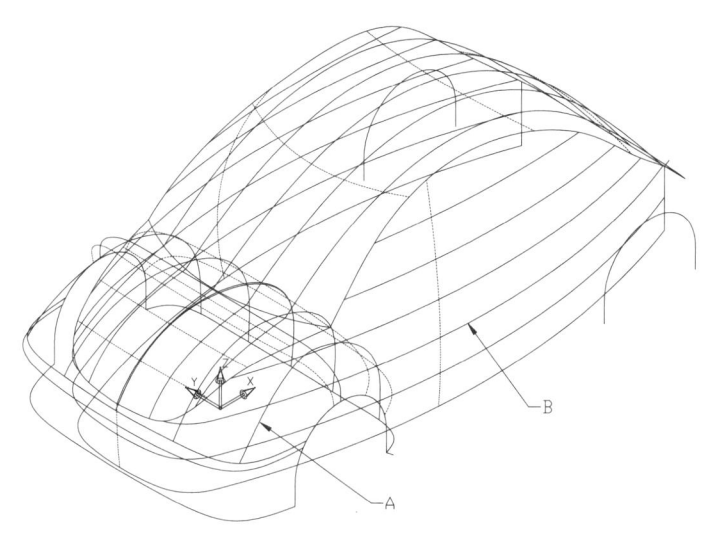

Figure 8–18 *Loft UV surface constructed*

89. Referring to Figure 8–19, trim the Loft UV surface and the quilt and stitch the trimmed surface and the quilt into a single quilt.
90. Select Surface>Create Surface>Sweep or select Swept Surface from the Surface Modeling toolbar.
91. Select A, B, and C (see Figure 8–19) one by one and press the ENTER key.
92. Select D and E (see Figure 8–19).
93. In the Sweep Surface dialog box, use scale transition and select the OK button.

A swept surface is constructed (see Figure 8–20).

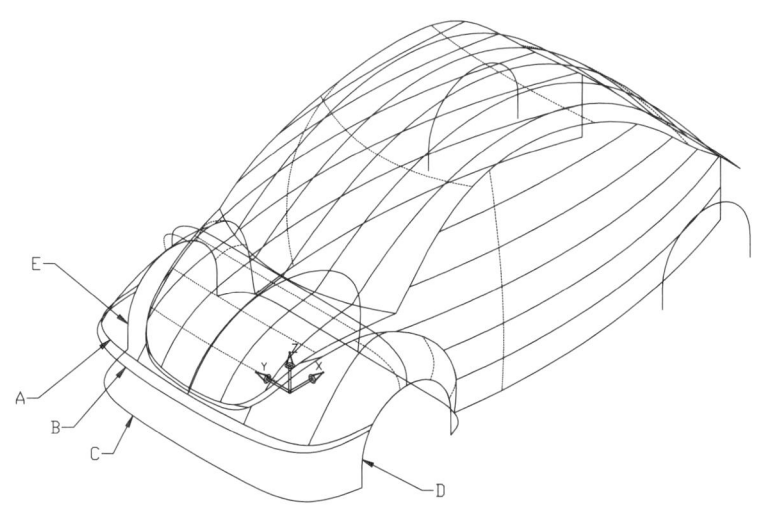

Figure 8–19 *Loft UV surface and the quilt trimmed and stitched*

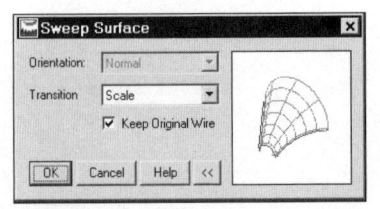

Figure 8–20 *Sweep Surface dialog box*

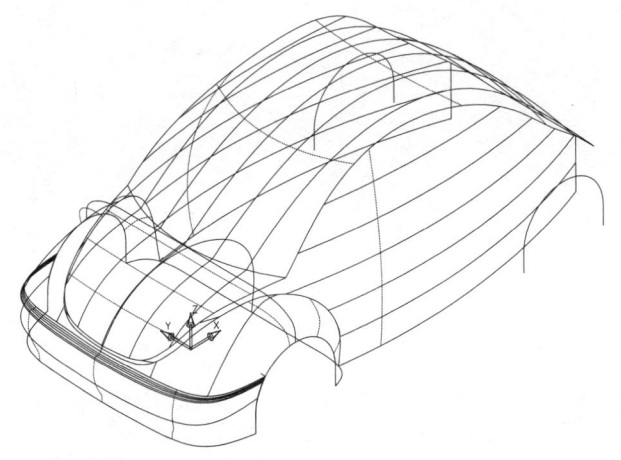

Figure 8–21 *A swept surface constructed*

Next you will construct six splines (see Figure 8–22).

    94. Set the display to an isometric view.

    95. Set the current layer to Wire.

    96. Referring to the information provided by the tables below, construct six splines.

Spline	1	2	3
First Point	124,–42	121,–42,16	120,–42,17.5
Next Point	129,–39	133.5,–39,16	129,–39,17.5
Next Point	133.5,–23	137.5,–23,16	133.5,–23,17.5
Next Point	134,0	138.5,0,16	134,0,17.5
Next Point	133.5,23	137.5,23,16	133.5,23,17.5
Next Point	129,39	133.5,39,16	129,39,17.5
Next Point	124,42	121,42,16	120,42,17.5
Next Point	ENTER	ENTER	ENTER
Start Tangent	ENTER	ENTER	ENTER
End Tangent	ENTER	ENTER	ENTER

Surface Modeling II 375

Spline	4	5		6
First Point	108,–42,23	92,–42	First Point	134,0,17.5
Next Point	108,–39,30	85,–39	Next Point	125,0,26
Next Point	108,–31,32	83,–31.5	Next Point	113.5,0,31
Next Point	108,0,31.5	83.5,0	Next Point	90,0,24
Next Point	108,31,32	83,31.5	Next Point	84.5,0,13
Next Point	108,39,30	85,39	Next Point	83.5,0,0
Next Point	108,42,23	92,42	Next Point	ENTER
Next Point	ENTER	ENTER	Start Tangent	ENTER
Start Tangent	ENTER	ENTER	End Tangent	ENTER
End Tangent	ENTER	ENTER		

Now you will construct a Loft UV surface (see Figure 8–23).

  97. Set the current layer to Surface.
  98. Select Surface>Create Surface>Loft UV or select Loft UV Surface from the Surface Modeling toolbar.
  99. Select A, B, and C (see Figure 8–22) one by one and press the ENTER key.
100. Select D, E, and F (see Figure 8–22) one by one and press the ENTER key.

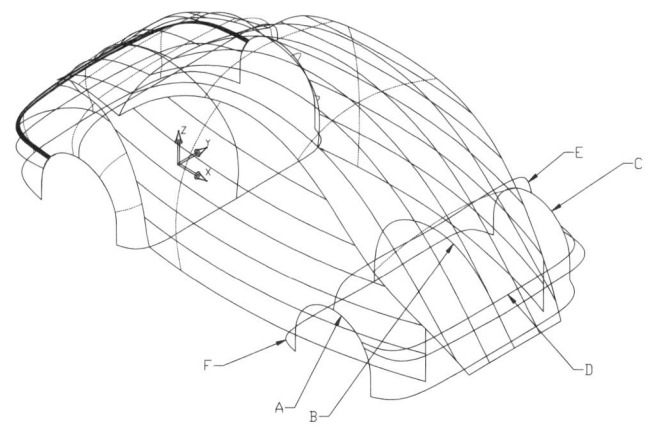

Figure 8–22 *Six splines constructed*

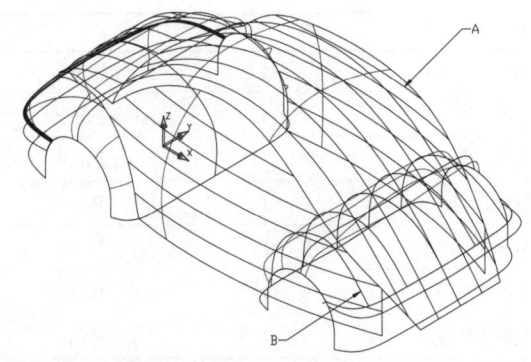

Figure 8–23 *Loft UV surface constructed*

Now you will trim the Loft UV surface and the quilt.

    101. Referring to Figure 8–23, trim the Loft UV surface and the quilt, and stitch the trimmed surface and the quilt into a single quilt.

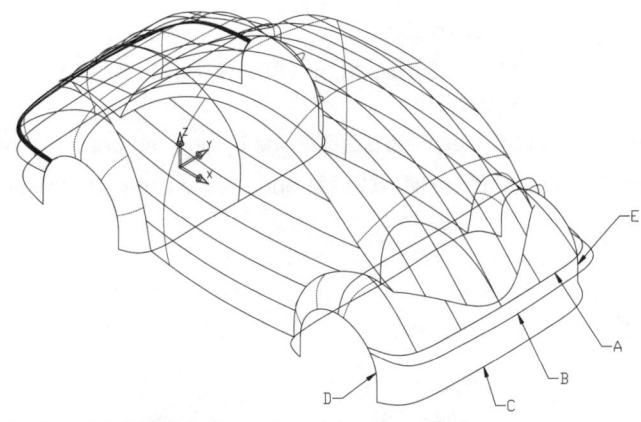

Figure 8–24 *Quilt and Loft UV surface trimmed and stitched*

Now you will construct a swept surface.

    102. Select Surface>Create Surface>Sweep or select Swept Surface from the Surface Modeling toolbar.

    103. Select A, B, and C (see Figure 8–24) one by one and press the ENTER key.

    104. Select D and E (see Figure 8–24).

    105. In the Sweep Surface dialog box, use scale transition and select the OK button.

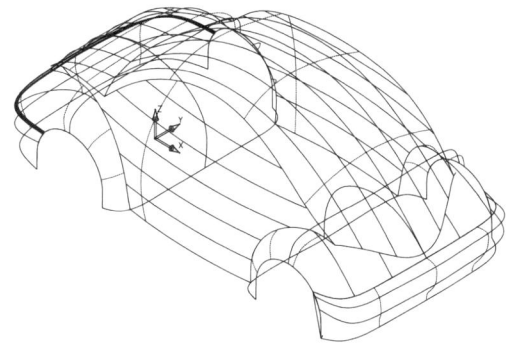

Figure 8–25 *Swept surface constructed*

Now you will construct a spline.

> 106. Set the current layer to Wire.
> 107. Referring to the information provided by the following table, construct a spline (see Figure 8–26).

First Point	16,–42,0	
Next Point	16,–39,4	
Next Point	16,–34,5	
Next Point	ENTER	
Start Tangent	ENTER	
End Tangent	ENTER	

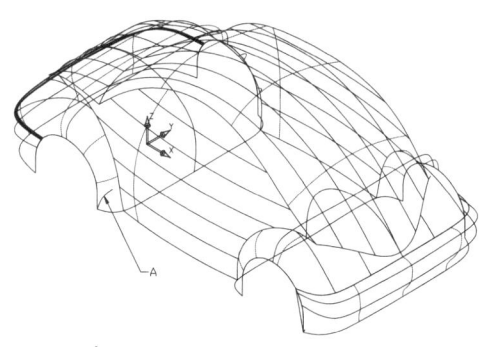

Figure 8–26 *Spline constructed*

Now you will construct an extruded surface.

> 108. Set the current layer to Surface.
> 109. Select Surface>Create Surface>Extrude or select Extruded Surface from the Surface Modeling toolbar.
>
>   Command: AMEXTRUDESF

110. Select A (see Figure 8–26) and press the ENTER key.
111. Type **X** to specify the direction of extrusion.
112. Type **76** to specify the extrusion distance.
113. Press the ENTER key to accept the direction.
114. Press the ENTER key to accept zero degrees taper.

An extruded surface is constructed (see Figure 8–27). Now you will construct a mirror copy of the extruded surface.

115. Select Modify>3D Operation>Mirror 3D

    Command: MIRROR3D

116. Select A (see Figure 8–27) and press the ENTER key.
117. Type **ZX** to specify the mirror plane orientation.
118. Type **0,0** to specify the mirror plane location.
119. Press the ENTER key (not to delete the original).
120. Turn off layer Wire.

The extrude surface is mirrored (see Figure 8–28).

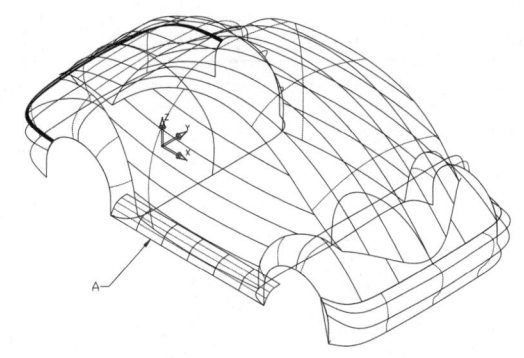

Figure 8–27 *Extrude surface constructed*

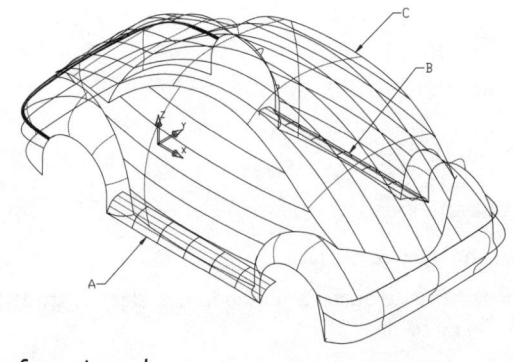

Figure 8–28 *Rule surface mirrored*

121. Referring to Figure 8–29, trim the extruded surfaces and the quilt, and stitch the trimmed surfaces and quilt into a single quilt.

The model is complete. Save your file (file name: *Model_A.dwg*).

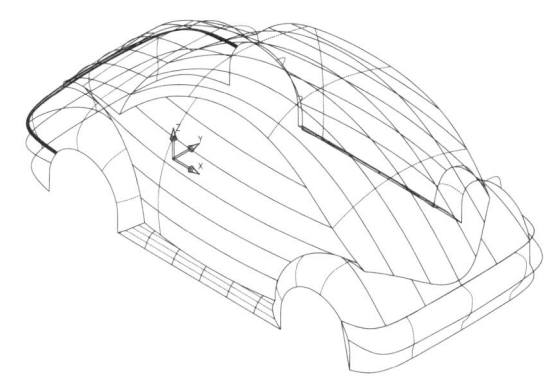

Figure 8–29 *Quilt and surfaces unhidden*

MODEL CAR B

Figure 8–30 shows the surface model of a car body. Like the car body you just constructed, this model has a number of surfaces. Before starting to make the models, spend some time thinking about the kind of surfaces that are required. Then think about how to construct the wires for the surfaces. After that, construct the wires and make the surfaces.

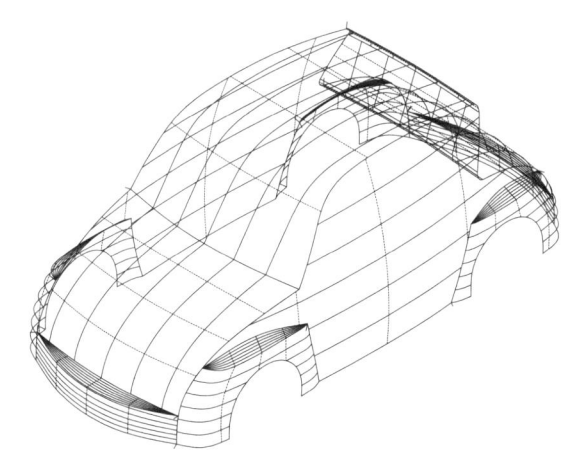

Figure 8–30 *Surface model of car model B*

To help you see more clearly what surfaces are required, Figure 8–31 shows an exploded view of the surfaces. The wires required to construct these surfaces are shown in Figure 8–32.

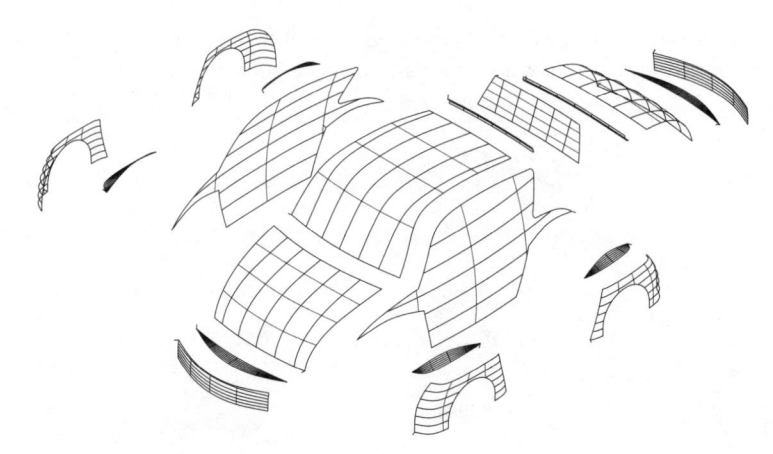

Figure 8–31 *Surfaces exploded apart*

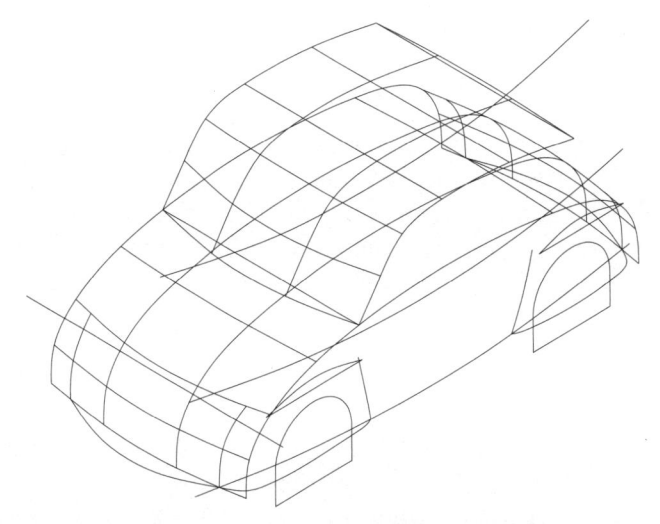

Figure 8–32 *Wires required for making the surfaces for the model car*

1. Start a new file. Use metric as the default.
2. Construct two layers: Wire and Surface.
3. Set layer Wire as the current layer.
4. Select Surface>Surface Options or select AutoSurf Options from the Surface Modeling toolbar.
5. In the Surface tab of the Mechanical Options dialog box, select Model Size.
6. In the Approximate Model Size dialog box, set the model size to 200 and select the OK button.
7. After returning to the Mechanical Desktop dialog box, select the OK button.

Now you will create a Loft UV surface as shown in Figure 8–33. To make this surface, you will construct a set of points and construct a number of splines passing through the points.

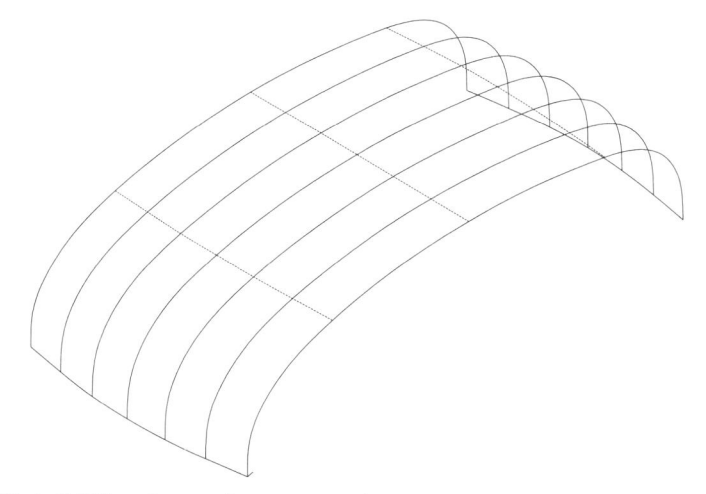

Figure 8–33 *Loft UV surfaces to be constructed*

    8. Select Assist>Format>Point Style

       Command: DDPTYPE

    9. Referring to Figure 8–34 (the Point Style dialog box), select the highlighted point style and select the OK button.

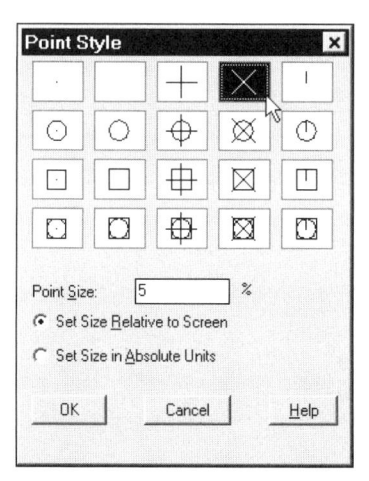

Figure 8–34 *Point Style dialog box*

    10. Set the display to an isometric view.

    11. Select Design>Point>Multiple Point

12. Referring to the information provided in the table below, key in the coordinates of the points one by one.

Point	X-coordinate	Y-coordinate	Z-coordinate
A	−22	0	0
B	−21	0	10
C	−14	0	19
D	1	0	26
E	15	0	29
F	78	0	28
G	100	0	19
H	105	0	10
J	106	0	0
K	−25	20	0
L	−24	20	10
M	−18	20	19
N	−1	20	27
P	14	20	30
R	78	20	29
S	102	20	20
T	108	20	10
U	109	20	0

Eighteen points are constructed (see Figure 8–35).

Figure 8–35 *Eighteen points constructed*

13. Select Modify>Mirror

    Command: MIRROR

14. Select all the points and press the ENTER key.
15. Type **0,32** to specify the first mirror point.
16. Type **1,32** to specify the second mirror point.
17. Press the ENTER key to accept the default (not to delete the original objects).

The points are mirrored (see Figure 8–36).

18. Select Assist>Drafting Settings>Drafting Settings
19. In the Object Snap tab of the Drafting Settings dialog box (see Figure 8–37), select Node (snap to point objects) and select the OK button.

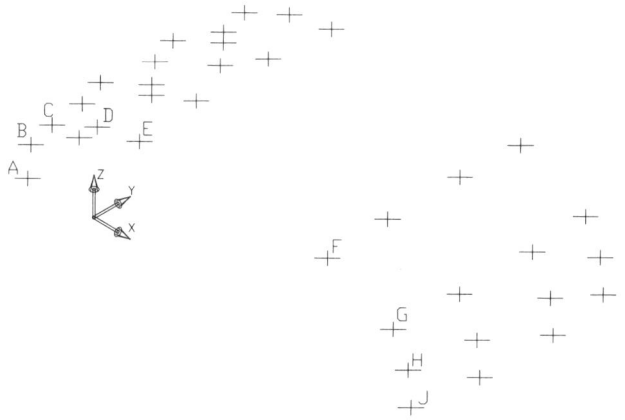

Figure 8–36 *Points mirrored*

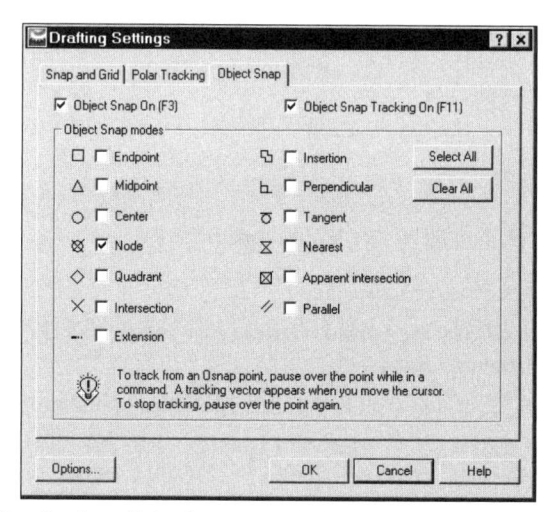

Figure 8–37 *Drafting Settings dialog box*

20. Referring to Figure 8–38, construct nine splines passing through the point objects.
21. Select View>3D Views>Front Left Isometric
22. Referring to Figure 8–39, construct four more splines.

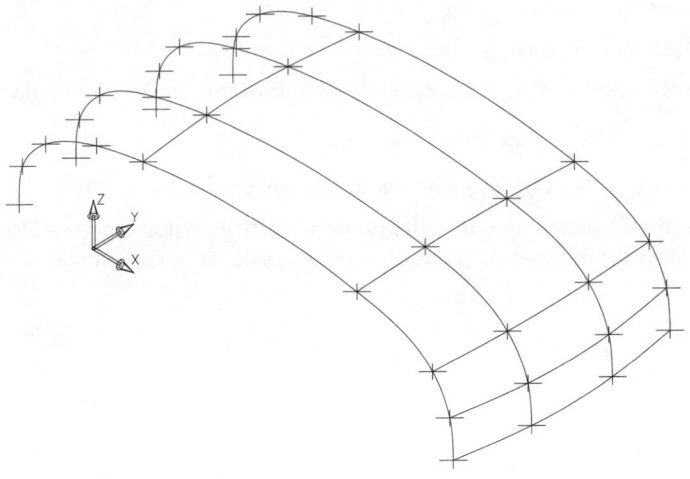

Figure 8–38 *Nine splines constructed*

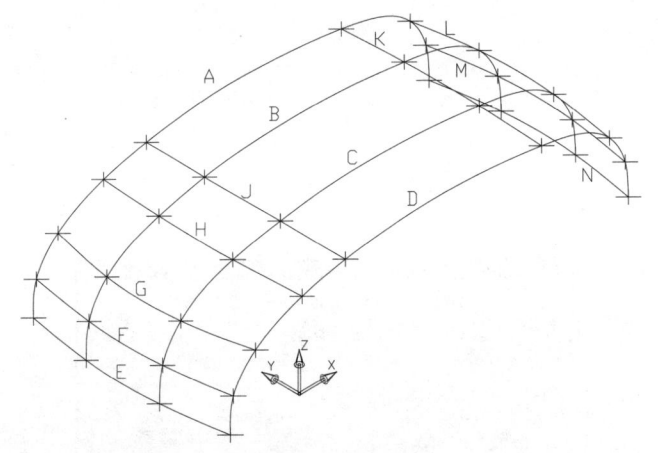

Figure 8–39 *Display set and four more splines constructed*

23. Set the current layer to Surface.
24. Select Surface>Create Surface>Loft UV or select Loft UV Surface from the Surface Modeling toolbar.
25. Select A, B, C, and D (see Figure 8–39) one by one and press the ENTER key.
26. Select E, F, G, H, J, K, L, M, and N (see Figure 8–39) one by one and press the ENTER key.

A Loft UV surface is constructed (see Figure 8–40).

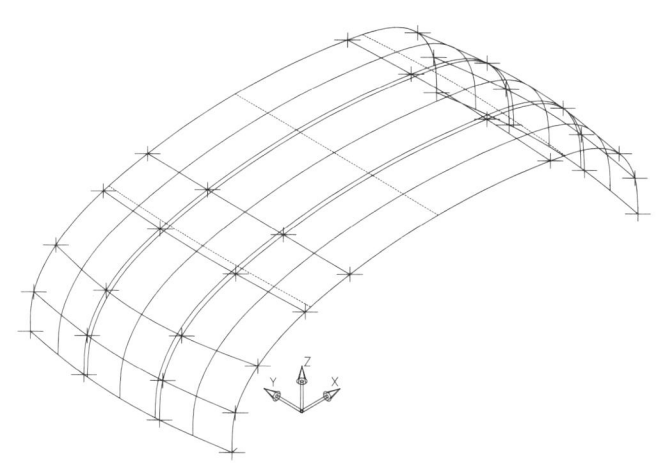

Figure 8–40 *Loft UV surface constructed*

27. Select Surface>Surface Visibility or select Object Visibility from the Surface Modeling toolbar.
28. In the Objects tab of the Desktop Visibility dialog box, select the Hide, Points, and Splines boxes, and select the OK button.

The points and splines are hidden (see Figure 8–41).

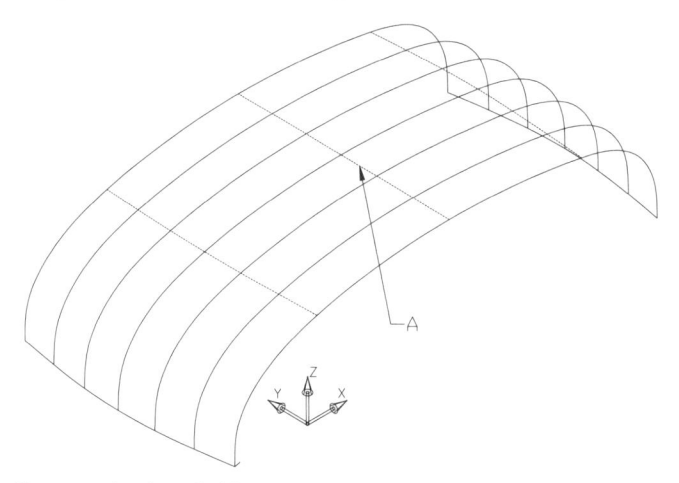

Figure 8–41 *Points and splines hidden*

Now you will break the Loft UV surface into two surfaces along a selected flow line.

29. Select Surface>Edit Surface>Break or select Break Surface from the Surface Modeling toolbar.

    Command: AMBREAK

30. Select the flow line A (see Figure 8–41).
31. Press the ENTER key.

The Loft UV surface is broken into two surfaces (50%/50%) along the selected U line (see Figure 8–42).

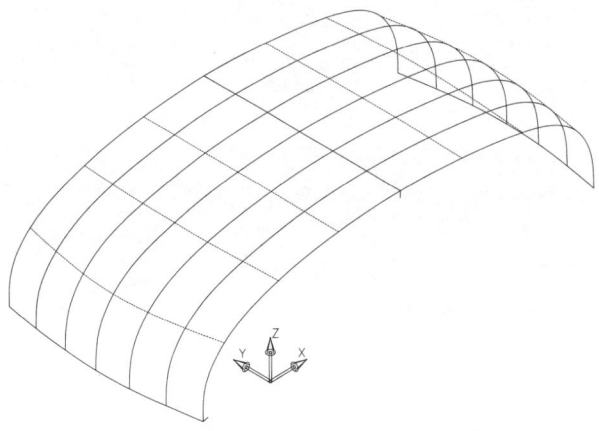

Figure 8–42 *Loft UV surface broken into two (50%/50%)*

Now you will construct a set of points, mirror the points, and construct a Loft UV surface.

32. Set the current layer to Wire.
33. Select Design>Point>Multiple Point
34. Referring to the information provided in the table below, key in the coordinates of each point one by one.

Point	X-coordinate	Y-coordinate	Z-coordinate
A	15	0	29
B	22	0	39
C	29	0	47
D	42	0	50
E	64	0	49
F	85	0	45
G	11	20	29
H	18	20	39
J	27	20	48
K	41	20	51
L	63	20	50
M	85	20	46

35. Select Modify>Mirror

    Command: MIRROR
36. Select all the points and press the ENTER key.
37. Type **0,32** to specify the first mirror point.
38. Type **1,32** to specify the second mirror point.
39. Press the ENTER key to accept the default (not to delete the original objects).

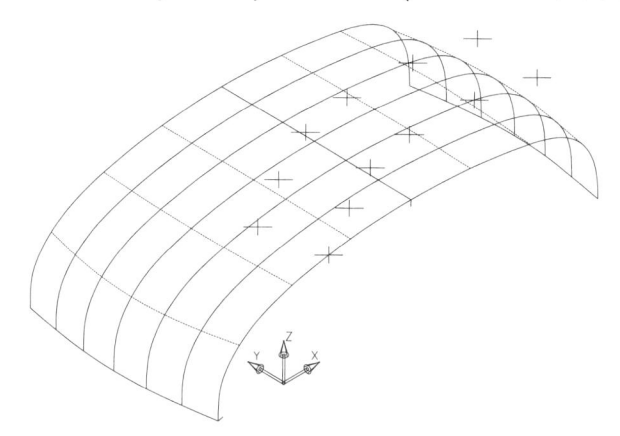

Figure 8–43 *Twelve points constructed*

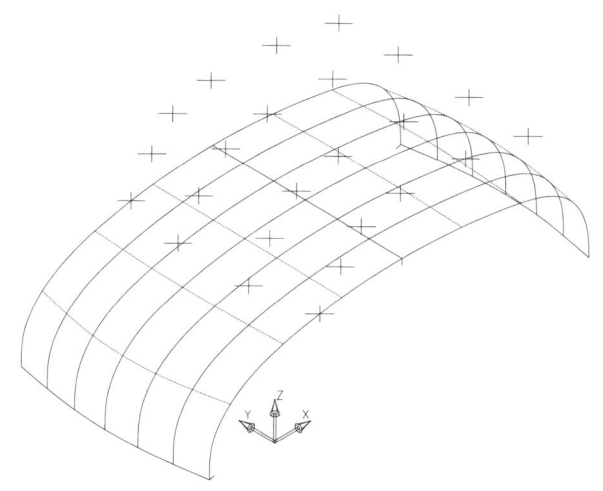

Figure 8–44 *Points mirrored*

40. Referring to Figure 8–45, construct 10 splines passing through the points.
41. Set the current layer to Surface.

42. Select Surface>Create Surface>Loft UV or select Loft UV Surface from the Surface Modeling toolbar.
43. Select A, B, C, and D (see Figure 8–45) one by one and press the ENTER key.
44. Select E, F, G, H, J, and K (see Figure 8–45) one by one and press the ENTER key.

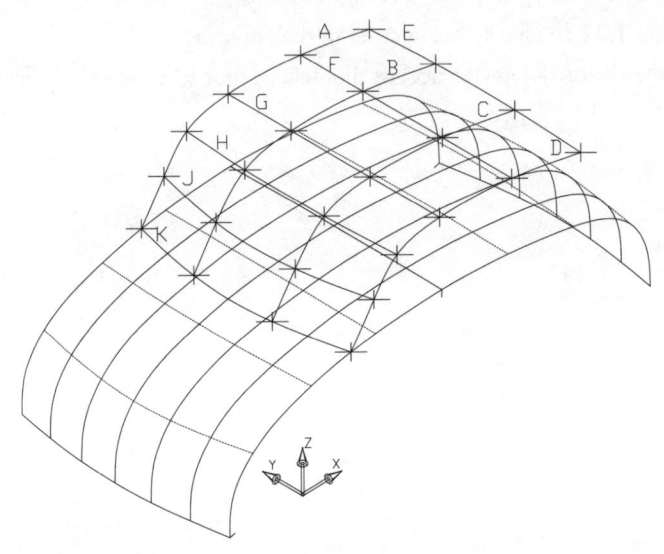

Figure 8–45 *Splines constructed*

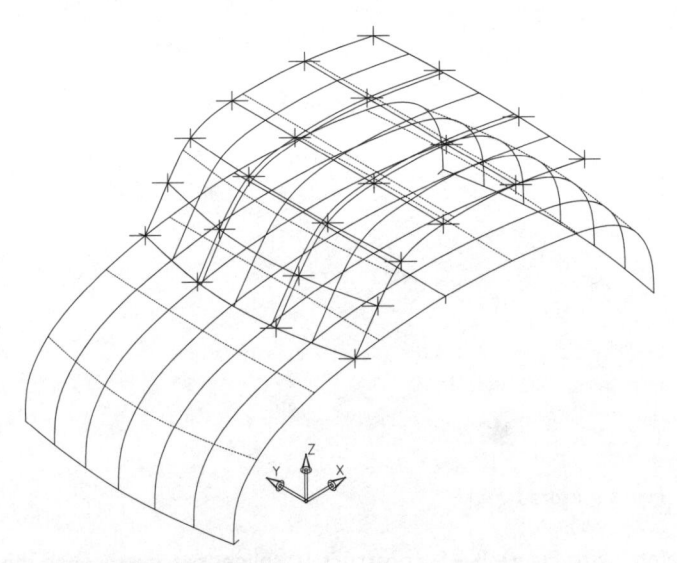

Figure 8–46 *Loft UV surface constructed*

45. Hide the splines and points.

Now you will lengthen a surface. (The surface is intentionally made a bit shorter in order to let you practice lengthening it.)

46. Select Surface>Edit Surface>Lengthen or select Lengthen Surface from the Surface Modeling toolbar.

    Command: AMLENGTHEN

47. Type **E** to use the Extend option.
48. Type **D** to use the Delta option.
49. Type **20** to set the incremental length to 20 mm.
50. Select A (see Figure 8–47).
51. Press the ENTER key.

The surface is lengthened a distance of 20 mm at the selected edge (see Figure 8–48).

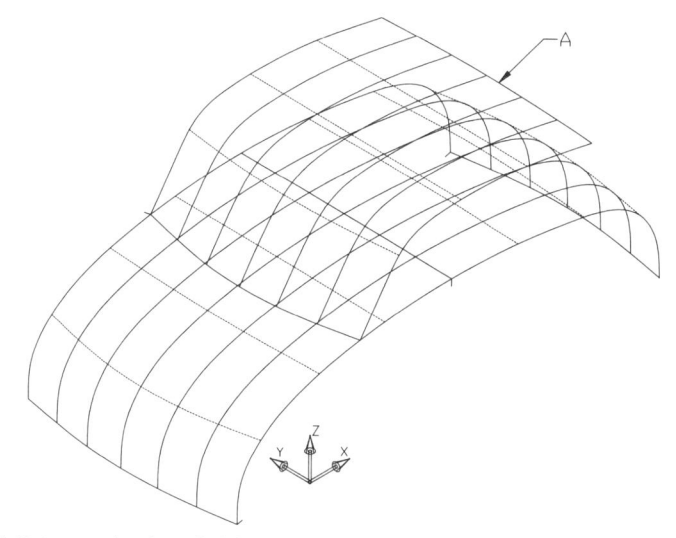

Figure 8–47 *Points and splines hidden*

Now you will trim two surfaces.

52. Select Surface>Edit Surface>Intersect Trim or select Intersect and Trim from the Surface Modeling toolbar.
53. Select A and B (see Figure 8–48).
54. In the Surface Intersection dialog box, select the OK button.

The intersecting surfaces are trimmed (see Figure 8–49).

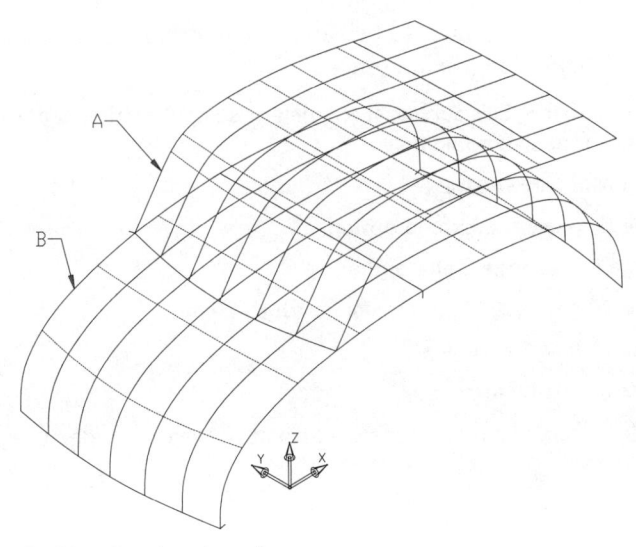

Figure 8–48 *Loft UV surface lengthened*

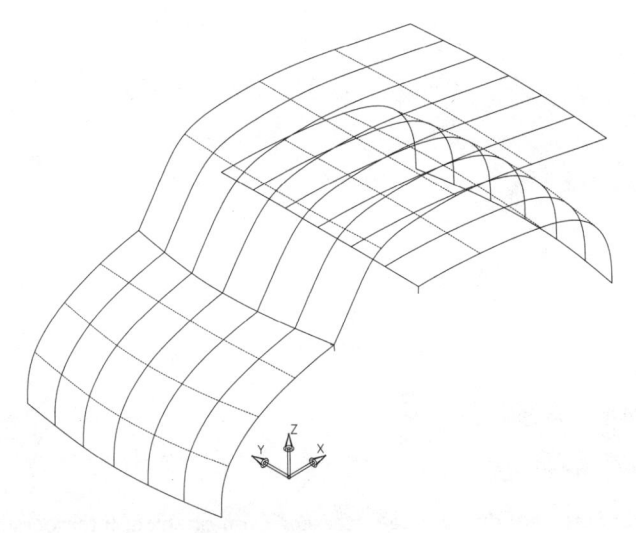

Figure 8–49 *Loft UV surfaces intersected and trimmed*

Now you will construct an extruded surface.

     55. Set the current layer to Wire.

     56. Select Design>Line

     57. Type **66,0** to specify the first point.

     58. Type **88,0,55** to specify the second point.

     59. Press the ENTER key.

A line is constructed (see Figure 8–50).

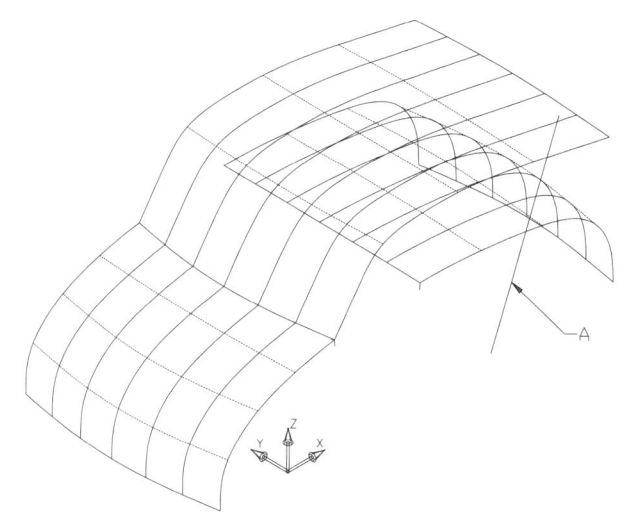

Figure 8–50 *Line constructed*

60. Set the current layer to Surface.
61. Select Surface>Create Surface>Extrude or select Extruded Surface from the Surface Modeling toolbar.

    Command: AMEXTRUDESF
62. Select A (see Figure 8–50) and press the ENTER key.
63. Type **Y** to specify the extrusion direction.
64. Type **64** to specify the extrusion distance.
65. Press the ENTER key to accept the default extrusion direction.
66. Press the ENTER key to accept the default zero degrees taper.

An extruded surface is constructed (see Figure 8–51). Now you will construct two fillet surfaces and stitch all the surfaces into a single quilt.

67. Select Surface>Create Surface>Fillet or select Fillet Surface from the Surface Modeling toolbar.

    Command: AMFILLETSF
68. Select A and B (see Figure 8–51).
69. In the Fillet Surface dialog box, select Constant fillet type, set the fillet radius to 2 mm, and select the OK button.

A fillet surface is constructed and the selected surfaces are trimmed (see Figure 8–52).

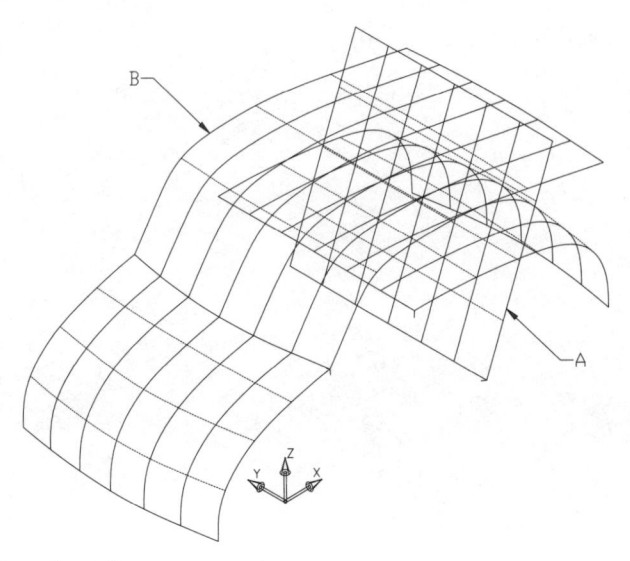

Figure 8–51 *Extrude surface constructed*

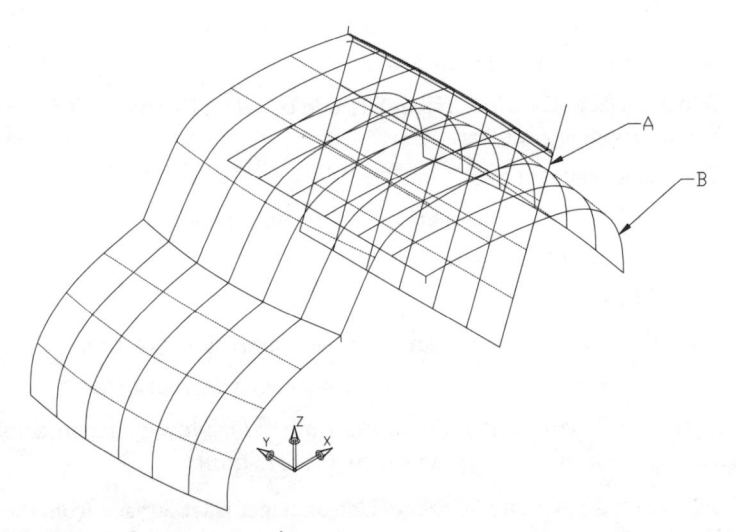

Figure 8–52 *Fillet surface constructed*

70. Repeat the AMFILLETSF command to construct a fillet surface between A and B (see Figures 8–52 and 8–53).
71. Select Surface>Surface Stitching or select Stitches Surfaces from the Surface Modeling toolbar.
72. Select all the surfaces and press the ENTER key.
73. Select the OK button from the Surface stitching dialog box.

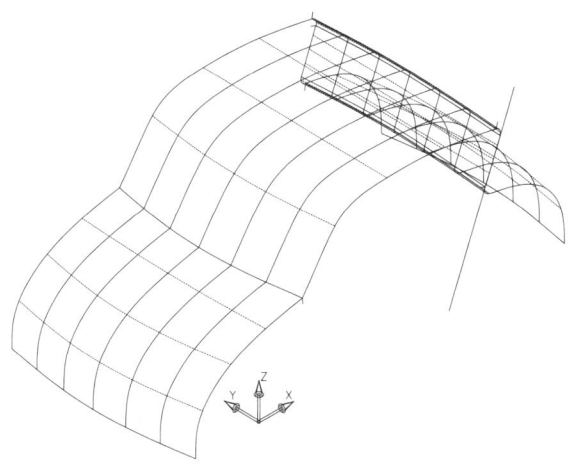

Figure 8–53 *Surfaces filleted*

Now you will construct a Loft U surface.

74. Set the current layer to Wire.
75. Referring to the information provided in the table below, construct a number of point objects (see Figure 8–54).

Point	X-coordinate	Y-coordinate	Z-coordinate
A	−30	9	0
B	−6	4	0
C	26	1	0
D	58	1	0
E	88	4	0
F	110	7	0
G	−30	11	25
H	−6	5	25
J	42	1	25
K	88	5	25
L	110	9	25
M	−30	20	55
N	0	12	55
P	32	10	55
Q	69	12	55
R	110	20	55

76. Construct three splines to pass through the points (see Figure 8–55).

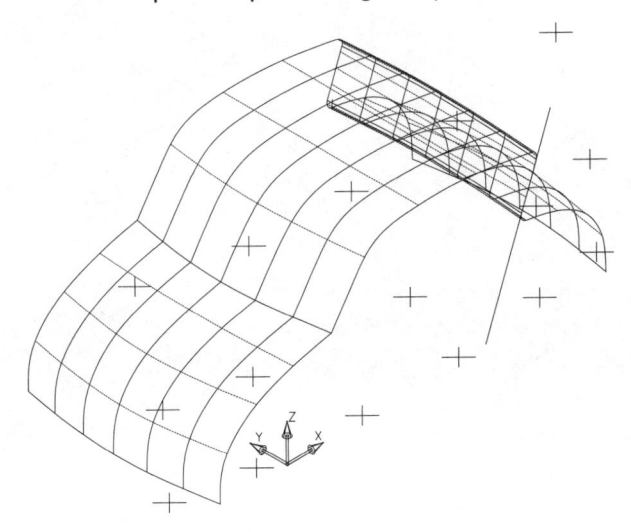

Figure 8–54 *Sixteen points constructed*

77. Set the current layer to Surface.
78. Select Surface>Create Surface>Loft U or select Loft U Surface from the Surface Modeling toolbar.
79. Select A, B, and C (see Figure 8–55) one by one and press the ENTER key.
80. Select the OK button in the Loft Surface dialog box.

A Loft U surface is constructed (see Figure 8–56).

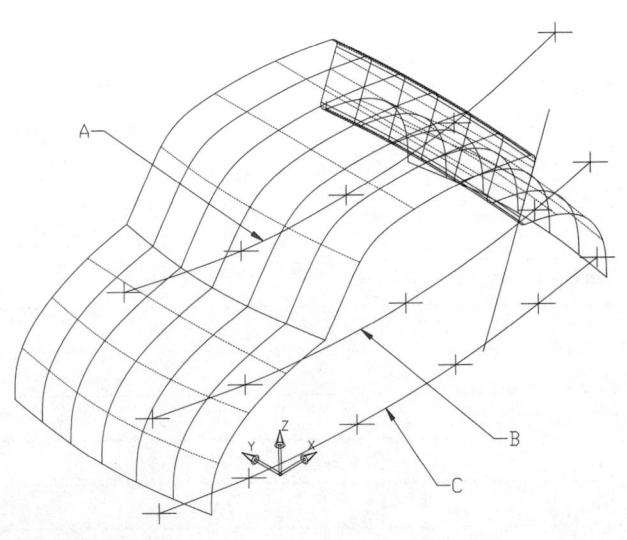

Figure 8–55 *Splines constructed*

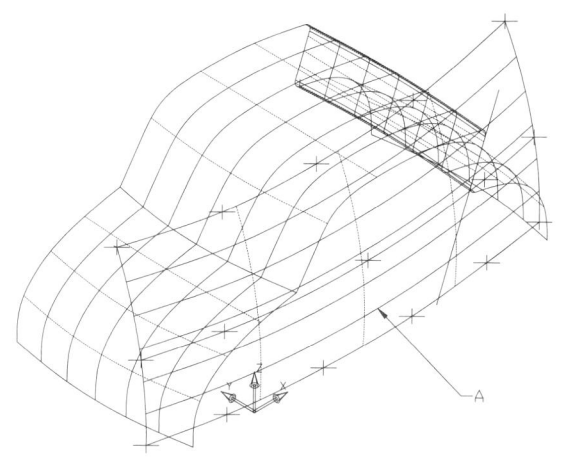

Figure 8–56 *Loft U surface constructed*

Now you will construct a mirror copy of the Loft U surface (see Figure 8–57).

    81. Select Modify>Mirror
    82. Select A (see Figure 8–56) and press the ENTER key.
    83. Type **0,32** to specify the first mirror point.
    84. Type **1,32** to specify the second mirror point.
    85. Press the ENTER key.

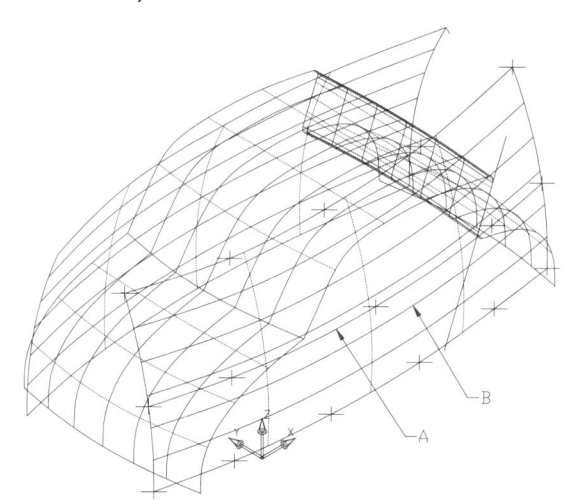

Figure 8–57 *Loft U surface mirrored*

Now you will check the fitness of a surface in comparison to the wires from which the surface is constructed. Because the complexity of a surface is optimized when it is made from a set of input wires, the surface may deviate from the input wires.

86. Select Surface>Utilities>Check Fit or select Check Fit from the Surface Modeling toolbar.

    Command: AMCHECKFIT
87. Select A, the wire (see Figure 8–57), and press the ENTER key.
88. Select B, the surface (see Figure 8–57).
89. Type **T** to display a table (see Figure 8–58).

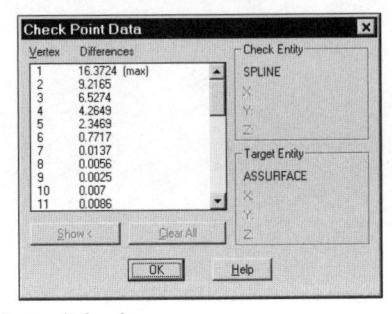

Figure 8–58 *Check Point Data dialog box*

The dialog box shows the deviation of the surface from the vertices of the wire.

90. In the Check Point Data dialog box, select the OK button.
91. Type **S** to use the scale option.
92. Type **10000** to set the scale factor.
93. Type **G** to display a graph to show the deviation.
94. Press the ENTER key.

Deviation of the surface from the selected wire is displayed graphically and magnified 10,000 times (see Figure 8–59).

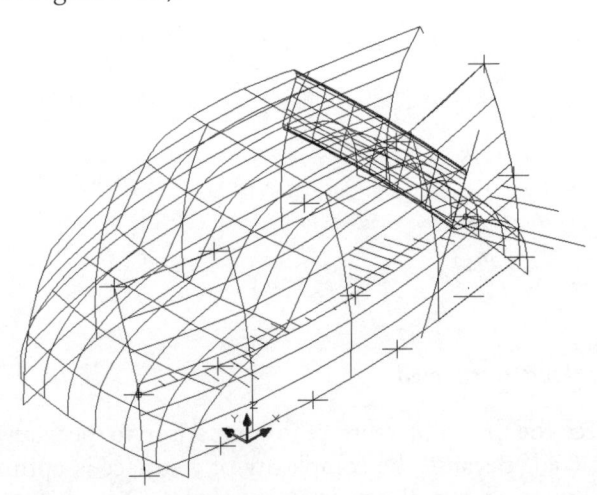

Figure 8–59 *Deviation shown graphically*

Now you will trim the surfaces and the quilt.

95. Hide the points and splines (see Figure 8–60).
96. Select Surface>Edit Surface>Intersect Trim or select Intersect and Trim from the Surface Modeling toolbar.
97. Select A and B (see Figure 8–60).
98. In the Surface Intersection dialog box, select the OK button.

The quilt and one of the Loft U surfaces are trimmed (see Figure 8–61).

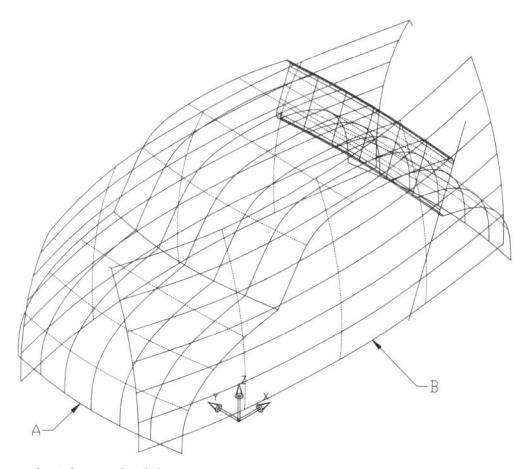

Figure 8–60 *Points and splines hidden*

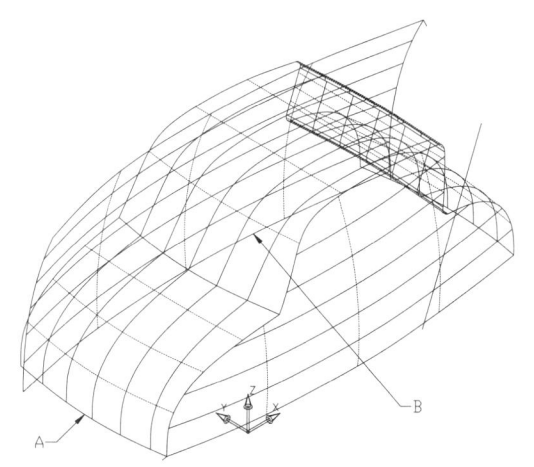

Figure 8–61 *Stitched quilt and a Loft U surface intersected and trimmed*

99. Repeat the AMINTERSF command.
100. Select A and B (see Figure 8–61).
101. Select the OK button.

The quilt and the second Loft U surface are trimmed (see Figure 8–62).

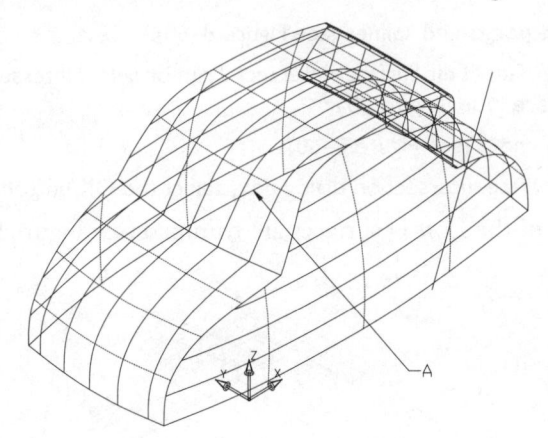

Figure 8–62 *Stitched quilt and the second Loft U surface intersected and trimmed*

Because the two sides of the model car are symmetrical, you will erase one side of the car, complete the remaining side of the car, and construct a mirror copy to complete the car.

    102. Select Modify>Erase>Erase

    103. Select A (see Figure 8–62) and press the ENTER key.

A surface is erased (see Figure 8–63).

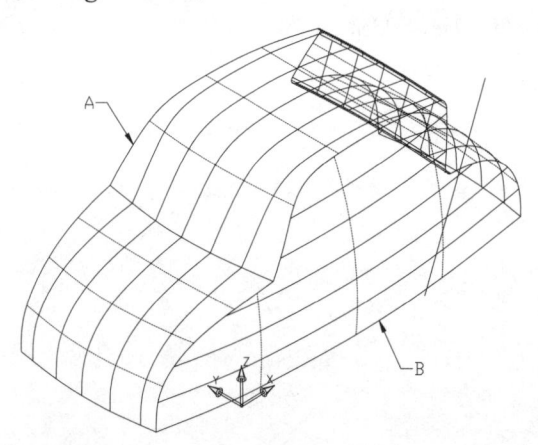

Figure 8–63 *A surface erased*

Because you need to work on the individual surfaces of the quilt, you have to convert the quilt back to individual surfaces.

    104. Select Create Surface>From ACAD or select Convert Face from the Surface Modeling toolbar.

        Command: AM2SF

Surface Modeling II 399

105. Select A (see Figure 8–63) and press the ENTER key.

After conversion, you will not notice any visual changes on your screen. However, the quilt is now broken down into a number of smaller individual surfaces.

106. Hide all the surfaces except B (see figures 8–63 and 8–64).

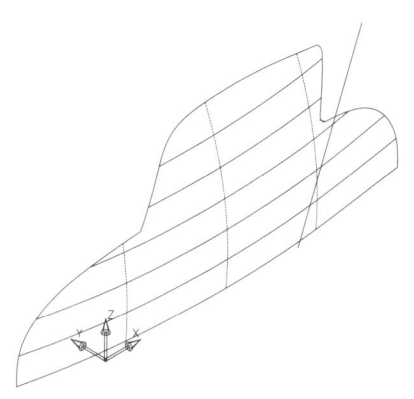

Figure 8–64 *Stitched quilt converted to surfaces and hidden*

Now you will construct a spline.

107. Referring to the information provided by the following table, construct a spline.

First Point	20,0,0
Next Point	16,0,19
Next Point	15,0,29
Next Point	ENTER
Start Tangent	ENTER
End Tangent	ENTER

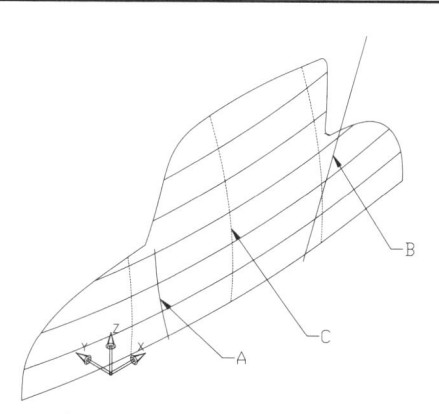

Figure 8–65 *Spline constructed*

Now you will project the spline and the line onto the surface.

108. Select Surface>Create Wireframe>Project Wire or select Projection Wire from the Surface Modeling toolbar.

    Command: AMPROJECT

109. Select A and B (see Figure 8–65) and press the ENTER key.
110. Select C (see Figure 8–65).

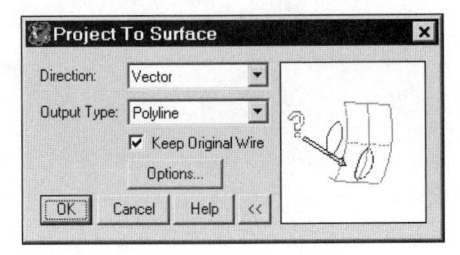

Figure 8–66 *Project To Surface dialog box*

The Project To Surface dialog box has a number of areas. The direction area enables you to project the wire normal to the surface, perpendicular to the UCS, in a direction specified by a vector, or perpendicular to the display view. The Output area enables you to output an augmented line or polyline, or trim or break the surface.

111. In the Project To Surface dialog box, select Vector and Polyline, and select the OK button.
112. Type **Y** to specify the direction of projection.

The wires are projected (see Figure 8–67).

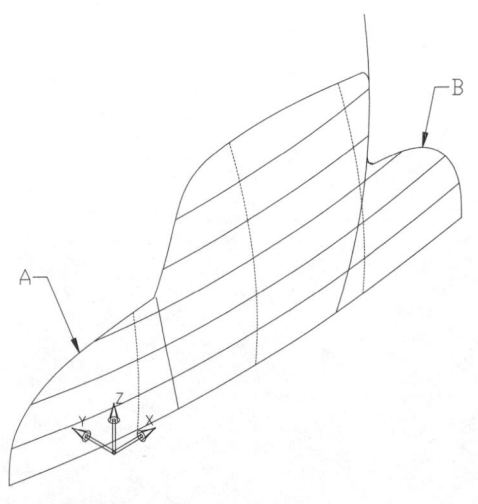

Figure 8–67 *Wires projected on the surface*

Now you will copy the edges of a surface and output it as a spline.

113. Select Surface>Edit Trim Edges>Copy Edge or select Copy Surface Edge from the Surface Modeling toolbar.

    Command: AMEDGE

    OUTPUT

114. Type **S** to output a spline.
115. Select A and B (see Figure 8–67) and press the ENTER key.
116. Hide the surface (see Figure 8–68).

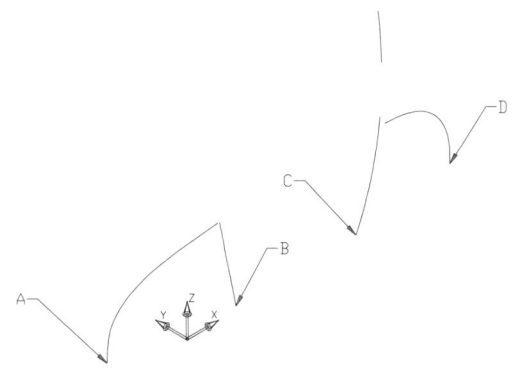

Figure 8–68 *Edges copied and surface hidden*

Now construct a line and two splines.

117. Select Design>Line
118. Type **–30,0,19** to indicate the first point.
119. Type **120,0,19** to indicate the second point.
120. Press the ENTER key.
121. Referring to the information provided by the table below, construct two splines.

First Point	END point of A (see Figure 8–68)	END point of C (see Figure 8–68)
Next Point	–20,2.2	67,1
Next Point	–13,0	70,0.2
Next Point	14,0	98,0.2
Next Point	18,0.5	104,2
Next Point	END point of B (see Figure 8–68)	END point of D (see Figure 8–68)
Next Point	ENTER	ENTER
Start Tangent	ENTER	ENTER
End Tangent	ENTER	ENTER

A line and two splines are constructed (see Figure 8–69).

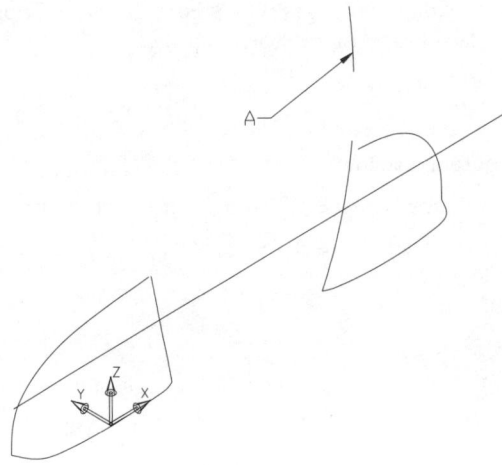

Figure 8–69 *Line and splines constructed*

Now you will erase a wire, set the display, and set the UCS (see Figure 8–70).

122. Select Modify>Erase>Erase
123. Select A (see Figure 8–69) and press the ENTER key.
124. Type **6** to set the display to the front view.
125. Select Assist>New UCS>View

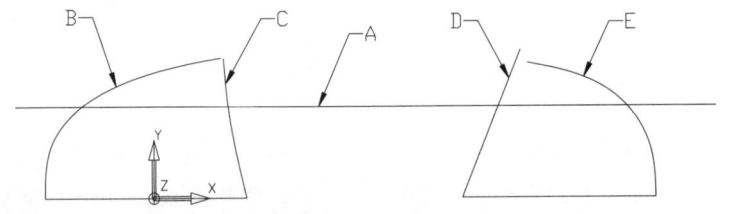

Figure 8–70 *Wire erased, display changed, and UCS set*

Now you will trim the wires. Before you can trim the projected edge, you will refine it and explode it.

126. Select Surface>Edit Wireframe>Refine

    Command: AMREFINE3D
127. Select D (Figure 8–70) and press the ENTER key.
128. Type **P** to specify the number of control points.
129. Type **2** to specify two control points.
130. Type **EXPLODE** at the command line area.

131. Select D (see Figure 8–70) and press the ENTER key.
132. Select Modify>Trim
133. Select A (see Figure 8–70) and press the ENTER key.
134. Select B, C, D, and E (see Figure 8–70) and press the ENTER key.
135. Repeat the TRIM command.
136. Select A, B, C, and D (see Figure 8–71) and press the ENTER key.
137. Select E, F, and G (see Figure 8–71) and press the ENTER key.
138. Type **8** to set the display to an isometric view (see Figure 8–72).

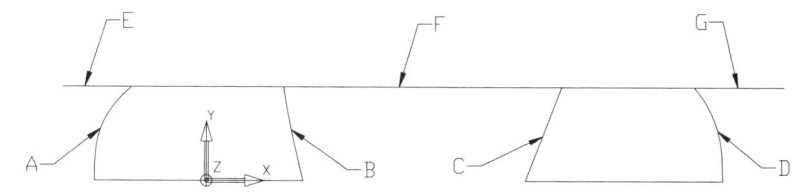

Figure 8–71 *Wires trimmed*

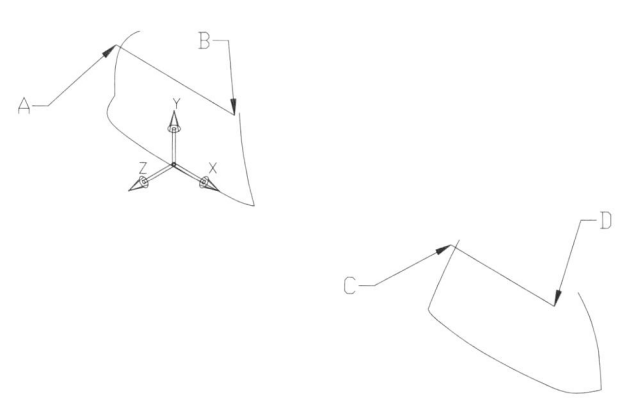

Figure 8–72 *Wires trimmed and display set*

Now you will construct two more splines.

139. Referring to the information provided by the table below, construct two splines (see Figure 8–73).

First Point	END point of A (see Figure 8–72)	END point of C (see Figure 8–72)
Next Point	0,22	84,22
Next Point	END point of B (see Figure 8–72)	END point of D (see Figure 8–72)
Next Point	ENTER	ENTER
Start Tangent	ENTER	ENTER
End Tangent	ENTER	ENTER

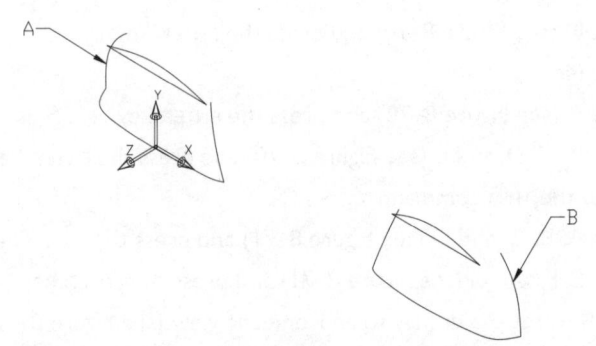

Figure 8–73 *Two splines constructed and display set to an isometric view*

Now you will construct a mirror copy of the two splines you just constructed.

140. Select Modify>3D Operation>Mirror 3D
141. Select A and B (see Figure 8–73) and press the ENTER key.
142. Type **XY** to specify the mirror plane.
143. Type **0,0,–32** to specify the mirror plane position.
144. Press the ENTER key.
145. Referring to the table below, construct two splines (see Figure 8–75).

First Point	END point of A (see Figure 8–74)	END point of C (see Figure 8–74)
Next Point	–29,0,–20	112,0,–20
Next Point	–29,0,–42	112,0,–42
Next Point	END point of B (see Figure 8–74)	END point of D (see Figure 8–74)
Next Point	ENTER	ENTER
Start Tangent	ENTER	ENTER
End Tangent	ENTER	ENTER

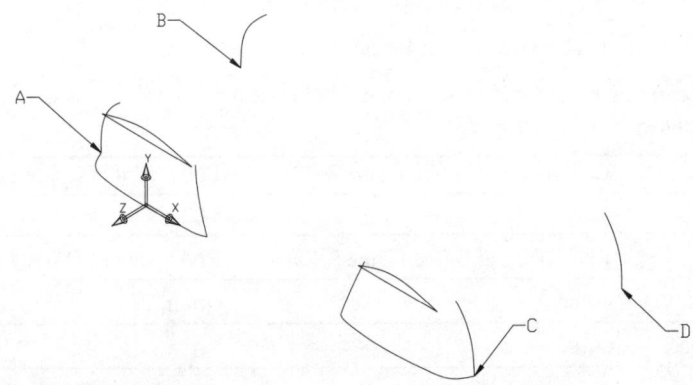

Figure 8–74 *Wires mirrored*

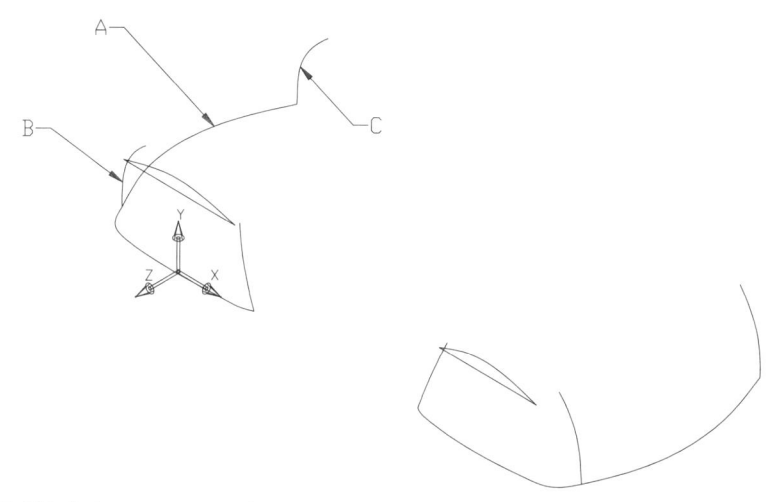

Figure 8–75 *Splines constructed*

Now you will construct four swept surfaces.

146. Set the current layer to Surface.
147. Select Surface>Create Surface>Sweep or select Swept Surface from the Surface Modeling toolbar.
148. Select A (see Figure 8–75) and press the ENTER key.
149. Select B and C (see Figure 8–75).

The Sweep Surface dialog box has two areas. The Orientation area enables you to determine the orientation of the cross sections (Normal, Parallel, or Direction). The Transition area enables you to determine how the cross-sections are swept along the paths (Scale or Stretch).

150. In the Sweep Surface dialog box, select Stretch and the OK button.
151. Repeat the AMSWEEPSF command.
152. Select A (see Figure 8–76) and press the ENTER key.
153. Select B and C (see Figure 8–76) and press the ENTER key.
154. Select the OK button from the Sweep Surface dialog box.
155. Repeat the AMSWEEPSF command.
156. Select A (see Figure 8–77) and press the ENTER key.
157. Select B and C (see Figure 8–77).
158. In the Sweep Surface dialog box, select Scale transition and the OK button.
159. Repeat the AMSWEEPSF command.
160. Select D (see Figure 8–77) and press the ENTER key.

161. Select E and F (see Figure 8–77).
162. Select the OK button from the Sweep Surface dialog box (see Figure 8–78).

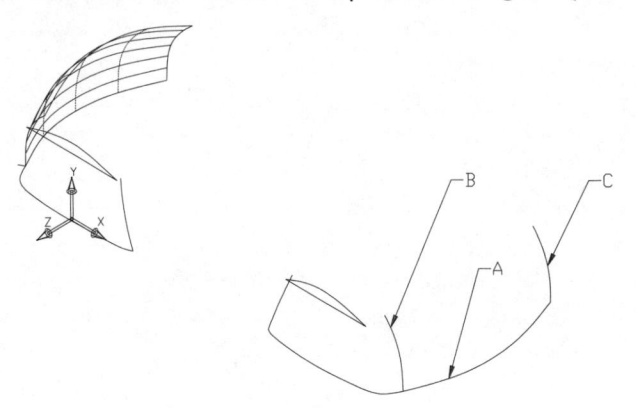

Figure 8–76 Stretch swept surface constructed

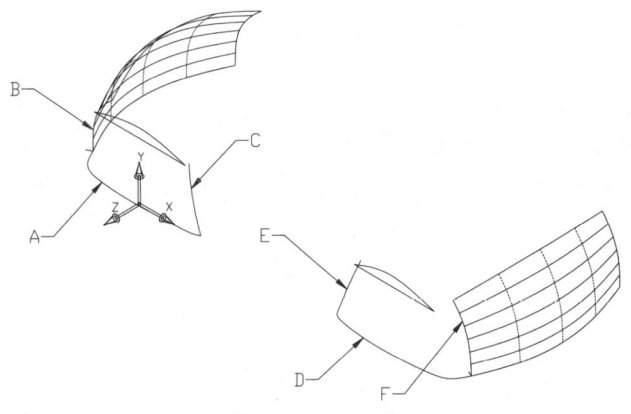

Figure 8–77 Second stretch swept surface constructed

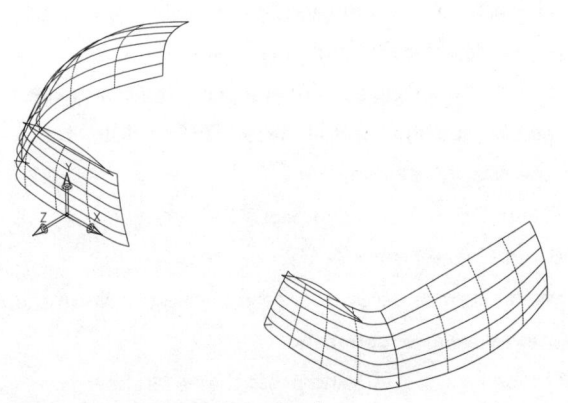

Figure 8–78 Two scale swept surfaces constructed

Now you will construct a polyline and refine its control points (see Figure 8–79).

163. Set the current layer to Wire.
164. Select Design>Polyline
165. Type **–12.5,–7** to specify the first point.
166. Type **–12.5,3** to specify the second point.
167. Type **A** to use the arc option.
168. Type **12.5,3** to specify the end point of the arc.
169. Type **L** to use the line option.
170. Type **12.5,–7** to specify the end point of the line.
171. Type **CL** to close the polyline.

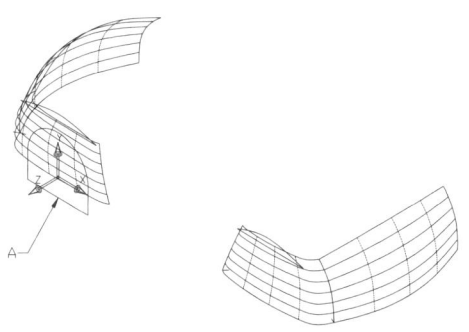

Figure 8–79 *Polyline constructed*

172. Select Surface>Edit Wireframe>Refine or select Refine Wire from the Surface Modeling toolbar.

    Command: AMREFINE3D
173. Select A (see Figure 8–79) and press the ENTER key.
174. Type **0.02** to specify the tolerance.
175. Select the polyline to display the control points (see Figure 8–80).

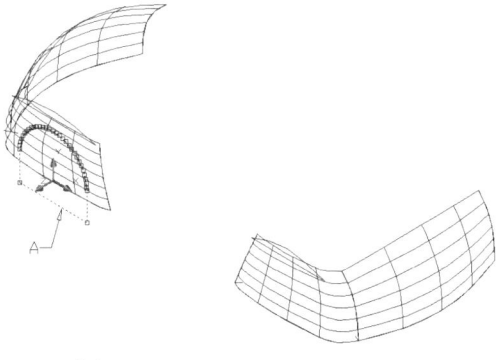

Figure 8–80 *Control points of the wire*

Now you will copy the polyline and construct a line (see Figure 8–81).

176. Select Modify>Copy
177. Select A (see Figure 8–80) and press the ENTER key.
178. Type **84<0** to specify the displacement.
179. Press the ENTER key.
180. Select Design>Line
181. Type **–10,8,0** to specify the first point.
182. Type **–10,8,–84** to specify the second point.
183. Press the ENTER key.

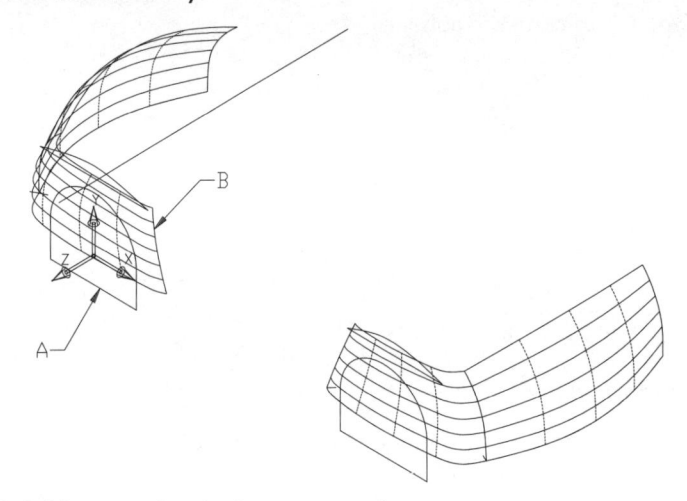

Figure 8–81 *Polyline copied and a line constructed*

Set the system variable DELOBJ to 0 so that objects are not deleted after being used for projection or offsetting. (You may also select Toggles Keep Original from the Surface Modeling Toolbar.)

184. Type **DELOBJ** at the command line area.
185. Type **0**.

Now you will trim the surfaces.

186. Select Surface>Edit Surface>Project Trim or select Project and Trim from the Surface Modeling toolbar.
187. Select A (see Figure 8–81) and press the ENTER key.
188. Select B (see Figure 8–81) and press the ENTER key.
189. In the Project to Surface dialog box, select Vector direction and select the OK button.
190. Type **Z** to specify the direction.

A surface is trimmed (see Figure 8–82).

191. Repeat the command.
192. Select A (see Figure 8–82) and press the ENTER key.
193. Select B (see Figure 8–82) and press the ENTER key.
194. Select Vector direction and select the OK button.
195. Type **Z** to indicate the projection direction.

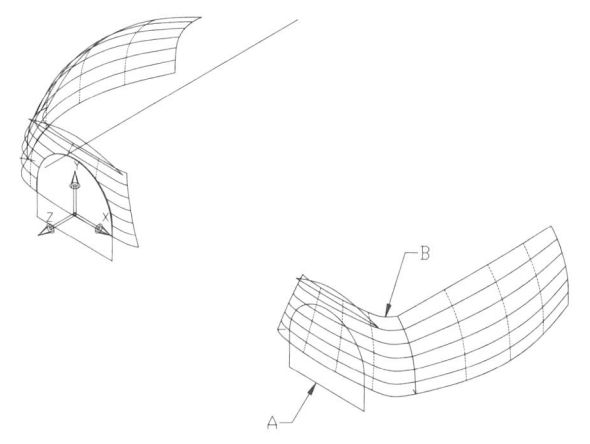

Figure 8–82 *Wire projected and swept surface trimmed*

The selected surface is trimmed (see Figure 8–83).

196. Repeat the command.
197. Select A (see Figure 8–83) and press the ENTER key.
198. Select B and C (see Figure 8–83) and press the ENTER key.
199. Select Vector direction and select the OK button.
200. Type **X** to specify the direction.

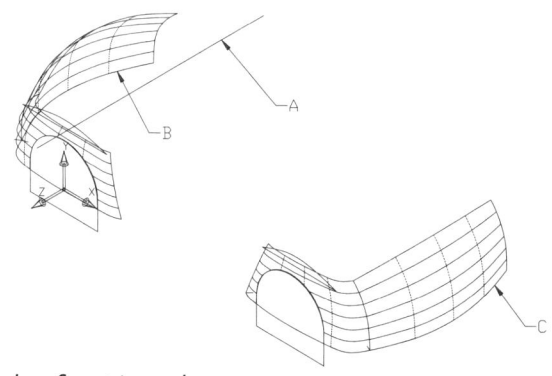

Figure 8–83 *Second surface trimmed*

Two surfaces are trimmed (see Figure 8–84).

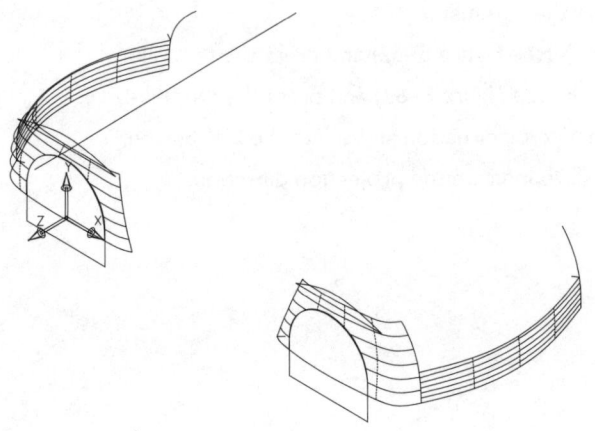

Figure 8–84 *Surfaces trimmed*

201. Hide and unhide surfaces in accordance with Figure 8–85.
202. Trim surfaces B and C (see Figure 8–85) by projecting the line A (see Figure 8–85).

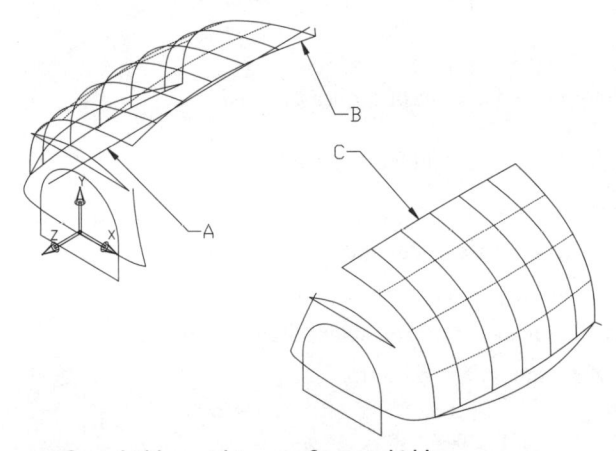

Figure 8–85 *Four surfaces hidden and two surfaces unhidden*

The surfaces are trimmed (see Figure 8–86).

203. Referring to Figure 8–87, hide the surfaces and unhide a surface.
204. Type **6** to set the display to the front view (see Figure 8–88).
205. Project the curves A, B, C, and D (see Figure 8–88) to trim the surface E (see Figures 8–88 and 8–89).

Surface Modeling II 411

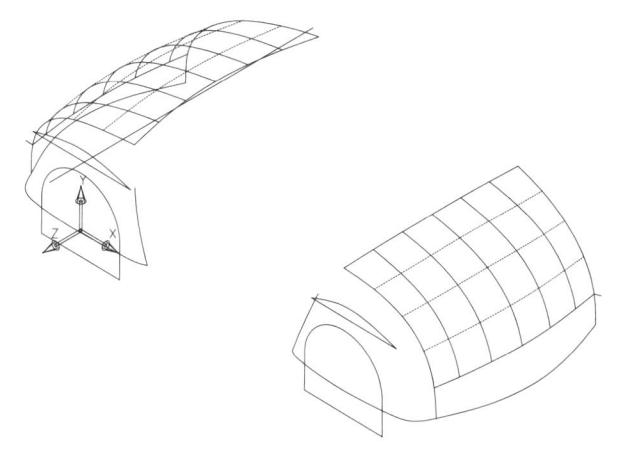

Figure 8–86 *Surfaces trimmed*

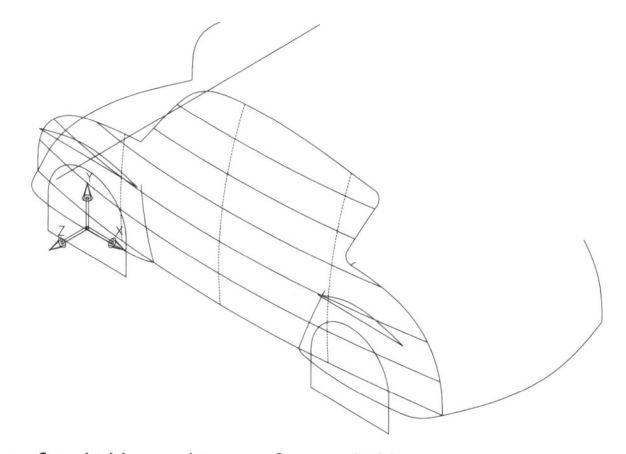

Figure 8–87 *A surface hidden and two surfaces unhidden*

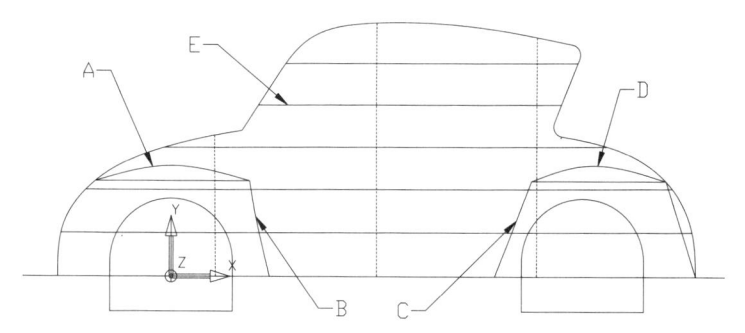

Figure 8–88 *Display set*

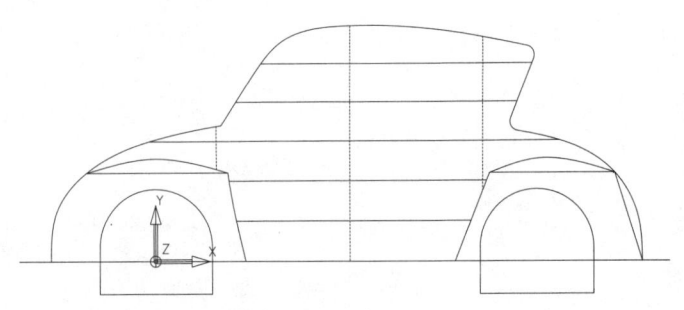

Figure 8–89 *Wire projected and surface trimmed*

Now you will construct four rule surfaces.

  206. Hide all the curves and unhide all the surfaces (see Figure 8–90).
  207. Set the current layer to Surface.
  208. Select Surface>Create Surface>Rule or select Ruled Surface from the Surface Modeling toolbar.

     Command: AMRULE

  209. Select A and B (see Figure 8–90).
  210. Repeat the AMRULE command.
  211. Select C and D (see Figure 8–90).

Two rule surfaces are constructed (see Figure 8–91).

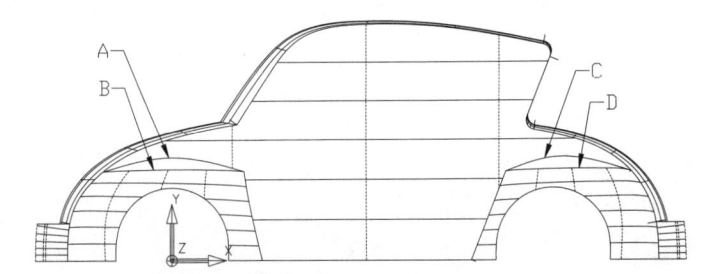

Figure 8–90 *Wires hidden and surfaces unhidden*

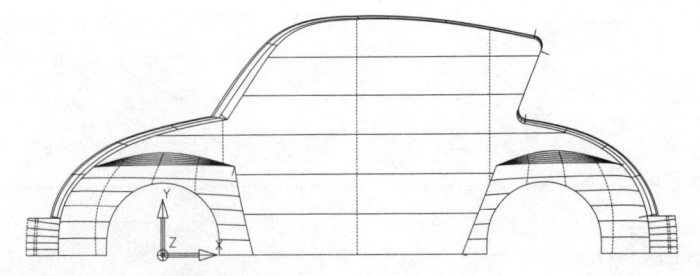

Figure 8–91 *Rule surfaces constructed*

Surface Modeling II 413

212. Select View>3D View>Front left Isometric (see Figure 8–92)
213. Select Surface>Create Surface>Rule or select Ruled Surface from the Surface Modeling toolbar.
214. Select A and B (see Figure 8–92).

A rule surface is constructed (see Figure 8–93).

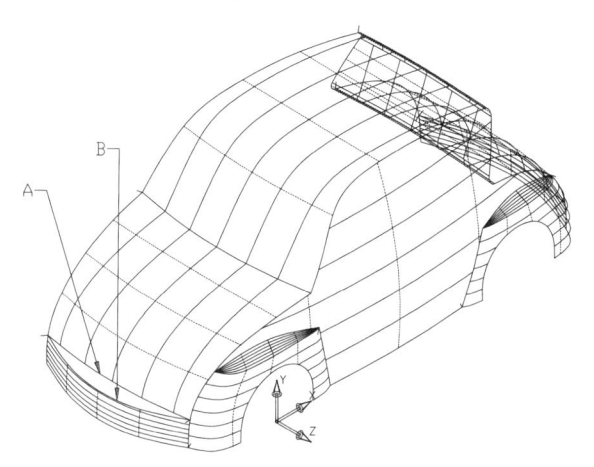

Figure 8–92 *Display changed*

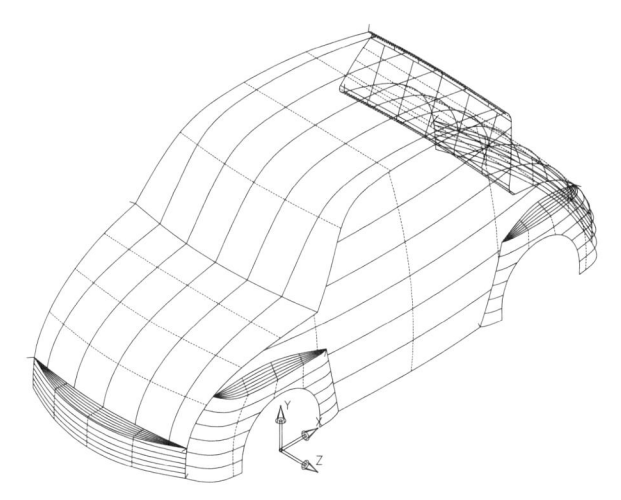

Figure 8–93 *Rule surface constructed*

215. Type **8** to set the display to an isometric view.
216. Select Surface>Create Surface>Rule or select Ruled Surface from the Surface Modeling toolbar.

217. Select A and B (see Figure 8–94).

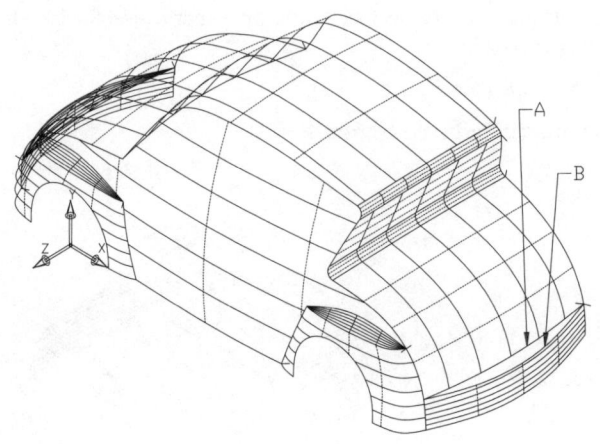

Figure 8–94 *Display set*

A rule surface is constructed (see Figure 8–95).

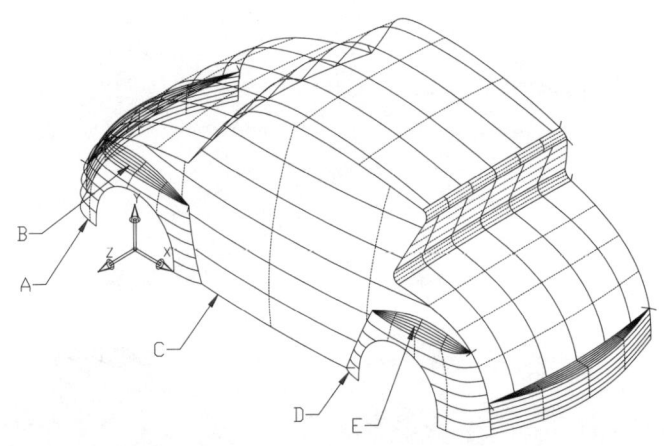

Figure 8–95 *Rule surface constructed*

Now you will perform a visual check of the normal direction of the surfaces by shading.

218. Select View>Shade>Gouraud Shaded (see Figure 8–96)

Because a surface on the computer screen has no thickness and no volume, a normal vector is assigned to the surface to tell which side of the surface has a volume (back side of the vector) and which side of the surface is void (front side of the vector). The direction of the normal vector is assigned according to the sequence in which the wires are selected when you make the surface. When a surface is rendered, the front side of the surface is rendered and the back side of the surface is ignored. Therefore, you find some of the surfaces missing.

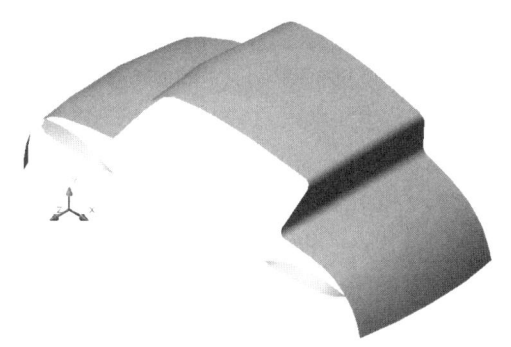

Figure 8–96 *Some surfaces missing in the shaded view*

    219. Select View>Shade>Gouraud Shaded, Edges On

Now the missing surfaces are represented by wires (see Figure 8–97).

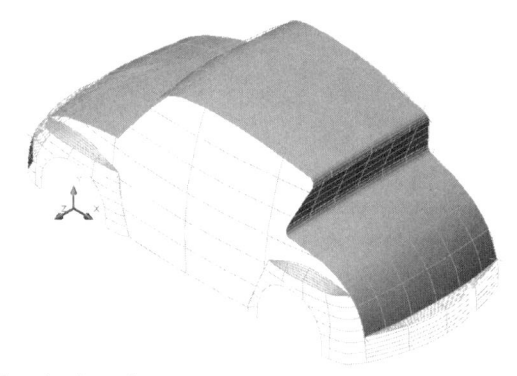

Figure 8–97 *Shaded and edges shown*

    220. Select Surface>Edit Surface>Adjust Normals or select Flip Surface Normal from the Surface Modeling toolbar.

    221. Select the surfaces that have a wrong direction of normal and press the ENTER key.

The surface normals are corrected (see Figure 8–98).

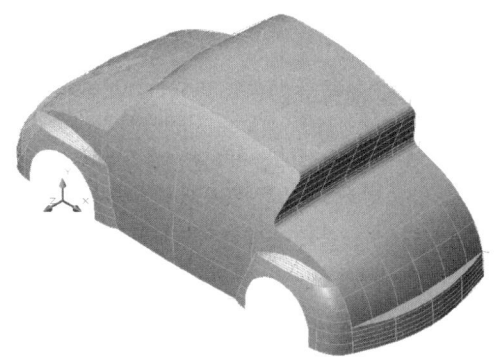

Figure 8–98 *Normal of the surfaces flipped*

Now you will construct the remaining surfaces of the model car.

222. Select View>Shade>3D Wireframe
223. Select Assist>New UCS>World
224. Select Modify>3D Operation>Mirror 3D
225. Select A, B, C, D, and E (see Figure 8–95) and press the ENTER key.
226. Type **ZX** to specify the mirror plane.
227. Type **0,32** to indicate the location of the mirror plane.
228. Press the ENTER key.

The model car is complete (see Figure 8–99.) Save your file (file name: *Model_B.dwg*).

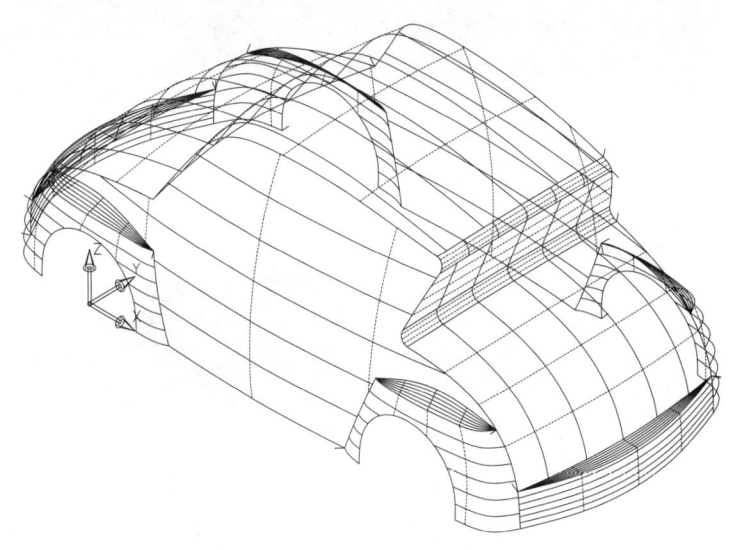

Figure 8–99 *Completed model*

MODEL CAR C

To illustrate how surfaces can be scaled, you will construct a surface model by scaling a set of surfaces.

1. Open the file *Model_B.dwg*, if you have already closed it.
2. Select File>SaveAs and specify a new file name (file name: *Model_C.dwg*).
3. Type **DELOBJ** at the command line area and type **1** so that objects are erased after they are operated.
4. Select Surface>Edit Surface>Scale

   Command: AMSCALE
5. Select all the surfaces and press the ENTER key.
6. Type **I** to use the independent option.

7. Type **1.25** to specify the X factor.
8. Type **1.2** to specify the Y factor.
9. Type **1.5** to specify the Z factor.
10. Type **0,0,0** to indicate the origin.

The surfaces are scaled (see Figure 8–100). Save and close your file.

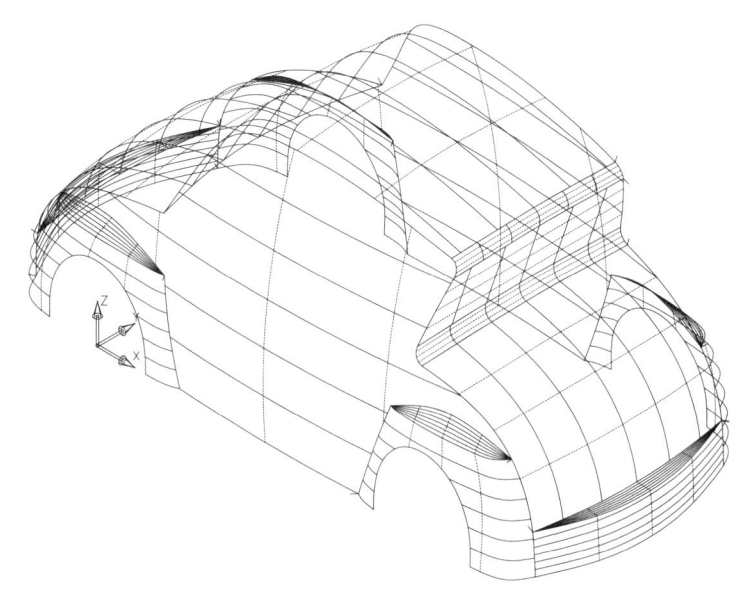

Figure 8–100 *Surfaces scaled*

REVIEW QUESTIONS

1. What are the two basic methods to represent a surface in the computer?

2. Use simple sketches to illustrate the various kinds of primitive surfaces, free-form surfaces, derived surfaces, and trimmed surfaces.

3. Write an account of the kinds of wires that you can construct by using Mechanical Desktop?

4. Explain how a solid model with free-form features can be constructed by using the surface modeling tool.

CHAPTER 9

Part Modeling III

OBJECTIVES

The aim of this chapter is to give you an opportunity to construct solid parts using various advanced techniques, including the use of a surface to cut a solid, the combination of two solid parts into a single solid part, splitting a solid part into two solid parts, applying a face draft to a solid, splitting a face of a solid, setting up a design table, scaling a solid part, mirroring a solid part, using an AutoCAD solid, and making a parametric solid static. After studying this chapter, you should be able to do the following:

- Use a surface to cut a solid
- Combine two solid parts to a single solid part
- Split a solid part into two solids
- Apply a face draft to a solid
- Split a face of a solid
- Use a table to define design parameters
- Modify a solid part by scaling
- Construct a new solid part by mirroring an existing solid part
- Use an AutoCAD solid as a static base solid feature
- Convert a parametric solid part to a static solid part

OVERVIEW

As the concluding chapter of the three chapters on part modeling, a number of advanced modeling techniques will be introduced here. First, you will learn more advanced placed solid features, including the surface cut, combine, part split, and face draft. Then you will learn about the face split (a special kind of sketched feature), use of design variables, scaling a part, mirroring a part, converting an AutoCAD solid to a Mechanical Desktop solid, and making a parametric solid static (non-parametric).

SURFACE CUT

The major drawback of using the solid modeling approach alone to construct a solid part is that forms and shapes are restricted by the four fundamental operations on sketches, namely, extrude, revolve, loft, and sweep. To widen the repertoire of form and shapes, you can add a surface cut feature to the solid by using a surface to cut a solid. Basically, there are four ways to use surfaces in solid modeling: you can adjust the thickness of a surface, stitch a set of surfaces to a Mechanical Desktop base solid, use a surface to cut an AutoCAD native solid, or use a surface to cut a Mechanical Desktop solid.

Now you will construct a solid part, a work point (as a reference point), and a surface, and use the surface to cut the solid.

1. Start a new part file. Use metric as the default.
2. Referring to Figure 9–1, construct a rectangle, resolve it to a profile, add parametric dimensions, and extrude it to a solid.

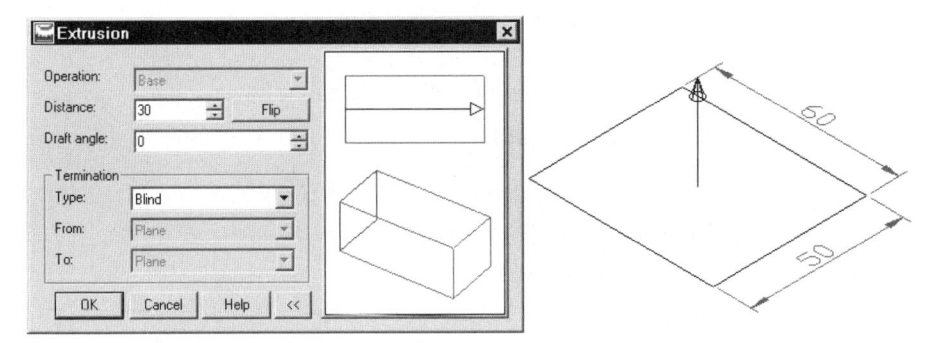

Figure 9–1 *Rectangle being extruded to a solid*

3. Construct a work point and add parametric dimensions (see Figure 9–2).
4. Select Assist>New UCS>Origin
5. Use the End point object snap and select the end point indicated in Figure 9–2.
6. Referring to the following table, construct three splines (see Figure 9–3).

Spline	A	B	C
First Point	−5,−5,15	65,−5,10	−5,25,20
Next Point	−5,25,20	65,25,25	30,25,20
Next Point	−5,55,15	65,55,10	65,25,25
Next Point	ENTER	ENTER	ENTER
Start Tangent	ENTER	ENTER	ENTER
End Tangent	ENTER	ENTER	ENTER

Now you will construct a sweep surface.

7. Select Surface>Create Surface>Sweep
8. Select A and B (shown in Figure 9–3) and press the ENTER key.
9. Select C (shown in Figure 9–3) and press the ENTER key.
10. In the Sweep Surface dialog box, select Normal orientation, and select the OK button. Figure 9–4 shows the results.

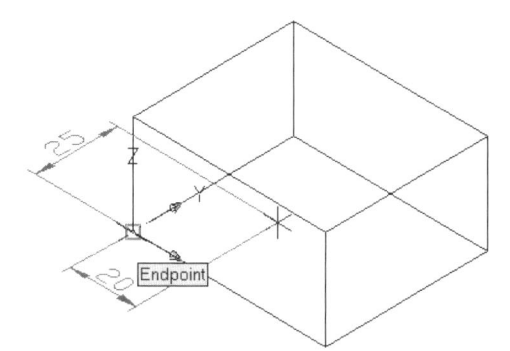

Figure 9-2 *Work point constructed*

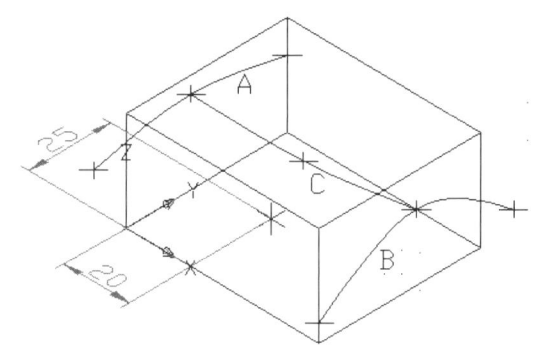

Figure 9-3 *Splines constructed*

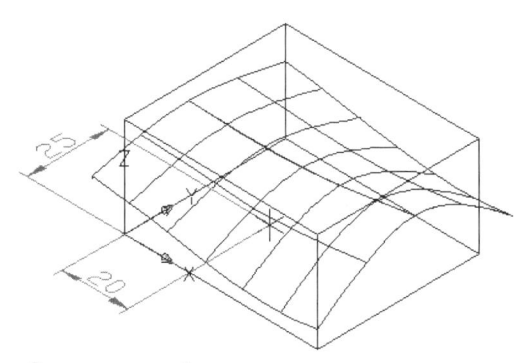

Figure 9-4 *Sweep surface constructed*

Before you use the surface to cut the solid, you will save two copies of the file.

11. Select File>Save (file name: *SurfaceSplit.dwg*).
12. Select File>SaveAs (file name: *SurfaceCut.dwg*).

Now you have two files, *SurfaceSplit.dwg* and *SurfaceCut.dwg*, and you work on the file *SurfaceCut.dwg*.

Now cut the solid by using the surface.

13. Select Part>Placed Features>Surface Cut

    Command: AMSURFCUT

14. Select the surface and then the work point.
15. Left-click to toggle the direction of the cut.
16. Referring to Figure 9–5, right-click to accept.
17. Hide the splines (see Figure 9–6).

The part is complete. Save and close your file.

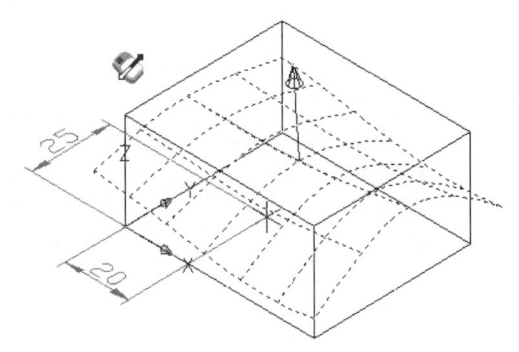

Figure 9–5 *Surface and work point selected*

Figure 9–6 *Surface cut feature placed*

COMBINATION OF SOLID PARTS

As you learned in Chapters 3 and 4, the way to construct a complex solid is to first make a sketched solid feature and then sequentially construct additional solid features. While making the additional sketched features, you need to decide how the features are to be combined. This way of solid construction imposes limitations on the shape of the object that you can construct because you must combine each and every sketched solid feature you make.

To overcome this limitation, you can start or import another solid part in the same part file and combine the solid parts to form a single solid part. Before you combine two solid parts, you need to properly position them in relation to each other. To maintain a parametric position relationship among the solid parts to be combined, apply assembly constraints to the solid parts in a way similar to what you would do on the components of an assembly. While combining the solid parts in pairs, you select one solid part to be the active solid and treat the other solid as a tool body.

Similar to constructing an assembly, there are three approaches to constructing a complex solid part by combining two or more solid parts: top-down, bottom-up, and hybrid.

Now you will construct two solid parts in a part file (top-down) and combine them to form a single solid part.

1. Start a new part file. Use metric as the default.
2. Referring to Figure 9–7, construct a rectangle, resolve it to a profile, add parametric dimensions, and extrude it a distance of 30 mm.
3. Construct a shell feature with a shell thickness of 2 mm, and remove the top face (see Figure 9–8).
4. Right-click in the Desktop Browser and select New Part/Toolbody (see Figure 9–9).
5. Press the ENTER key to accept the part name.
6. Referring to Figure 9–10, construct a rectangle, resolve it to a profile, add parametric dimensions, and extrude it a distance of 25 mm.

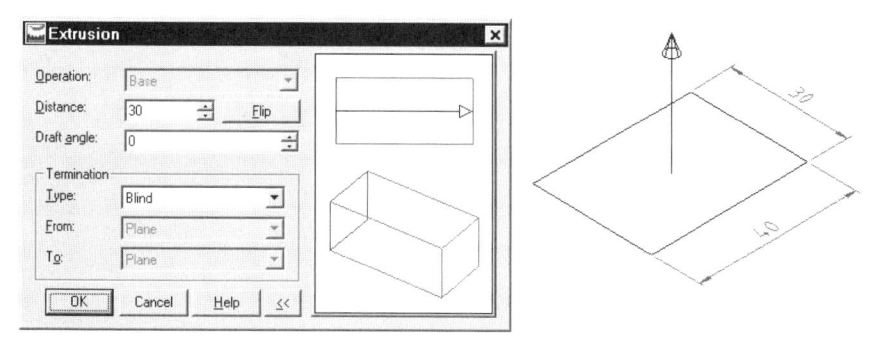

Figure 9–7 *Extruded solid being constructed*

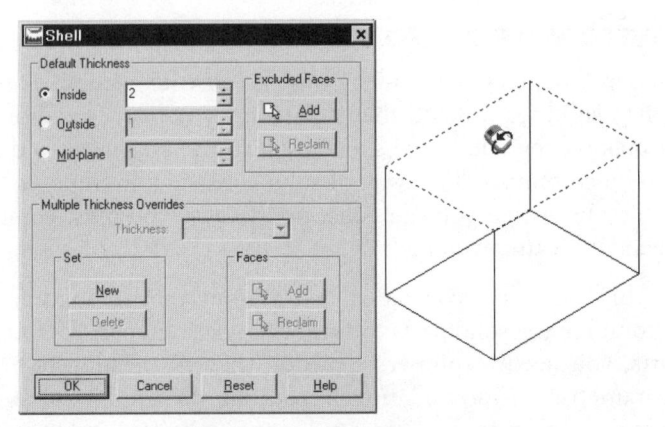

Figure 9-8 *Shell feature being placed*

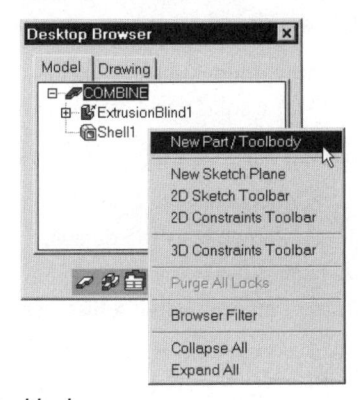

Figure 9-9 *Starting a new tool body*

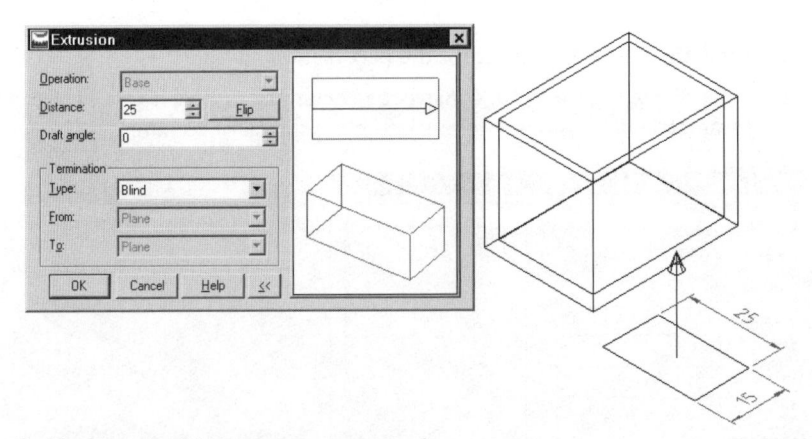

Figure 9-10 *Second solid part being constructed*

7. Construct a shell feature (2 mm thick) and remove the top face (see Figure 9-11).

Now you have two solid parts in a single solid part file. Before you combine them, you will add assembly constraints to control the position relationship between them.

8. Select Toolbody>3D Constraints>Flush
9. Referring to Figure 9–12, select the faces and press the ENTER key.

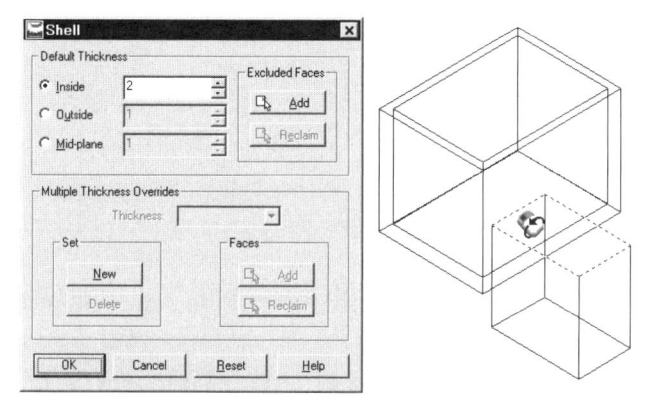

Figure 9–11 *Shell feature being placed on the second solid part*

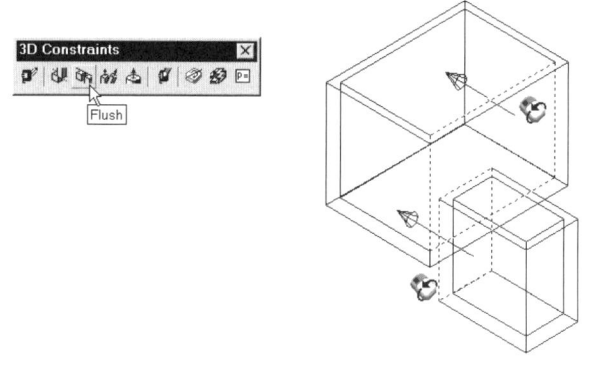

Figure 9–12 *Faces being flushed*

10. Repeat the AMFLUSH command twice on the pair of faces highlighted in Figures 9–13 and 9–14.

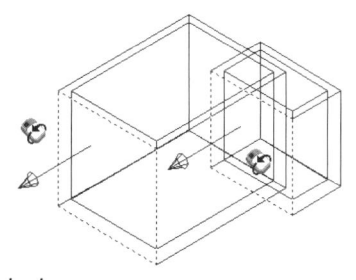

Figure 9–13 *Faces being flushed*

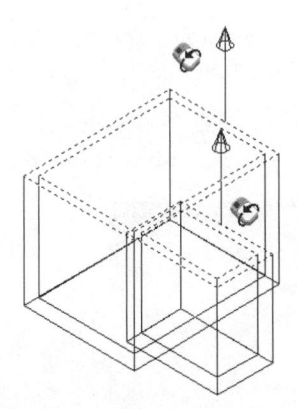

Figure 9–14 *Faces being flushed*

There are three ways to combine two solid parts: join, cut, and intersect. Now you will combine the solid parts by joining.

11. Double-click the first solid part in the Browser to activate it.
12. Select Part>Placed Features>Combine
13. Type **J** to use the join operation.
14. Select the toolbody indicated in Figure 9–15.

The solid parts are combined. Now you will import a solid part to the part file and combine them into a single solid. Importing a solid part is equivalent to the bottom-up approach in an assembly. Because you first use the top-down approach and then the bottom-up approach, you are, in essence, using a hybrid approach.

To import a file, you can either use the part catalog, insert a part and localize it, or you can copy in a solid part. Here we will use the part catalog. (To copy in a solid part, you select Part>Part>Copy In and select a file from the File to Load dialog box.)

15. Select Toolbody>Catalog or the Catalog button from the desktop Browser.

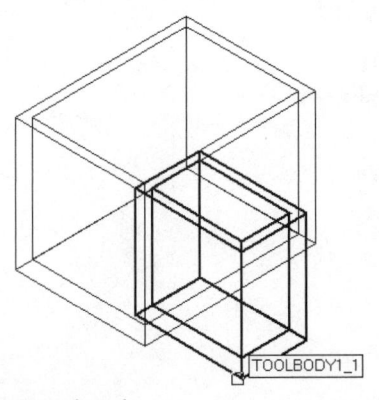

Figure 9–15 *Solid parts being combined*

The Part Catalog dialog box is similar to the Assembly Catalog dialog box.

16. In the Directories box, right-click and select Add Directory (see Figure 9–16).

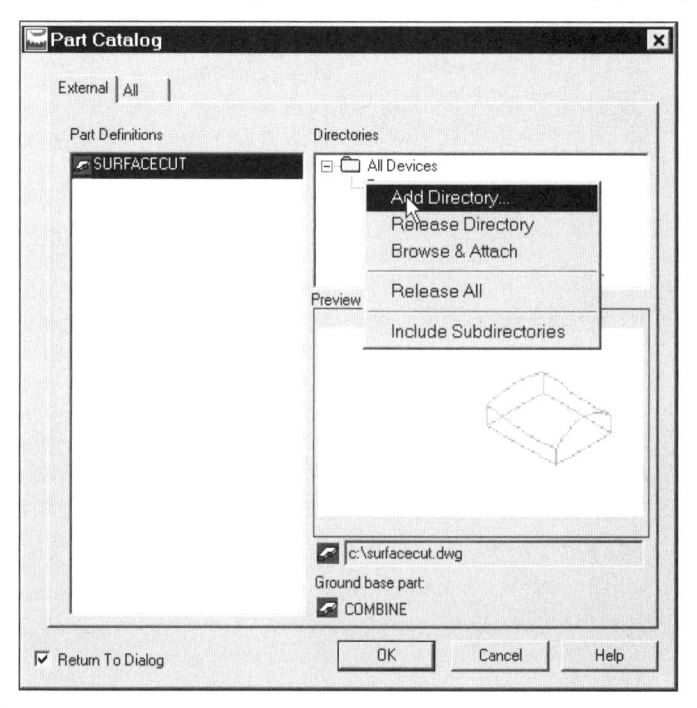

Figure 9–16 *Part Catalog dialog box*

17. In the Browser for folder dialog box, select the folder where you saved the file *SurfaceCut.dwg* and select the OK button.
18. On returning to the Part Catalog dialog box, double-click *SurfaceCut.dwg* in the Part Definition box, select a location in the screen, and right-click.
19. In the All tab of the Part Catalog dialog box (shown in Figure 9–17), select *SurfaceCut.dwg*, right-click, and select Localize.
20. Select the OK button.

Similar to using the assembly catalog, localizing an external part is equivalent to copying in the part. There will be no relationship between the local definition and the external solid part.

21. Referring to Figure 9–18, apply the AMFLUSH command three times on the solid parts.

Figure 9-17 *Localizing a solid part*

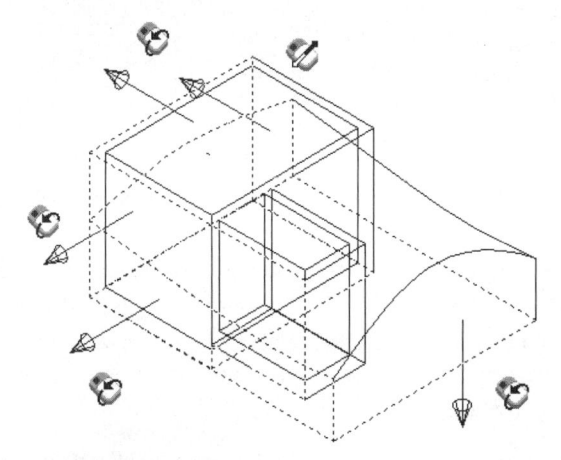

Figure 9-18 *Solid parts assembled*

Now combine the solid parts by intersecting.

    22. Select Part>Placed Features>Combine

    23. Type **I** to use the intersect option.

    24. Select the imported solid part (see Figure 9–19).

The solid part is complete (see Figure 9–20). Save and close your file (file name: *Combine.dwg*).

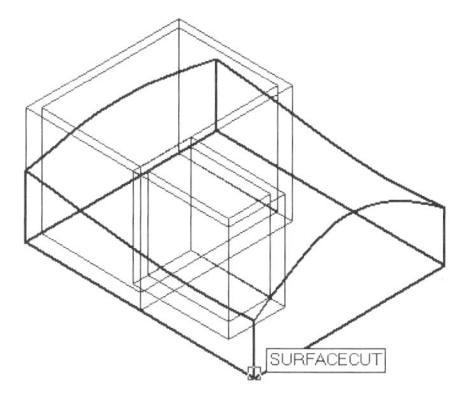

Figure 9–19 *Solid parts being combined*

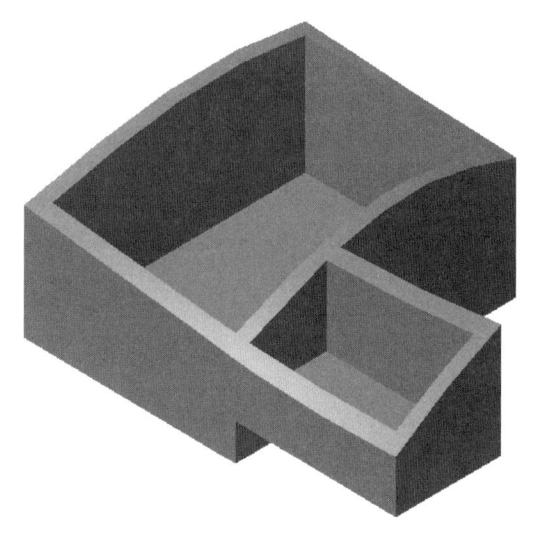

Figure 9–20 *Solid parts combined*

PART SPLITTING

In contrast to combining two solid parts to form a single solid part, you can split a single solid into two solid parts along a planar face of the solid, a work plane, a surface, or a split line. After splitting, you can externalize the solids to form individual solid parts.

Now you will split a part along a surface.

1. Open the file *SurfaceSplit.dwg*.
2. Select Part>Placed Features>Part Split
3. Select the surface and then the work point.

4. Left-click to flip the direction for the new solid (see Figure 9–21).
5. Right-click to accept.
6. Press the ENTER key to accept the default name.
7. Hide the splines.

The solid is split into two (see Figure 9–22).

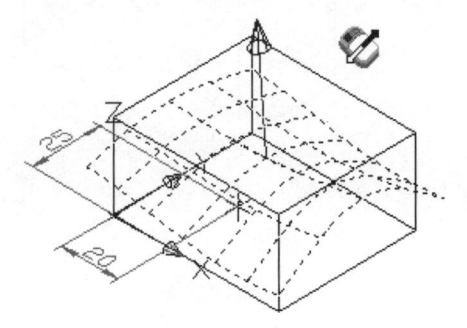

Figure 9–21 *Solid being split*

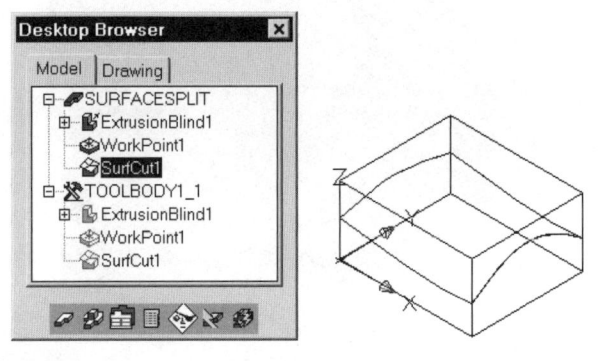

Figure 9–22 *Two solids*

Two solids are constructed in a single part file. They are both local definitions. To export one of them to become a separate part file, you can either select Part>Part>Copy Out or use the Part Catalog to externalize it.

Now you will copy out the solid part definition.

8. Select Part>Part>Copy Out (see Figure 9–23).

   Command: AMCOPYOUT

9. In the Part/Subassembly Out dialog box, select TOOLBODY1 and select the File button.

10. In the Output File dialog box, select a folder, specify a file name (Upper), and select the Save button.

11. After returning to the Part/Subassembly Out dialog box, select the OK button.

Figure 9–23 *Part/Subassembly Out dialog box*

The selected solid part is copied to a file. Because the local definition is not required any longer, you will remove it.

12. Select Toolbody>Catalog
13. In the All tab of the Part Catalog dialog box, select TOOLBODY1 from the Local Toolbody Definitions area, right-click, and select Remove.
14. Select the OK button.

The toolbody is removed (see Figure 9–24).

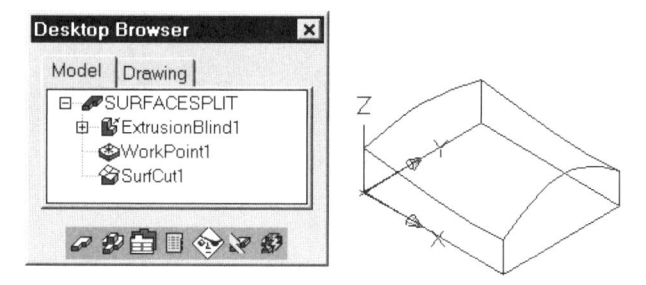

Figure 9–24 *Toolbody removed*

Now you will construct a work plane and split the solid along the work plane.

15. Select Part>Work Features>Work Plane
16. In the Work Plane dialog box, select On Edge/Axis in the 1st Modifier area and On Edge/Axis in the 2nd Modifier area, and select the OK button.
17. Select the edges indicated in Figure 9–25.

A work plane is constructed.

18. Select Part>Placed Features>Part Split

19. Select the work plane.
20. Right-click to accept the direction of the split (see Figure 9–26).
21. Press the ENTER key to accept the default new part name.

The solid is split (see Figure 9–27). Save and close your file.

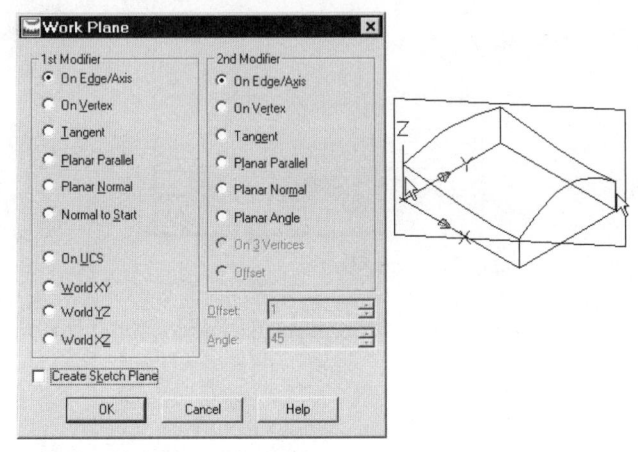

Figure 9–25 *Work plane being constructed*

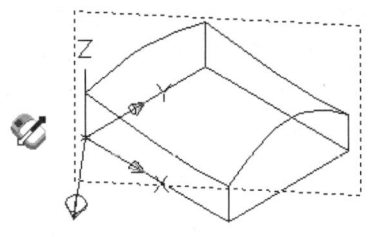

Figure 9–26 *Part being split along the work plane*

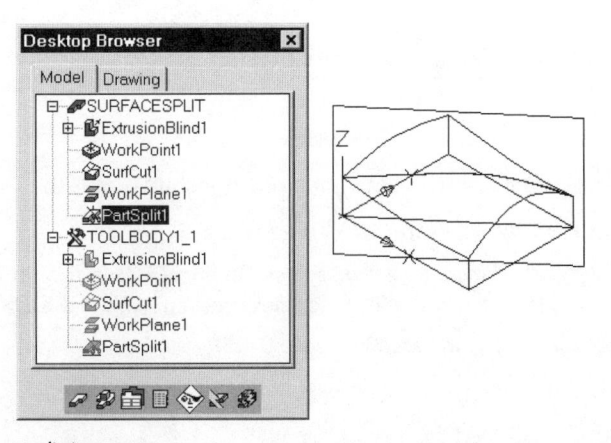

Figure 9–27 *Part split into two*

Now you will split a part along a planar face.

1. Start a new file. Use metric as the default.
2. Referring to Figure 9–28, construct a sketch, resolve it to a profile, add parametric dimensions, and extrude it a distance of 30 mm.
3. Select Part>Placed Features>Part Split
4. Select the planar face highlighted in Figure 9–29.
5. Right-click to accept the direction of the split.
6. Press the ENTER key to accept the default new part name.

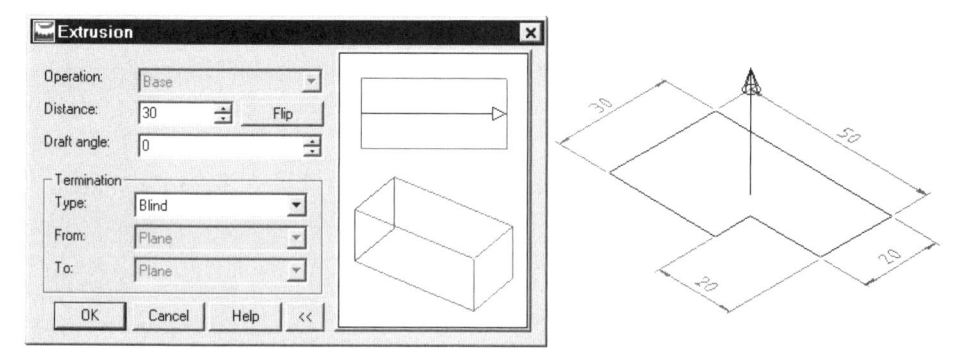

Figure 9–28 *Sketch being extruded*

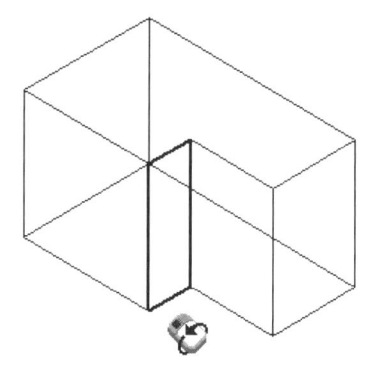

Figure 9–29 *Part being split*

The solid is split along a planar face (see Figure 9–30). Save your file (file name: *PlaneSplit.dwg*).

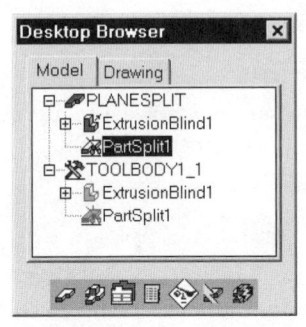

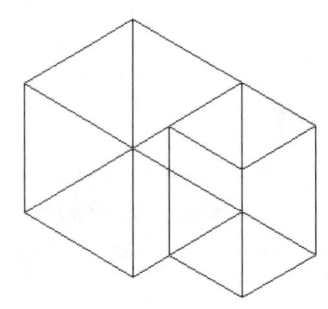

Figure 9–30 *Solid part split into two parts*

Now you will split a part along a split line. (A split line is a special kind of parametric sketch. You can use it to cut a solid into two solids or the selected face of a solid into two faces.)

    7. Select the main solid part from the Browser and double-click to activate it, if it is not activated.

    8. Referring to Figure 9–31, construct an arc.

Now you will resolve the sketch to a split line.

    9. Select Part>Sketch Solving>Split Line

        Command: AMSPLITLINE

    10. Select the arc and press the ENTER key.

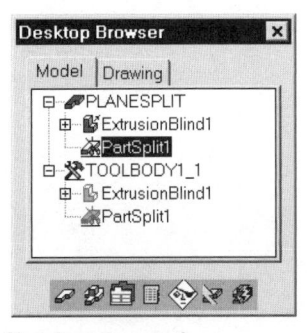

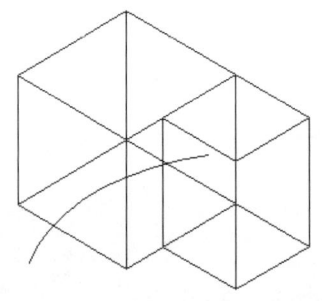

Figure 9–31 *Sketch constructed*

    11. Referring to Figure 9–32, add parametric dimensions.

Now you will split the solid along the split line.

    12. Select Part>Placed Features>Part Split

    13. Select the split line.

    14. Right-click to accept the direction of the split (see Figure 9–33).

    15. Press the ENTER key to accept the default new part name.

Part Modeling III 435

The part is split along the split line (see Figure 9–34). Save and close your file.

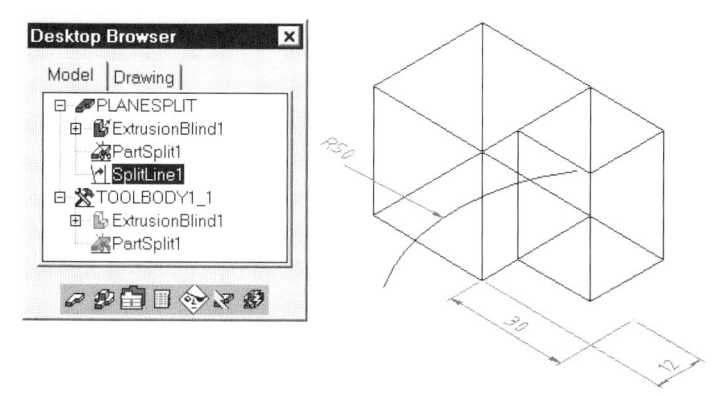

Figure 9–32 *Dimensions added*

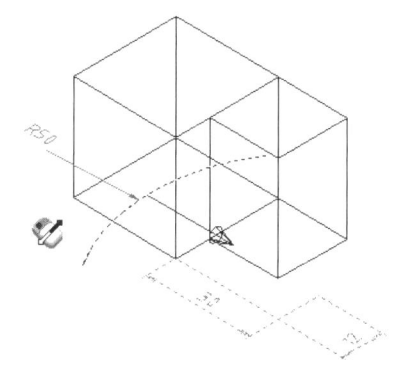

Figure 9–33 *Part being split along the split line*

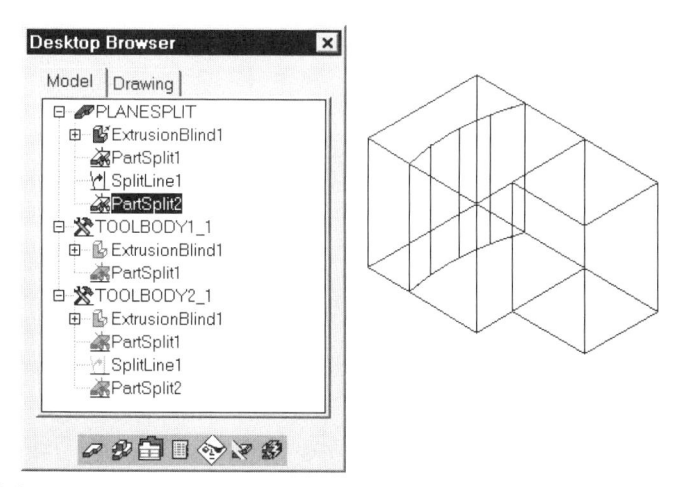

Figure 9–34 *Part split along the split line*

FACE DRAFT

In mold and die design, a draft angle is applied to the vertical faces to facilitate removal of the product from the die or mold. To add a draft angle to a solid, you place the face draft feature.

Now you will construct an extruded solid and place a face draft on its faces.

1. Start a new part file. Use metric as the default.
2. Referring to Figure 9–35, construct a sketch, resolve it to a profile, and add parametric dimensions.
3. Set the display to an isometric view.
4. Extrude the profile a distance of 20 mm.
5. Select Part>Placed Features>Face Draft (see Figure 9–36).

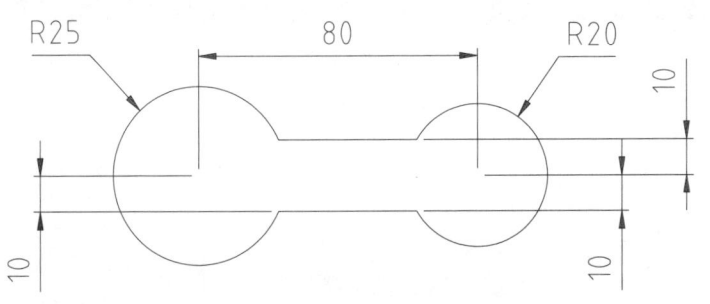

Figure 9–35 *Resolved sketch*

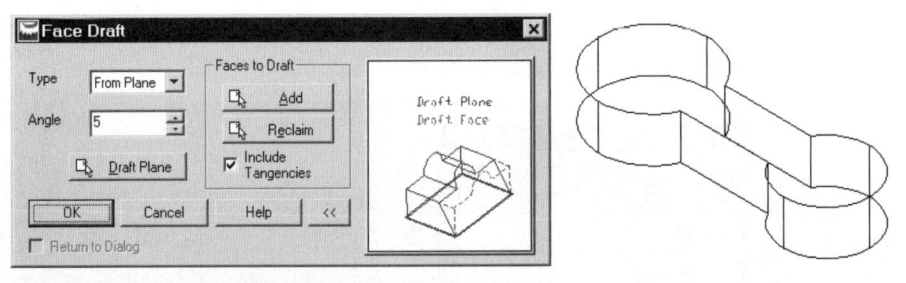

Figure 9–36 *Face Draft being placed*

Using the Face Draft dialog box, specify a face draft in three ways: From Plane, From Edge, and Shadow. Using the From Plane option, you construct a face draft from a selected work plane or planar face. Using the From Edge option, you construct a face draft from a selected edge. Using the Shadow option, you construct a face draft from a selected plane to a point tangent to a selected cylindrical face.

6. Select the From Plane option, set Angle to **5**, and select the Draft Plane button.
7. Referring to Figure 9–37, select the upper face of the solid.

8. Right-click to accept the direction of the face draft, or left-click to flip.
9. On returning to the Face Draft dialog box, select the Add button.
10. Select all the faces perpendicular to the draft plane and press the ENTER key.
11. Select the OK button.

The face draft feature is placed (see Figure 9–38). Save your file (file name: *FaceDraft.dwg*).

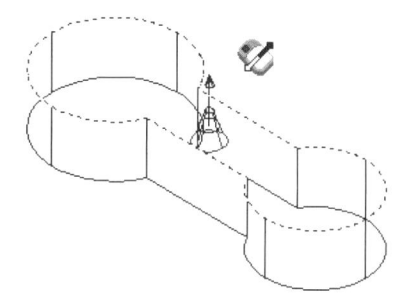

Figure 9–37 *Top face selected*

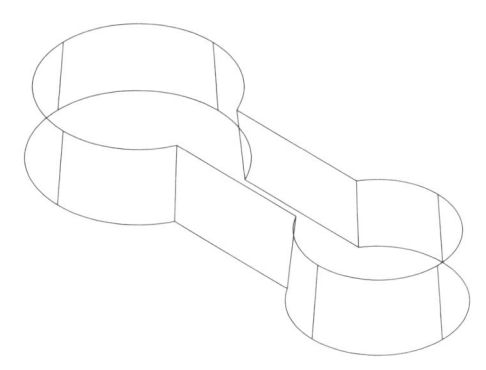

Figure 9–38 *Face draft placed*

FACE SPLITTING

Because the face draft feature applies to the entire selected face of a solid, you need to split a face into two faces if you want to apply face draft to a face of a part along a line in two directions. To split faces, you construct a split line and use the line to split the selected faces or you use a plane to split selected faces.

1. Open the file *FaceDraft.dwg*, if you have already closed it.
2. Select File>SaveAs and specify a new file name (file name: *FaceSplitDraft.dwg*).
3. Select the face draft feature from the Browser, right-click, and select Delete.
4. Referring to Figure 9–39, establish a sketch plane.

5. Construct a horizontal line, resolve it to a split line, and add a parametric dimension (see Figure 9–40).

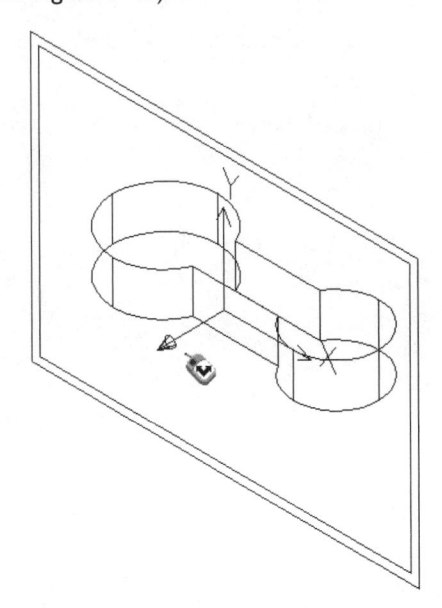

Figure 9–39 *Sketch plane being constructed*

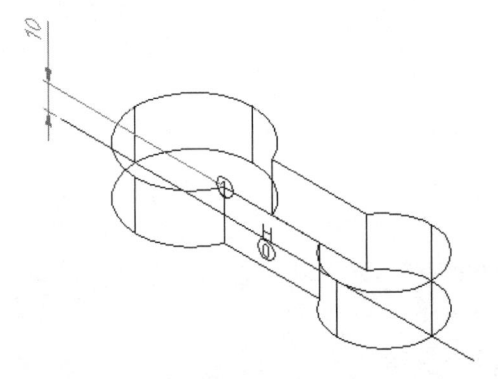

Figure 9–40 *Split line constructed*

6. Select Part>Sketch Features>Face Split

   Command: AMFACESPLIT

7. Press the ENTER key to accept (if the default option is R) or type **R** to use the projected option.

8. Type **A** to select all the faces.

9. Press the ENTER key.

All the faces are split by the split line (see Figure 9–41).

Now you will construct a work plane and use the work plane to split the faces.

10. Select the Face Split feature from the Browser, right-click, and select Suppress.
11. Referring to Figure 9–42, construct a work plane parallel to the top face of the solid part and offset a distance of 10 mm.

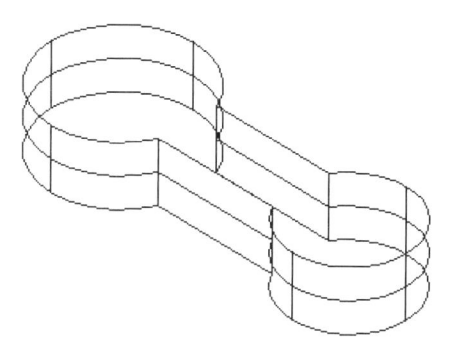

Figure 9–41 *Faces split*

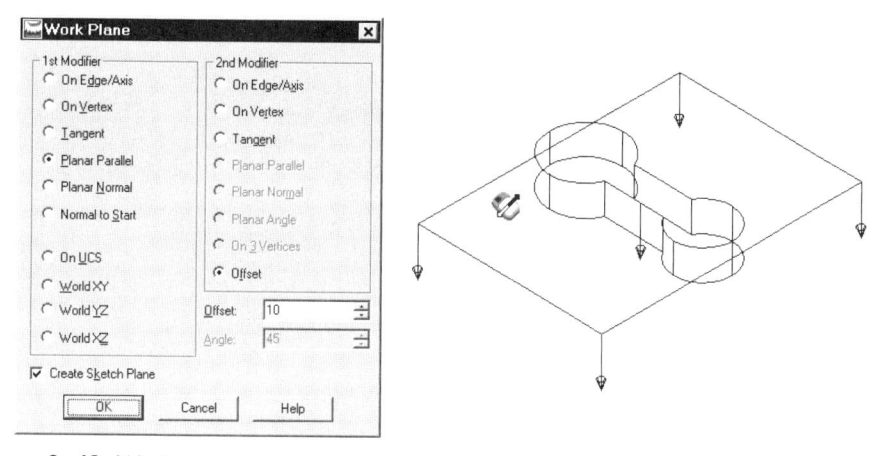

Figure 9–42 *Work plane being constructed*

12. Select Part>Sketch Features>Face Split.
13. Type **P** to use the planar option.
14. Type **A** to split all the faces.
15. Press the ENTER key.
16. Select the work plane.

The faces are split (see Figure 9–43). Now you will place a face draft feature.

17. Select Part>Placed Features>Face Draft

18. Set the draft angle to 5 degrees.
19. Select the Draft Plane button and select the work plane (see Figure 9–43).
20. Select the Add button.
21. Select the faces highlighted in Figure 9–44 and press the ENTER key.
22. Select the OK button.

The face draft feature is placed (see Figure 9–45). Save and close your file.

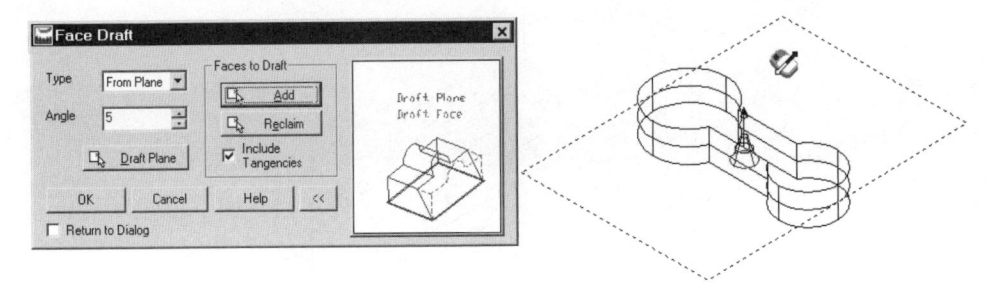

Figure 9–43 *Work plane selected and face draft direction accepted*

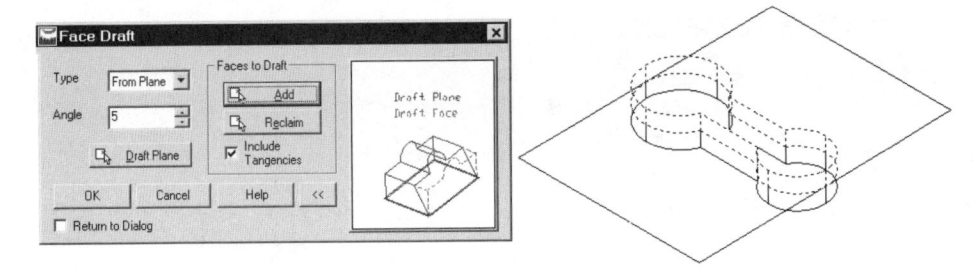

Figure 9–44 *Faces selected*

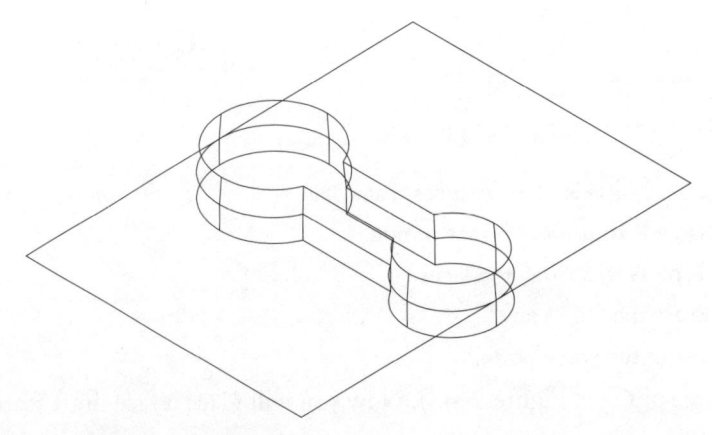

Figure 9–45 *Face draft feature placed*

DESIGN VARIABLES

To better control the parametric dimensions of the solids in a single file and across a set of files, you maintain a set of design variables by setting up a design variable file or constructing an Excel spreadsheet.

DESIGN VARIABLE FILE

Now you will learn how to use a design variable file to control a set of files.

1. Open the part file *Toybase.dwg* that you constructed in Chapter 5.
2. Select Part>Design Variables

    Command: AMVARS

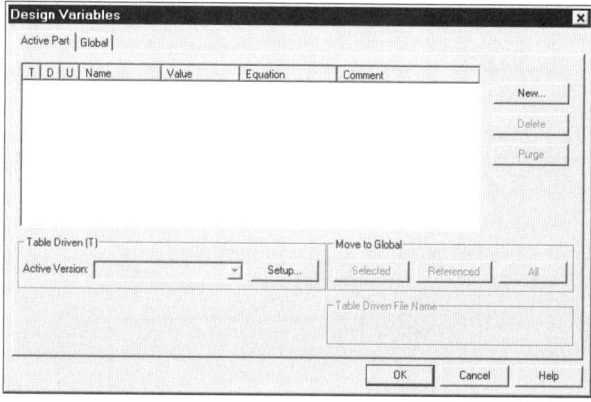

Figure 9–46 *Design Variables dialog box*

In the Design Variables dialog box, there are two tabs: Active Part and Global. The Active Part tab concerns the variables of the active solid part, and the Global tab concerns the variables for a set of solid parts in the file.

3. Select the Global tab (see Figure 9–47).

Figure 9–47 *Global tab*

4. In the Global tab, select the New button to display the New Part Variable dialog box (see Figure 9–48).

The New Part Variable dialog box has three fields: Name, Equation, and Comment. Name is the name of the variable. Equation is a mathematical expression that is represented by the variable name. Comment is any explanation that you add.

5. Input data in accordance with the following table and select the OK button.

Name	Equation	Comment
WheelBase	84	Distance between two axles

6. Repeat steps 4 and 5 to input two more design variables as follows (see Figure 9–49):

Name	Equation	Comment
Length	100	Length of the base plate
Diameter	2	Diameter of the axle

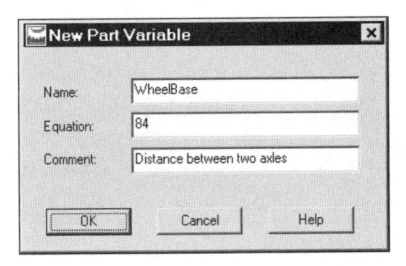

Figure 9–48 *New Part Variable dialog box*

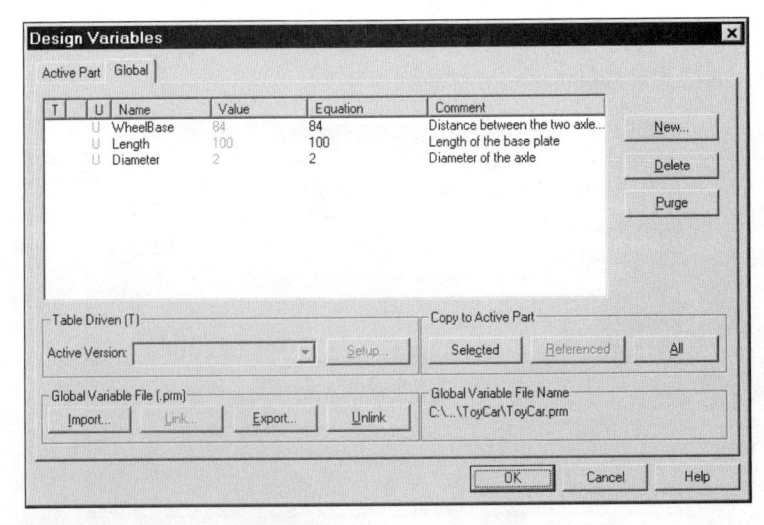

Figure 9–49 *Design variable constructed*

Three design variables are constructed in the solid part file. The U letter along each variable stands for unused. In order for these variables to be usable by other part files, you will export them to an external design variable file.

7. Select the Export button.
8. In the Export dialog box, select a directory, specify a file name (*ToyCar.prm*) and select the OK button. (The default file extension for a design variable file is .prm.)

To reference the design variables in the solid part to a design variable file, you will link to the file.

9. Select the Link button.
10. In the Link dialog box, select the design variable file that you exported.
11. Select the Open button.

Once the solid part is linked to a design variable file, any change to the file will be reflected in the solid part.

To gain a better understanding of the design variable file, you will now open it with the Notepad application.

12. Select the Start button of your Windows application.
13. Select Programs>Accessories>Notepad
14. In the Notepad application window, select File>Open.
15. In the Open dialog box, select the file *ToyCar.prm* from the folder where you saved the file and select the Open button (see Figure 9–50).

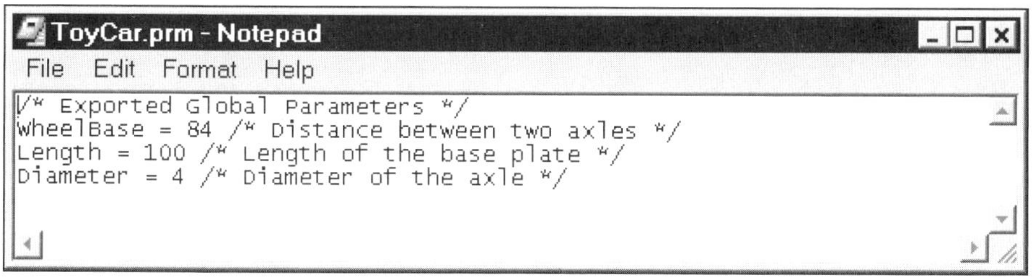

```
/* Exported Global Parameters */
WheelBase = 84 /* Distance between two axles */
Length = 100 /* Length of the base plate */
Diameter = 4 /* Diameter of the axle */
```

Figure 9–50 *Design variable file opened in Notepad*

The design variable file has a number of lines. The first line tells you that it is a set of exported global parameters. Each of the second and the subsequent lines delineate the name, equation, and comment of a global parameter. Because the design variable file is a text file, you can add or modify design variables by using any text editor.

16. Select File>Exit to close the design variable file.

Now you will continue to work on the solid part.

17. Select Part>Dimensioning>Dimensions as Equation

18. Referring to Figure 9–51, select the profile of the extruded feature from the Browser, right-click, and select Edit Sketch.
19. Select the dimension indicated in Figure 9–51 and double-click.
20. In the Power Dimensioning dialog box, change the expression to Length (a design variable name that you constructed) and select the OK button.
21. Select the Update button from the Browser.

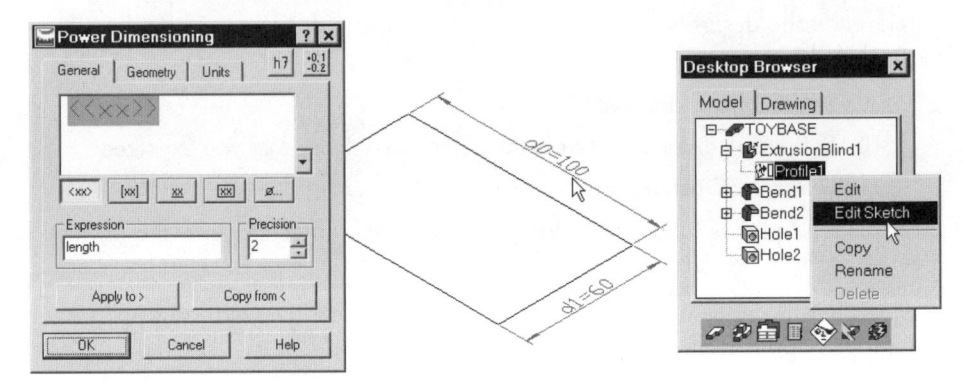

Figure 9–51 *Dimension being modified*

22. Referring to Figure 9–52, double-click the hole feature from the Browser.
23. In the Hole dialog box, change the hole size to Diameter (a design variable) and select the OK button of the Hole dialog box (see Figure 9–52).

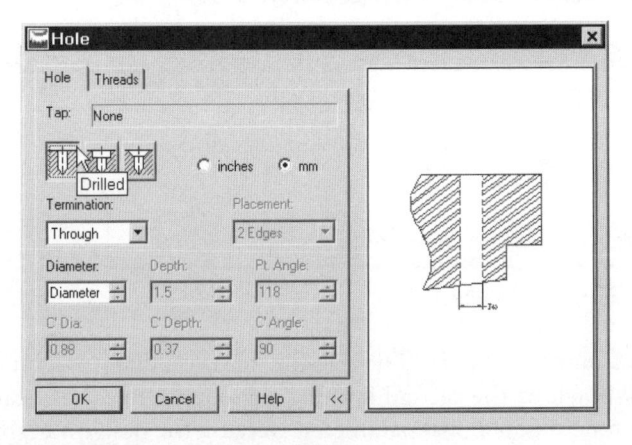

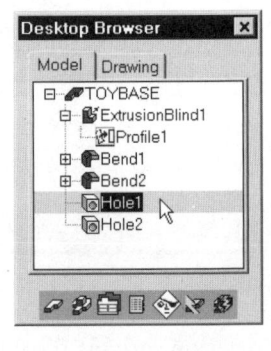

Figure 9–52 *Hole size being changed*

24. Select the dimension indicated in Figure 9–53.
25. Type **(Length-WheelBase)/2** at the command line area to change the dimension to an expression.
26. Press the ENTER key.

27. Select the Update button from the Browser.
28. Repeat steps 21 through 26 to modify the size and location of the other hole (see Figure 9–54).

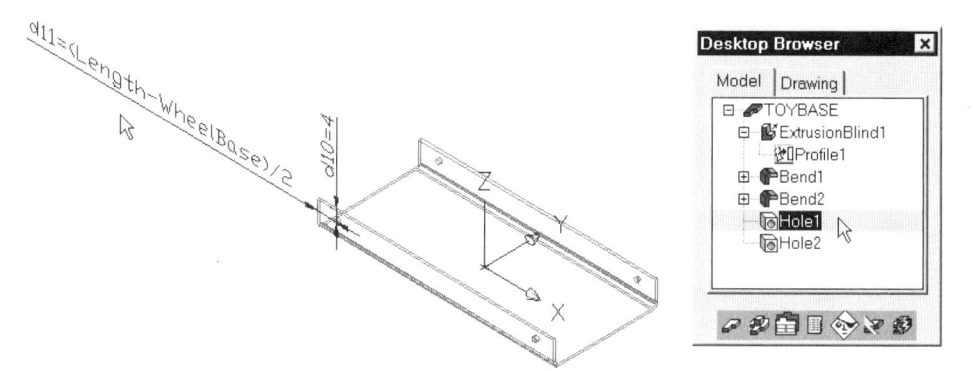

Figure 9–53 *Hole location being changed*

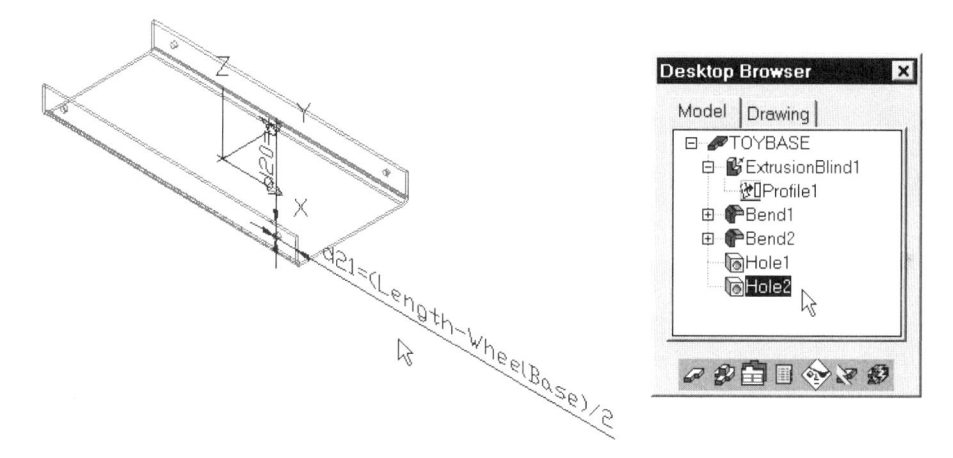

Figure 9–54 *Hole size and location changed*

Five dimensions of the solid part are changed. They are referenced to the design variables depicted in a linked variable file. If you open the Design Variable dialog box, the U prefix to the design variables will disappear because they are now referenced.

Save and close your file. Now you will modify another solid part.

29. Open the file *ToyShaft.dwg*.
30. Select Part>Design Variables
31. In the Design Variable dialog box, select the Global tab, and select the Link button.
32. In the Link dialog, select the design variable file *ToyCar.prm*.
33. On returning to the Design Variable dialog box, select the OK button.

34. Referring to Figure 9–55, modify the diameter of the sketch of the extruded feature.
35. In the Power Dimensioning dialog box, select the OK button.
36. Select the Update button from the Browser.

A dimension is modified. Save and close your file.

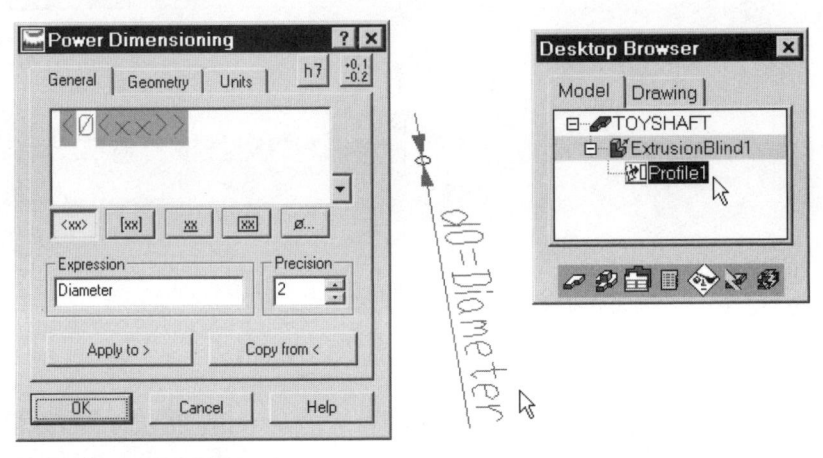

Figure 9–55 *Dimension changed*

Now you will modify another solid part.

1. Open the file *ToyWheel.dwg*.
2. Following the steps delineated above, change a dimension indicated in Figure 9–56.
3. Update the change.

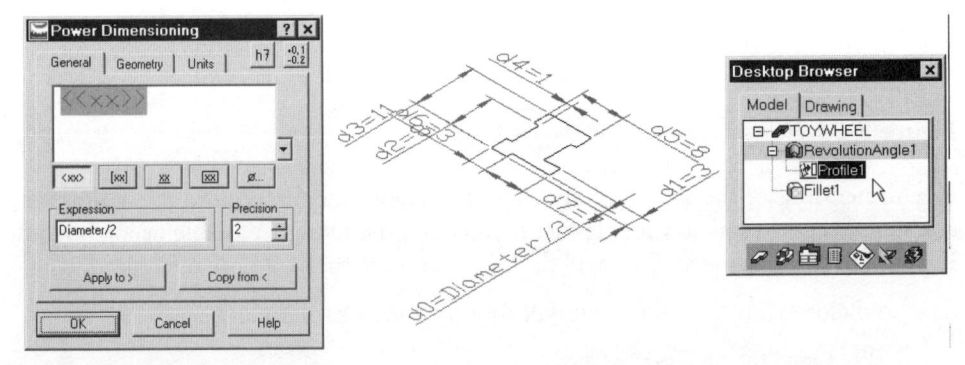

Figure 9–56 *Dimension changed*

Save the file. Now three components of the toy car assembly are referenced to the same design variable file. If the design variable file changes, all the referenced dimensions will change.

There are two ways to edit the design variable file:

- Use Notepad or any text editor to edit the text file.
- Use the Design Variable dialog box.

Now you will edit a design variable in the Design Variable dialog box.

4. Select Part>Design Variables
5. Select the Equation column of the "Diameter" variable and double-click.
6. Change the value to 3 (see Figure 9–57).

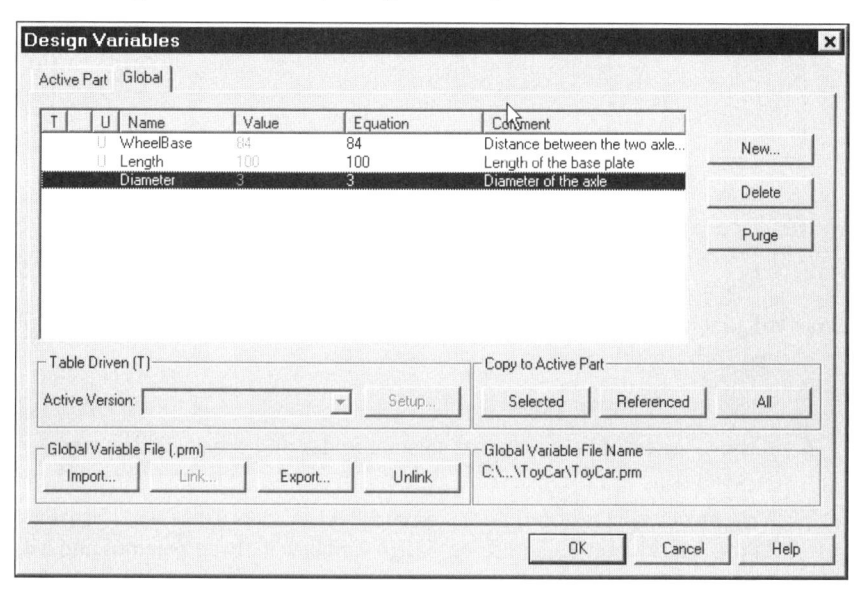

Figure 9–57 *Design variable changed*

7. Select the Export button.
8. In the Export dialog box, select the file *ToyCar.prm* to overwrite it.

Now the design variable file is modified. All three files linked to the design variable file will be affected. Open the solid parts one by one to observe the changes.

TABLE-DRIVEN PARTS

Now you will learn how to use an Excel spreadsheet to construct a set of similar parts. (You need to have Excel properly installed in your computer in order to construct table driven parts.) In the columns of the Excel table, you will define a set of design variables. In the rows, you will define a set of versions of design variables. Each design variable will have a different value in each of the versions.

1. Open the file *Nut.dwg* that you constructed in Chapter 3.
2. Select File>SaveAs and specify a new file name (file name: *HexNut.dwg*).

3. Select Part>Design Variables
4. In the Active Part tab of the Design Variables dialog box, select the Setup button to display the Table Driven Setup dialog box (see Figure 9–58).

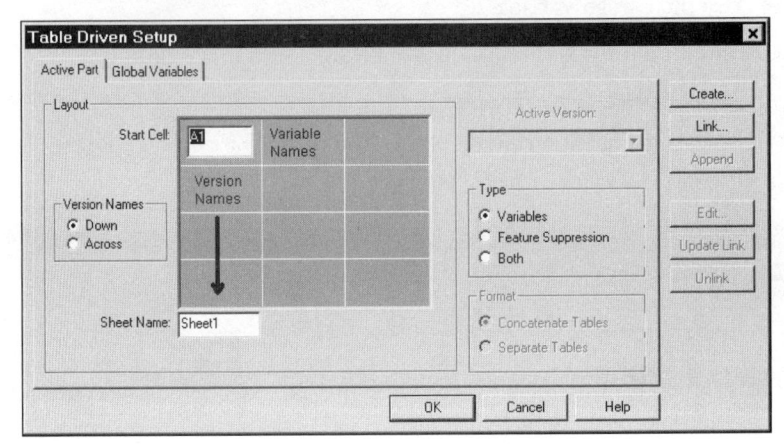

Figure 9–58 *Table Driven Setup dialog box*

Here you will decide the start cell of the Excel table to read and the directions of version names and variable names.

5. Select the Create button.
6. In the Create Table dialog box, select a folder and specify a file name (*HexNut.xls*).
7. After invoking Excel, construct a spreadsheet in accordance with Figure 9–59. (In the spreadsheet, define three design variables in three columns and five versions in five rows.)

Figure 9–59 *Excel spreadsheet*

8. In the Microsoft Excel window, select File>Save and then File>Close.
9. On returning to the Table Driven Setup dialog box, select the Update Link button and then the OK button.

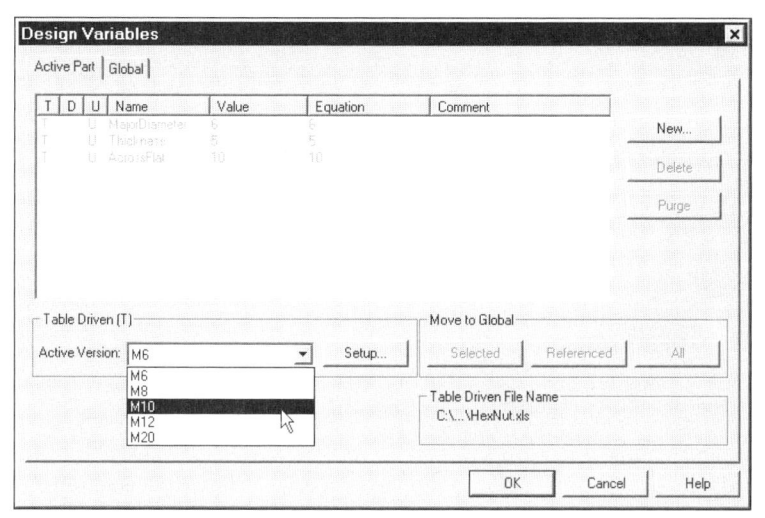

Figure 9–60 *Table-driven design variables constructed*

An Excel spreadsheet is constructed and is linked to the solid part. In the Design Variable dialog box (shown in Figure 9–60), you will find three design variables prefixed by the letters T and U. (As mentioned before, U stands for unused.) Here, T stands for table-driven. In the pull-down box in the Table Driven area, you will find five versions of design variables.

10. Select M10 from the pull-down box from the Table Driven area.
11. Select the OK button.

Now you will use the design variables in the dimensions of the solid part.

12. Referring to Figures 9–61 through 9–63, use the design variables on three dimensions.

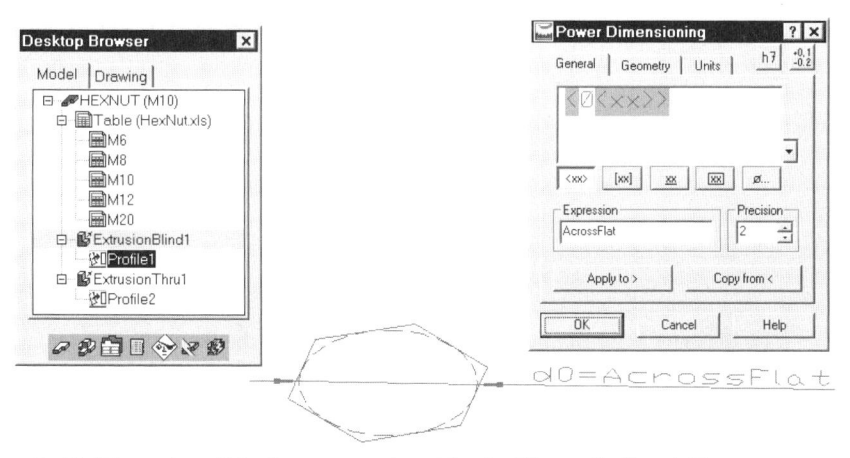

Figure 9–61 *Dimension of the hexagon replaced by the "AcrossFlat" variable*

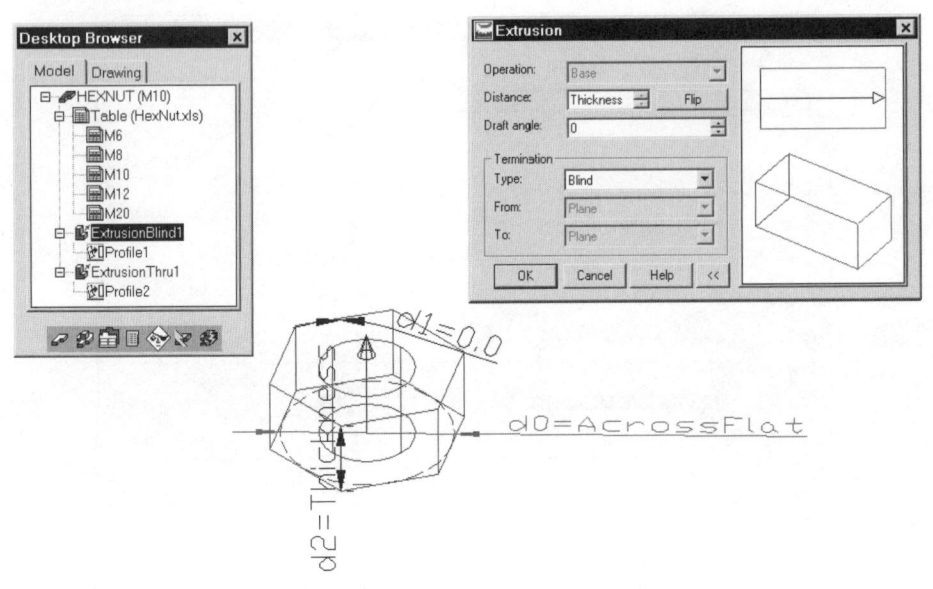

Figure 9-62 *Extrusion distance replaced by the "Thickness" variable*

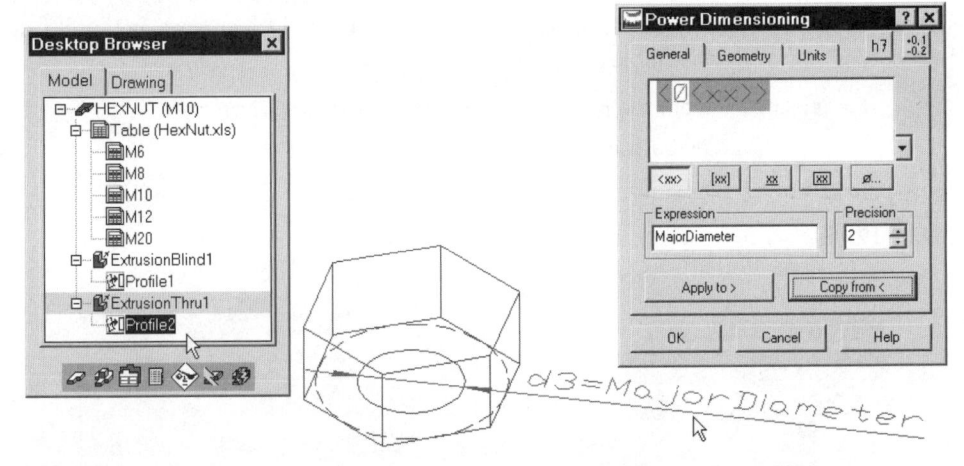

Figure 9-63 *Diameter of the sketch replaced by the "MajorDiameter" variable*

13. Update the changes.

The design variables are applied to the dimensions of the solid part. Save your file.

14. To discover how the Excel spreadsheet controls the version of the solid part, select the other versions one by one from the Browser and double-click (see Figure 9-64).

Close your file.

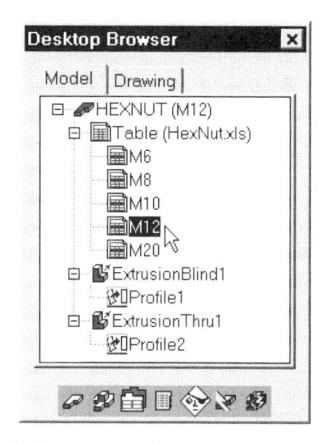

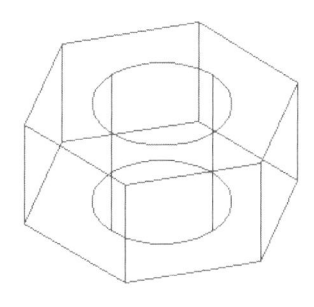

Figure 9–64 *M12 version selected*

SCALE

After an object manufactured by injection molding or casting cools down, it will shrink in size. To account for this amount of shrinkage, you must scale up the mold cavity. Apart from mold making, there may be other reasons that require you to scale a model up or down. Now you will learn how to scale a solid part.

1. Open the file *Seat.dwg* that you constructed in Chapter 4.
2. Select File>SaveAs and specify a new file name (file name: *SeatScale.dwg*).
3. Select Modify>Scale.
4. Select the solid part and press the ENTER key.
5. Type **0,0** to specify the base point.
6. Type **1.05** to specify the scale factor.

The model is scaled, as shown in Figure 9–65. Save your file.

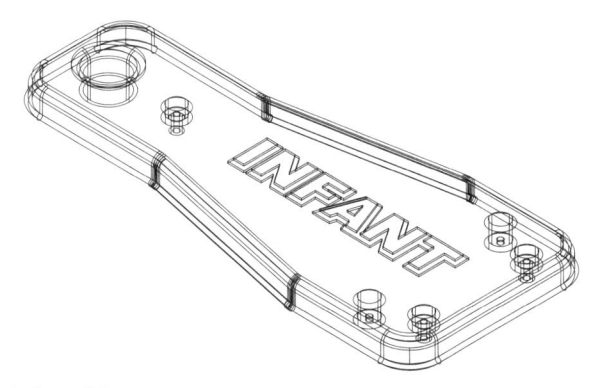

Figure 9–65 *Scaled model*

MIRROR

Our left hand is a mirror copy of our right hand. In designing the handle of a bicycle for example, you construct a right side handle and make a mirror copy of the solid part.

Now you will construct a mirror part of the seat of the infant scooter.

1. Open the file *Seat.dwg*.
2. Select File>SaveAs and specify a new file name (file name: *SeatMirror.dwg*).
3. Type at the command line AMMIRROR.
4. Select the solid part.
5. Type **L** to use the line option.
6. Type **0,0** to specify the first mirror line.
7. Type **0,1** to specify the second mirror line.
8. Press the ENTER key to create a new part.
9. Press the ENTER key to accept the default new part.

A new part is constructed (see Figure 9–66). Save and close your file.

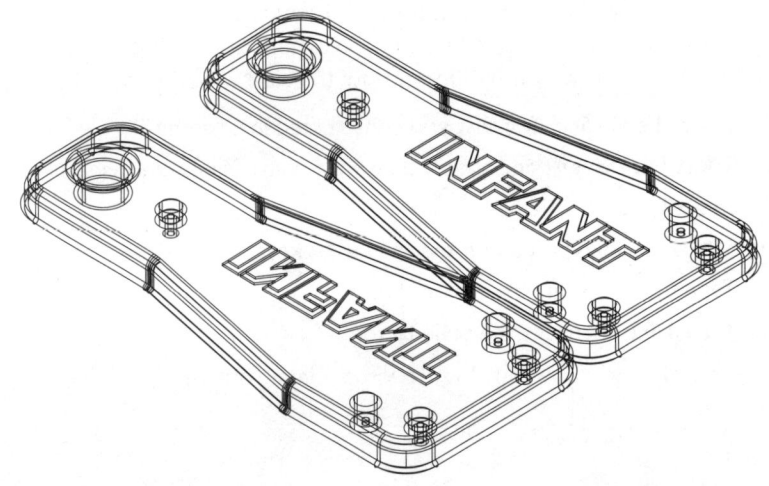

Figure 9–66 *Mirror part constructed*

MAKE BASE PART

Sometimes it may be necessary to freeze the parameters of a solid part so that it cannot be modified. To make a part not modifiable, you remove the parametric history and make it a static part. Select Part>Part>Make Base Part. Figure 9–67 shows the Browser of a base part.

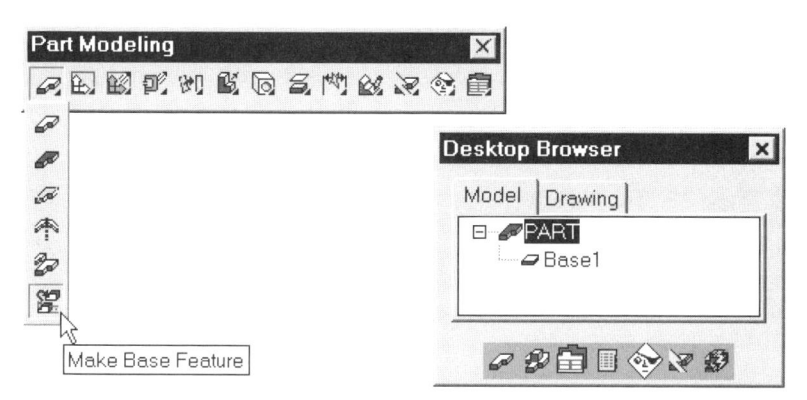

Figure 9–67 *Browser showing a static base part*

Remember that once a parametric solid part is made static, its parametric history delineated in the Browser will be permanently lost! So proceed with caution when using this command.

CONVERT AUTOCAD SOLID

If you already have an AutoCAD solid or a set of solids in a file (including inserted blocks), you can convert them into a set of base solid parts. Basically, a converted solid is static and is non-parametric. However, the solid features (sketched, placed, and work features) that you add to it subsequently are parametric. To convert an AutoCAD solid, you select Part>Part>Convert Solid and select the solid that you want to convert.

Now you will construct an AutoCAD native solid and convert it to a Mechanical Desktop base feature.

1. Start a new part file. Use metric as the default.
2. Set the display to an isometric view.
3. Select Design>Solids>Box

    Command: BOX

4. Select any point on the screen to specify the first corner of the solid box.
5. Type **@40,30,20** to indicate the opposite upper corner of the box.

A native solid box is constructed (see Figure 9–68). Now you will construct a solid cylinder.

6. Select Design>Solids>Cylinder

    Command: CYLINDER

7. Press the SHIFT key and right-click.
8. In the right-click menu, select Reference From.
9. Press the SHIFT key and right-click.

10. Select Endpoint from the right-click menu.
11. Select the end point indicated in Figure 9–69 as the reference point.
12. Type **@20,15** to specify a location from the reference point.
13. Type **10** to specify the radius of the cylinder.
14. Type **35** to specify the height of the cylinder.

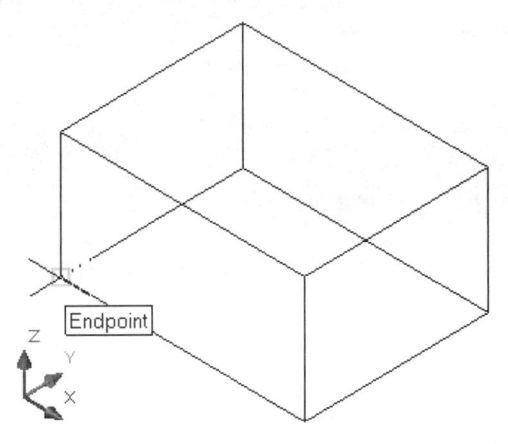

Figure 9–68 *Solid box constructed*

A native solid cylinder is constructed (see Figure 9–69). Now you will combine them.

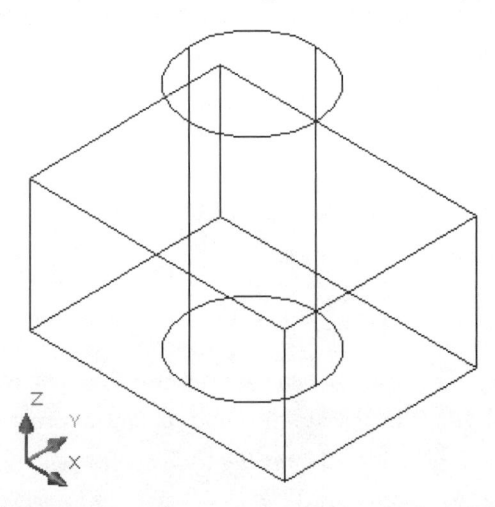

Figure 9–69 *Solid cylinder constructed*

15. Select Modify>Solid Editing>Union

    Command: UNION

16. Select the solid box and the solid cylinder, and press the ENTER key.

The solids are united (see Figure 9–70). Now you will convert the solid to a Mechanical Desktop base feature.

17. Select Part>Part>Convert Solid>Single Part
18. Select the solid.
19. Press the ENTER key to accept the default part name.

The solid is converted to a Mechanical Desktop base part (see Figure 9–71). Save your file (file name: *NativeBoxCylinder.dwg*).

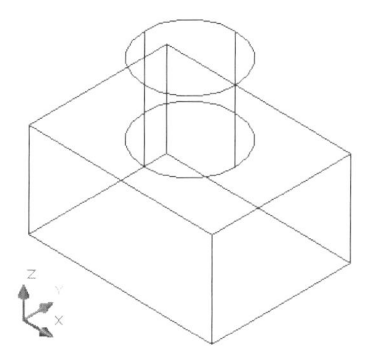

Figure 9–70 *Solid box and solid cylinder united*

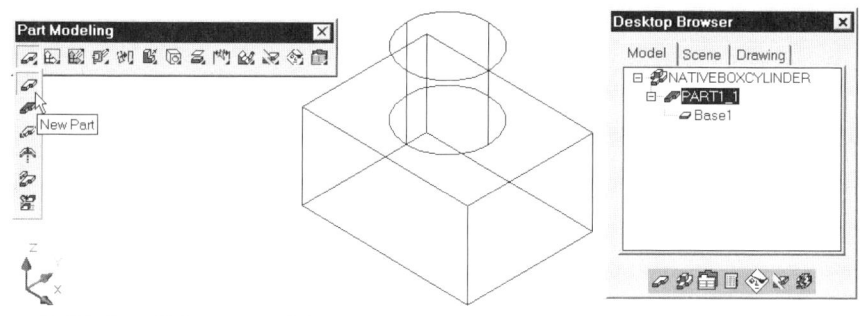

Figure 9–71 *AutoCAD native solid converted to Mechanical Desktop base part*

Now you will construct a more complex AutoCAD drawing and convert it to a Mechanical Desktop file. First you will construct a box and a number of cylinders and subtract the cylinders from the box.

1. Start a new part file. Use metric as the default.
2. Set the display to an isometric view.
3. Select Design>Solids>Box

    Command: BOX

4. Select any point on the screen to specify the first corner of the solid box.
5. Type **@100,80,10** to indicate the opposite upper corner of the box.

A solid box is constructed.

6. Select Design>Solids>Cylinder

   Command: CYLINDER
7. Press the SHIFT key and right-click.
8. Select Reference From on the right-click menu.
9. Press the SHIFT key and right-click.
10. Select Endpoint from the right-click menu.
11. Select the end point indicated in Figure 9–72 as the reference point.
12. Type **@10,10** to specify a location from the reference point.
13. Type **5** to specify the radius of the cylinder.
14. Type **20** to specify the height of the cylinder.

A solid cylinder is constructed (see Figure 9–73).

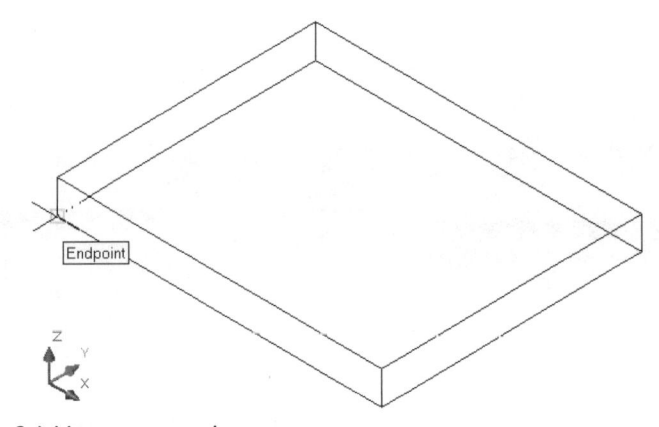

Figure 9–72 *Solid box constructed*

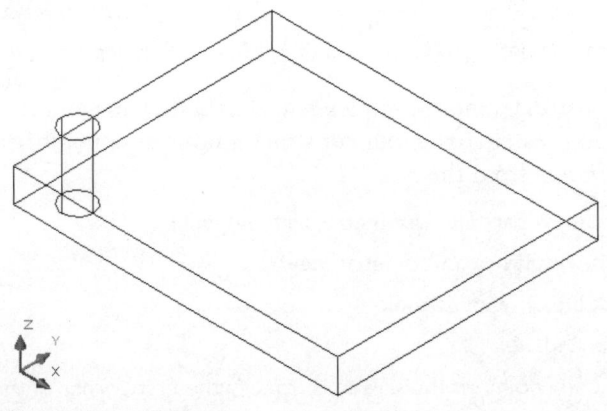

Figure 9–73 *Solid cylinder constructed*

15. Select Modify>Array

    Command: ARRAY

16. In the Array dialog box (shown in Figure 9–74), select the Rectangular array button, if it is not already selected.
17. Set the number of rows to 4 and the number of columns to 5.
18. Set the row offset distance to 20 mm and the column offset distance to 20 mm.
19. Select the Select objects button.
20. Select the cylinder and press the ENTER key.
21. On returning to the Array dialog box, select the OK button.

An array of cylinder is constructed (see Figure 9–75).

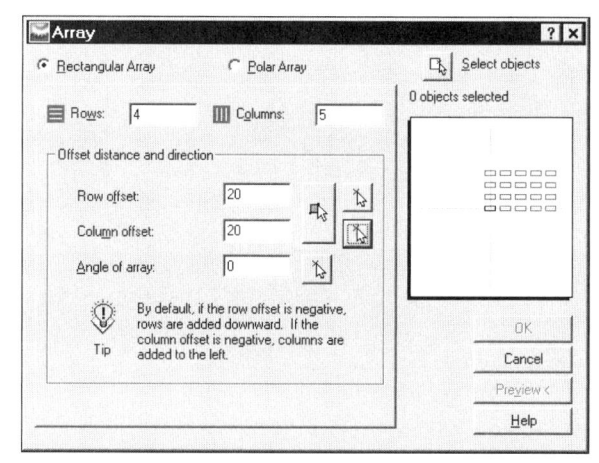

Figure 9–74 *Array dialog box*

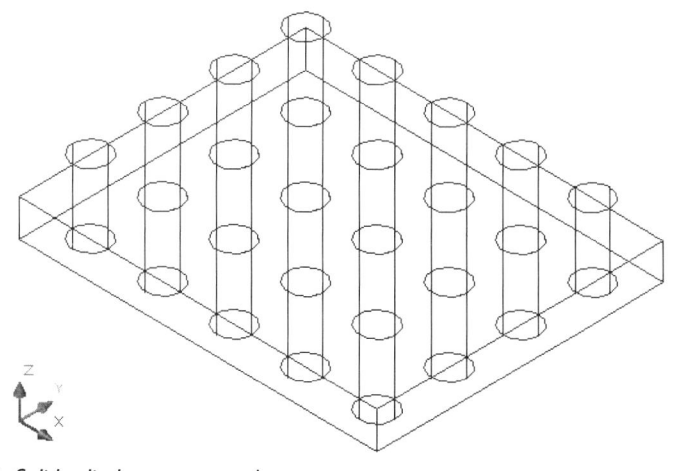

Figure 9–75 *Solid cylinder constructed*

22. Select Modify>Solid Editing>Subtract

    Command: SUBTRACT

23. Select the solid box and press the ENTER key.
24. Select all the solid cylinders and press the ENTER key.

The cylinders are subtracted from the box (see Figure 9–76). Save your file (file name: *NativePlate.dwg*).

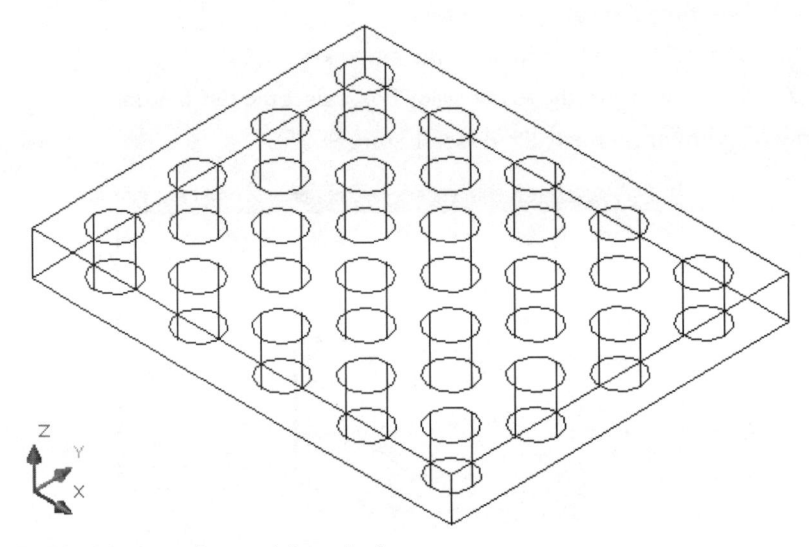

Figure 9–76 *Cylinders subtracted from the box*

Now you will construct another native AutoCAD solid in a separate file.

25. Start a new part file. Use metric as the default.
26. Set the display to an isometric view.
27. Design>Solids>Cylinder

    Command: CYLINDER

28. Select a point on your screen.
29. Type **5** to specify the radius of the cylinder.
30. Type **10** to specify the height of the cylinder.

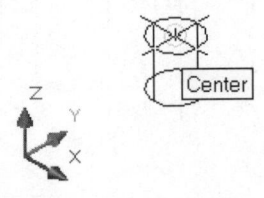

Figure 9–77 *Cylinder constructed*

31. Repeat the CYLINDER command.
32. Press the SHIFT key and right-click.
33. In the right-click menu, select Center.
34. Select the center indicated in Figure 9–77.
35. Type **6** to specify the radius.
36. Type **3** to specify the height.

Two cylinders are constructed (see Figure 9–78).

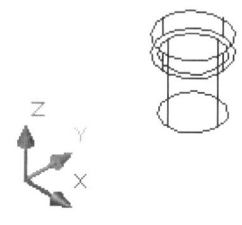

Figure 9–78 *Cylinders constructed*

37. Select Modify>Solid Editing>Union

    Command: UNION

38. Select the two cylinders and press the ENTER key.

The cylinders are united. Now you will set the insertion base point.

39. Select Design>Block>Base

    Command: BASE

40. Press the SHIFT key and right-click.
41. Select Center from the right-click menu.
42. Select the center indicated in Figure 9–79.

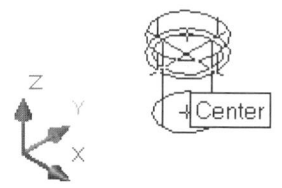

Figure 9–79 *Base point selected*

Save and close your file (file name: *NativePlug.dwg*).

Now you will insert the file *NativePlug.dwg* into the file *NativePlate.dwg*.

43. Select Insert>Block

    Command: INSERT

44. In the Insert dialog box (shown in Figure 9–80), select the Browse button.
45. In the Select Drawing File dialog box, select the file NativePlug.
46. On returning to the Insert dialog box, select the OK button.
47. Press the SHIFT key and right-click.
48. Select Center from the right-click menu.
49. Select the center point indicated in Figure 9–81.

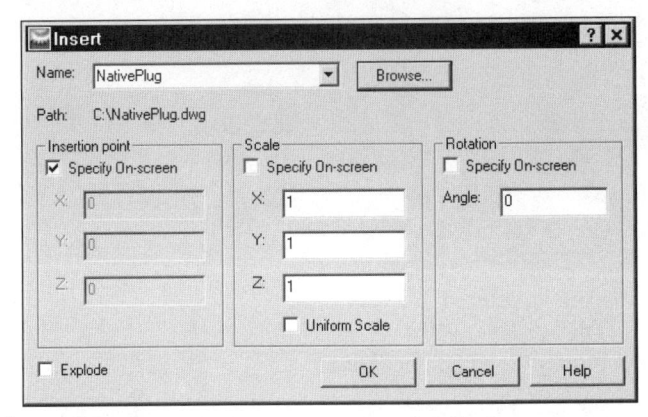

Figure 9–80 *Insert dialog box*

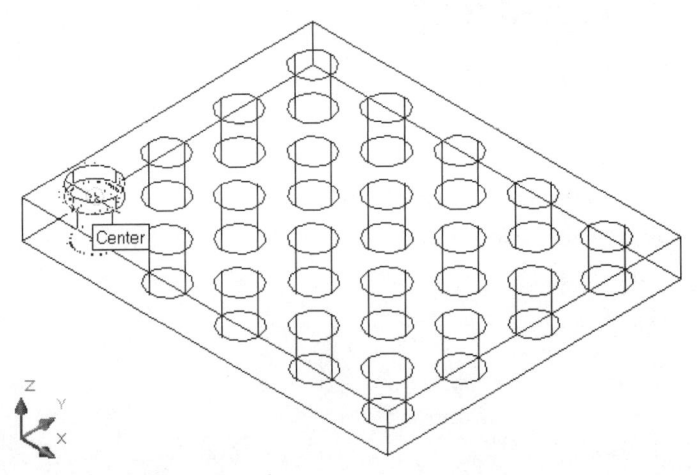

Figure 9–81 *Insertion position selected*

A block is inserted. Now you will construct an array of the inserted block.

50. Select Modify>Array

    Command: ARRAY

51. In the Array dialog box, select the Rectangular array button, if it is not already selected.

52. Set the number of rows to 4 and the number of columns to 5.
53. Set the row offset distance to 20 mm and the column offset distance to 20 mm.
54. Select the Select objects button.
55. Select the block and press the ENTER key.
56. On returning to the Array dialog box, select the OK button.

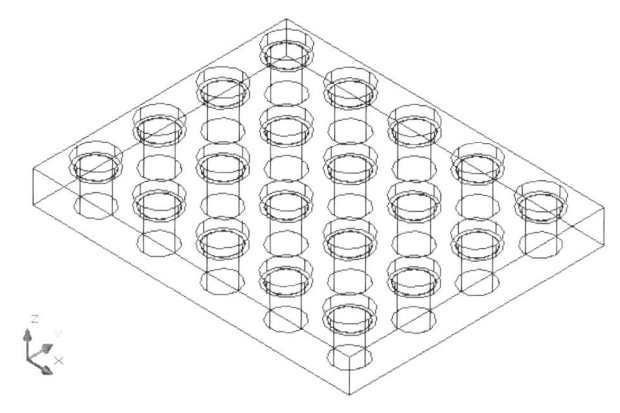

Figure 9–82 *Multiple insertions constructed*

Now you will convert the solid and the instances of the block into Mechanical Desktop solid parts.

57. Select Part>Part>Convert Solid>Multiple Parts
58. Press the ENTER key to convert local objects.
59. Select all the objects and press the ENTER key.

The solid and the block instances are converted (see Figure 9–83).

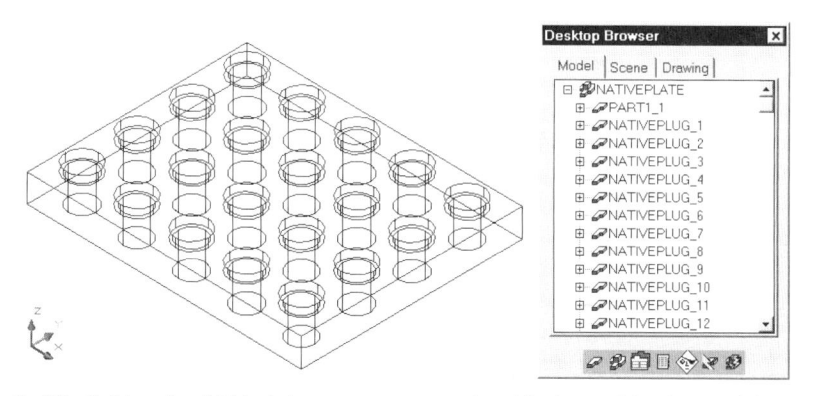

Figure 9–83 *Solid and solid block instances converted to Mechanical Desktop solid parts*

Conversion is complete. Save your file.

BASE FEATURE EDITING

Basically, a base solid feature is non-parametric and static. By leveraging AutoCAD's solid editing commands, you can modify a base solid feature in various ways.

Now you will edit a base solid feature.

1. Open the file *NativeBoxCylinder.dwg*, if you have already closed it.
2. Select the base feature from the Browser and double-click.

The Solid Editing State toolbar displays (see Figure 9–84).

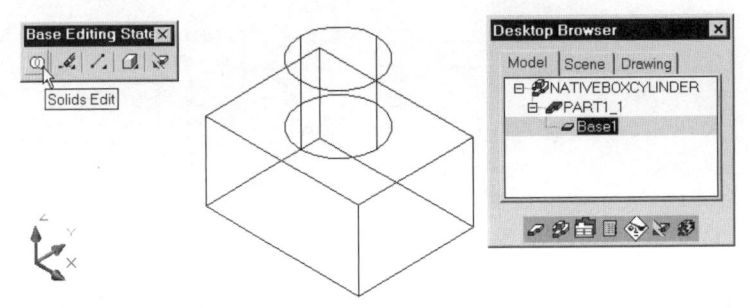

Figure 9–84 *Base feature selected*

The Solid Edits fly-out toolbar has 18 icons:

Union	enables you to construct an AutoCAD solid and unite it to the base solid.
Subtract	enables you to construct an AutoCAD solid and subtract it from the base solid.
Intersect	enables you to construct an AutoCAD solid and intersect it with the base solid.
Extrude Faces	enables you to select planar faces from the base solid and extrude them to a specified height or along a path
Move Faces	enables you to select faces from the base solid and move them to a specified height or distance
Offset Faces	enables you to offset faces from the base solid by a specified distance or through a specified point
Delete Faces	enables you to remove faces (including chamfer and fillet) from the base solid
Rotate Faces	enables you to rotate selected faces from the base solid around an axis
Taper Faces	enables you to taper faces from the base solid
Color Faces	enables you to change the color of the faces of the base solid
Copy Faces	enables you to copy faces from the base solid
Color Edges	enables you to change the color of the edges of the base solid
Copy Edges	enables you to copy edges from the base solid
Imprint	enables you to imprint curves onto faces of the base solid

Clean	enables you to remove shared edges or vertices from the base solid
Separate	enables you to separate solids with disjoint volumes into separate solids
Shell	enables you to make the base solid hollow
Check	enables you to validate the base solid

Now you will move a face of the base solid.

    3. Select Move Faces from the Solid Edit State toolbar.

    4. Select the faces highlighted in Figure 9–85 and press the ENTER key.

    5. Type **15<-15** at the command line area.

    6. Press the ENTER key.

The selected faces are moved a distance of 15 mm in the −15 degree direction (see Figure 9–86).

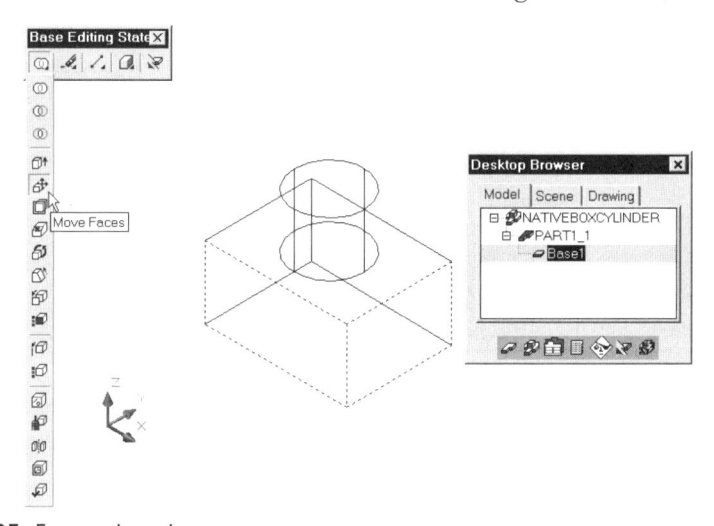

Figure 9–85 *Faces selected*

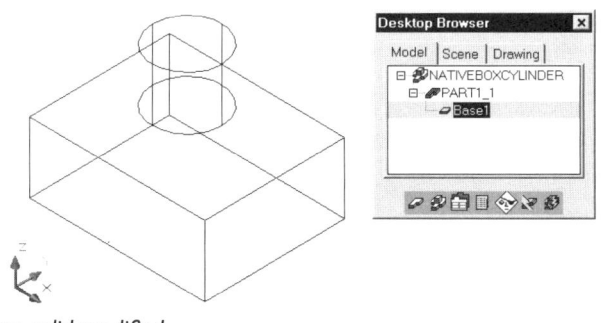

Figure 9–86 *Base solid modified*

    7. Select the Update button from the Browser to update the solid part.

The solid is complete. Save and close your file. Now you will make hollow a base solid part.

1. Open the file *NativePlate.dwg*, if you have already closed it.
2. Select an instance of the NativePlug definition from the Browser and double-click to activate it.
3. Select the base feature and double-click (see Figure 9–87).

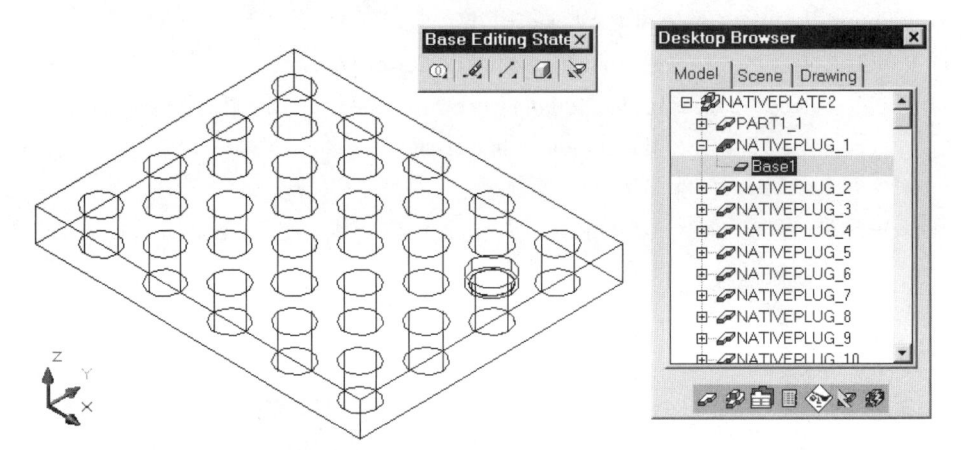

Figure 9–87 *An instance of the NativePlug being edited*

4. Zoom in on the display.
5. Select Shell from the Solid Edit State toolbar (see Figure 9–88).

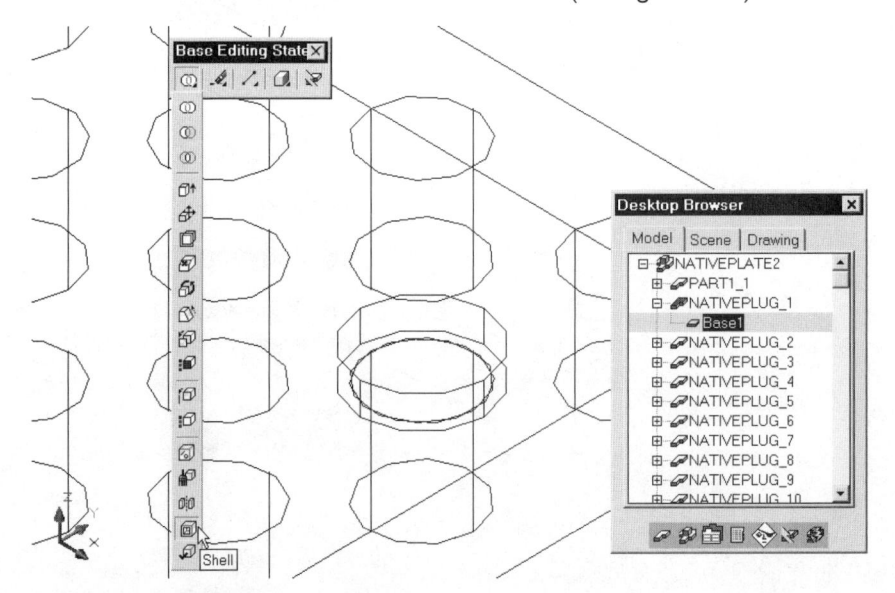

Figure 9–88 *Base solid selected for editing*

6. Referring to Figure 9–89, select the instance to make hollow and select the edge to indicate the faces to be removed while shelling.

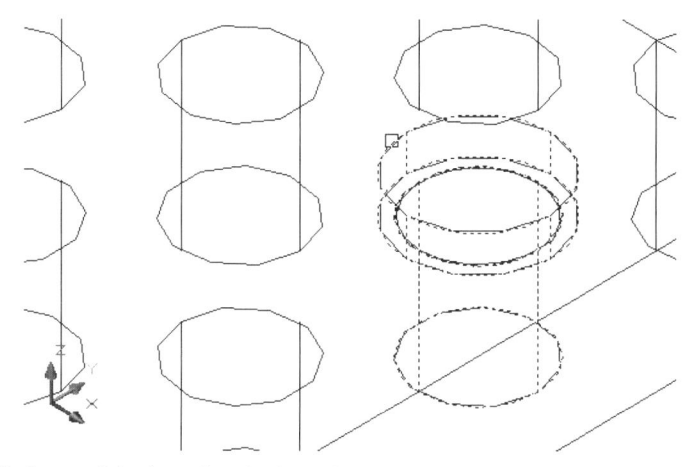

Figure 9–89 *Base solid selected and edge selected*

Because the selected edge is shared by two faces, two faces are selected for removal. Now you will reclaim a face.

7. Type **A** at the command line area.
8. Select the edge indicated in Figure 9–90.
9. Press the ENTER key.

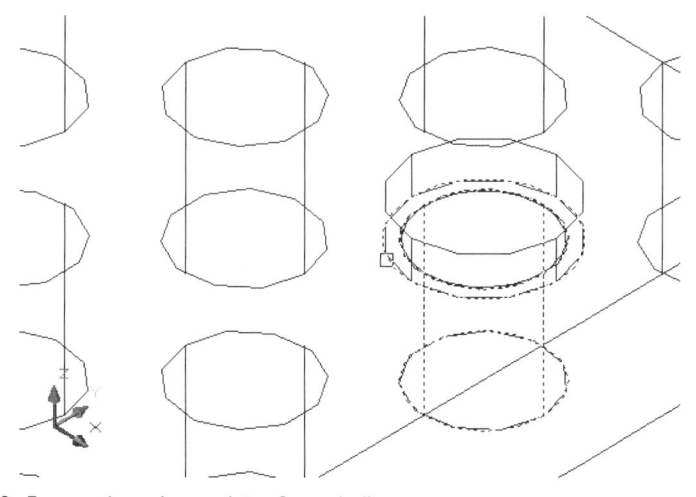

Figure 9–90 *Faces selected to reclaim from shelling*

10. Type **1** to specify the shell thickness.
11. Press the ENTER key to exit.

The base solid is made hollow (see Figure 9–91).

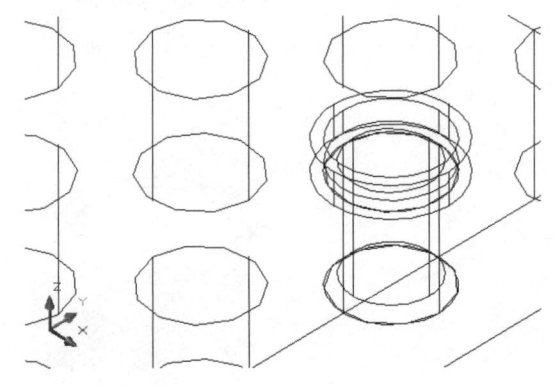

Figure 9–91 *Base solid made hollow*

12. Select the Update button from the Browser.
13. Shade the display and zoom to extent.

The solid part is updated and all the instances of the solid modified (see Figure 9–92). Save and close your file.

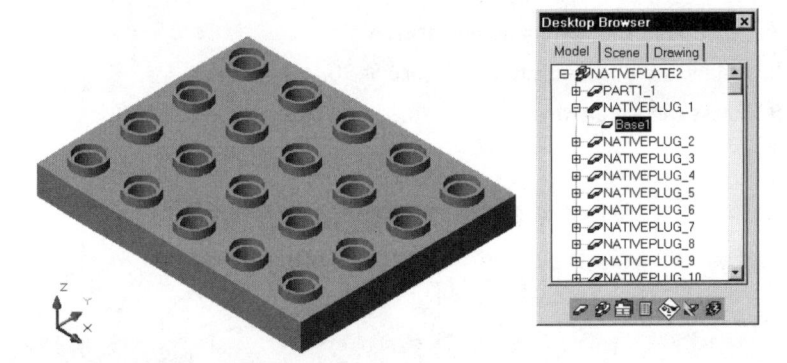

Figure 9–92 *All instances modified*

SOLID PART EXPORT

To cope with the different data structure requirements of downstream computerized operations, you can output a solid part to various formats. You may export to ACIS, VRML, STL, IGES, and STEP formats.

ACIS

"Save As Text (SAT)" is a file format of the ACIS object-oriented 3D geometric modeling engine by Spatial Technology, Inc. It is a format for translating 3D wires, surfaces, and solids.

1. Open the file *Frame_R.dwg* that you constructed in Chapter 4.
2. To export an ACIS file, select File>Export>Desktop ACIS
   Command: AMACISOUT
3. Select the solid part and press the ENTER key.
4. In the Save file dialog box, specify a file name and select the Save button. (file name: *Feature_1.sat*).

VRML

Virtual Reality Modeling Language (VRML) is a kind of file format that enables 3D objects to be viewed and manipulated on a Web page. To save selected parts in VRML format, you can use the AMVRMLOUT command. This command outputs a file with an extension of .wrl.

You can insert a WRL file into a Web page written in HTML format. To view the WRL file, you need VRML plug-in applications.

5. To export a VRML file, select File>Export>Desktop VRML
   Command: AMVRMLOUT
6. Select the solid part and press the ENTER key.
7. In the Save file dialog box, specify a file name and select the Save button (file name: *Frame_R.wrl*).

STL

"Stereolithography (STL)" is a standard file format for use in most rapid prototyping machines. An STL file is a list of triangular surfaces that depict the 3D model. A 3D solid model saved in STL format is downgraded into a 3D model approximated by a set of triangular flat surfaces.

8. To export STL file, type amstlout at the command line area.
   Command: AMSTLOUT
9. Select the solid part and press the ENTER key.
10. In the Save file dialog box, specify a file name and select the Save button (file name: *Frame_R.stl*).

IGES

"Initial Graphics Exchange Specification (IGES)" format is an American standard started by the American National Standards Institute in 1979. Because it is a format for translating 3D wires and surfaces, a solid that you output in IGES format will exhibit as a surface model. Naturally, volume data will be lost. Figure 9–93 shows the IGESOUT Translation dialog box.

11. To export an IGES file, select File>Export>IGES
    Command: IGESOUT

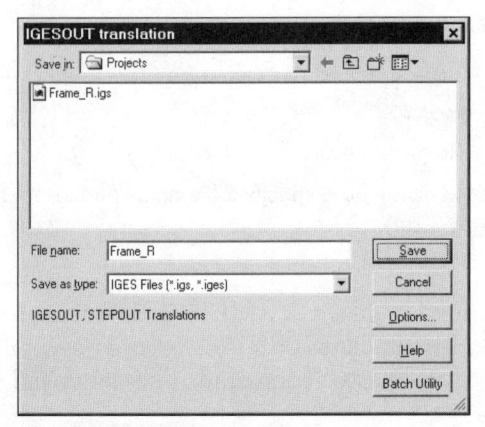

Figure 9-93 *IGESOUT Translation dialog box*

12. In the IGESOUT Translation dialog box, specify a file name and select the Save button (file name: *Frame_R.igs*).

STEP

"STandard for the Exchange of Product Model Data (STEP)" is a product model data exchange standard that was initially developed by the International Organization for Standardization. Like SAT files, it is used for translating wires, surfaces, and solids. Figure 9-94 shows the STEPOUT Translation dialog box.

13. To export a STEP file, select File>Export>STEP

    Command: STEPOUT

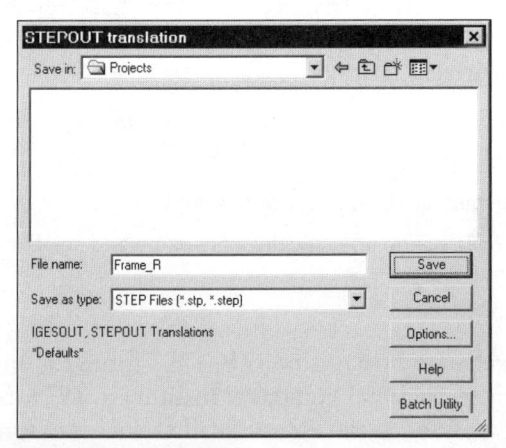

Figure 9-94 *STEPOUT Translation dialog box*

14. In the STEPOUT Translation dialog box, specify a file name and select the Save button (file name: *Frame_R.stp*).
15. Close the file.

FEATURE RECOGNITION

Basically, solids imported from other computer applications are static base solids. Apart from leveraging AutoCAD solid editing commands to edit these kinds of solids, you can also use the feature exchange add-on utility to transform solid bodies into fully parametric parts with features.

Using this application inside Mechanical Desktop, common part features are detected and then constructed from any 3D solid. A Browser style list of detected features are provided after the solid part is analyzed. Figure 9–95 shows the Feature Recognition pull-down menu and toolbar.

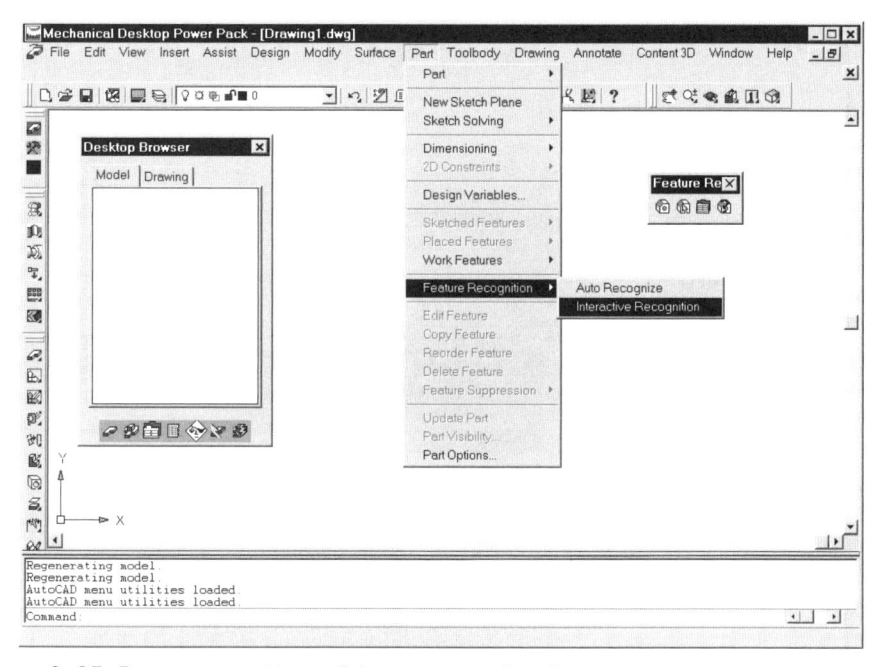

Figure 9–95 *Feature recognition pull-down menu and toolbar*

In the Feature Recognition dialog box, there are four buttons:

Auto Recognize	starts the recognition process automatically
Interactive Recognition	starts the recognition process interactively
Feature Recognition Options	sets recognition options
Feature Recognition Help	displays the feature recognition help menu

Now you will insert an ACIS solid and convert it into a parametric solid part.

1. Start a new part file. Use metric as the default.
2. Select Insert>ACIS File

3. In the select ACIS file dialog box, select the file (*Feature_1.sat*) that you saved earlier.
4. Use the shortcut key 8 to set the display to an isometric view.
5. Select Feature Recognition Help from the Feature Recognition dialog box (see Figure 9–96).

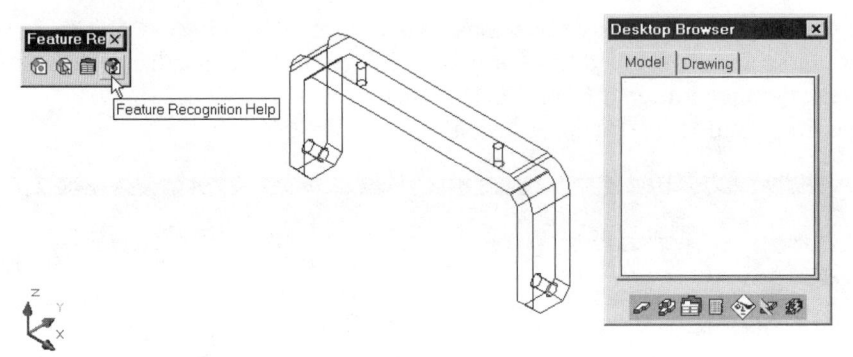

Figure 9–96 *Feature Recognition help button*

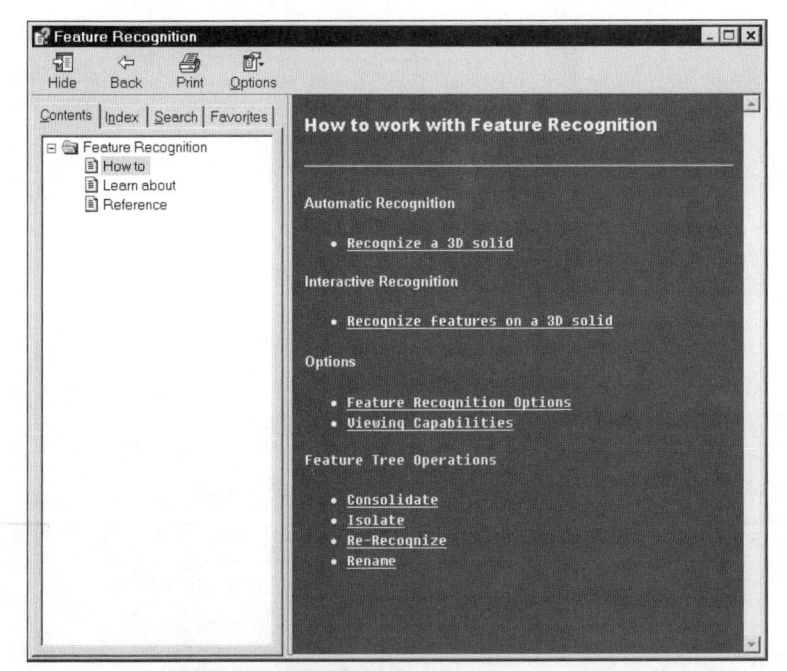

Figure 9–97 *Feature Recognition help dialog box*

6. Close the Feature Recognition Help dialog box.
7. Select Feature Recognition Options from the Feature Recognition dialog box (see Figure 9–98).

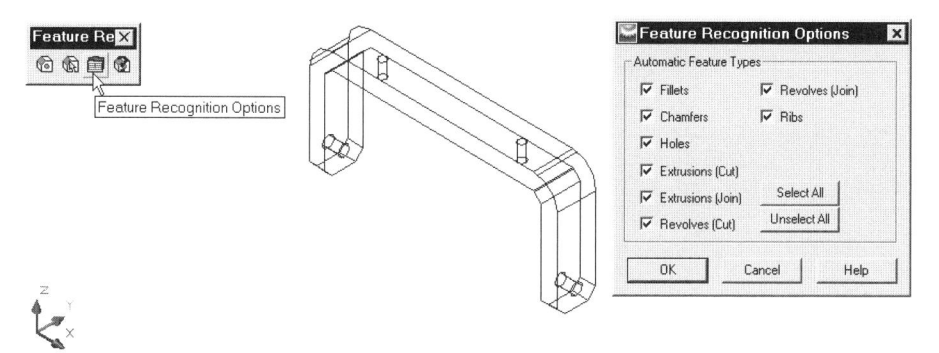

Figure 9–98 *Feature Recognition Options dialog box*

In the Feature Recognition Options dialog box, select the types of features that the automatic recognition process will attempt to identify.

8. Select the OK button to close the Feature Recognition Options dialog box.
9. Select Auto Recognize from the Feature Recognition toolbar (see Figure 9–99).

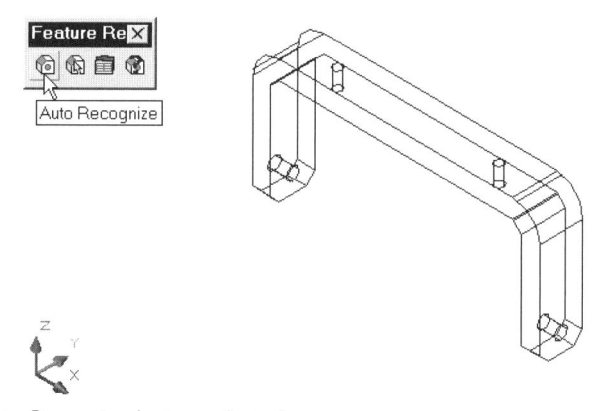

Figure 9–99 *Auto Recognize button selected*

10. Select the imported ACIS solid.

The Feature displays, showing you the process of feature recognition (see Figure 9–100). After the recognition process is complete, the Feature Recognition – [Automatic] dialog box displays. It shows you the recognition result (see Figure 9–101).

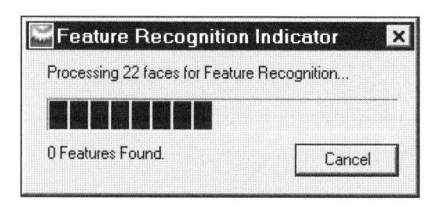

Figure 9–100 *Feature Recognition Indicator dialog box*

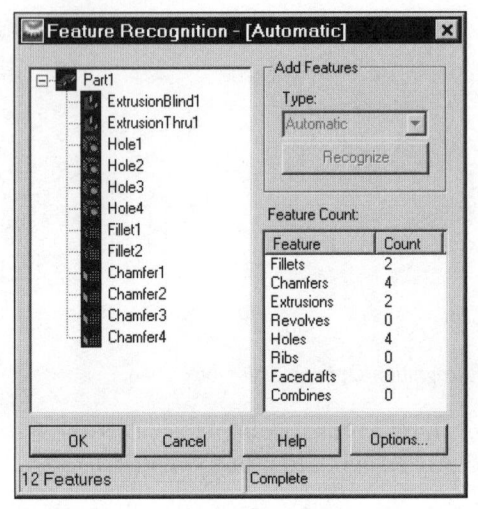

Figure 9–101 *Feature Recognition – [Automatic] dialog box*

11. Select the OK button.

Recognition is complete. The ACIS solid is converted into a parametric feature-based solid part with its features delineated in the Browser (see Figure 9–102). Save and close your file (file name: *Feature_1.dwg*).

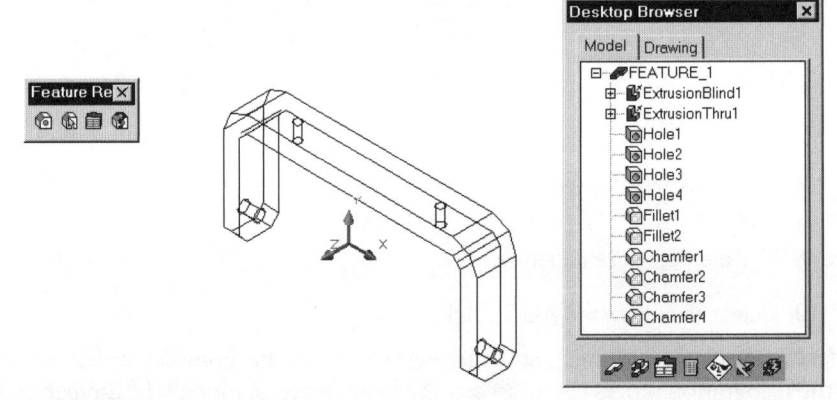

Figure 9–102 *Recognition completed*

Now you will use interactive recognition process.

1. Start a new part file. Use metric as the default.
2. Insert the ACIS file (*Feature_1.sat*).
3. Set the display to an isometric view.
4. Select Interactive Recognize from the Feature Recognize toolbar (see Figure 9–103).

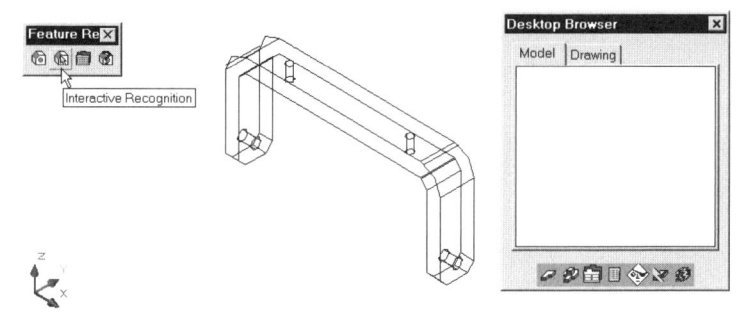

Figure 9–103 *Interactive Recognize button selected*

5. Select the ACIS solid.

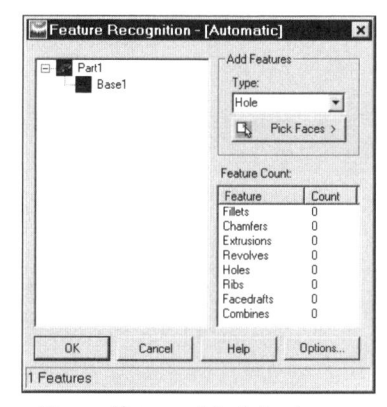

Figure 9–104 *Feature Recognition – [Automatic] dialog box*

6. In the Type pull-down box of the Feature Recognition dialog box, select Hole.
7. Select the Pick Faces button.
8. Referring to Figure 9–105, select the hole.
9. Left-click to cycle and right-click to accept.
10. Press the ENTER key.

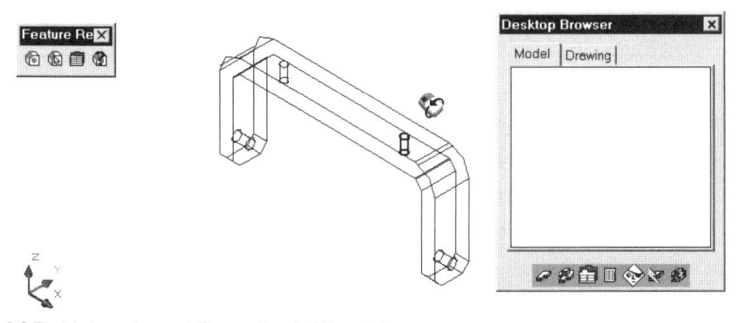

Figure 9–105 *Hole selected from the ACIS solid*

A hole feature is recognized interactively (see Figure 9–106).

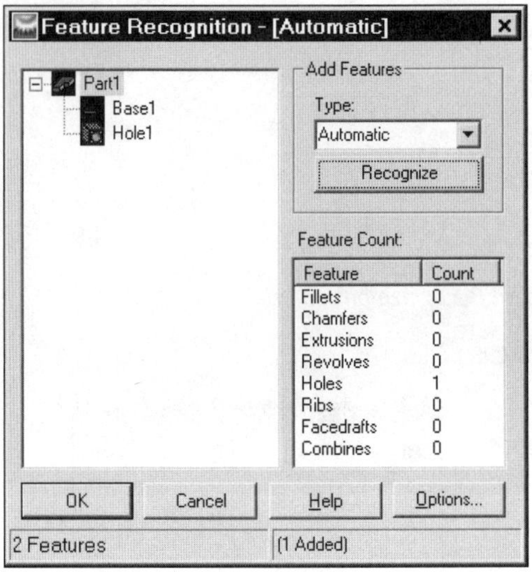

Figure 9–106 *Hole feature recognized*

Now you will complete the recognition by using the automatic mode.

11. Select Automatic from the Types pull-down box and select the Recognize button.
12. Select the OK button.

Recognition is complete. Save and close your file (file name: *Feature_2.dwg*).

COLLABORATION

Besides exporting to the kinds of data explained above, you can also export data in XGL and ZGL format:

XGL	"X Windows Graphics Library (XGL)" is a standard file format that captures all the 3D information that can be rendered by the OpenGL rendering library.
ZGL	"Compressed XG (ZGL)" is a compressed XGL format that is about 10 times smaller than XGL files.

Using ZGL file, you can upload to Autodesk Streamline, which is a hosted service for sharing design data across the extended manufacturing enterprise. You can obtain more information about how to upload Mechanical Desktop files to Autodesk Streamline by visiting http://www.autodesk.com.

SUMMARY

Using basic operations on sketches and the fundamental approach of building feature by feature in a sequential way, the repertoire of shapes that you can construct is limited. To include free-form features in a solid part, you construct a surface and use it to cut a solid. To combine solids in a more flexible way, you construct additional solid parts or import an external solid part in a part file, apply assembly constraints to them to establish a relationship among them, and combine them in pairs.

In contrast to combining two solid parts into a single solid, you can split a solid into two and export the parts to become individual solid parts.

To facilitate removal of objects from a mold in the manufacturing process, draft angles are applied on vertical faces of an object. To split a solid or a face of a solid into two, you can use a split line. To better control the dimensions of a solid part or a set of solid parts, you can use design variables. You can construct design variable files to control a set of solid parts and use Excel spreadsheets to control a set of table-driven parts.

To cater for shrinkage, you can scale a solid part. To make a mirror copy of the solid, you can make a mirror part. If you already have an AutoCAD solid, you can convert it to a base solid part. If you want to remove the parametric history of a solid part, you can convert it to a base feature. Although a base feature is non-parametric, you can still modify it by leveraging AutoCAD solid editing tools or using the feature exchange add-on utility. To facilitate downstream computerized manufacturing, you can export a solid part to various file formats.

PROJECTS

Now you will enhance your part modeling knowledge by working on the following projects.

CAMERA CASING

Figure 9–107 shows two exploded views (front-right isometric and back-left isometric) of the component parts of a camera casing. To make these component parts, you will construct two solid parts with the surface cut feature; combine them into a single solid part; add surface cut, extruded, and shell features to the combined solid; and split the solid into two solid parts.

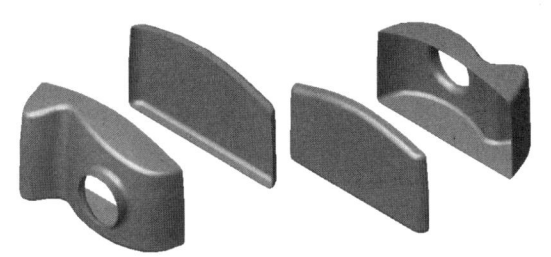

Figure 9–107 *Camera casing*

Now you will use the top-down approach to construct the assembly of two solid parts.

1. Start a new part file. Use metric as the default.
2. Referring to Figure 9–108, construct a rectangle, resolve it to a profile, add parametric dimensions, and extrude it a distance of 90 mm.

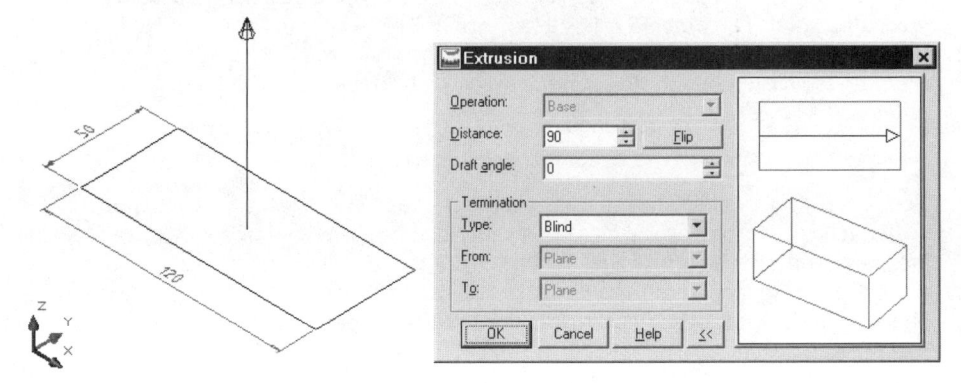

Figure 9–108 *Profile being extruded*

3. Construct a work point and add two parametric dimensions (see Figure 9–109).
4. Select Assist>New UCS>Origin
5. Press the SHIFT key and right-click.
6. Select Endpoint from the right-click menu.
7. Select the end point indicated in Figure 9–110.

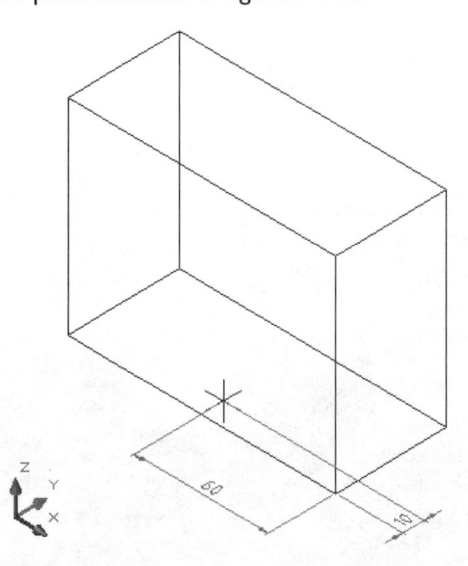

Figure 9–109 *Work point constructed*

Part Modeling III 477

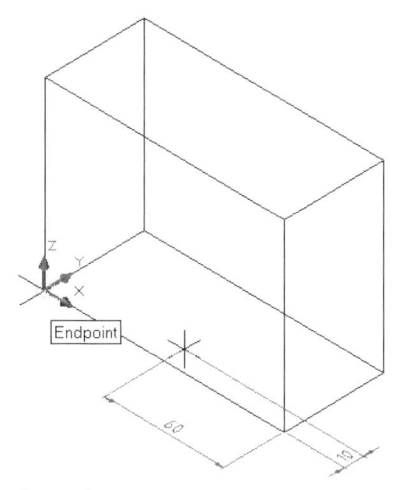

Figure 9–110 *UCS origin relocated*

8. Construct two new layers: Wire and Surface.
9. Set layer Wire the current layer.
10. Construct two splines (shown in Figure 9–111) as follows:

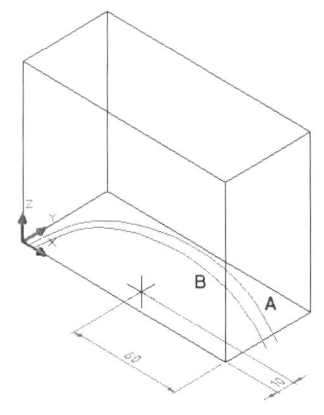

First point	3,0	6,0
Next point	10,30	12,25
Next point	60,48	70,40
Next point	125,25	125,19
Next point	ENTER	ENTER
Start tangent	ENTER	ENTER
End tangent	ENTER	ENTER

Figure 9–111 *Two splines constructed*

Now you will copy spline A (Figure 9–111) a distance of 40 mm in the Z direction and move spline B (Figure 9–111) a distance of 90 mm in the Z direction (see Figure 9–112).

11. Select Modify>Copy
12. Select spline A (shown in Figure 9–111) and press the ENTER key.
13. Type **0,0,40** to specify a distance.
14. Press the ENTER key to use the first point as displacement.
15. Select Modify>Move
16. Select spline B (shown in Figure 9–111) and press the ENTER key.
17. Type **0,0,90** to specify a distance.
18. Press the ENTER key to use the first point as displacement.
19. Set current layer to Surface.

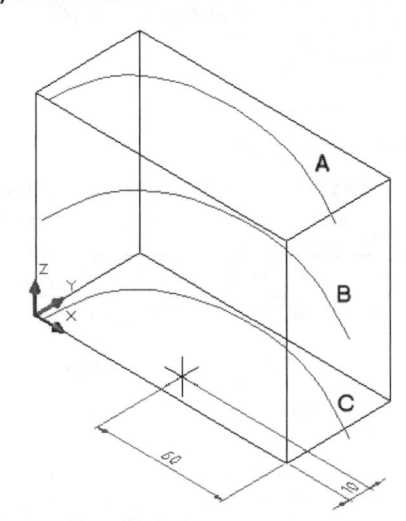

Figure 9–112 *A spline copied and a spline moved*

20. Select Surface>Create Surface>Loft U
21. Select A, B, and C (shown in Figure 9–112) one by one and press the ENTER key.
22. In the Loft Surface dialog box, select the OK button (see Figure 9–113).
23. Set current layer to 0.
24. Select Part>Placed Features>Surface Cut
25. Select the surface and then the work point.
26. Referring to Figure 9–114, right-click to accept.

An extruded solid part with a surface cut feature is constructed (see Figure 9–115).

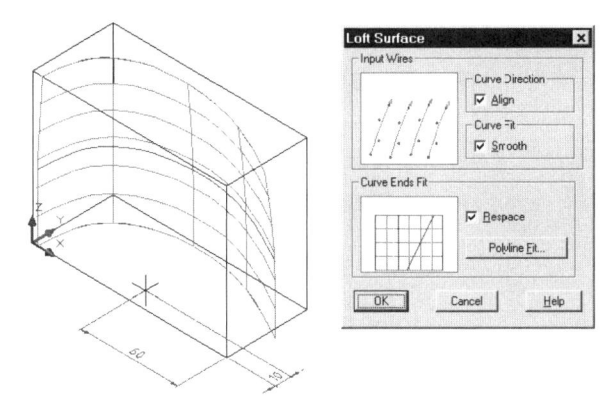

Figure 9–113 *Loft U surface constructed*

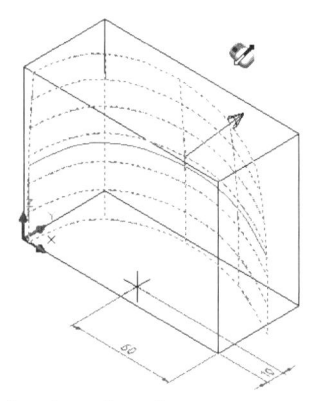

Figure 9–114 *Surface and work point selected*

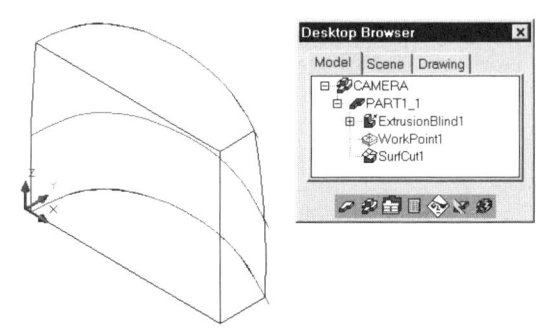

Figure 9–115 *Surface cut feature constructed*

Now you will construct another extruded solid part with a surface cut feature.

27. Start another new solid part in the file.
28. Referring to Figure 9–116, construct a rectangle, resolve it to a profile, add parametric dimensions, and extrude it a distance of 90 mm.

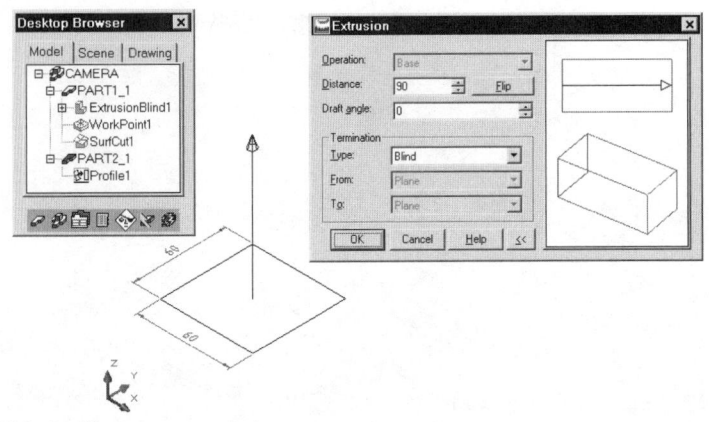

Figure 9–116 *Profile being extruded*

29. Construct a work point (see Figure 9–117).
30. Set current layer to Wire.
31. Set the UCS origin to a corner of the new solid part indicated in Figure 9–118.

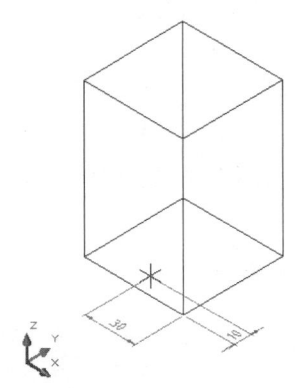

Figure 9–117 *Work point constructed*

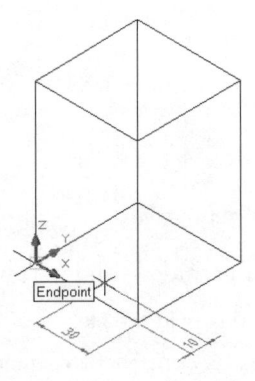

Figure 9–118 *UCS origin relocated*

32. Construct two splines (shown in Figure 9–119) as follows:

First point	−8,0	−4,0
Next point	8,40	8,32
Next point	30,55	22,50
Next point	55,0	50,0
Next point	ENTER	ENTER
Start tangent	ENTER	ENTER
End tangent	ENTER	ENTER

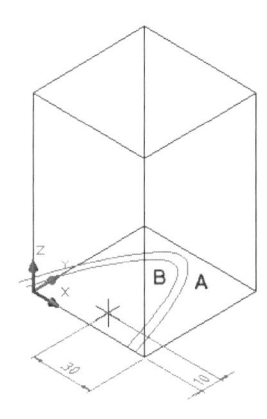

Figure 9–119 *Spline constructed*

33. Copy spline A a distance of 40 mm in the Z direction and move spline B a distance of 90 mm in the Z direction (see Figure 9–120).
34. Set the current layer to Surface.
35. Construct a Loft U surface (see Figure 9–121).

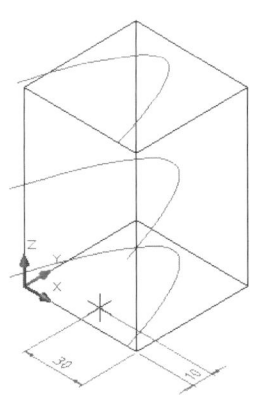

Figure 9–120 *A spline copied and a spline moved*

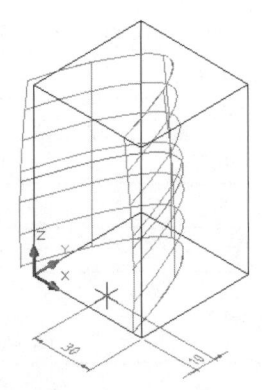

Figure 9–121 *Loft U surface constructed*

36. Set the current layer to 0.
37. Using the surface and the work point, construct a surface cut feature (see Figure 9–122).

Figure 9–122 *Surface cut feature placed*

You have successfully constructed two solid parts with the surface cut feature. Now you will apply assembly constraints to put them together and combine them into a single solid.

38. Hide the splines.
39. Referring to Figure 9–123, apply a mate constraint with –10 mm offset.

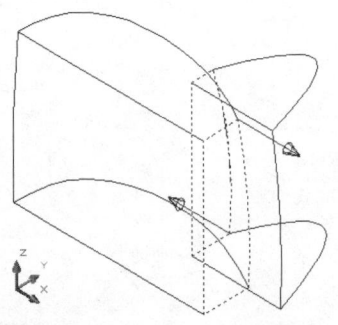

Figure 9–123 *Mate constraint with –10 mm offset applied*

40. Apply two flush constraints (one by one) on the faces indicated in Figure 9–124.
41. Select Part1 from the Browser and double-click to activate it.
42. Select Part>Placed Features>Combine
43. Type **J** to use the join option.
44. Select Part2.

The solid parts are combined into a single solid (see Figure 9–125).

45. Referring to Figure 9–126, construct a work point and add parametric dimensions.

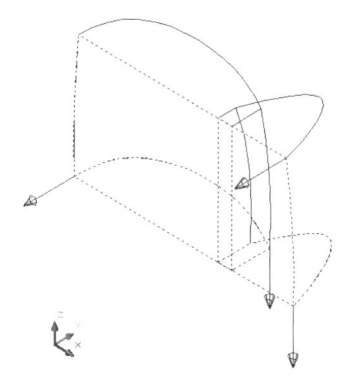

Figure 9–124 *Flush constraints applied*

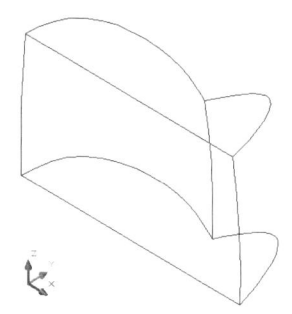

Figure 9–125 *Solid parts combined*

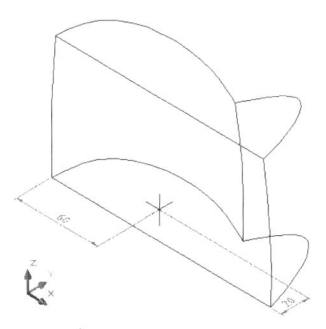

Figure 9–126 *Work point constructed*

46. Set the UCS origin to the end point of the solid indicated in Figure 9–127.
47. Set current layer to Wire.
48. Construct three splines as follows:

First point	−13,−40,60	−13,10,65	−13,60,60
Next point	80,−40,82	80,10,88	80,60,82
Next point	167,−40,70	167,10,72	167,60,70
Next point	ENTER	ENTER	ENTER
Start tangent	ENTER	ENTER	ENTER
End tangent	ENTER	ENTER	ENTER

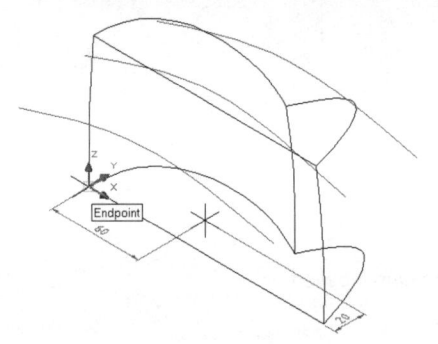

Figure 9–127 *UCS origin relocated and splines constructed*

49. Set current layer to Surface.
50. Construct a Loft U surface (see Figure 9–128)
51. Using the Loft U surface and the work point, add a surface cut feature to the solid.

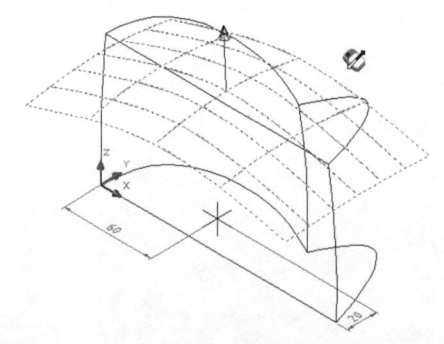

Figure 9–128 *Loft U surface constructed and being used to cut the solid part*

52. Set current layer to 0 and turn off layer Wire.
53. Referring to Figure 9–129, set up a sketch plane and construct a profile.
54. Extrude the profile a distance of 55 mm to join the solid part (see Figure 9–130).

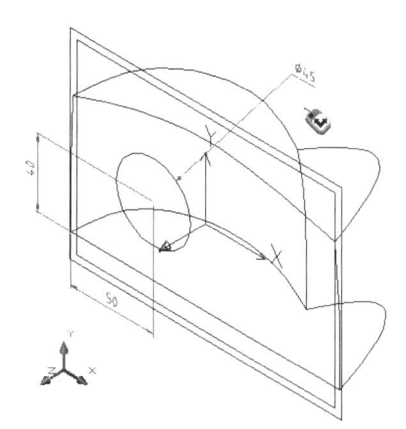

Figure 9–129 *Profile constructed*

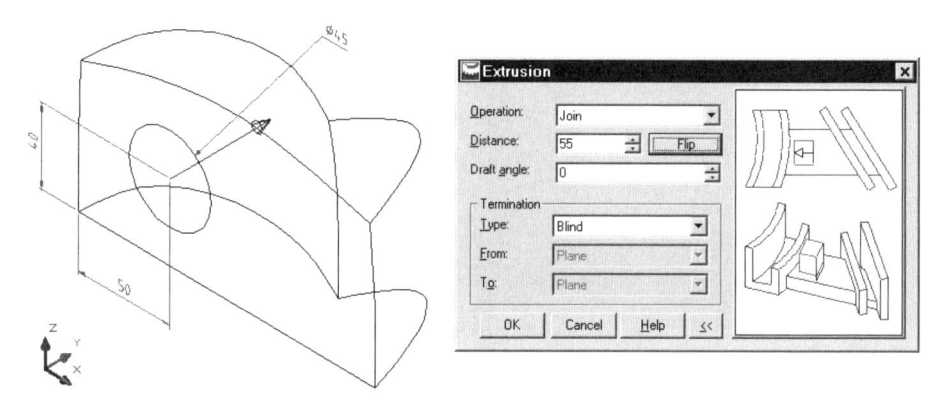

Figure 9–130 *Profile being extruded*

55. Construct a fillet with a constant radius of 5 mm in accordance with Figure 9–131.
56. Construct three fillets with a constant radius of 10 mm (see Figure 9–132).
57. Construct two fillets with a constant radius of 5 mm (see Figure 9–133).

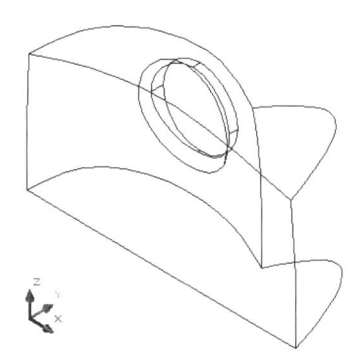

Figure 9–131 *Fillet constructed*

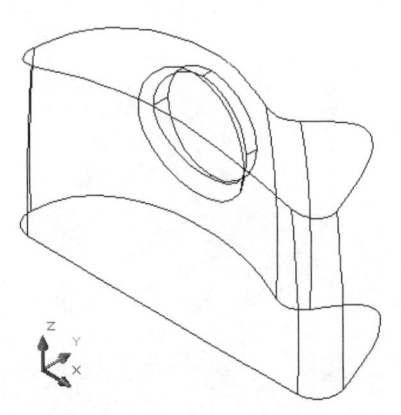

Figure 9–132 *Three fillets constructed*

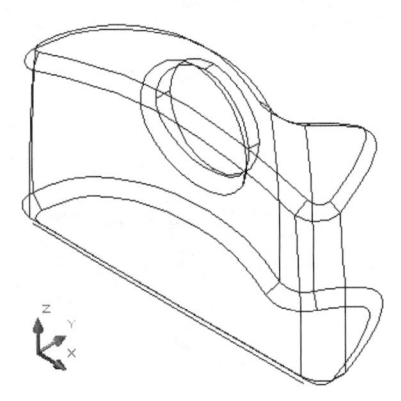

Figure 9–133 *Two constant fillets constructed*

58. Construct an offset work plane in accordance with Figure 9–134.
59. Referring to the shaded image shown Figure 9–135, set the display to back left isometric view and add a shell feature with a thickness of 1 mm.
60. Set the display to 3D wireframe.
61. Select Part>Placed Features>Part Split
62. Select the work plane as the split plane.
63. Right-click to accept the direction of cutting.
64. Press the ENTER key to accept the default part name.

The solid part is split into two solids (see Figure 9–136).

The solid parts for the camera casing are complete. You may further improvise the design by adding ribs, bosses, and other details to the casings. After finishing the design, you should add assembly constraints to assemble the casings and externalize them to form individual solid part files.

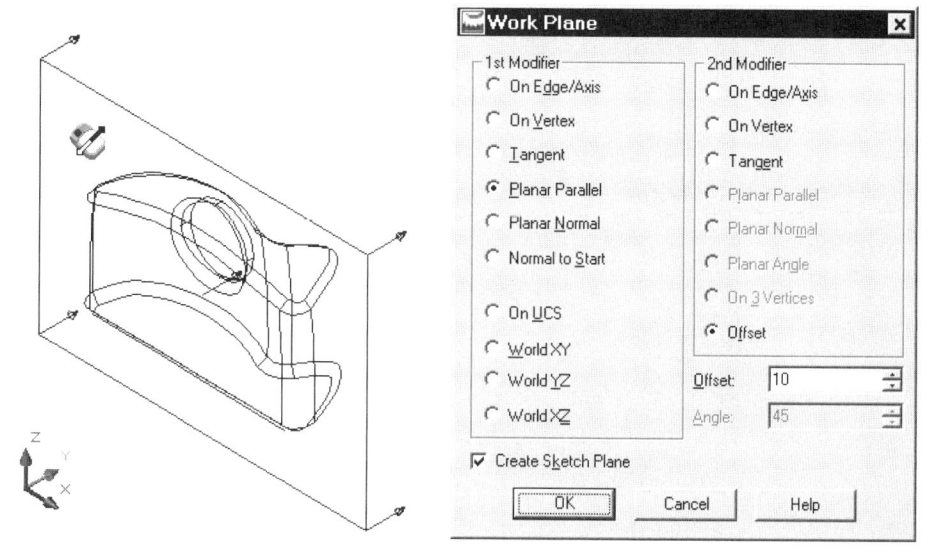

Figure 9–134 Work plane constructed

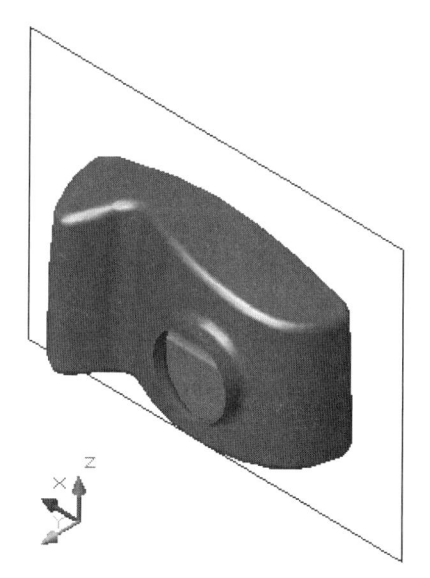

Figure 9–135 Shell feature constructed

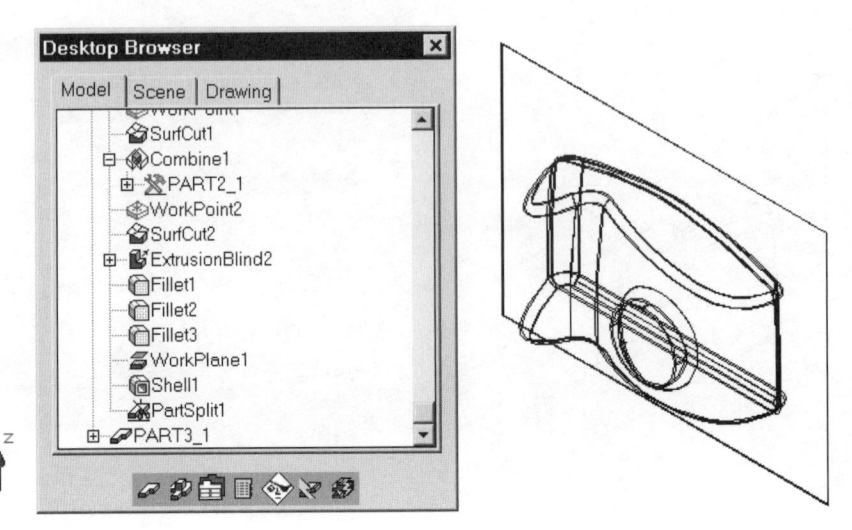

Figure 9–136 *Part split into two solids*

CAMERA LENS HOUSING

Now you will construct a series of lens housings for the camera body that you constructed.

1. Start a new part file. Use metric as the default.
2. Select Part>Design Variables
3. In the Design Variables dialog box, select the Setup button.
4. In the Table Drive Setup dialog box, select the Create button.
5. Specify a file name in the Create Table dialog box (file name: *Lens.xls*).
6. Referring to Figure 9–137, construct an Excel spreadsheet.

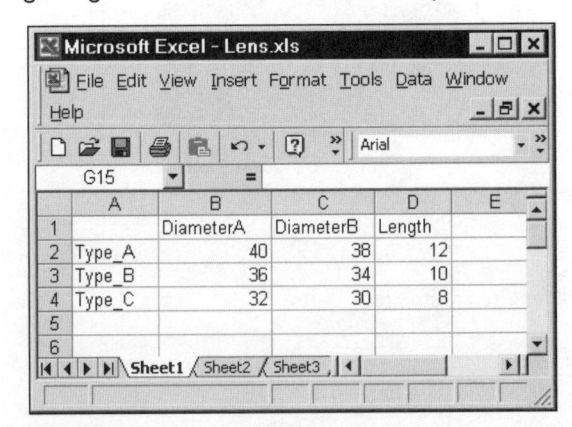

Figure 9–137 *Variables defined in a spreadsheet*

7. Save and close the spreadsheet.

8. After returning to the Table Driven Setup dialog box, select the Update Link button and then the OK button.
9. Select the OK button in the Design Variable dialog box.
10. Referring to Figure 9–138, construct a sketch, resolve the sketch to a profile, and add parametric dimensions by using the variable names depicted in the spreadsheet. Note that the dimension names (d1, d2, etc.) may not be the same as yours because the name suffix is in accordance with your dimensioning sequence, which is different.

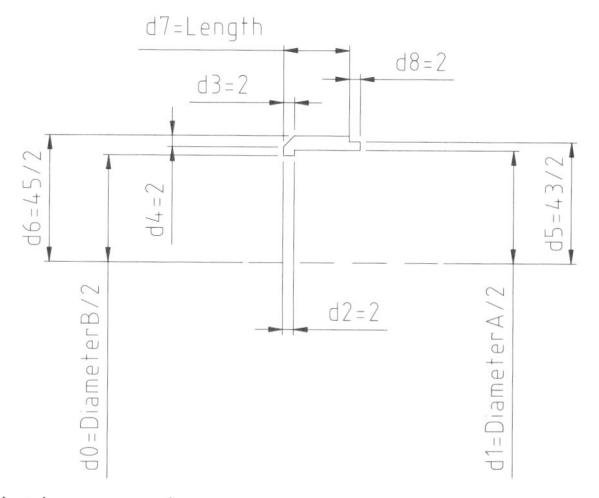

Figure 9–138 *Sketch constructed*

11. Revolve the profile around the hidden line for 360 degrees (see Figure 9–139).

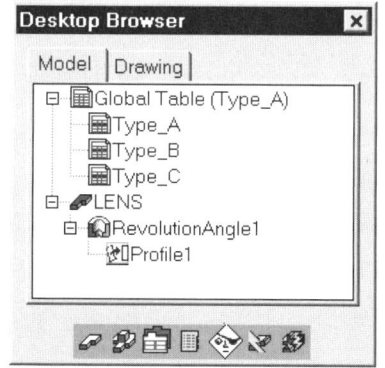

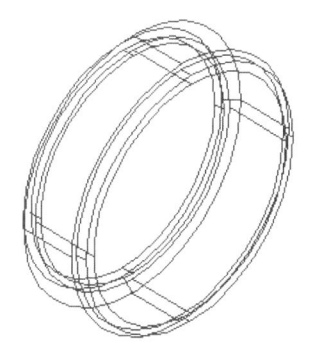

Figure 9–139 *Table driven part constructed*

The solid part with three versions is complete. Save and close your file (file name: *Lens.dwg*). To complete the camera casing project, open the assembly file *Camera.dwg*, put the lens housing in the assembly, and add assembly constraints.

REVIEW QUESTIONS

1. What are the three methods to construct a solid part of free-form shape?

2. Outline the procedure of combining two solid parts.

3. Explain the ways to split a solid part into two solid parts.

4. How will you split the face of a solid part?

5. State the steps to construct design tables.

6. You must include a work point as a reference when using a surface to cut a solid part. True or false?

7. A surface feature is a parametric feature. True or false?

8. Split operation is not available when combining two solid parts in a part file. True or false?

9. You can combine two solid parts even if you do not apply assembly constraints between them. True or false?

10. You must externalize or copy out the split part of a solid. True or false?

11. Base features are static. Therefore, they cannot be modified. True or false?

CHAPTER 10

Engineering Drafting I

OBJECTIVES

The aims of this chapter are to introduce the use of drawing mode and the key concepts of associative engineering drafting; to delineate the ways to construct various kinds of 2D engineering drawing views from 3D solid parts and surfaces that you already constructed; to explain how to construct a cutting line sketch for making an offset sectional view and a break line sketch for making breakout section views; and to familiarize you with the methods that allow you to include annotations to an engineering drawing. After studying this chapter, you should be able to do the following:

- Describe the key concepts of associative drafting
- Construct various kinds of engineering drawing views for solid parts and surfaces
- Apply annotations to an engineering drawing of a solid part

OVERVIEW

Apart from exporting electronic data for downstream computerized manufacturing systems, it is sometimes necessary to output 2D engineering drawing for conventional production processes. By using the set of drawing tools, you can output 2D engineering drawing from surfaces, solid parts, and assemblies, and include annotations to the drawing. In the drawing, you can construct orthographic, isometric, auxiliary, detailed, and broken drawing views.

Engineering drawings and their solid parts are bidirectionally associative. If you change the solid part, the drawing changes to reflect the modification. On the other hand, if you change a parametric dimension in the drawing, the model also changes. In this chapter, you will learn how to construct engineering drawings for solid parts and surfaces. In the next chapter, you will work on engineering drawings for assemblies.

ENGINEERING DRAFTING CONCEPTS

A Mechanical Desktop part file has two working environments in which you place entities: model mode and drawing mode. Model mode is the environment in which you construct the main constituents of the drawing: 3D NURBS surfaces, 3D parametric solids, or assembly of solid parts. Drawing mode is the environment in which you construct an engineering drawing document from 3D Mechanical Desktop objects. By selecting the Model or the Drawing tab of the Browser, you can toggle between model mode and drawing mode. If you have already constructed 3D solids and assembled 3D solids in the computer, constructing a 2D engineering drawing is very simple. You just switch to drawing mode, construct a layout, let the computer project orthographic views from the objects you constructed in the model mode, and add annotations to the drawing views.

ENGINEERING DRAWING CONSTRUCTION

To construct an engineering drawing of objects that you have already constructed in model mode, you use drawing mode by selecting the Drawing tab from the Browser. Because a 2D drawing needs to be plotted in a plotting device and the drawing has to comply with standard engineering practice, you must perform several tasks before you construct any drawing views: select a plotting device and paper size, insert a title block, and set relevant standard and drawing options.

With a drawing layout defined and options set, you can construct engineering drawings of various kinds and add annotations to the drawing.

LAYOUT

Construction of the drawing layout consists of two tasks: configuring the layout to cope with the printer that you will use to print the drawing, and construction of an engineering title block to comply with standard engineering practice, such as ANSI, DIN, ISO, and so on.

Page Setup

Because a 2D engineering drawing needs to be plotted on a piece of paper by using a plotting device, you have to set up the page by selecting a plotter and a paper size.

1. Open the file *Frame_R.dwg* that you constructed in Chapter 4.
2. Select File>Save As and specify a new file name (file name: *Frame-R1.dwg*).
3. Select the Drawing tab from the Browser.
4. In the Drawing tab, select Layout1, right-click, and select Page Setup (see Figure 10–1).

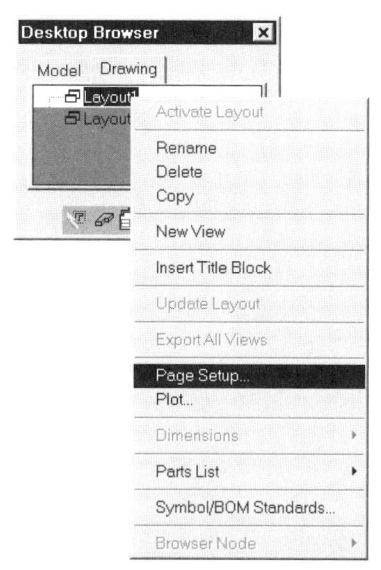

Figure 10–1 *Selecting a layout from the Drawing tab of the Browser*

5. In the Plot Device tab of the Page Setup dialog box, select a plotting device that is already connected to your computer (see Figure 10–2).
6. In the Layout Settings tab, select A3 (or another paper size that is supported by your plotting device) and select the OK button (see Figure 10–3).

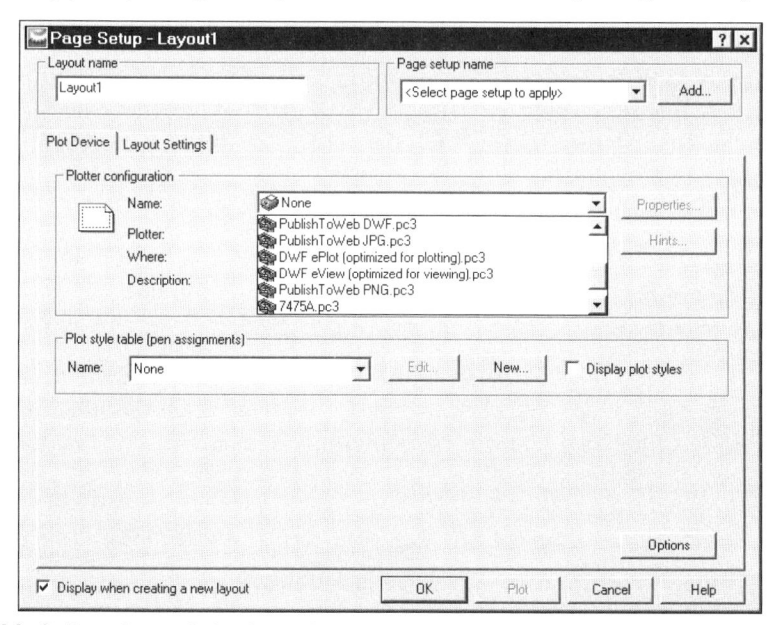

Figure 10–2 *Page Setup dialog box – Plot Device tab*

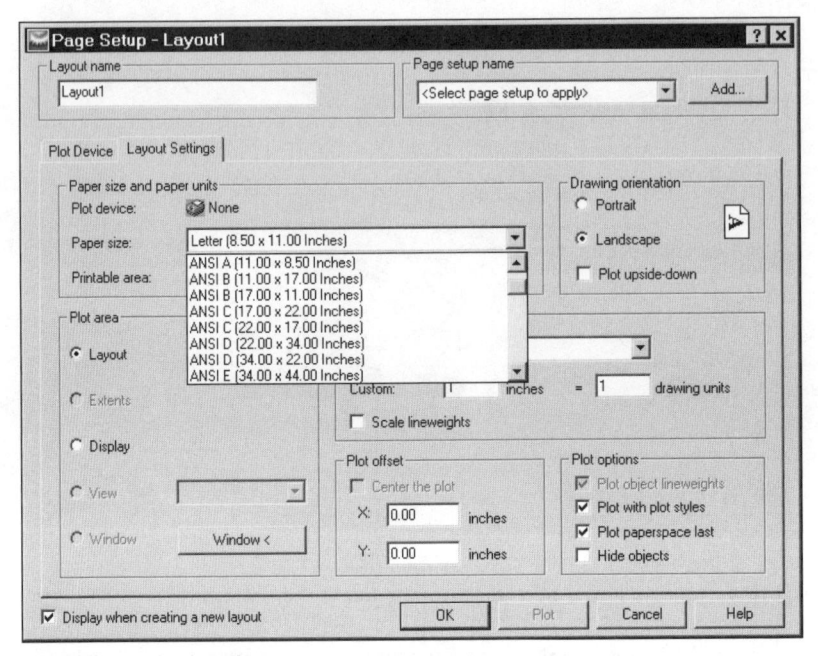

Figure 10–3 *Page Setup dialog box – Layout Settings tab*

Title Block

A title block is a rectangle, or four border lines, with margins around the edge of the drawing paper. Within the title block, in addition to the engineering drawing itself, you need to include general engineering information. Typically, it should have the name of the company, possibly with the company logo, and textual data. You need to state who constructed the drawing, who checked the drawing, who approved the drawing, the date, the plotting scale, the material, the surface finish requirement, the tolerances, and so on.

You should have a title block ready before you start constructing engineering drawings. If you do not have one, you can either make one now and save it in the computer for insertion into the engineering drawing, or use the default title blocks.

Whichever title block you use, it is very important to ensure that the printable area depicted in the Page Setup dialog box is equal to or greater than the size of the rectangular title block. Otherwise, you will encounter problems in printing the drawings.

Now you will use the default title blocks.

7. Select Layout1 from the Browser, right-click, and select Insert Title Block.
8. In the Drawing Border with Title Block dialog box (see Figure 10–4), select A3 paper format (or another format corresponding to your page setup) and ISO Title Block A title block, and select the OK button.

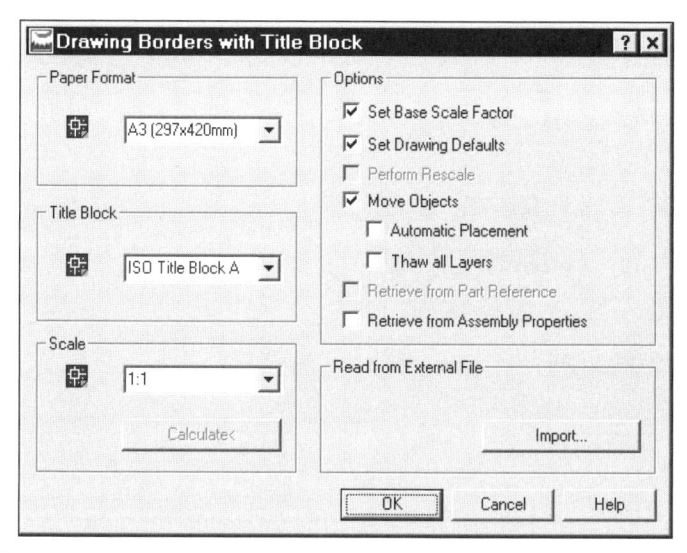

Figure 10–4 *Drawing Border with Title Block dialog box*

9. Select the OK button in the Page Setup dialog box.
10. Select a location near the lower-left corner of the layout.
11. In the Change Title Block Entry dialog box, fill in the information to be printed on the drawing title block (see Figure 10–5).

A title block is inserted (see Figure 10–6). Here, the X and Y scale of the title block has been slightly modified to match the printable area of the selected plotting device.

Figure 10–5 *Change Title Block Entry dialog box*

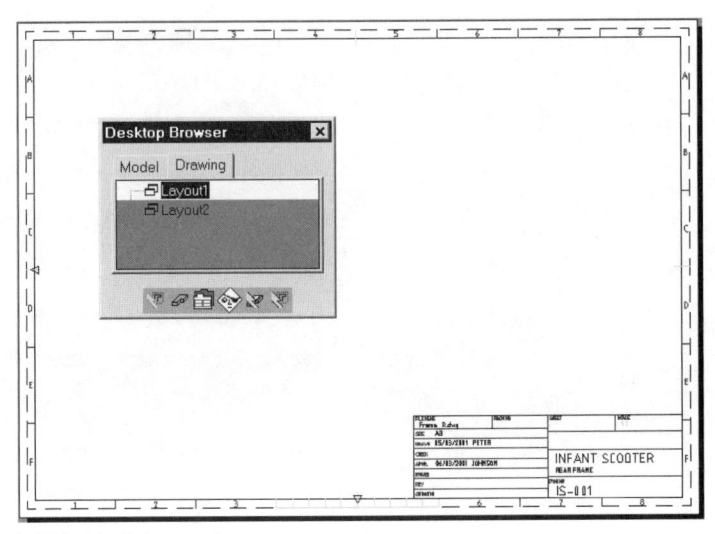

Figure 10–6 *Title block inserted*

ENGINEERING STANDARDS

Before you construct engineering drawing views, you should spend some time setting the relevant drawing options to meet the requirements of relevant engineering standards.

12. Select Drawing>Drawing Options to display the Mechanical Options dialog box (see Figure 10–7).

The Drawing tab of the Mechanical Options dialog box has five areas (shown in Figure 10–7).

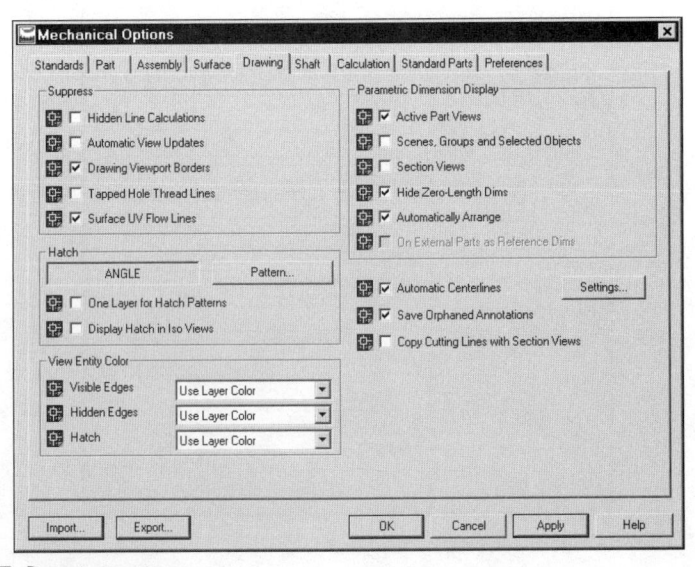

Figure 10–7 *Drawing options*

The Suppress area has five check boxes:

Hidden Line Calculation	suppresses hidden line calculation
Automatic View Updates	suppresses automatic update of drawing views
Drawing Viewport Borders	suppresses the display of the drawing viewport borders
Tapped Hole Thread Lines	suppresses the display of the thread lines of tapped holes
Surface UV Flow Lines	suppresses the display of U and V flow lines of NURBS surfaces

The Hatch area has two check boxes and one button:

One Layer for Hatch Patterns	puts the hatch patterns in a separate layer
Display Hatch in Iso Views	displays hatch patterns in isometric views

13. Select the Pattern button (see Figure 10–8).
14. Select a hatch pattern and select the OK button.

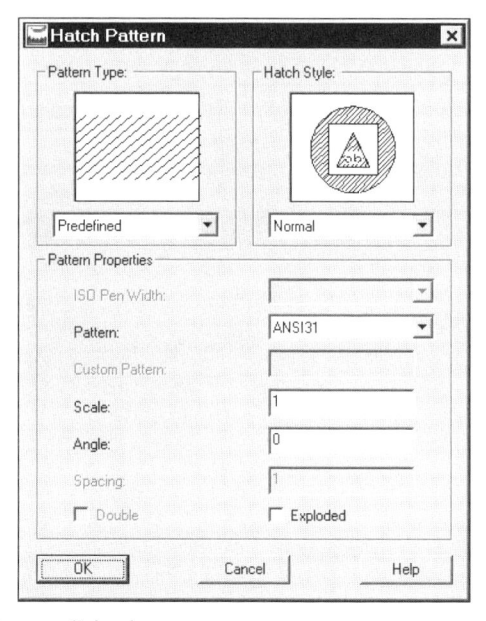

Figure 10–8 *Hatch Pattern dialog box*

The View Entity Color area has three pull-down list boxes:

Visible Edges	enables you to use layer, part, or part and feature color
Hidden Edges	enables you to use layer, part, or part and feature color
Hatch	enables you to use layer or part color

The Parametric Dimension Display area has six check boxes:

Active Part Views	displays parametric dimensions in drawing views of the active part
Scenes, Groups, and Selected Objects	displays parametric dimensions in drawing views of scenes, groups, and selected objects
Section Views	displays parametric dimensions in section views
Hide Zero-Length Dims	hides parametric dimensions with a zero value (zero value dimensions are used to align edges in the solid part)
Automatically Arrange	arranges the parametric dimensions automatically
On External Parts as Reference Dims	displays parametric dimensions of external definitions in an assembly drawing as reference dimensions

The fifth area of the Drawing tab of the Mechanical Options dialog box has three check boxes and a button.

Automatic Centerlines	constructs centerlines in the drawing views automatically
Save Orphaned Annotations	saves orphaned annotations instead of removing them
Copy Cutting Lines with Section Views	copies cutting lines when section views are copied

15. Select the Settings button to display the Centerlines dialog box (see Figure 10–9).

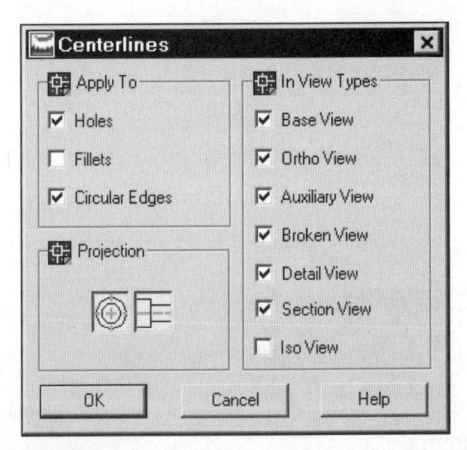

Figure 10–9 *Centerlines dialog box*

The Centerlines dialog box enables you to apply center lines to holes, fillets, and circular edges in various kinds of drawing views, as well as to axial and profile projections.

16. Select the OK button to accept the default.
17. On returning to the Mechanical Options dialog box, select the Standards tab (see Figure 10–10).

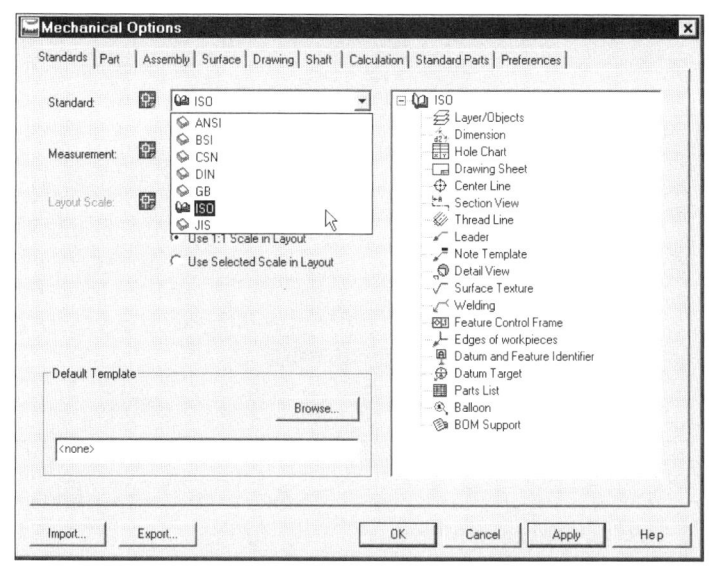

Figure 10-10 *Standard tab of the Mechanical Options dialog box*

A number of national and international standards are available in the Standards tab of the Mechanical Options dialog box.

18. In the Standards tab, select ISO, metric, and 1:1 scale, and select the OK button.

DRAWING VIEWS

Construction of engineering drawing views is partially automated; you simply select the objects to be displayed in the drawing views and specify the kind of view and parameters of the views. The engineering drawing constructed is associative to the objects residing in the model environment of the computer file.

Now you will construct drawing views of the solid part residing in model mode.

19. Select Layout1 from the Browser, right-click, and select New View to display the Create Drawing View dialog box (see Figure 10-11).

    Command: AMDWGVIEW

In the Create Drawing View dialog box, select view type, data set, layout, and drawing scale. There are also three tabs that enable you to control hidden line display, select section view, and control representation of the part in the 2D drawing.

If this is the first drawing view on the drawing layout, there are three kinds of views:

Base	is the first drawing view
Multiple	is a set of drawing views, including a base view and a number of orthographic and isometric views
Broken	is the first drawing view broken into two parts

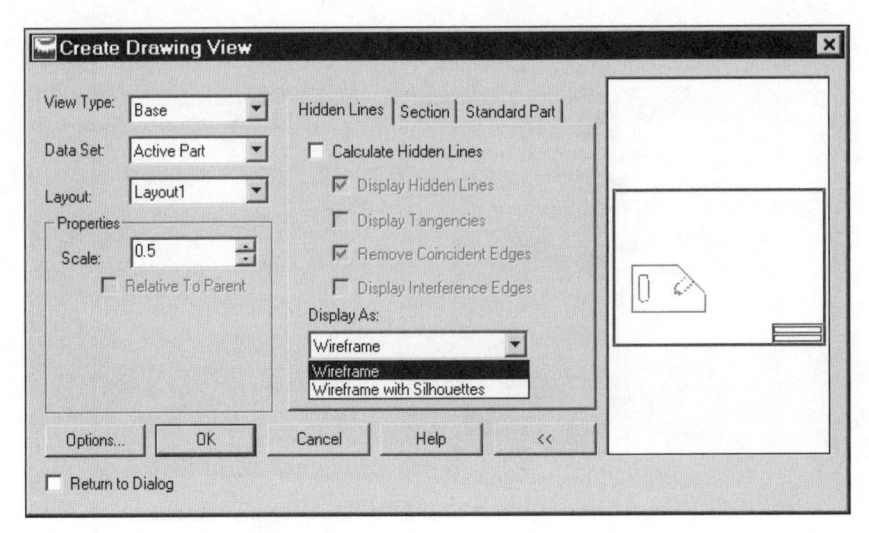

Figure 10–11 *Create Drawing View dialog box*

Data Set refers to the kind of objects to be included in the drawing view. In a part file, there are two kinds of data set:

Active Part	enables you to construct drawing views of the active solid part
Select	enables you to construct drawing views of selected objects (including NURBS surfaces)

The Hidden Lines tab has five check boxes and a pull-down box:

Calculate Hidden Lines	determines whether hidden lines are calculated or not
Display Hidden Lines	controls the display of hidden lines (if hidden lines are calculated)
Display Tangencies	controls the display of tangency edges (if hidden lines are calculated)
Remove Coincident Edges	removes coincident hidden lines (if hidden lines are calculated)
Display Interference edges	displays the edges between intersecting objects
Display As	sets the display to wireframe or wireframe together with silhouettes

20. Select the Section tab to find out what you can do (see Figure 10–12).

The Section tab enables you to construct sectional views and set section view options. For a base view, there are three kinds of section views: full, offset, and breakout. (You will learn these and other section views later in this chapter.) If you have decided to construct a section view, you will need to provide a label to the section, select a hatch pattern, and decide whether you will hide the obscured hatch.

21. Select None in the Section Type pull-down list box.

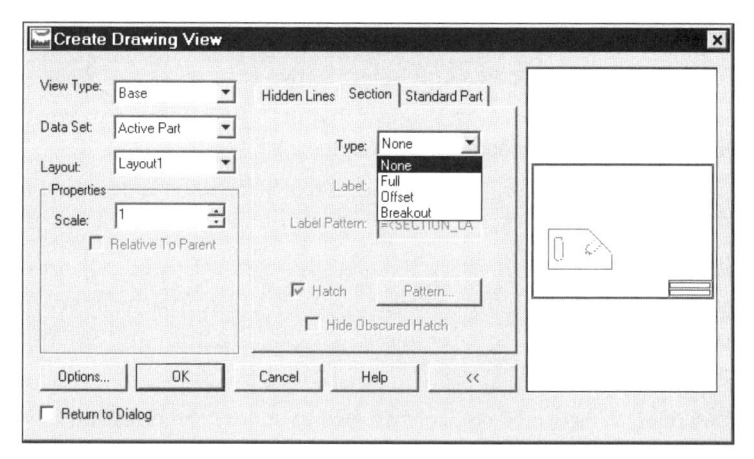

Figure 10–12 *Section tab of the Create Drawing View dialog box*

22. Select the Standard Part tab (see Figure 10–13).

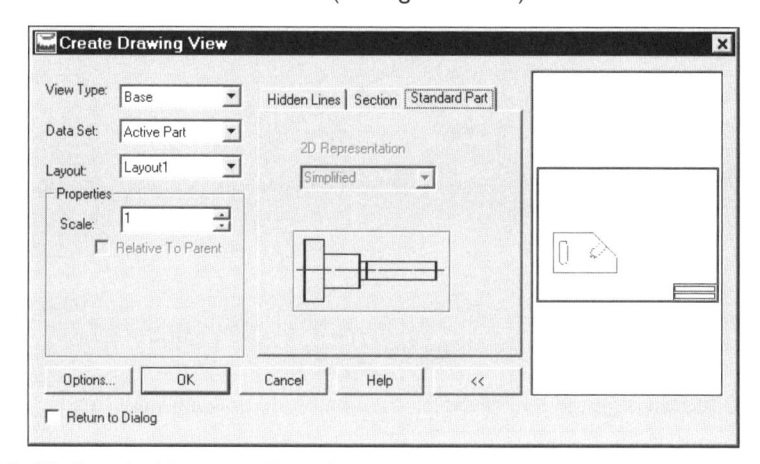

Figure 10–13 *Standard Part tab of the Create Drawing View dialog box*

The options of the Standard tab of the Create Drawing View dialog box are not available for the base view. For other engineering views, you can set 2D representation to three levels of detail:

True	enables you to construct a projected view of the part with all the details
Standard	enables you to construct a section view with details displayed
Simplified	enables you to construct a simplified section view without details

23. Select the Hidden Lines tab.
24. In the Hidden Lines tab, check the Calculate Hidden Lines, Display Hidden Lines, Remove Coincident Edges, and the Display Interference Edges boxes.

25. In the View Type pull-down box, select Base.
26. In the Data Set pull-down box, select Active Part.
27. In the Layout pull-down box, select Layout1. (Remember that you already set up layout1 and inserted a title block there.)
28. In the Properties area, set the scale to 0.5. (This value is based on an estimation of the size of the paper and the size of the component.)
29. Select the OK button.

After you have selected the OK button, the display will automatically switch to model mode for you to select a face and an edge to set the direction and orientation of the drawing view.

30. Referring to Figure 10–14, select a face to specify the direction of viewing and right-click to accept.

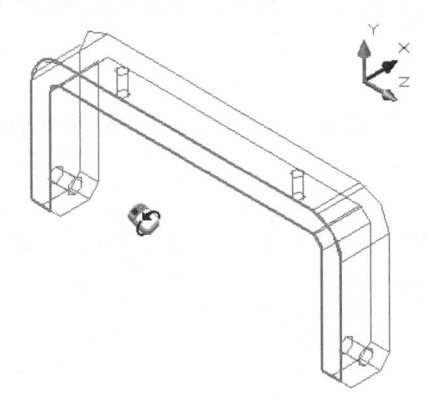

Figure 10–14 *Face selected*

31. Select an edge to indicate the orientation of the drawing view (see Figure 10–15).
32. Right-click to accept.

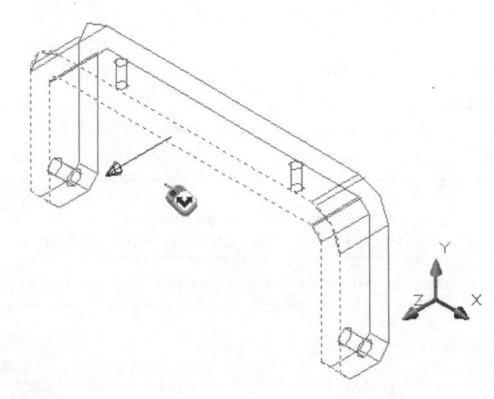

Figure 10–15 *Edge selected*

After you have selected a face and an edge, the display returns to drawing mode to let you specify the location of the drawing view on the paper.

33. Referring to Figure 10–16, select a point and press the ENTER key.

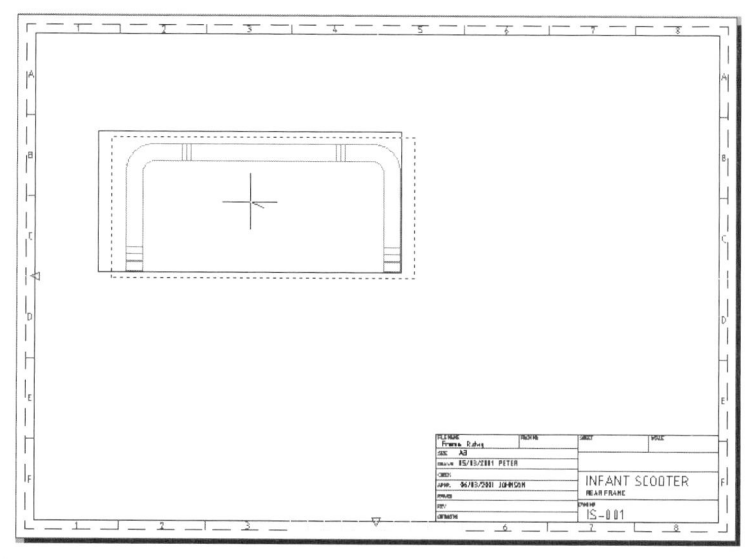

Figure 10–16 *Location of drawing view selected*

A drawing view is constructed (see Figure 10–17). Note that parametric dimensions and centerlines are displayed automatically and a base view object is listed in the Browser.

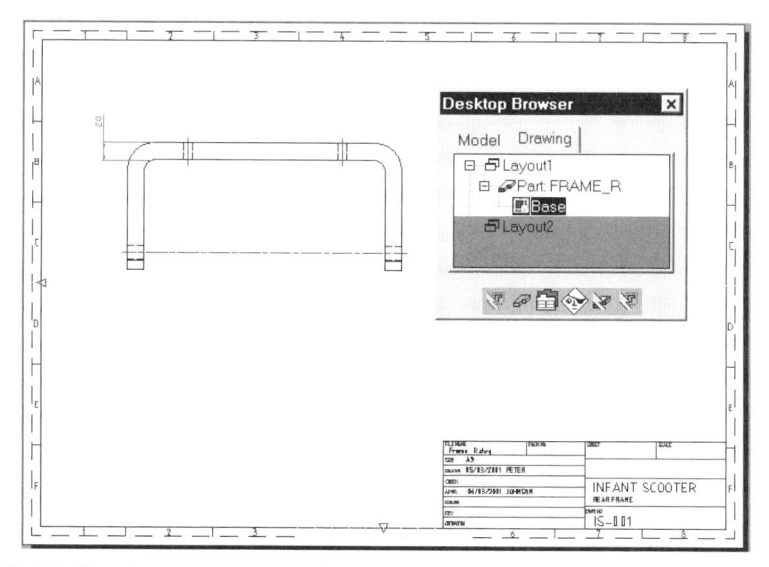

Figure 10–17 *Drawing view constructed*

Now you will construct an orthographic view and an isometric view.

34. Select the base view from the Browser, right-click, and select **New View**.

After you have constructed a drawing view in the layout and continue to construct engineering drawing views, the type of view that you can construct are orthographic, auxiliary, isometric, and detail views, in addition to base, multiple, and broken views.

35. In the Create Drawing View dialog box, select Ortho from the View Type pull-down box and the OK button (see Figure 10–18).

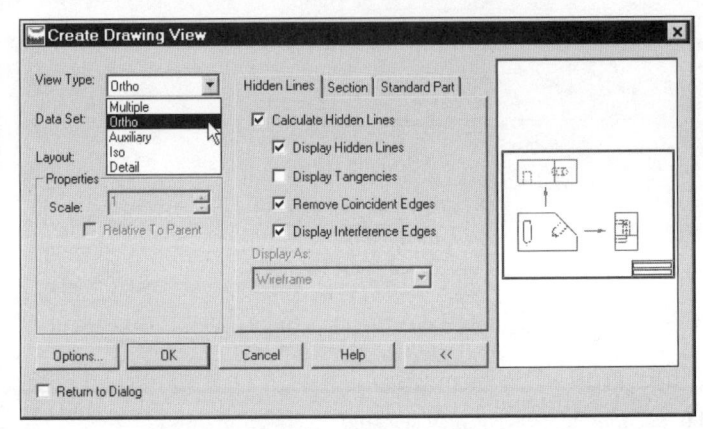

Figure 10–18 *Orthographic view selected*

36. Referring to Figure 10–19, select a location to indicate the position of the orthographic view, right-click, and select ENTER.

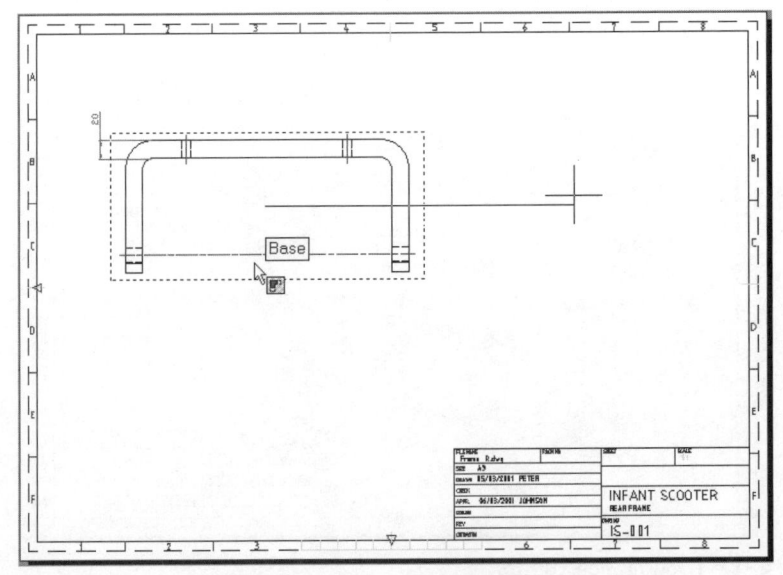

Figure 10–19 *Parent view and location of new view selected*

An orthographic view is constructed (see Figure 10–20).

Figure 10–20 *Orthographic view constructed*

Now you will construct an isometric view.

37. Select the base view again, right-click, and select New View.
38. In the Create Drawing View dialog box, select Iso from the View Type pull-down box and select the OK button.
39. Referring to Figure 10–21, select a location of the isometric view, right-click, and select ENTER.

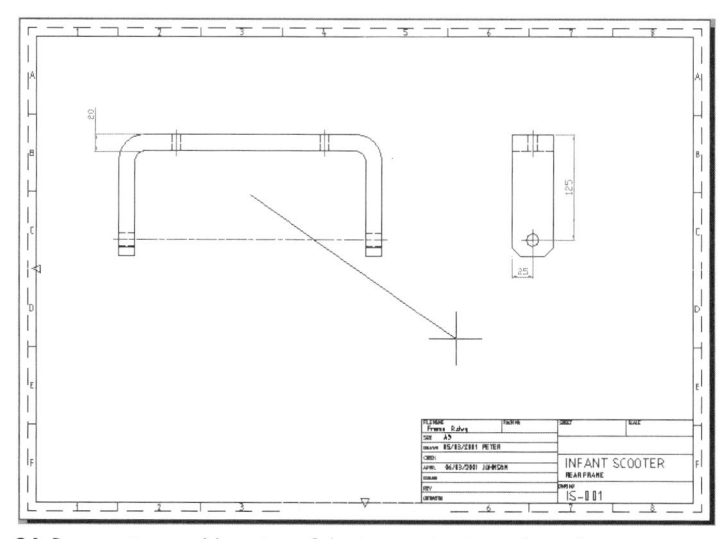

Figure 10–21 *Parent view and location of the isometric view selected*

An isometric view is constructed (see Figure 10–22).

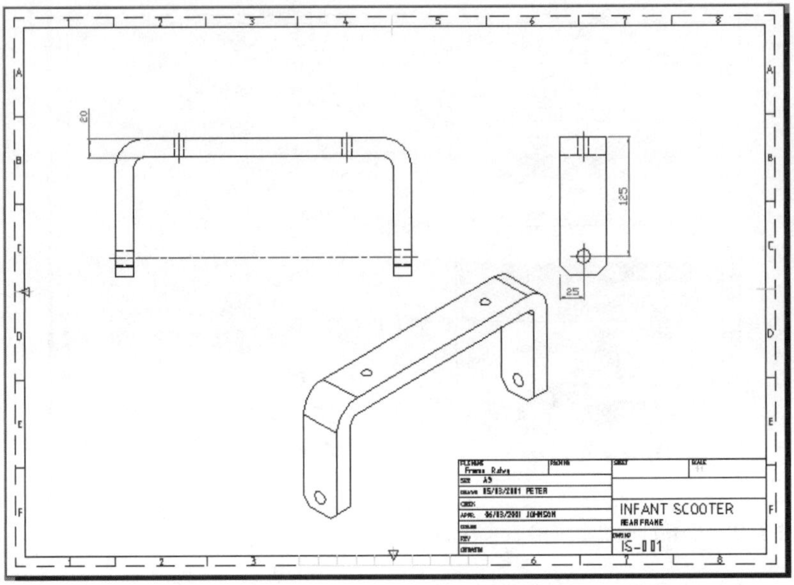

Figure 10–22 *Isometric view constructed*

A base view, an orthographic view, and an isometric view are complete. Save your file.

MANIPULATING DRAWING VIEWS

Now you will learn how to move, copy, delete, and edit a drawing view.

40. Select the base view from the Browser, right-click, and select Move (see Figure 10–23).
41. Move the cursor to a new location, right-click, and select ENTER.

Because the orthographic and isometric views are dependent views of the selected base view, they move as well.

42. Select the isometric view from the Browser, right-click, and select Move.
43. Move the cursor to a new location and right-click.

Only the isometric view moves because it does not have any dependent view.

44. Select the isometric view from the Browser, right-click, and select Copy (see Figure 10–24).
45. Select a location, right-click, and select ENTER.

Engineering Drafting I 507

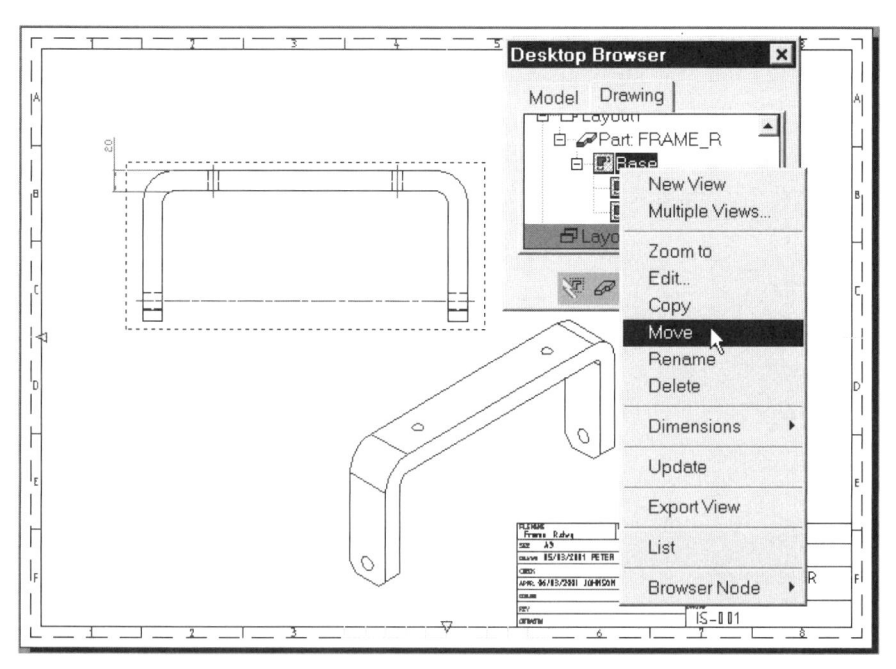

Figure 10–23 *Base view together with its dependent views being moved*

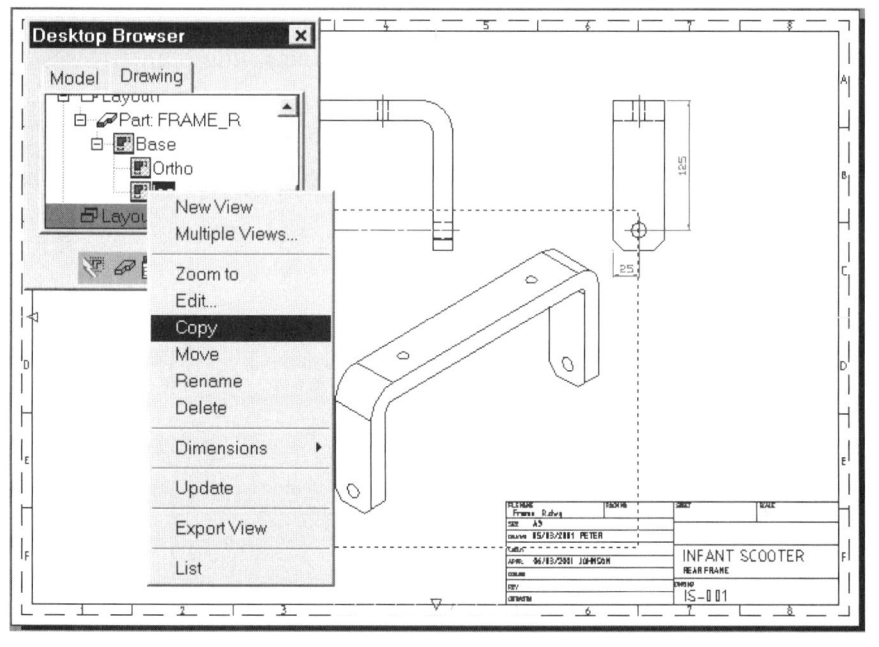

Figure 10–24 *Isometric view being copied*

The isometric view is copied (see Figure 10–25). Because the purpose of copying a view is simply to let you know that drawing views can be copied, you will delete one of the isometric views.

46. Select an isometric view from the Browser, right-click, and select Delete.

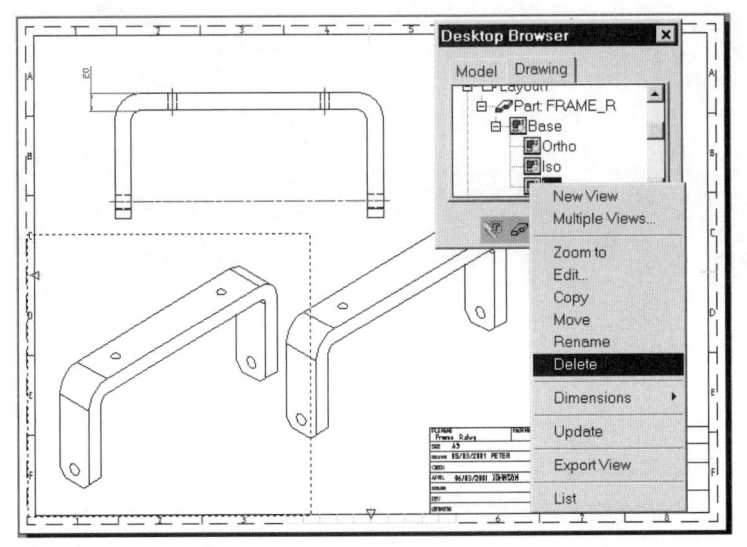

Figure 10–25 *Isometric copied and being deleted*

An isometric view is deleted. Now you will edit the remaining isometric view.

47. Select the remaining isometric view from the Browser, right-click, and select Edit to display the Edit Drawing View dialog box (see Figure 10–26).

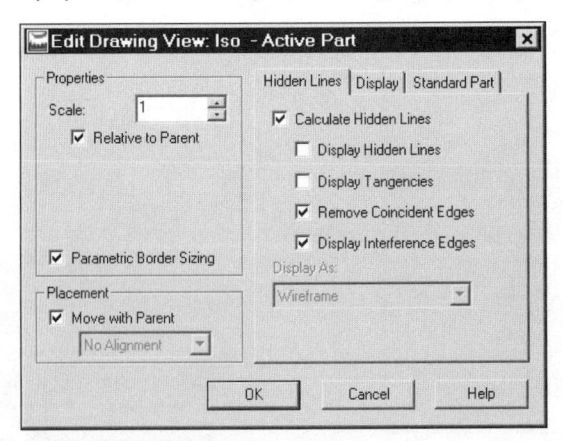

Figure 10–26 *Edit Drawing View dialog box*

48. In the Edit Drawing View dialog box, deselect the Display Tangencies box and select the OK button.

Tangency lines of the isometric view are suppressed (see Figure 10–27). Save and close your file.

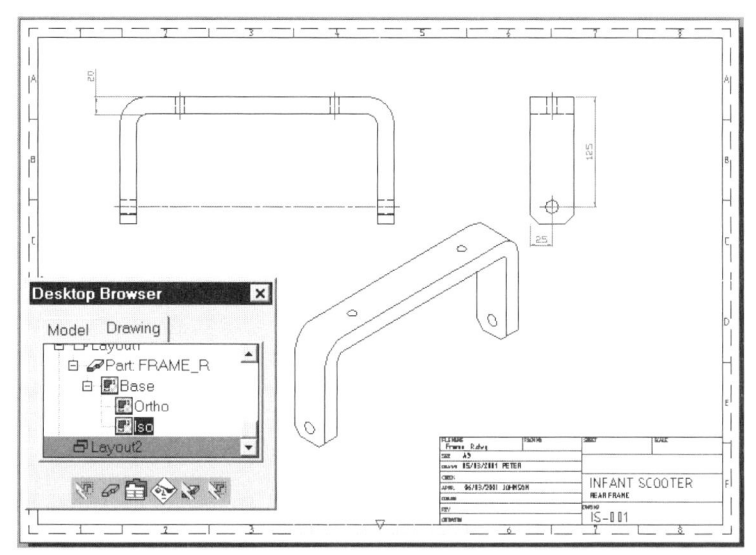

Figure 10–27 *Isometric view edited*

ENGINEERING DRAWING VIEWS

After learning how to construct base views, orthographic views, and isometric views of a solid part, you will now learn how to construct other kinds of drawing views including the multiple view, auxiliary view, detail view, broken view, and section view. In addition, you will learn how to construct engineering drawing views for a surface model.

MULTIPLE VIEWS FOR A SURFACE MODEL

Now you will learn how to use the multiple view option to construct a number of drawing views without returning to the Create View dialog box and learn how to construct engineering drawings for a surface model. Basically, the procedure for making the drawing views of the surface model is similar to that for a solid part. The only exception is that you use the Select option in the data set area of the Create Drawing View dialog box.

1. Open the file *RemoteControl1.dwg* that you constructed in Chapter 7.
2. Select the Drawing tab from the Browser.
3. Select Layout1 from the Drawing tab of the Browser, right-click, and select Page Setup.
4. In the Page Setup dialog box, select a plotting device and a paper size, and select the OK button.

5. Insert a title block (see Figure 10–28).
6. Select Layout1 from the Browser, right-click, and select New View.
7. In the Create Drawing View dialog box, select Multiple from View Type box and Select from Data Set box, set scale to 1, deselect Calculate Hidden Lines, select Wireframe with Silhouettes, and select the OK button (see Figure 10–29).

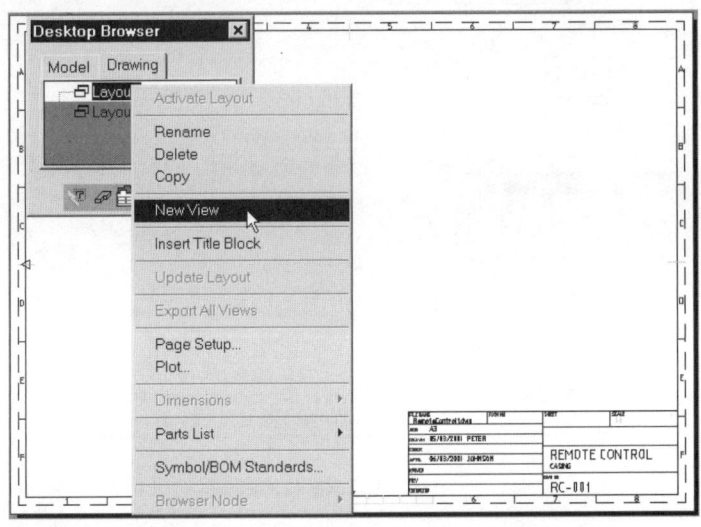

Figure 10–28 *Title block inserted*

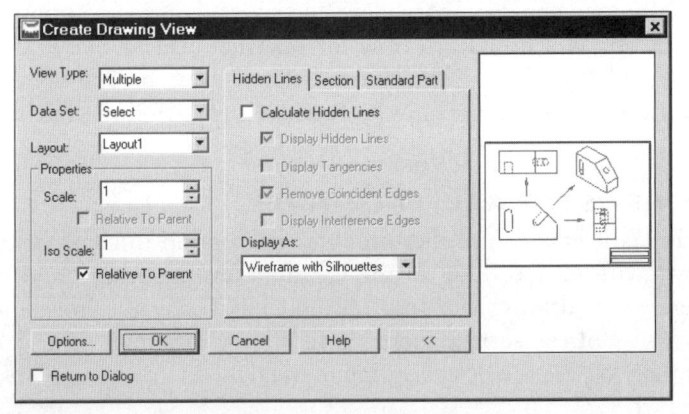

Figure 10–29 *Multiple views selected*

Now the display switches to Model mode.

8. Select all the surfaces and press the ENTER key.
9. Type **Z** to use world ZX plane.
10. Type **X** to use world X axis.
11. Press the ENTER key to accept.

12. After returning to the layout, select a location to indicate the position of the base view (see Figure 10–30).
13. Press the ENTER key.

A base view is constructed.

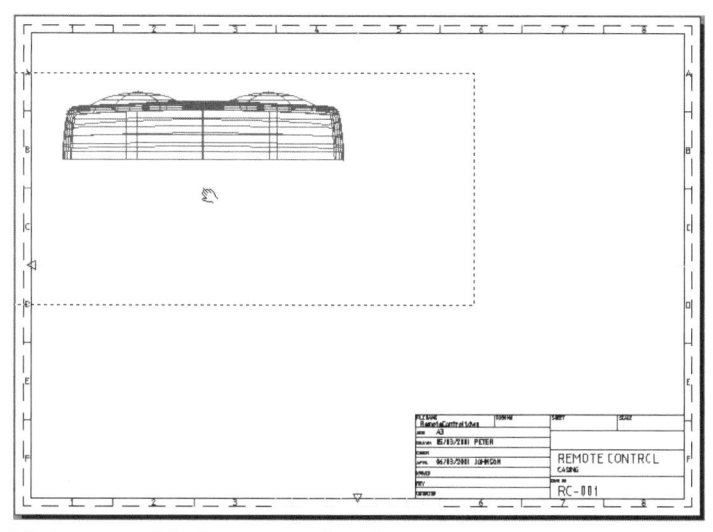

Figure 10–30 *Location of the base view selected*

14. Select a location to indicate the position of the projected view and press the ENTER key (see Figure 10–31).

An orthographic view is constructed.

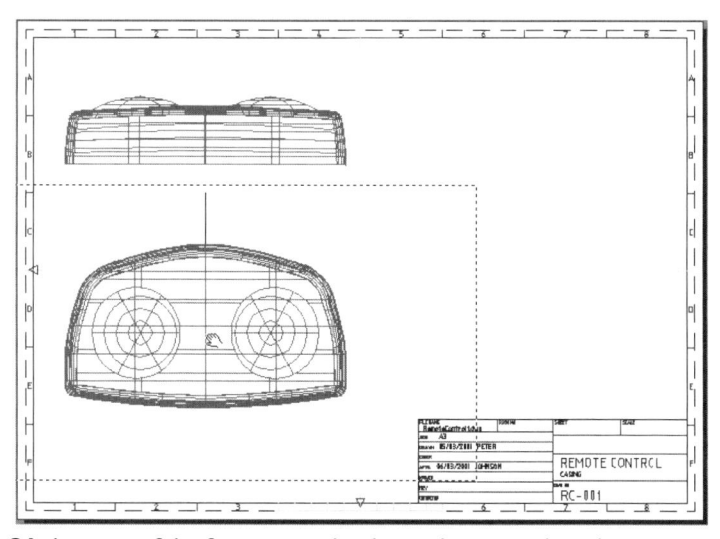

Figure 10–31 *Location of the first projected orthographic view selected*

15. Select a location to indicate the position of the projected side view and press the ENTER key (see Figure 10–32).
16. Referring to Figure 10–33, select a location to indicate the position of the isometric view and press the ENTER key.
17. Press the ENTER key again to exit.

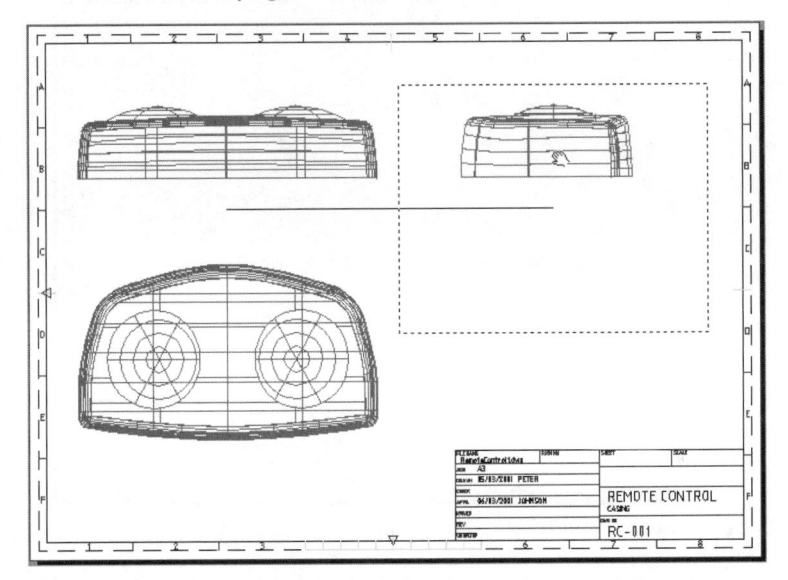

Figure 10–32 *Location of the side view selected*

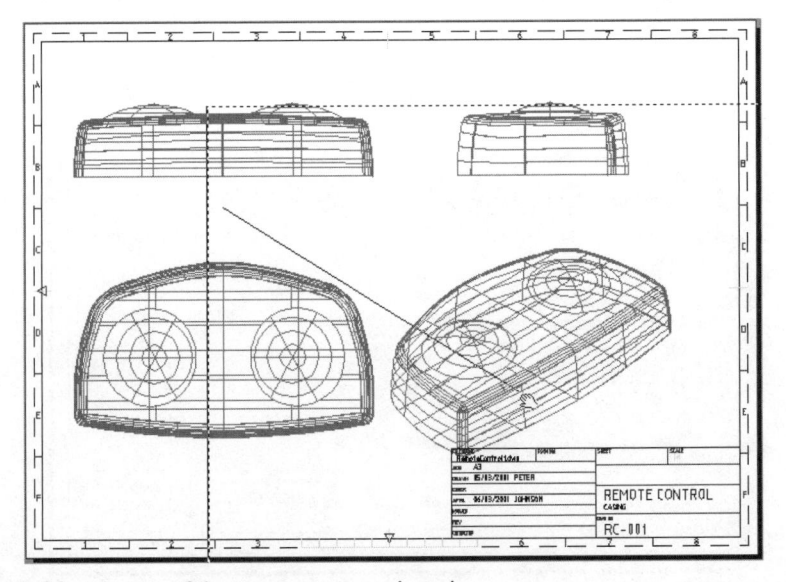

Figure 10–33 *Location of the isometric view selected*

The drawing is complete (see Figure 10–34). Save and close your file.

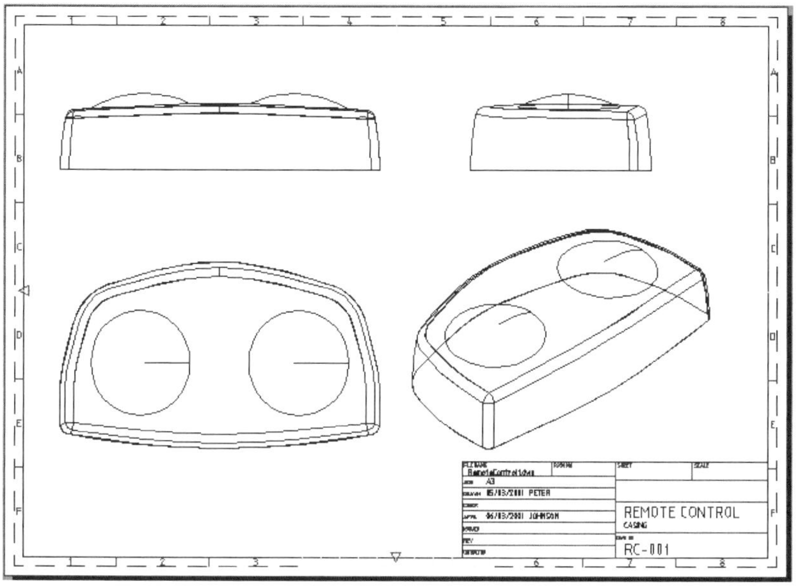

Figure 10-34 *Multiple views constructed*

AUXILIARY VIEW

Now you will learn how to construct an auxiliary view. An auxiliary view is a special kind of orthographic view in which the direction of projection is not orthogonal to the parent view. To indicate the direction of projection, you either specify two points or an edge.

1. Open the file *Frame_F.dwg* that you constructed in Chapter 4.
2. Select the Drawing tab from the Browser.
3. Select Layout1 from the Drawing tab of the Browser, right-click, and select Page Setup.
4. In the Page Setup dialog box, select a plotting device and a paper size, and select the OK button.
5. Insert a title block.
6. Referring to Figure 10–35, construct a front view, side view, and a top view. (Use a drawing scale of 0.5.)
7. Select Layout1 from the Browser, right-click, and select New View.
8. In the Create Drawing View dialog box, select Auxiliary view and select the OK button.
9. Select an edge indicated in Figure 10–36.

10. Press the ENTER key to use the edge.
11. Referring to Figure 10–36, select a location to position the auxiliary view and press the ENTER key.

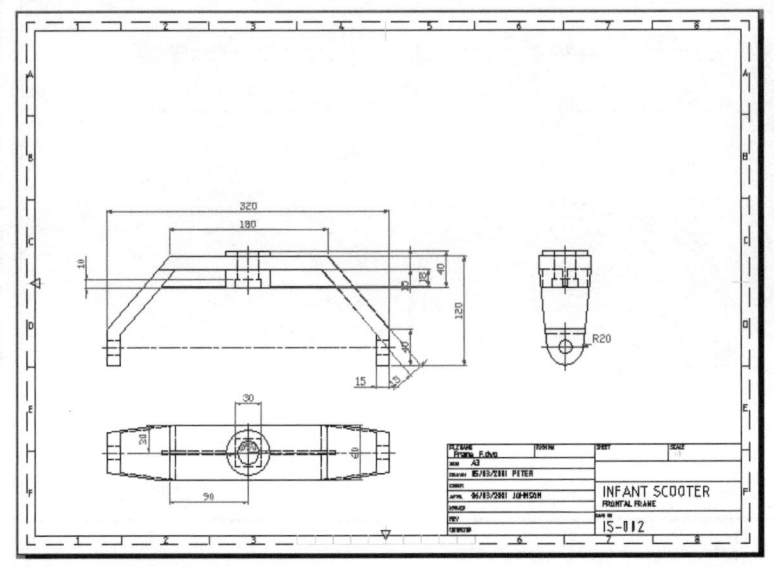

Figure 10–35 *Multiple views constructed*

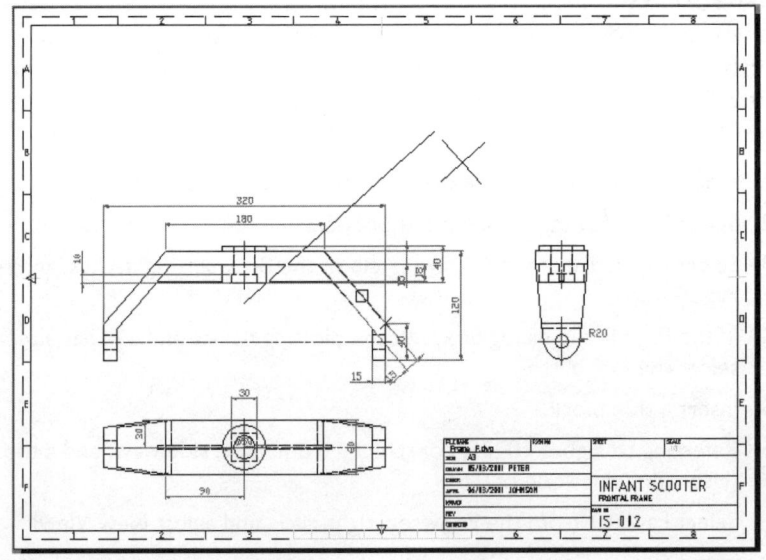

Figure 10–36 *Edge and location selected*

An auxiliary view with three orthographic views is constructed (see Figure 10–37). Save your file.

Engineering Drafting I 515

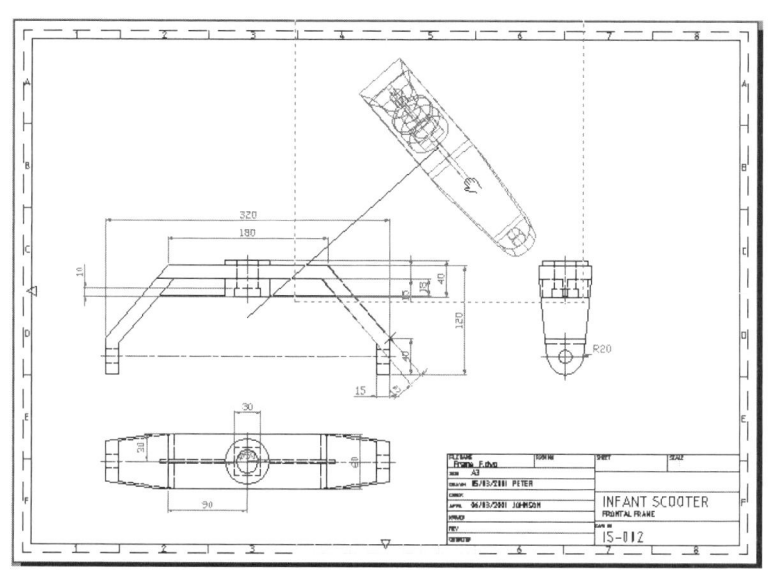

Figure 10-37 *Auxiliary view constructed*

DETAIL VIEW

Now you will learn how to construct a detail view. A detail view shows an enlarged portion of a drawing view. By using a detail view, you can maintain a drawing scale for all the views of the drawing and provide a more detailed closer look at a portion of the component.

12. Open the file *Frame_F.dwg*, if you have already closed it.
13. Select the ortho side view from the Browser, right-click, and select New View.
14. In the Create Drawing View dialog box, select Detail view (see Figure 10-38).

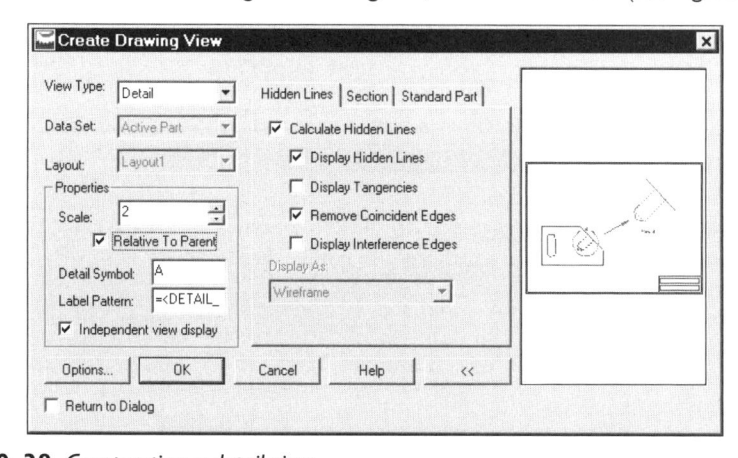

Figure 10-38 *Constructing a detail view*

In the Properties area of the Create Drawing View dialog box, you can specify the parameters of the detail view, including scale, symbol, label, and dependency of the view.

15. Specify a scale of 2 relative to the parent view.
16. Specify the letter "A" as the symbol name.
17. Use the default label pattern.
18. Check the Independent view display button. This way, the detail view is independent of the parent view's degree of detail delineated in the Standard Part tab of the Create Drawing View dialog box. For example, you can specify a simplified 2D representation for the parent view and use a true 2D representation in the detail view.
19. Select the OK button.
20. Referring to Figure 10–39, select a vertex.

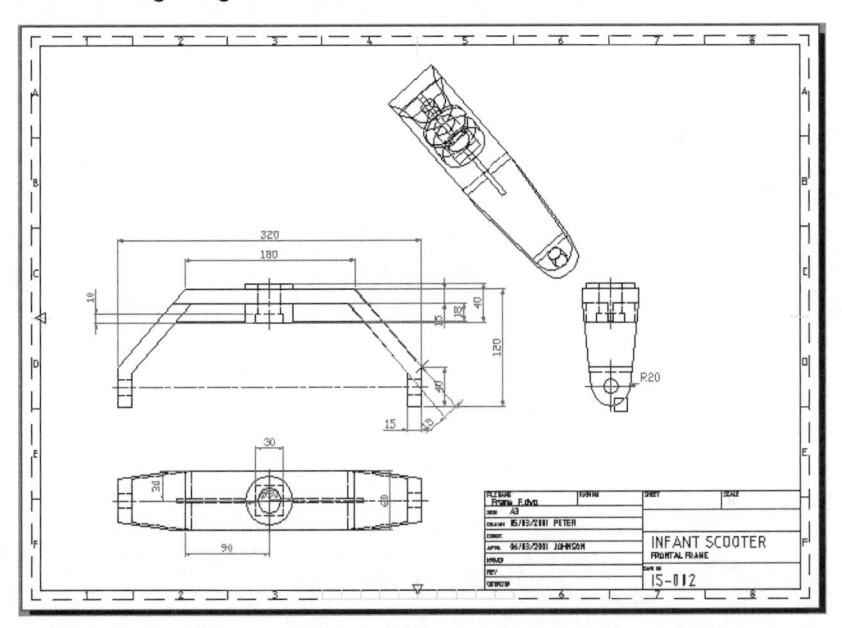

Figure 10–39 *Vertex of a parent view selected*

21. Select a point to indicate the center and a point to indicate the circumference of the circular area to be enlarged in the detail view (see Figure 10–40).

Engineering Drafting I 517

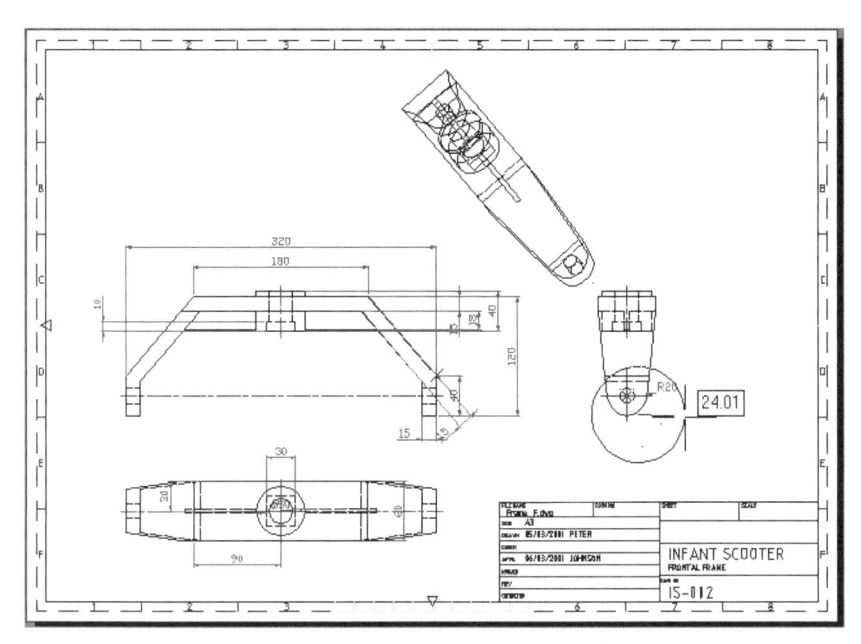

Figure 10–40 *Selection of a center and a circumferential point of the circular area to be enlarged*

22. Referring to Figure 10–41, select a location for the detail view.

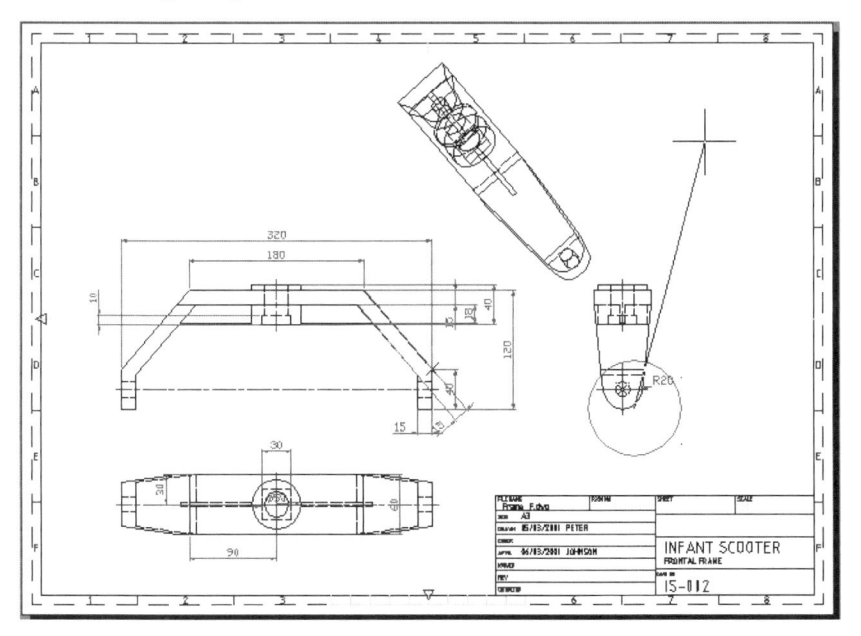

Figure 10–41 *Location of the detail view selected*

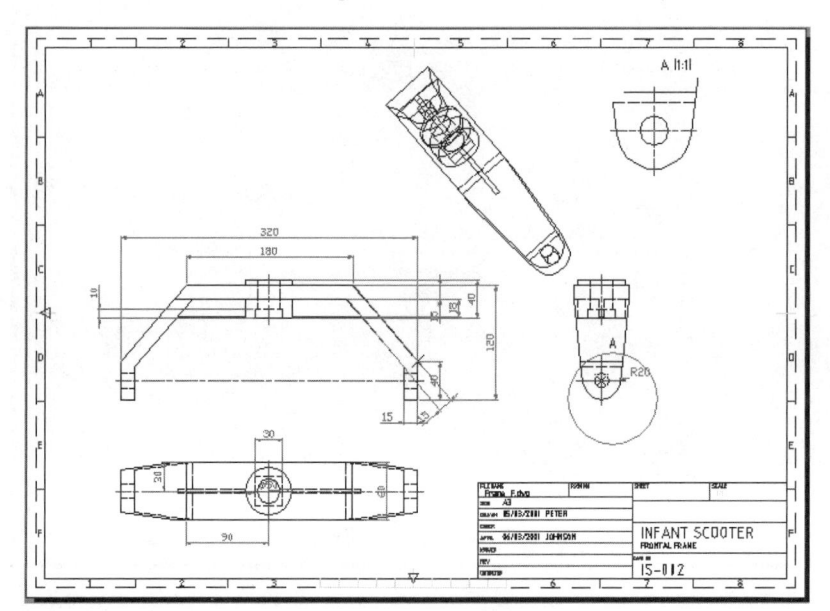

A detail view is constructed (see Figure 10–42). Save and close your file.

Figure 10–42 *Detail view constructed*

BROKEN VIEW

Now you will learn how to construct a broken view. When considering the scale of engineering drawing views, a dilemma exists if the component is a long and thin object. It is a normal engineering practice to remove the central portion of the drawing view of a very long object and move the remaining portions of the drawing view toward each other to minimize the space requirement of the drawing view. This is called a broken view.

1. Open the file *Axle.dwg* that you constructed in Chapter 3.
2. Select the Drawing tab from the Browser.
3. Select Layout1 from the Drawing tab of the Browser, right-click, and select Page Setup.
4. In the Page Setup dialog box, select a plotting device and a paper size, and select the OK button.
5. Insert a title block.
6. Select Layout1 from the Browser, right-click, and select New View.
7. In the Create Drawing View dialog box (shown in Figure 10–43), select Broken view, set the scale to 1, set the distance between the subviews (break gap) of a broken view to 20 mm, and select the OK button.
8. Type **Y** to use the world XY plane.

9. Type **X** to use the world X axis.
10. Right-click to accept.

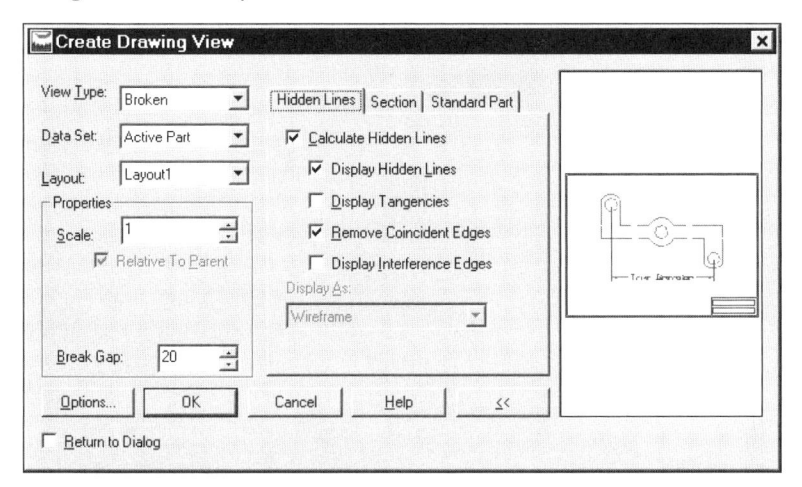

Figure 10–43 *Creating a broken view*

11. Referring to Figure 10–44, select a location to position the broken view.

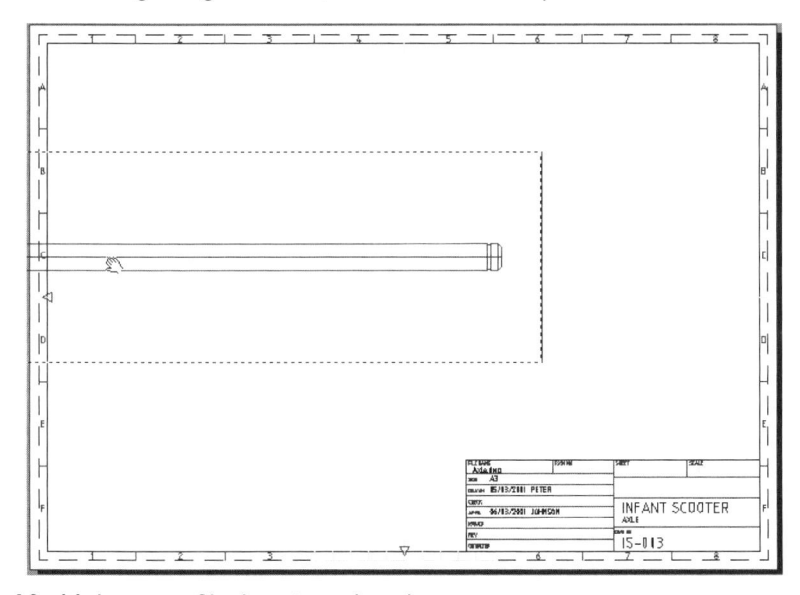

Figure 10–44 *Location of broken view selected*

When a temporary view is displayed on your screen, select a vertex of this view. After selecting a vertex on the temporary view, you are required to drag a rectangular box to define a subview.

12. Select a vertex to position the subview and select two points to describe a rectangle to define the subview (see Figure 10–45).

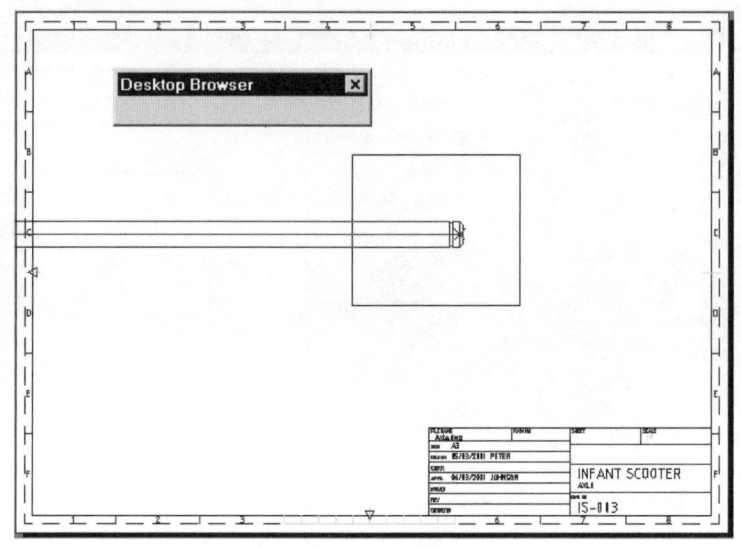

Figure 10–45 *Vertex and rectangular subview defined*

A broken view has at least two subviews. Now you will select a vertex and drag a rectangle to define the second subview.

13. Referring to Figure 10–46, select a vertex and select two points to define the second subview.

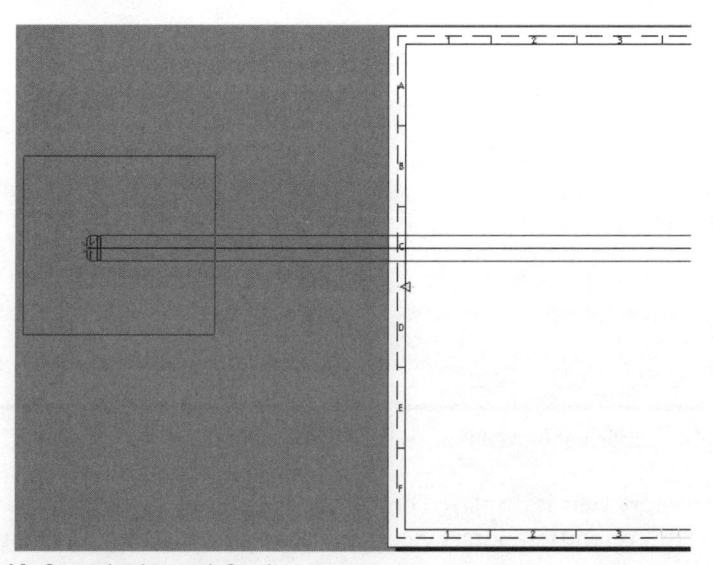

Figure 10–46 *Second subview defined*

Two subviews are defined. You may continue to define the third subview or press the ENTER key to exit.

14. Press the ENTER key.

A broken view with two subviews is constructed (see Figure 10–47). Save and close your file.

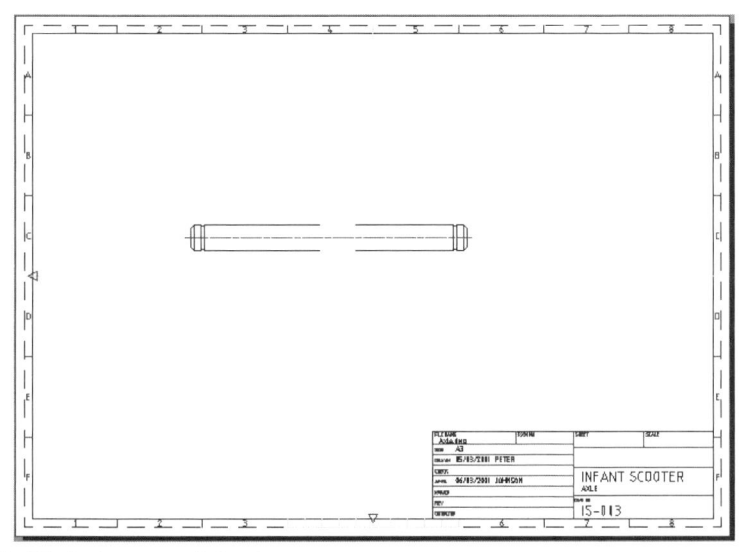

Figure 10–47 *Broken view defined*

SECTION VIEW

There are six major kinds of section views: full section, half section, offset section, aligned section, breakout section, and radial section. Now you will learn how to construct section views one by one.

Full Section

Construction of a full section view requires a plane or work plane on the model to define a cutting plane. Now you will construct a full section view.

1. Open the file *ScrewCap.dwg* that you constructed in Chapter 3.
2. Referring to Figure 10–48, construct a work axis.

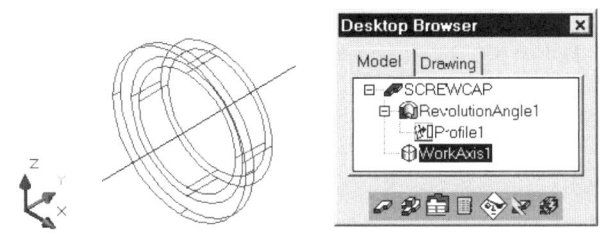

Figure 10–48 *Work axis constructed*

3. Construct a work plane on the world YZ plane (see Figure 10–49).

Figure 10–49 *Work plane on the world YZ plane constructed*

4. Construct a work plane parallel to the work plane on the world YZ plane and passing through the work axis (see Figure 10–50).

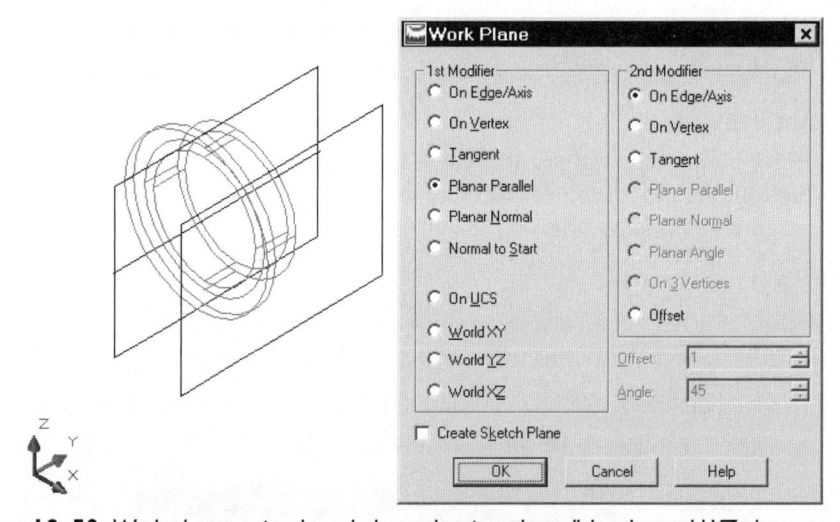

Figure 10–50 *Work plane passing through the work axis and parallel to the world YZ plane constructed*

5. Select the Drawing tab from the Browser.
6. Select Layout1 in the Drawing tab of the Browser, right-click, and select Page Setup.
7. In the Page Setup dialog box, select a plotting device and a paper size, and select the OK button.

8. Insert a title block.
9. Referring to Figure 10–51, construct a base view with a scale of 2.

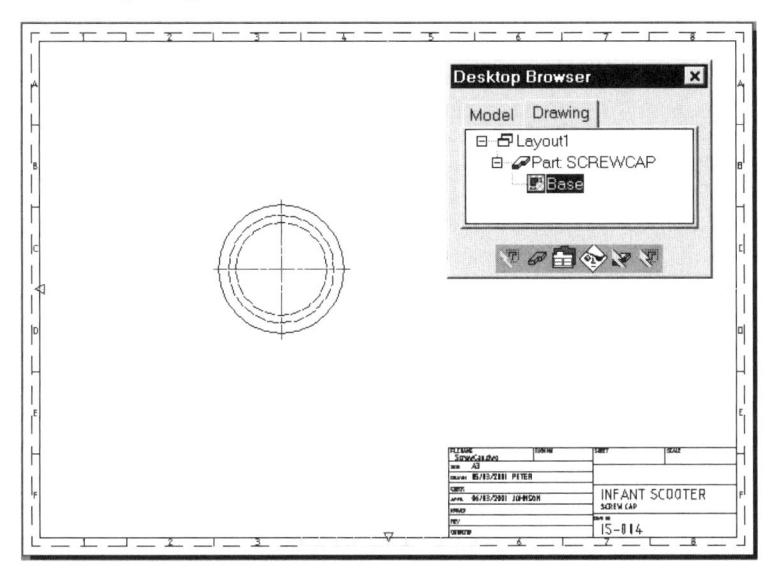

Figure 10–51 *Base view constructed*

10. Select the base view from the Browser, right-click, and select New View.
11. In the Create Drawing View dialog box (shown in Figure 10–52), select Ortho view, select the Section tab, select Full type, and select the Pattern button.

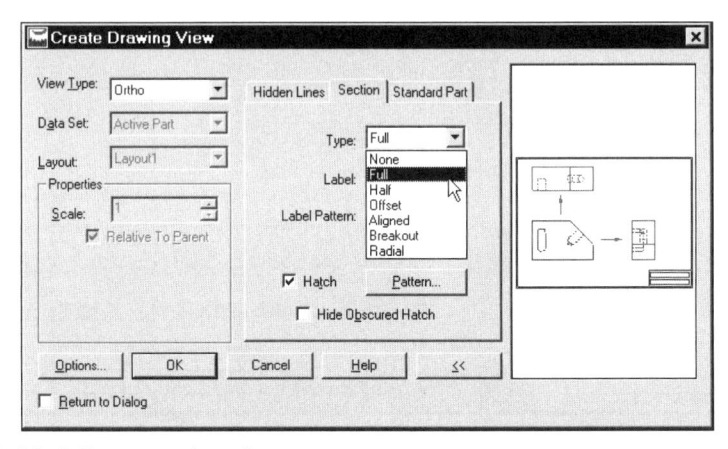

Figure 10–52 *Full section selected*

12. In the Hatch Pattern dialog box (shown in Figure 10–53), select Line pattern, set angle to 45, and select the OK button.
13. After returning to the Create Drawing View dialog box, select the OK button.

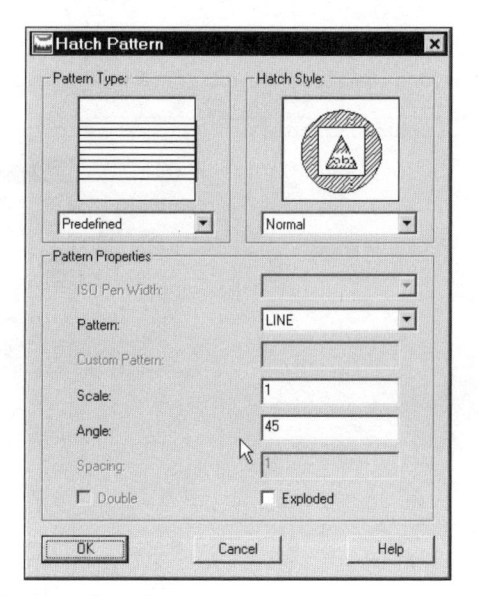

Figure 10–53 *Hatch Pattern selected*

14. Select a location for the section view (see Figure 10–54).

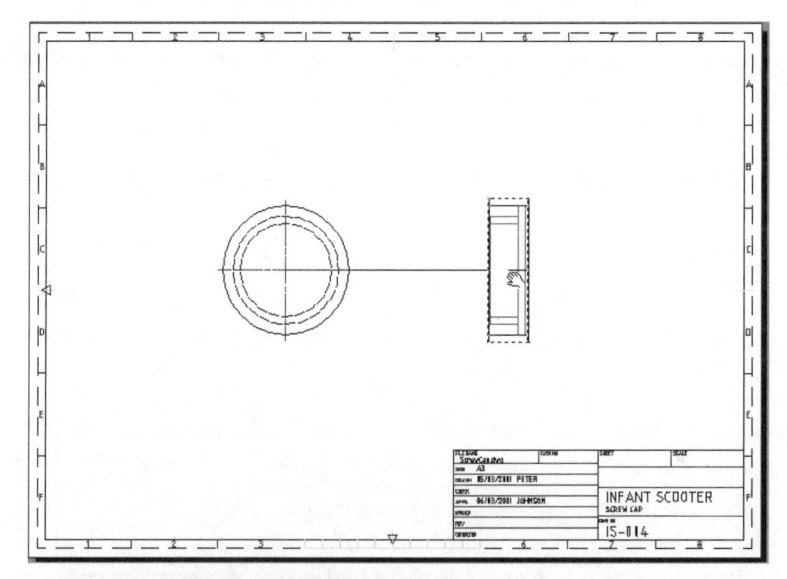

Figure 10–54 *Location of the section view selected*

15. Press the ENTER key to use the Workplane option.
16. Select the work plane from the parent view (see Figure 10–55).

A full section view is constructed (see Figure 10–56). Save and close your file.

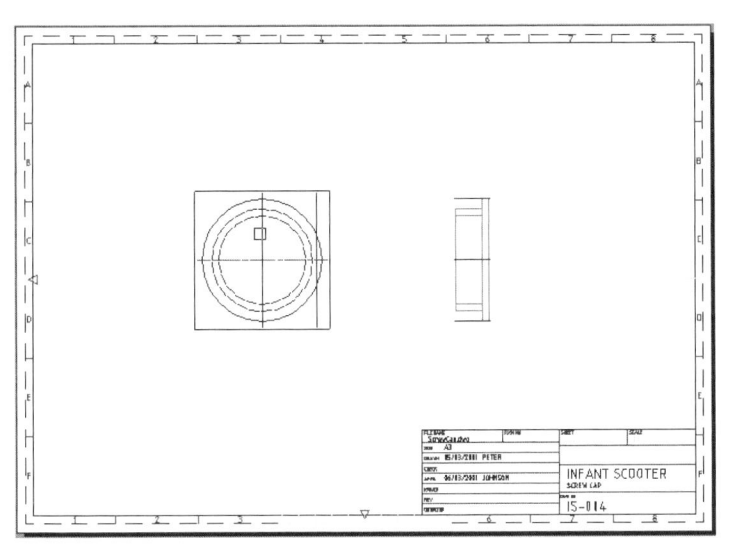

Figure 10–55 *Work plane being selected*

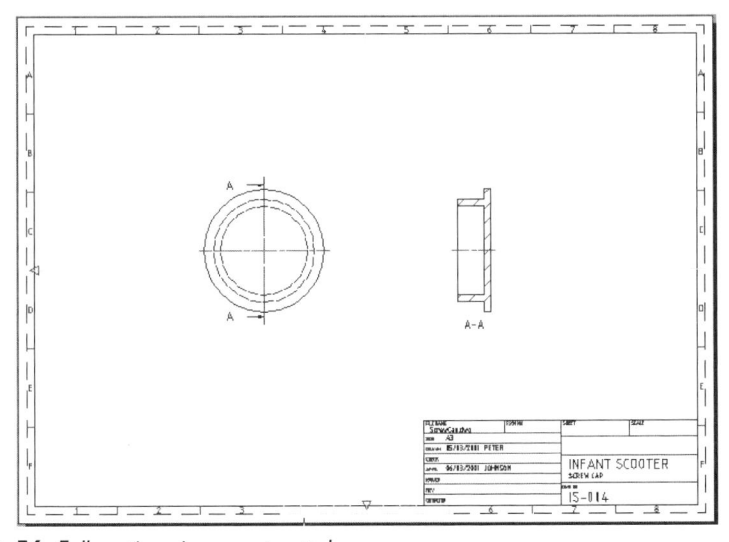

Figure 10–56 *Full section view constructed*

Half Section

Now you will construct a half-section view. Construction of a half-section view requires two work planes.

1. Open the file *Wheel.dwg* that you constructed in Chapter 4.
2. Referring to Figure 10–57, construct a work axis, a work plane passing through the work axis and parallel to the world XY plane, and a work plane passing through the work axis and parallel to the world ZX plane.

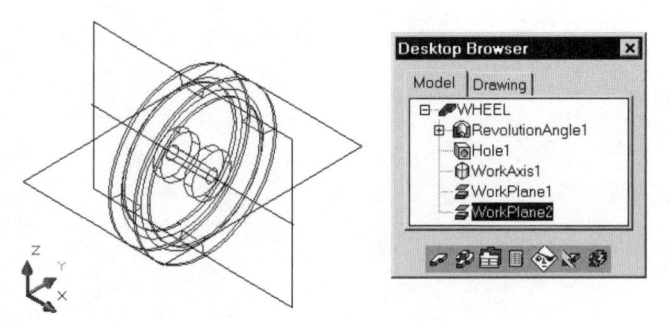

Figure 10–57 Work axis and work planes constructed

3. Select the Drawing tab from the Browser.
4. Select Layout1 from the Drawing tab of the Browser, right-click, and select Page Setup.
5. In the Page Setup dialog box, select a plotting device and a paper size, and select the OK button.
6. Insert a title block.
7. Using a scale of 1, construct a base view with reference to Figure 10–58.

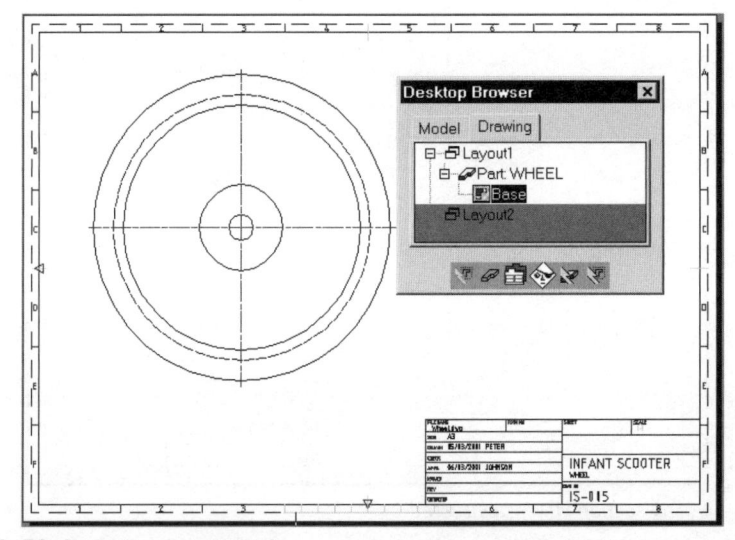

Figure 10–58 Base view constructed

8. Select the base view from the Browser, right-click, and select New View.
9. In the Create Drawing View dialog box, select Ortho view and select the Section tab.
10. In the Section tab, select Half type and select the Pattern button.
11. In the Hatch Pattern dialog box, select Line, set angle to 45, and select the OK button.

12. Select the OK button from the Create Drawing View dialog box.
13. Referring to Figure 10–59, select a location for the half section view.
14. Press the ENTER key to use the work plane option.
15. Select the work planes one by one (see Figure 10–60).

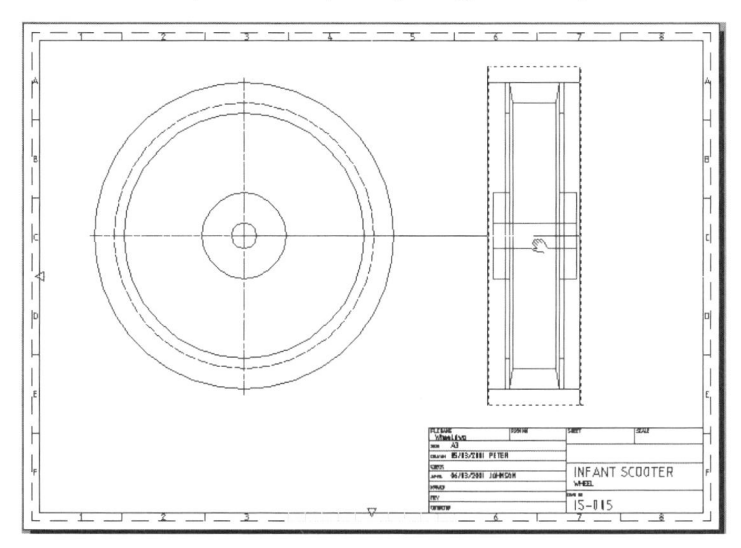

Figure 10–59 *Location of the half section view selected*

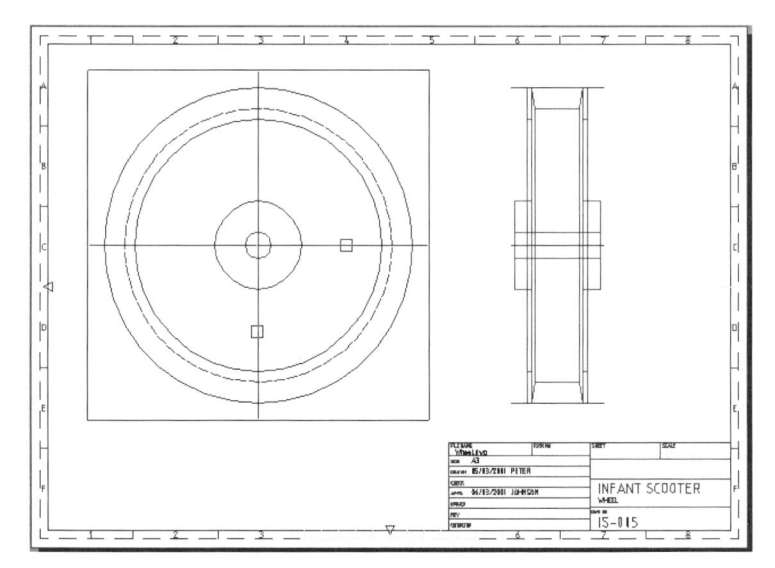

Figure 10–60 *Work planes selected*

16. Press the ENTER key to accept.

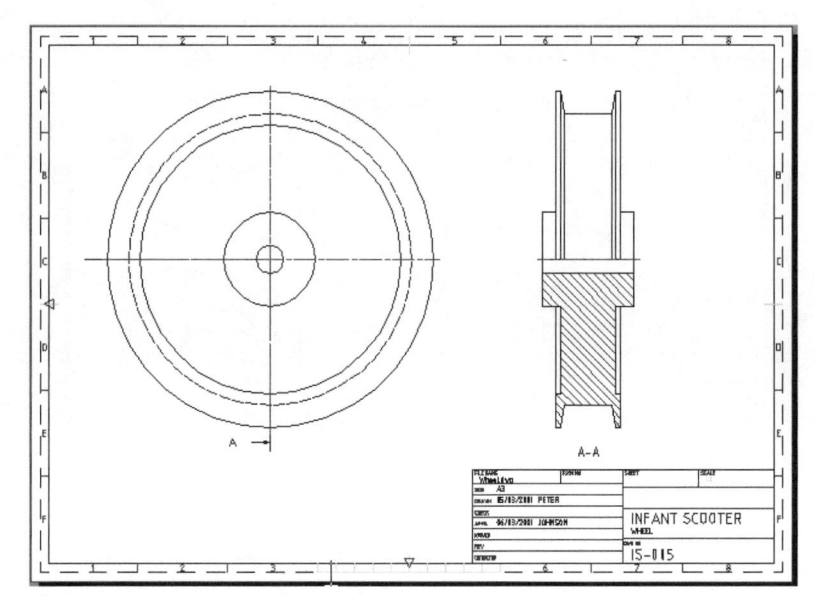

A half section view is constructed (see Figure 10–61). Save and close your file.

Figure 10–61 *Half section view constructed*

Offset Section

Now you will construct an offset-section view. Construction of an offset-section view requires a cut line sketch in the solid part.

1. Open the file *Combine.dwg* that you constructed in Chapter 9.
2. Set the display to the top view.
3. Referring to Figure 10–62, construct three line segments.

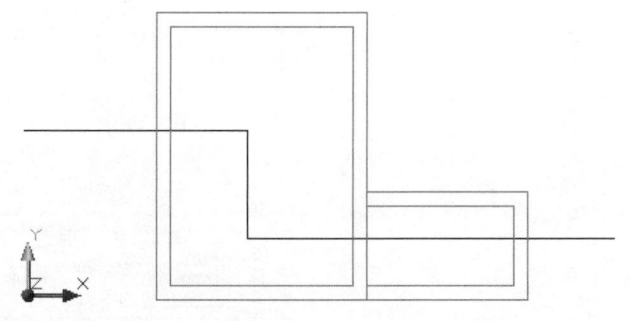

Figure 10–62 *Three line segments constructed*

4. Select Part>Sketch Solving>Cut Line
   Command: AMCUTLINE

5. Select the lines and press the ENTER key.
6. Add parametric dimensions to the cut line sketch in accordance with Figure 10–63.

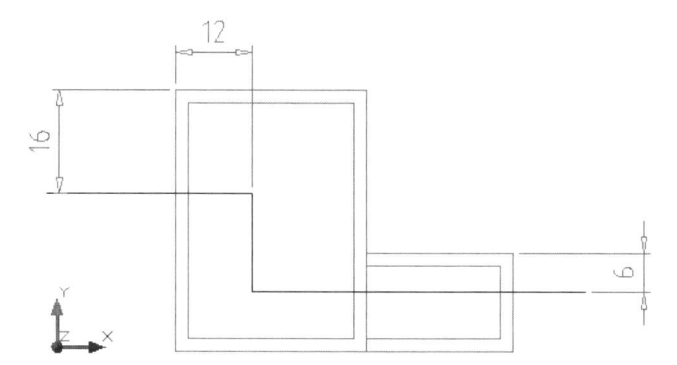

Figure 10–63 *Sketch resolved to a cut line sketch and parametric dimensions added*

7. Referring to Figure 10–64, setup the layout, insert a title block, and construct a base view with a scale of 2.

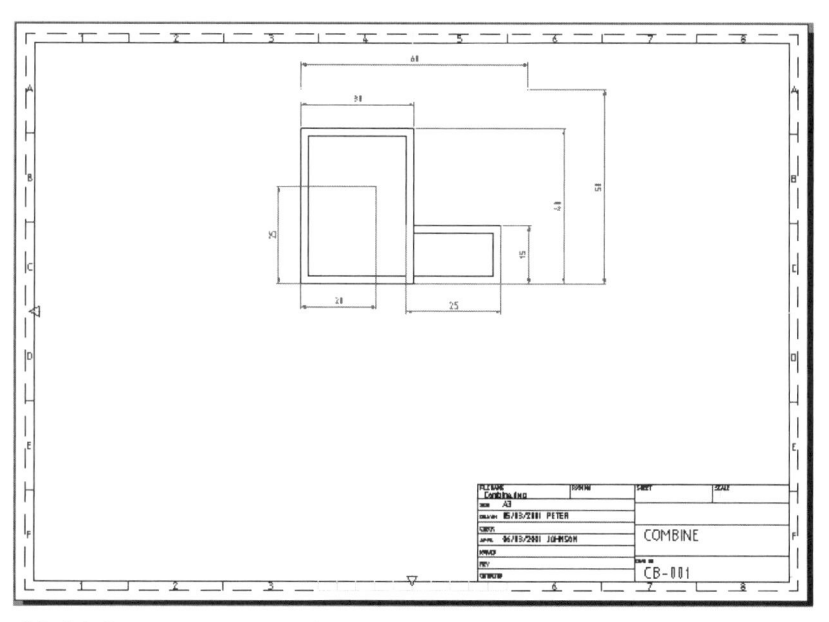

Figure 10–64 *Base view constructed*

8. Select the base view from the Browser, right-click, and select New View.
9. In the Create Drawing View dialog box, select Ortho type and select the Section tab.
10. In the Section tab, select Offset type, set hatch pattern, and select the OK button.

11. Referring to Figure 10–65, select a location for the offset section view.
12. In model mode, select the cut line sketch.

An offset section is constructed (see Figure 10–66). Save and close your file.

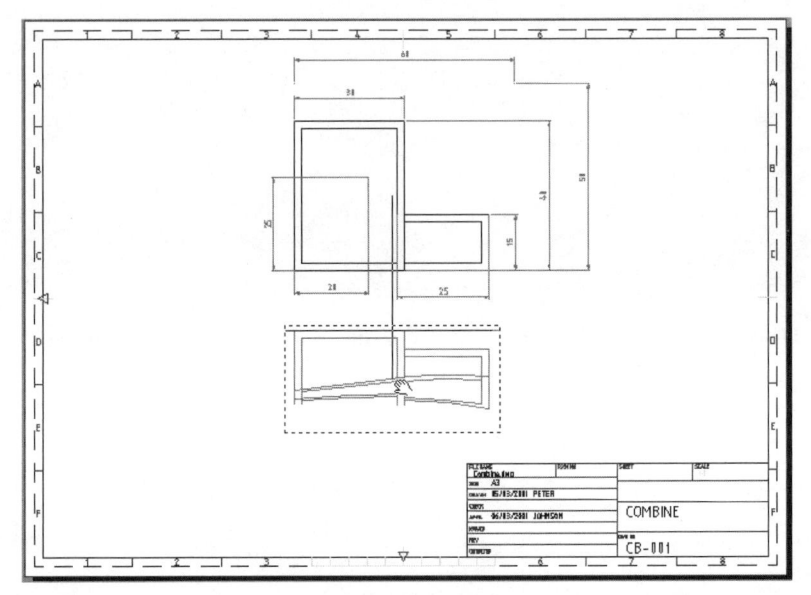

Figure 10–65 *Location of the offset section view selected*

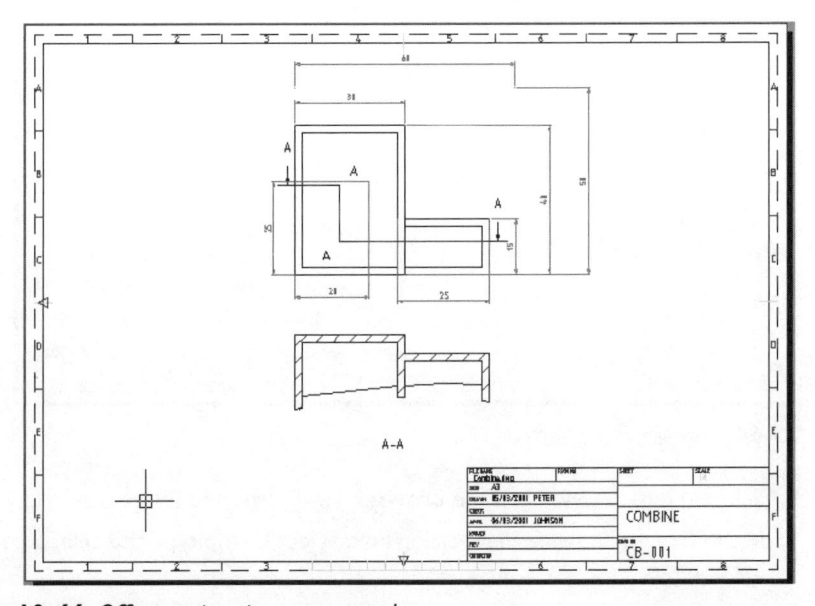

Figure 10–66 *Offset section view constructed*

Engineering Drafting I 531

Aligned Section

Now you will construct an aligned section view. Similar to an offset section, you will construct a cut line sketch to define the aligned section plane.

1. Open the file *Tire.dwg* that you constructed in Chapter 4.
2. Set the display to the front view.
3. Referring to Figure 10–67, establish a sketch plane, construct two line segments, resolve them to a cut line sketch, and add parametric dimensions.

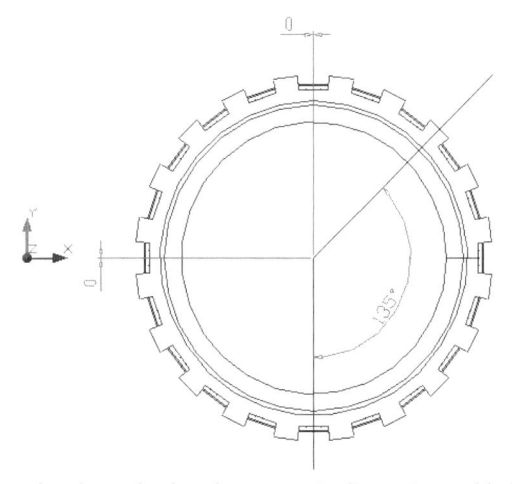

Figure 10–67 *Cut line sketch resolved and parametric dimensions added*

4. Referring to Figure 10–68, setup the layout, insert a title block, and construct a base view with a scale of 0.5.

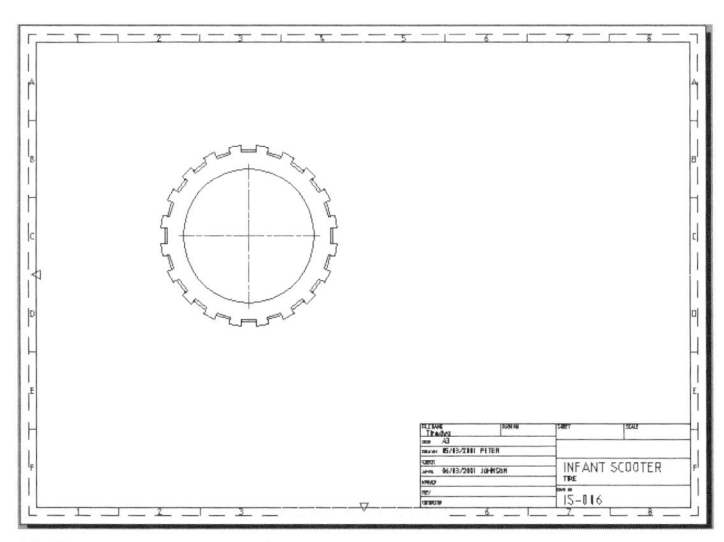

Figure 10–68 *Base view constructed*

5. Select Layout1, right-click, and select New View.
6. In the Create Drawing View dialog box, select Ortho type and select the Section tab.
7. In the Section tab, select Aligned type, set hatch pattern, and select the OK button.
8. Select the base view as the parent view and select a location for the aligned section view (see Figure 10–69).
9. Select the cut line in model mode.

An aligned section view is constructed (see Figure 10–70). Save and close your file.

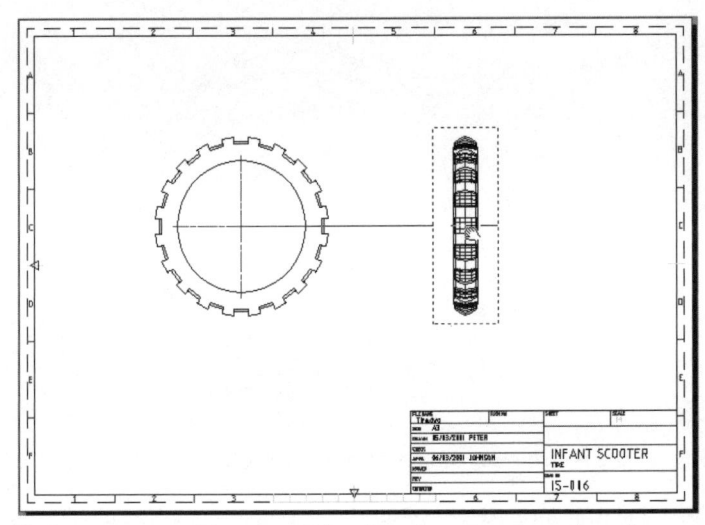

Figure 10–69 *Parent view and location of the aligned section view selected*

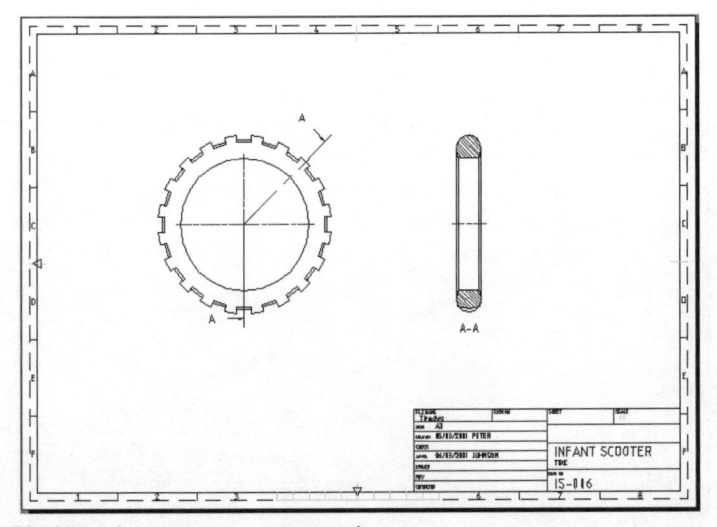

Figure 10–70 *Aligned section view constructed*

Breakout Section

Now you will construct a breakout section. To construct a breakout section view, construct a sketch in the solid part and resolve it to a break line sketch.

1. Open the file *Seat.dwg*.
2. Set the display to an isometric view.
3. Referring to Figure 10–71, construct a work axis.
4. Construct a work plane parallel to a vertical face and passing through the work axis (see Figure 10–72).

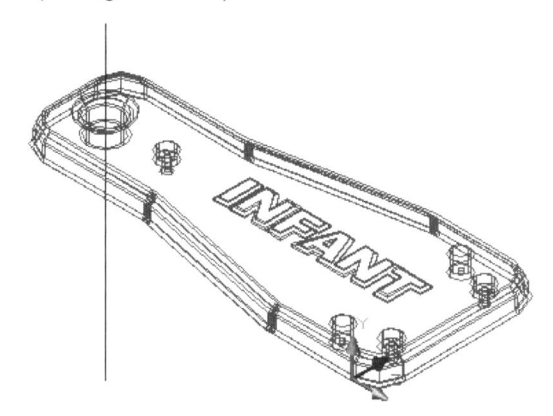

Figure 10–71 *Work axis constructed*

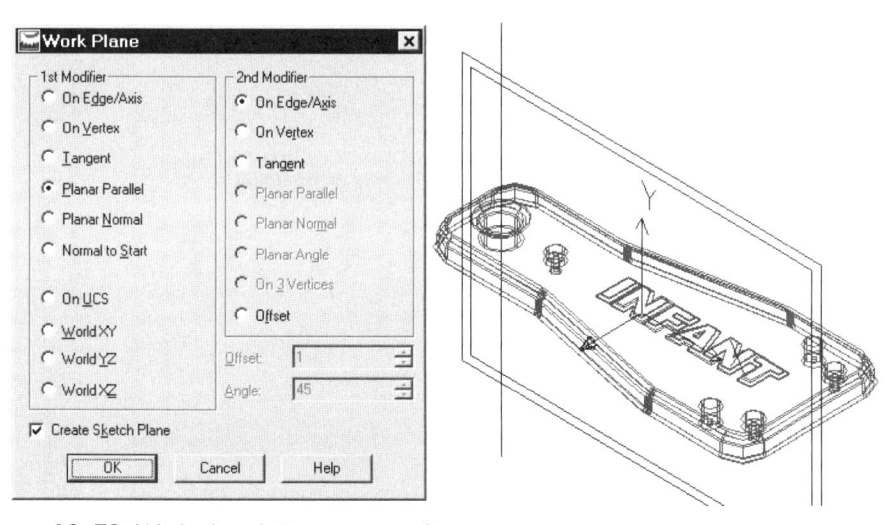

Figure 10–72 *Work plane being constructed*

5. Set the display to the front view.

6. Construct a circle (see Figure 10–73).
7. Select Part>Sketch Solving>Break Line

   Command: AMBREAKLINE
8. Select the circle and press the ENTER key.
9. Referring to Figure 10–74, add parametric dimensions.
10. Select the Drawing tab from the Browser.

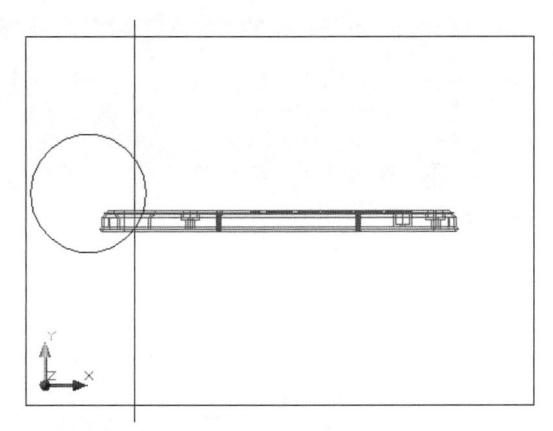

Figure 10–73 *Circle constructed*

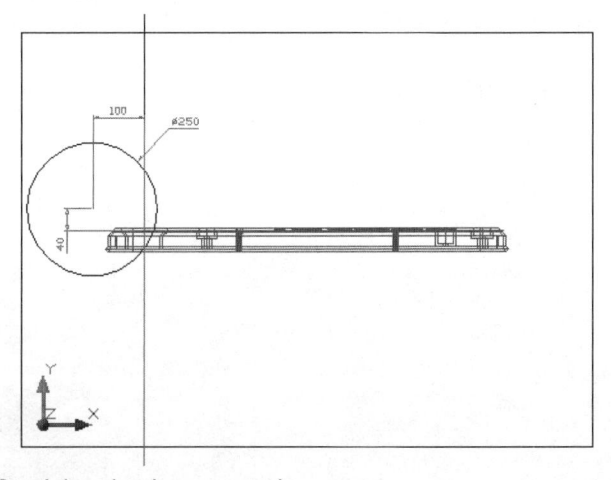

Figure 10–74 *Break line sketch constructed*

11. Referring to Figure 10–75, set up the layout, insert a title block, and construct a base view with a scale of 0.2.
12. Select the base view from the Browser, right-click, and select New View.
13. In the Create Drawing View dialog box, select Ortho type and select the Section tab.

Engineering Drafting I 535

14. Select Breakout type in the Section tab and select the OK button.
15. Select a location for the breakout section view and press the ENTER key (see Figure 10–76).
16. Select the break line in the model and press the ENTER key.

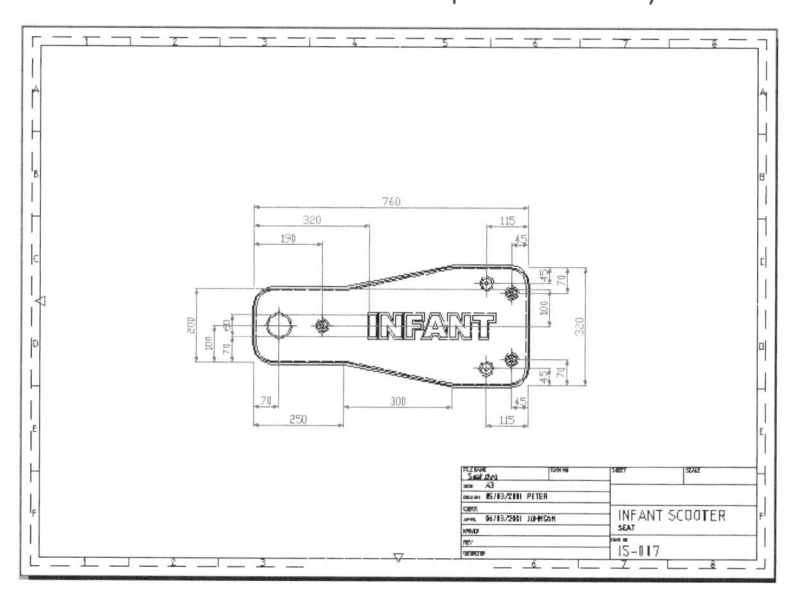

Figure 10–75 *Base view constructed*

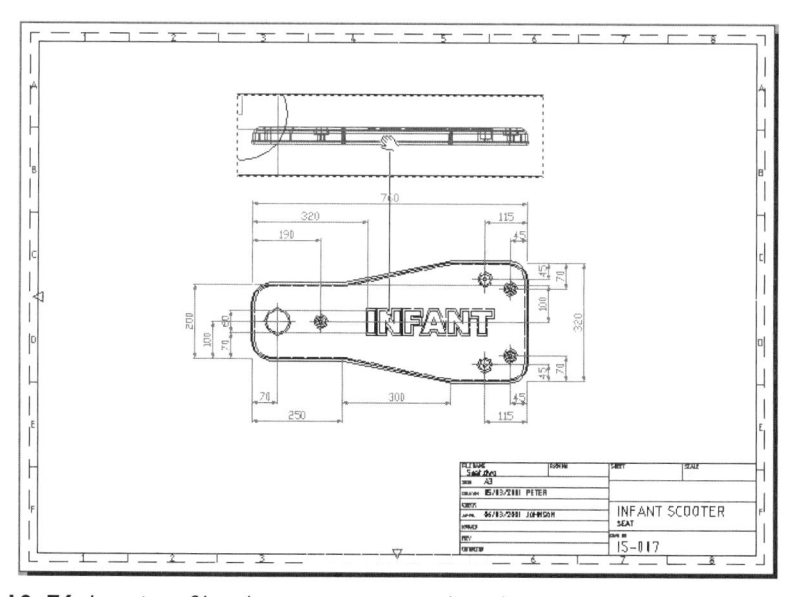

Figure 10–76 *Location of break out section view selected*

A breakout section view is constructed (see Figure 10–77). Save and close your file.

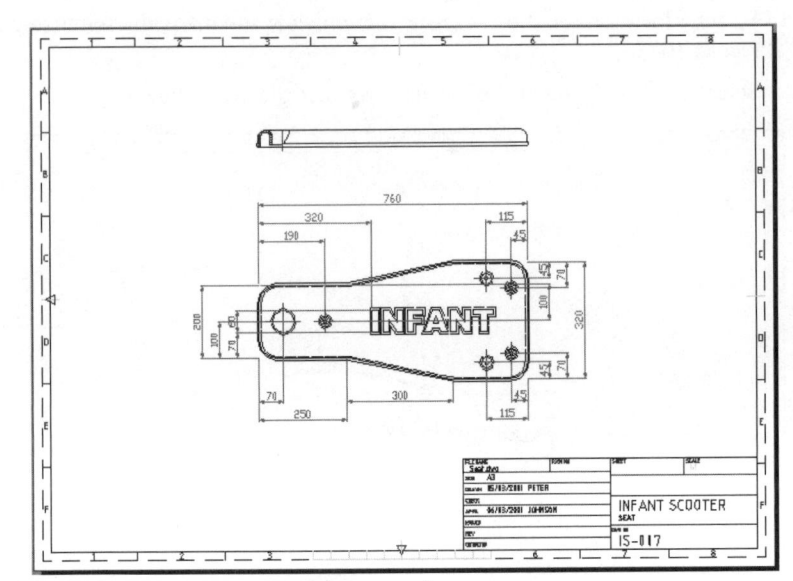

Figure 10–77 *Break out section view constructed*

Radial Section

A radial section view rotates an auxiliary section view around a selected point in the view. Now you will construct a radial section view.

1. Open the file *SteeringWheel.dwg* that you constructed in Chapter 4.
2. Referring to Figure 10–78, unhide four work planes.

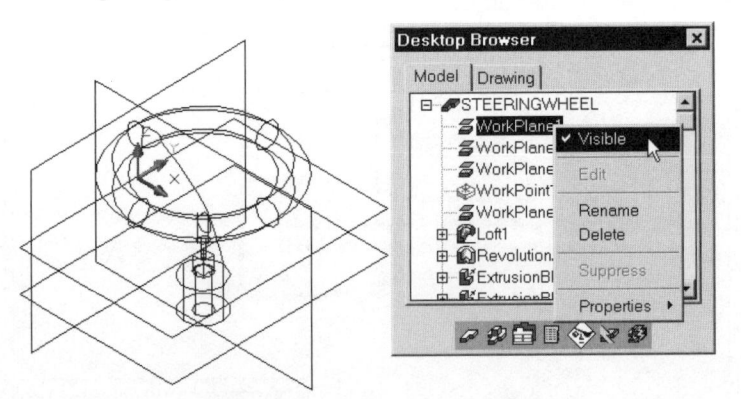

Figure 10–78 *Work planes unhidden*

3. Select the Drawing tab from the Browser.
4. Referring to Figure 10–79, set up the layout, insert a title block, and construct a base view with a scale of 0.5.

Engineering Drafting I 537

5. Select the base view from the Browser, right-click, and select New View.
6. In the Create Drawing View dialog box, select Ortho type and select the Section tab.
7. Select Radial type in the Section tab and select the OK button.
8. Select a location to position the radial section view (see Figure 10–80).

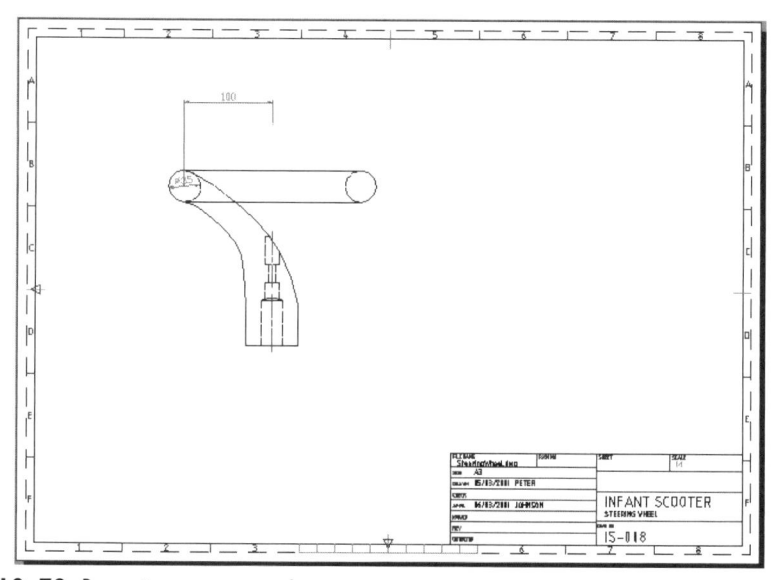

Figure 10–79 *Base view constructed*

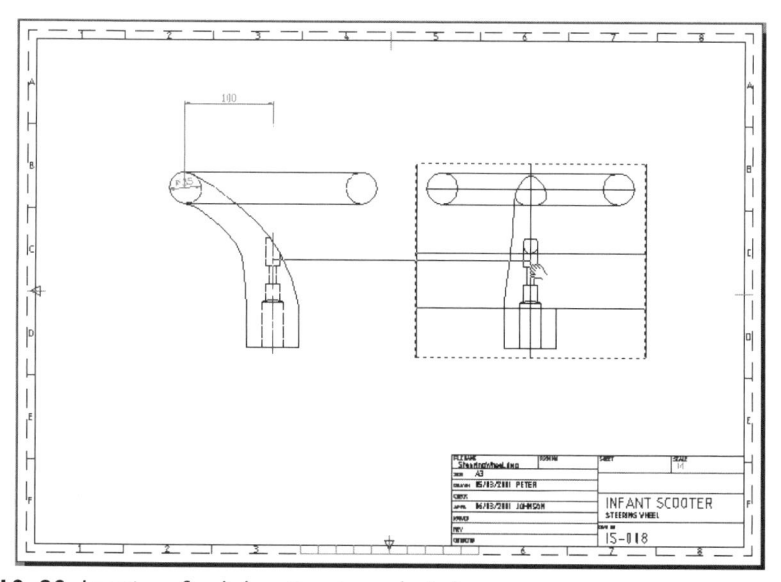

Figure 10–80 *Location of radial section view selected*

9. Select the work plane indicated in Figure 10–81 to specify the section plane.
10. Press the ENTER key to align the section view with the mid-point of the section.

A radial section view is constructed (see Figure 10–82). Save your file.

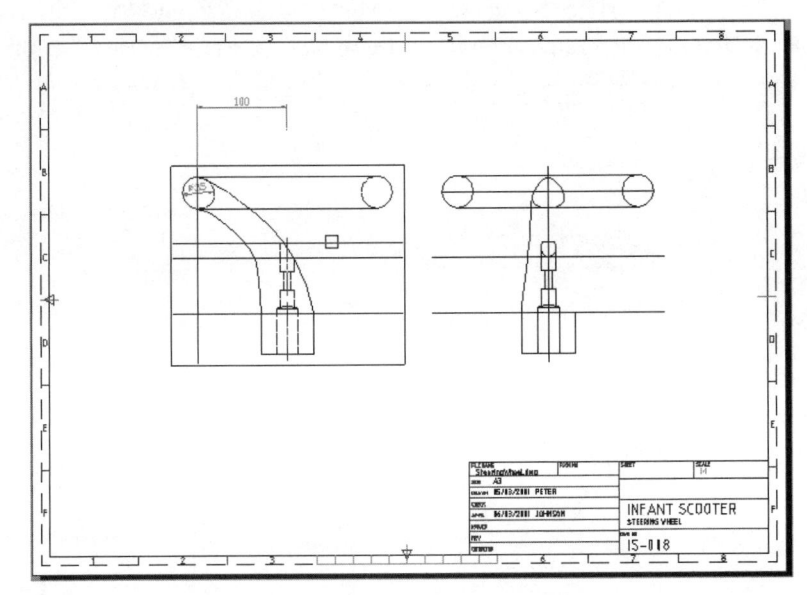

Figure 10–81 *Work plane selected*

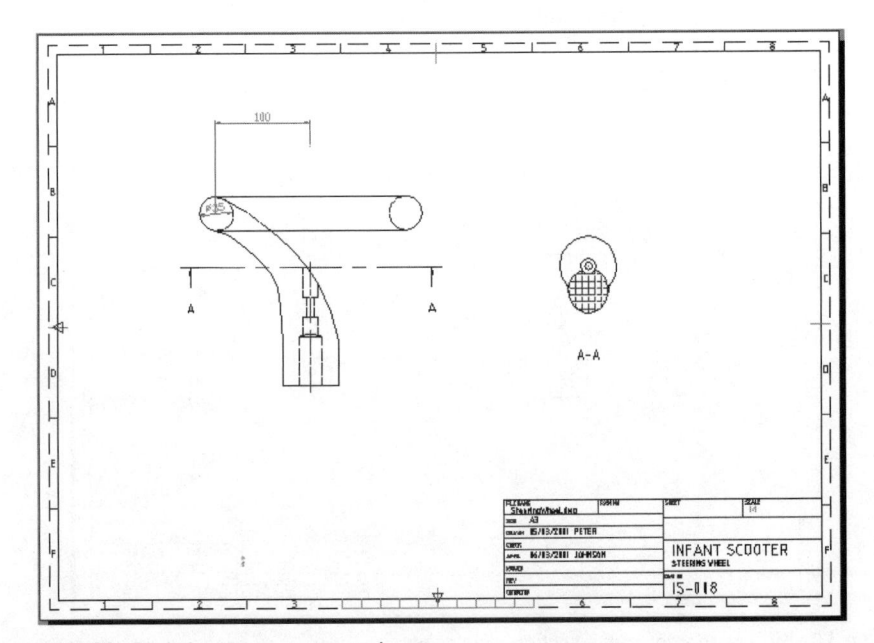

Figure 10–82 *Radial section constructed*

ANNOTATIONS

An engineering drawing is a 2D graphical representation of the 3D object. To better elaborate design ideas, intents, and requirements, annotations are required. Annotations that you add to a part drawing are center lines, textual information, dimensions, tolerances, and surface finishes. Before you add annotations to a drawing, it is important to select the relevant national or international standard by using the Standard tab of the Mechanical Options dialog box (refer back to Figure 10–10).

CENTERLINES

Centerlines can be added to drawing views automatically or manually. Basically, they are automatically placed to cylindrical features of a solid part as the drawing views are constructed. To add centerlines manually, you select a circular object and press the ENTER key or select two straight edges and specify the end points of the centerline.

Now you will add centerlines manually.

1. Open the file *SteeringWheel.dwg*, if you have already closed it.
2. Select Annotate>Annotation>Centerline

    Command: AMCENLINE
3. Select the circular object X (Figure 10–83) and press the ENTER key.

A pair of center lines is constructed (see Figure 10–84).

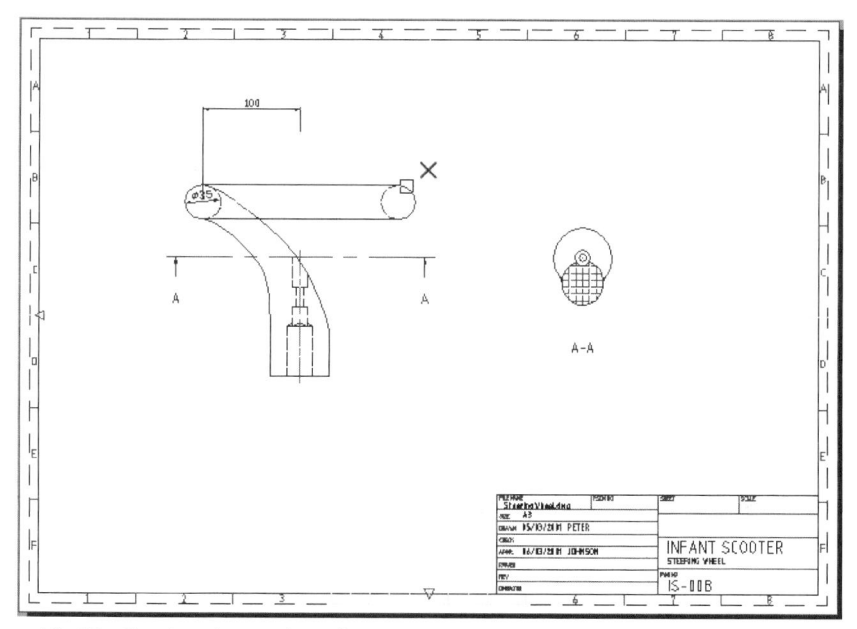

Figure 10–83 *Circular object selected*

4. Repeat the AMCENLINE command.

5. Referring to Figure 10–84, select two edges (X and Y) one by one.

A centerline with infinite length is placed temporarily. Now you will specify the end points of the centerline.

6. Select two points to indicate the end points of the center line (see Figure 10–85).

A center line is added between the selected edges.

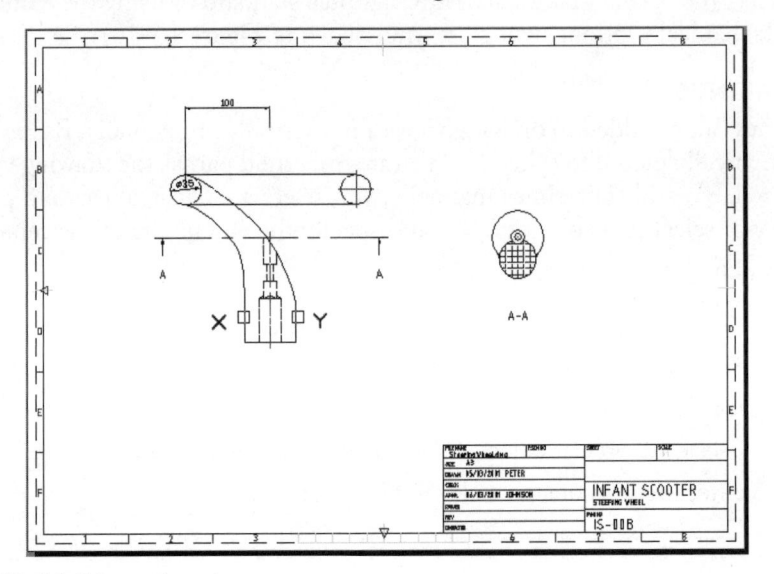

Figure 10–84 *Edges selected*

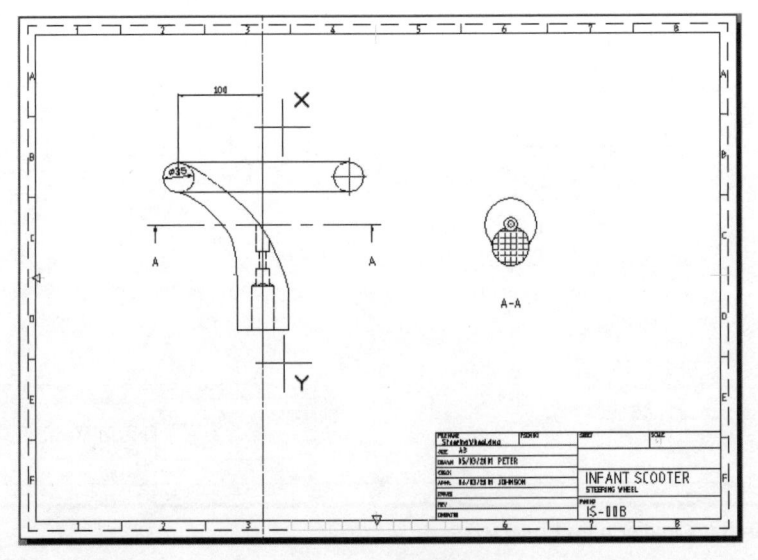

Figure 10–85 *Length of center line being determined*

TEXT

Construction of text objects in the drawing involves two steps: constructing a text string and associating the text string to one of the drawing views.

7. Select Annotate>Text>Paragraph Text
8. Referring to Figure 10–86, select two points to describe a rectangular area to place the paragraph text.

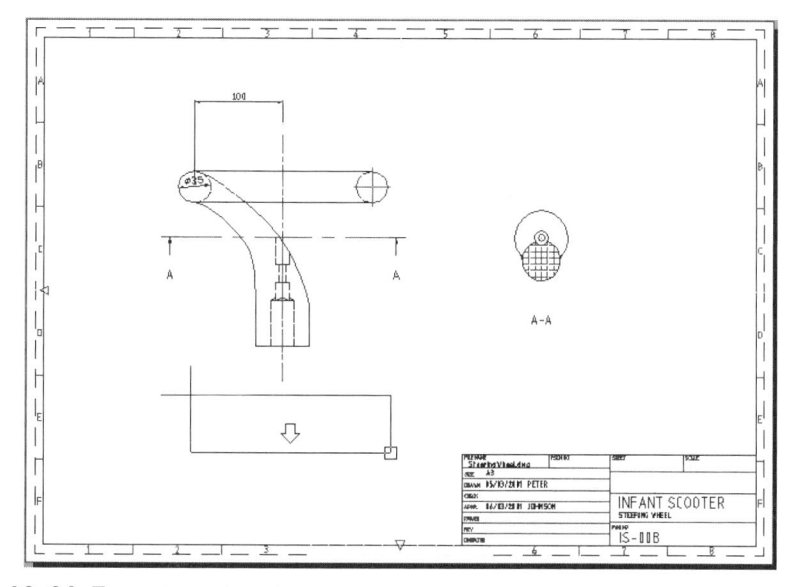

Figure 10–86 *Two points selected*

9. In the Multiline Text Editor, input a paragraph of text string and select the OK button.

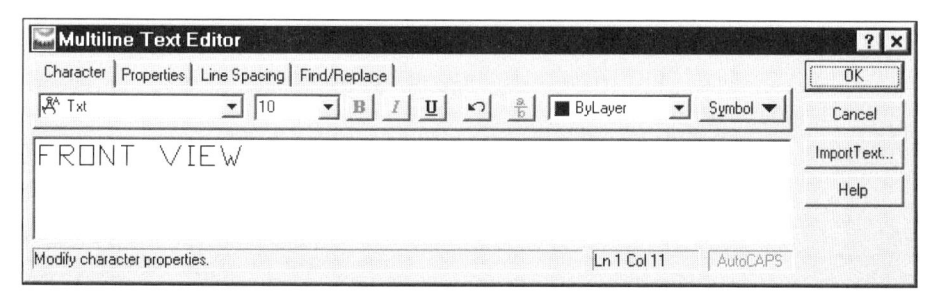

Figure 10–87 *Text string typed into the Multiline Text Editor dialog box*

A paragraph text is constructed. Now you will associate the text with a drawing view.

10. Select Annotate>Annotation>Create Annotation
11. Select the text and press the ENTER key.

12. Referring to Figure 10–88, select a vertex.

The text is associated with the drawing view. If you move the drawing view now, the associated text string will move as well.

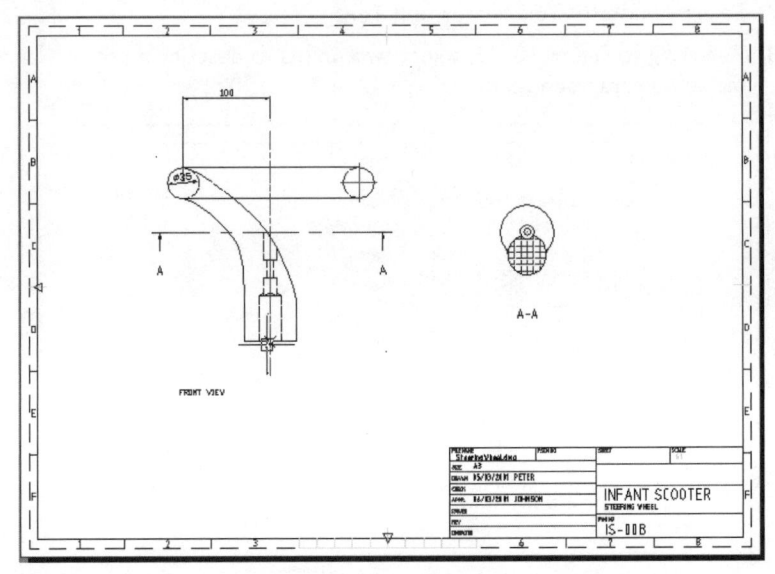

Figure 10–88 *Text string associated with a drawing view*

LEADER

A leader is a text string together with an arrow. Now you will add a leader to the drawing. Like working on text objects, you should also associate the leader with a drawing view.

13. Select Annotate>Annotation>Leader

    Command: QLEADER

14. Referring to Figure 10–89, select two points to indicate the first and second leader point, and press the ENTER key.
15. Press the ENTER key to accept the default text width.
16. Type the text string at the command line and press the ENTER key.

A leader is constructed (see Figure 10–90). Now you will associate the leader with the drawing view.

17. Select Annotate>Annotation>Create Annotation
18. Select the leader and press the ENTER key.
19. Select a vertex from a drawing view.

The leader is associated. Save and close your file.

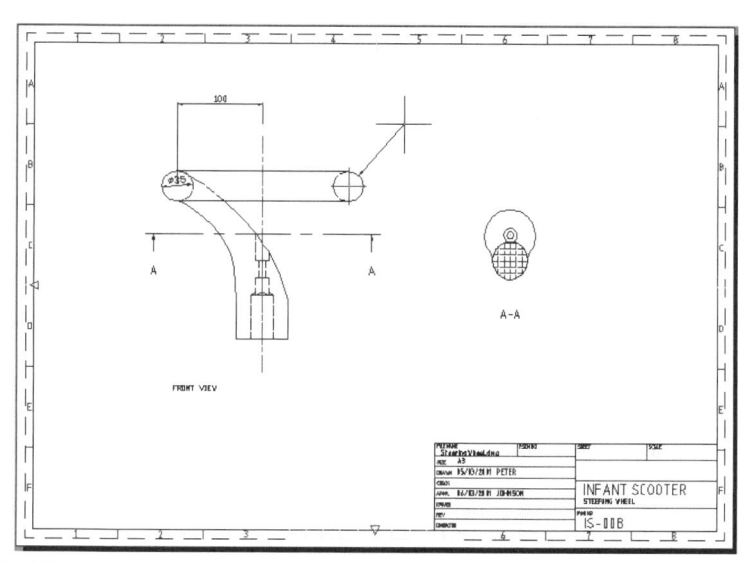

Figure 10-89 *Leader's first and second points selected*

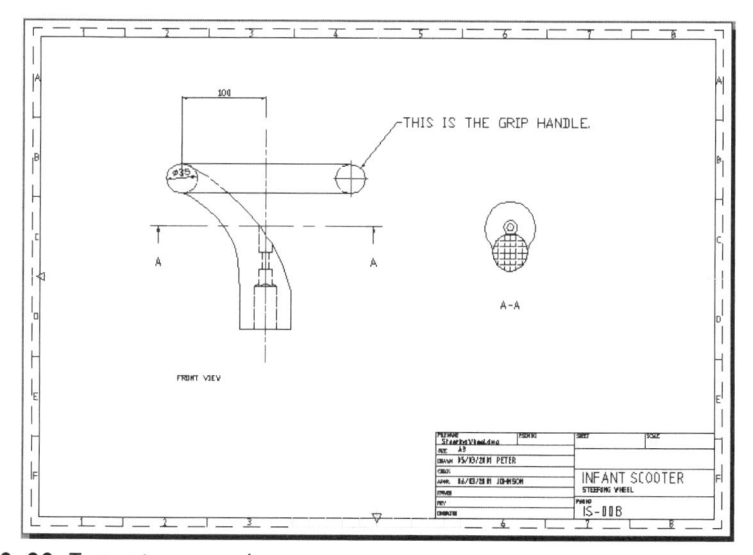

Figure 10-90 *Text string entered*

HOLE NOTE

A hole note is fully parametric textual information that states the parameters of the selected hole feature.

1. Open the file *Frame_F.dwg*.
2. Select Annotate>Annotation>Hole Note

    Command: AMNOTE

3. Referring to Figure 10–91, select the hole in the detail view and select a point to indicate the end point of the hole note leader.

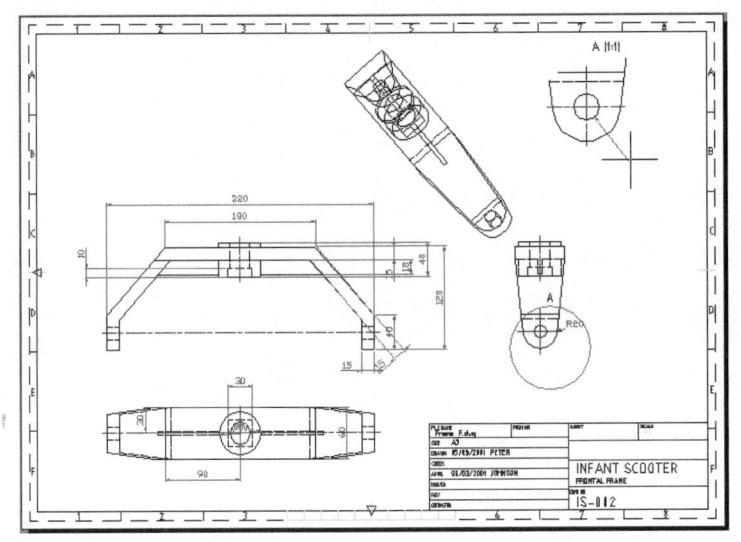

Figure 10–91 *Hole feature and end point of hole note selected*

4. Press the ENTER key (see Figure 10–92).

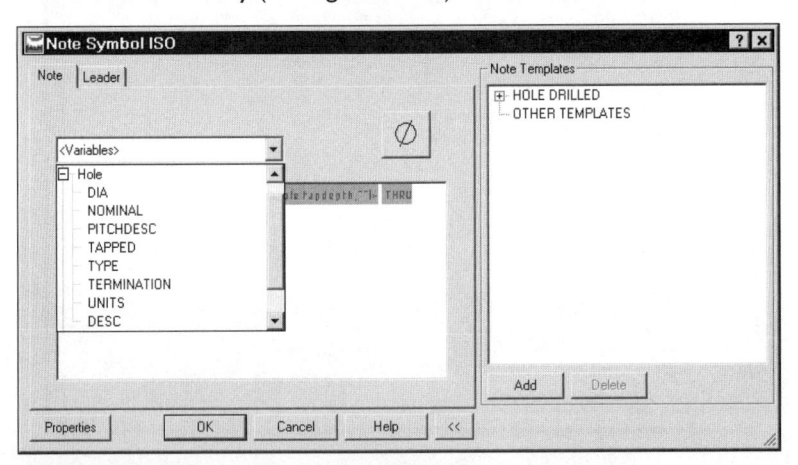

Figure 10–92 *Note tab of the Note Symbol dialog box*

The Note Symbol dialog box has two tabs: Note and Leader. In the Note tab, you select the variables from the list and insert them to the note text area.

5. Select the Leader tab (see Figure 10–93).

Here you specify the orientation, justification, and arrowhead of the leader.

6. Select the OK button.

A hole note is constructed (see Figure 10–94). Save and close your file.

Figure 10–93 *Leader tab of the Note Symbol dialog box*

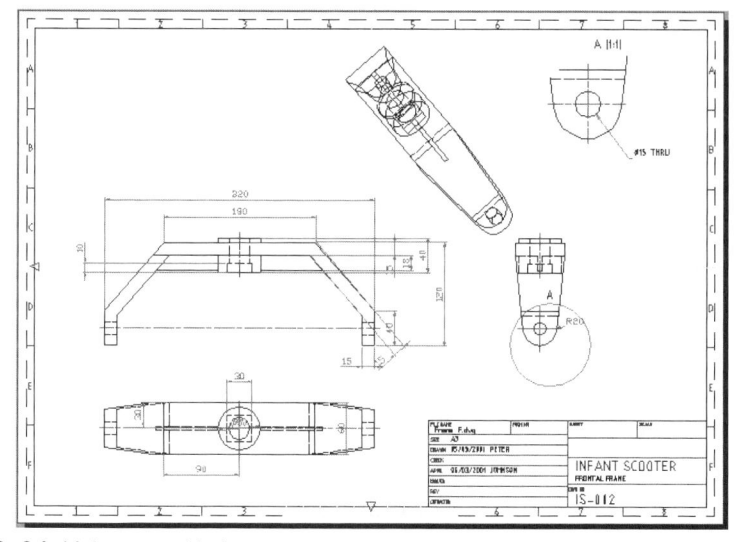

Figure 10–94 *Hole note added*

HOLE CHART

Design of complex components involves the use of a lot of hole features of various types. To save time in annotating these hole features, you can list them in a hole chart. Now you will construct a hole chart.

1. Open the file *NativePlate.dwg* that you constructed in Chapter 9.
2. Referring to Figure 10–95, insert a title block and construct a base view.
3. Select Annotate>Hole Charts

   Command: AMHOLECHART

4. Press the SHIFT key and right-click.
5. Select Endpoint from the right-click menu.
6. Select the end point indicated in Figure 10–96 to specify the origin.

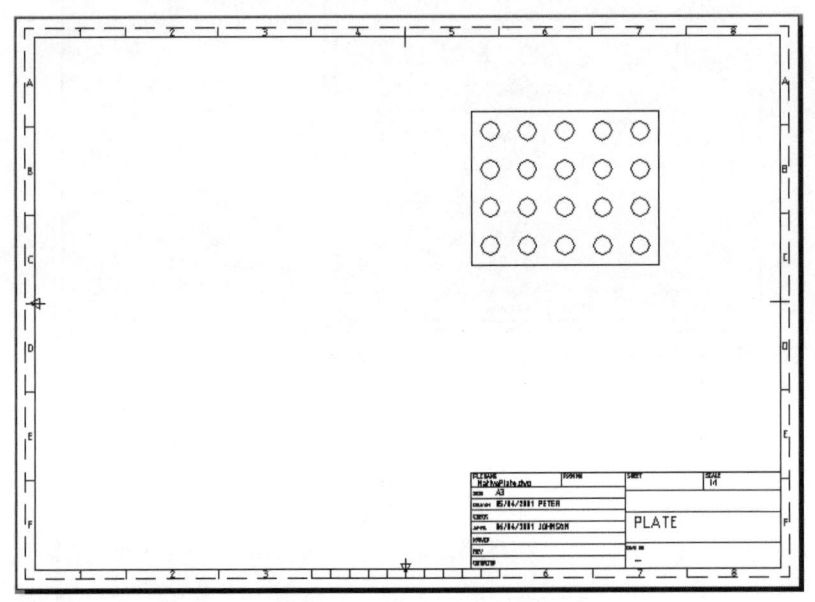

Figure 10–95 *Drawing view constructed*

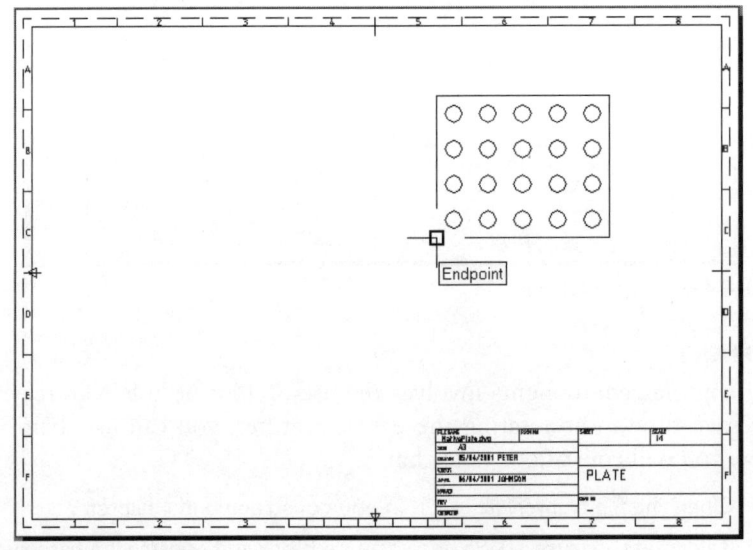

Figure 10–96 *Origin selected*

7. Press the ENTER key to use the default rotation angle.

8. Press the ENTER key to use the default origin name.
9. Select the circular objects in the drawing view and press the ENTER key.
10. Press the SHIFT key, right-click, and select Endpoint.
11. Select the end point indicated in Figure 10–97 to specify the location of the hole chart.

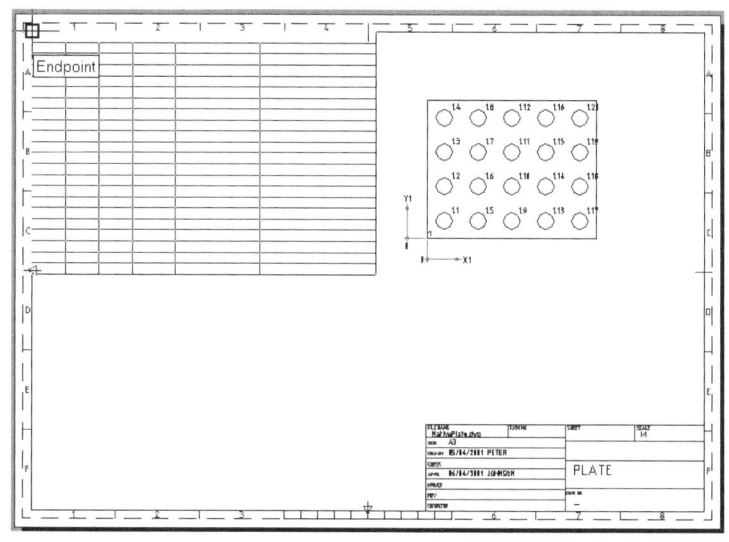

Figure 10–97 *Insertion point of hole chart selected*

A hole chart is constructed (see Figure 10–98). Save and close your file.

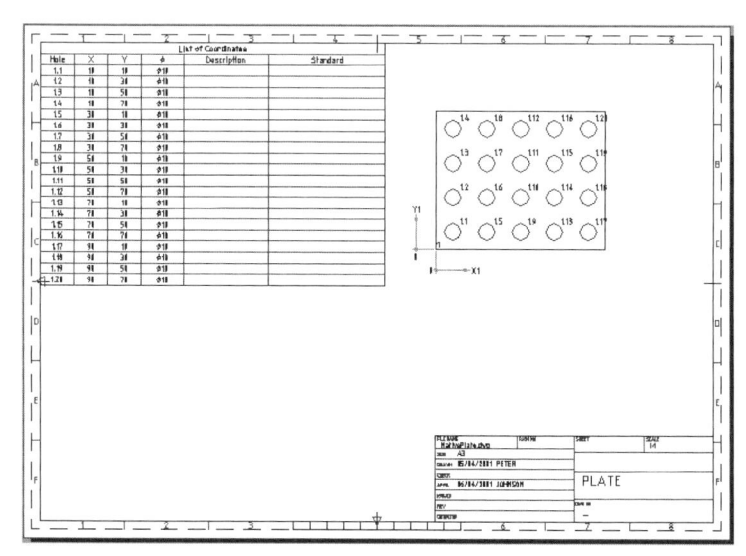

Figure 10–98 *Hole chart constructed*

DIMENSIONS

There are two kinds of dimensions in an engineering drawing generated from parametric 3D solids: parametric dimensions and reference dimensions.

Parametric Dimensions

Parametric dimensions are dimensions that you use to construct the features of the parametric solid part. Depending on the settings of the Mechanical Options dialog box, they appear automatically in the engineering drawing. Because the 2D engineering drawing in drawing mode is associative to the 3D objects in model mode, changes in the 3D objects cause corresponding changes in the 2D drawings. In particular, the 3D parametric solids and their drawings are bidirectionally associative. Therefore changing the parametric dimensions in the 2D drawing also causes changes in the 3D solid.

Reference Dimensions

The reference dimensions are dimensions that report the size. If you change a reference dimension, the solid part will not change. Hence, it may be detrimental to modify the values of the reference dimensions because the actual size of the solid part will not be affected. As a result, the assembly will still reference to a solid part of wrong size. As a rule of thumb, never modify a reference dimension. If you need to modify the solid part, change the parametric dimensions.

Dimensions in an Engineering Drawing

Because some parametric dimensions may not be the dimensions required for manufacturing the solid part, you will make adjustments to the drawing by hiding those parametric dimensions that may cause confusion and adding reference dimensions to help clarify the shape and size of the component.

Now you will learn how to manipulate parametric dimensions and reference dimensions in an engineering drawing.

1. Open the file *Frame_R.dwg*.

Now you will modify a parametric dimension of a drawing.

2. Select the dimension indicated in Figure 10–99 and double-click.
3. In the Power Dimensioning dialog box, change the dimension value from 125 to 50 and select the OK button.
4. Select the Update button from the Browser.
5. Select the Model tab from the Browser.

The solid part in model mode updates automatically.

6. Select the Hole indicated in Figure 10–100, right-click, and select Edit Sketch.
7. Select the dimension indicated in Figure 10–100 and double-click.

8. In the Power Dimensioning dialog box, change the dimension value from 50 to 125 and select the OK button.
9. Select the Update button.
10. Select the Drawing tab from the Browser.

The drawing updates automatically.

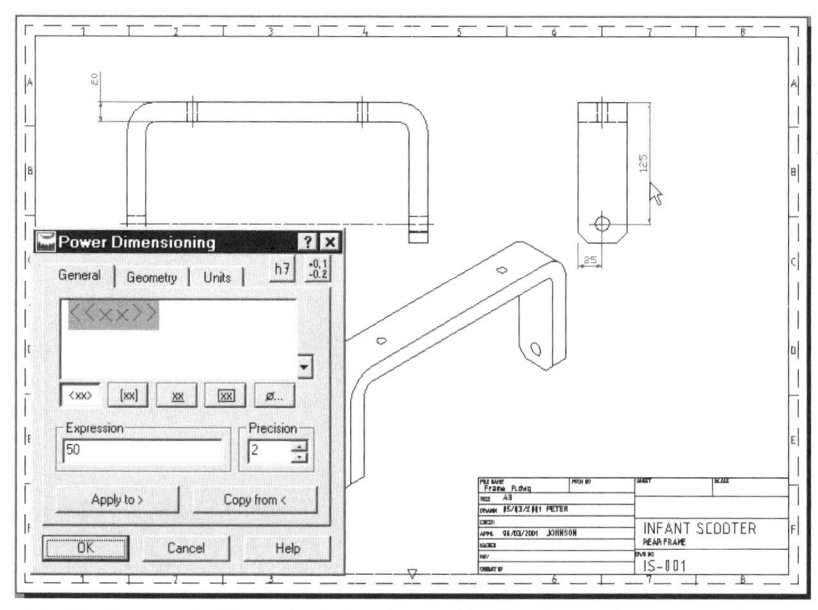

Figure 10–99 *Parametric dimension being changed in drawing mode*

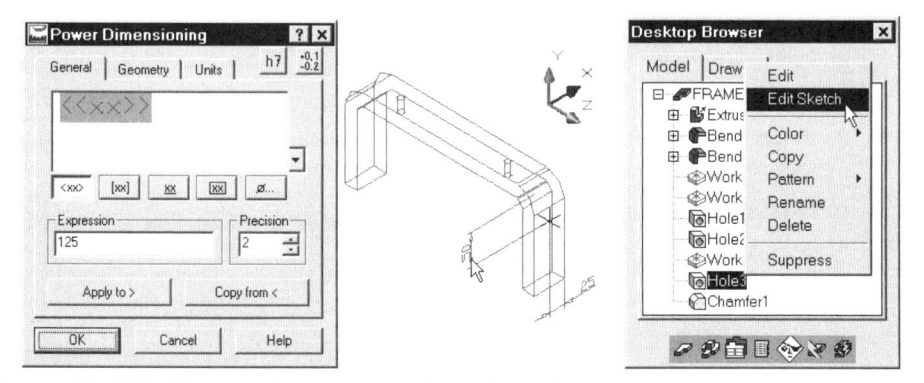

Figure 10–100 *Dimension being changed in model mode*

Change to the parametric in drawing mode causes the solid part in model mode to change. Similarly, change of the dimensions in model mode also causes the drawing to change. This way of association in two directions is called bidirectional associativity.

Now you will hide a parametric dimension.

11. Select the Visibility button from the Browser.
12. In the Desktop Visibility dialog box, check the Hide box and select the Select button.
13. Select the dimension indicated in Figure 10–101 and press the ENTER key.
14. On returning to the Desktop Visibility dialog box, select the OK button.

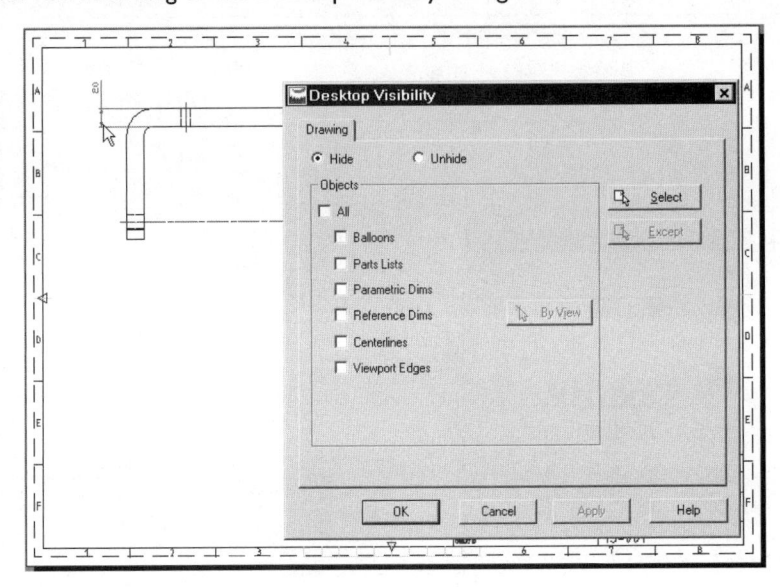

Figure 10–101 *Hiding a parametric dimension*

The selected dimension is hidden. Now you will add reference dimensions to the drawing.

15. Select Annotate>Reference Dimension

    Command: AMREFDIM

16. Referring to Figure 10–102, select a fillet corner to dimension and select a location to indicate the position of the dimension.
17. Press the ENTER key.

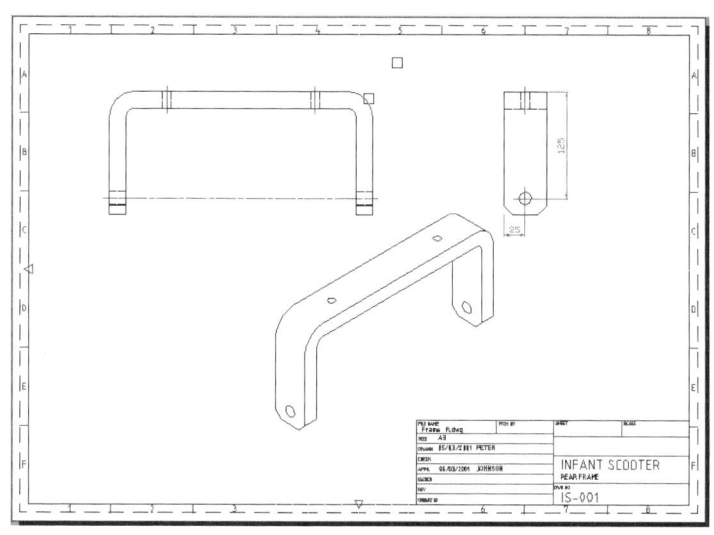

Figure 10–102 *A reference radial dimension being added*

A reference radial dimension is added.

    18. Repeat the AMREFDIM command.

    19. Select the edges indicated in Figure 10–103 and a location to indicate the position of the dimension.

    20. Press the ENTER key to confirm the location of the dimension.

    21. Press the ENTER key to exit.

A reference linear dimension is added (see Figure 10–104).

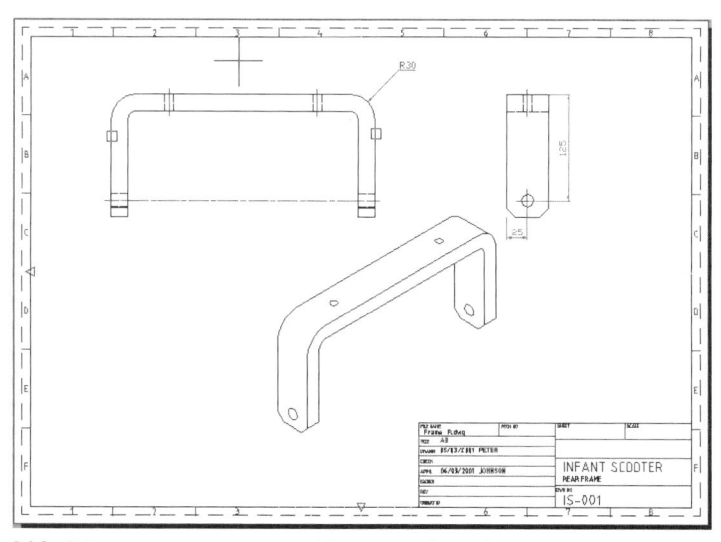

Figure 10–103 *Dimension extensions and location selected*

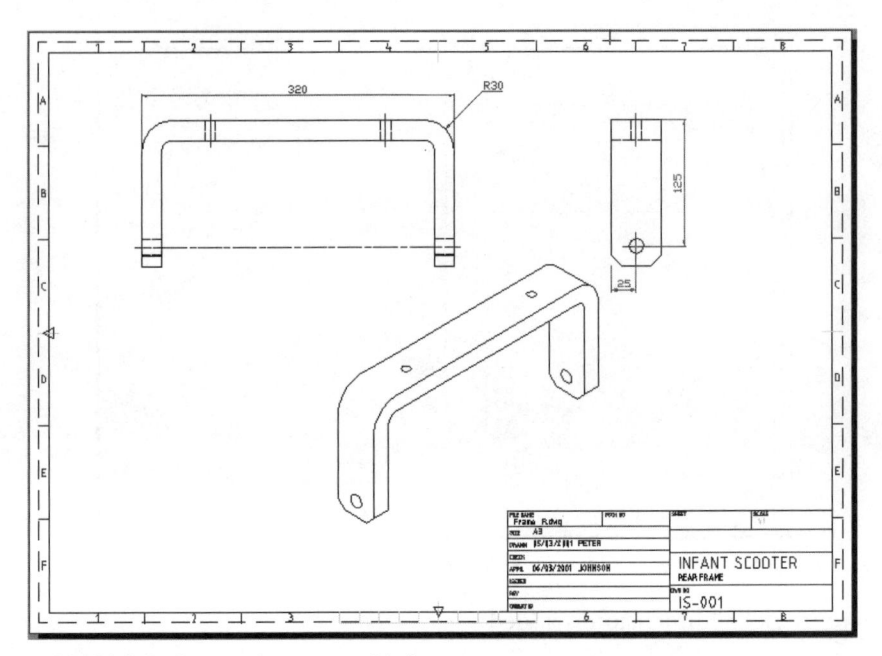

Figure 10–104 *Reference dimension added*

22. Referring to Figure 10–105, add nine more reference dimensions.

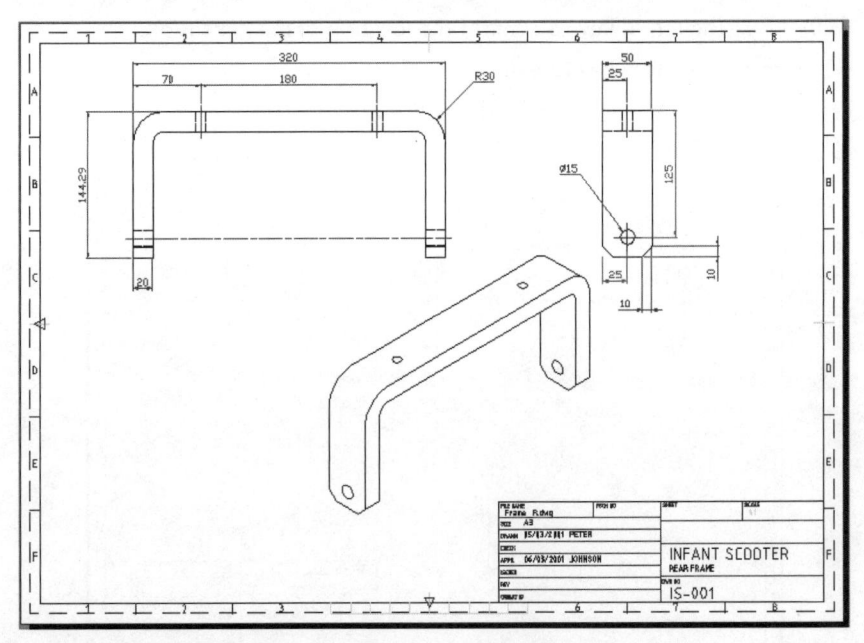

Figure 10–105 *Reference dimensions added*

Now you will change the displayed precision of a dimension.

23. Select the dimension indicated in Figure 10–106 and double-click.
24. In the Power Dimensioning dialog box, change the precision value to 0 and select the OK button.

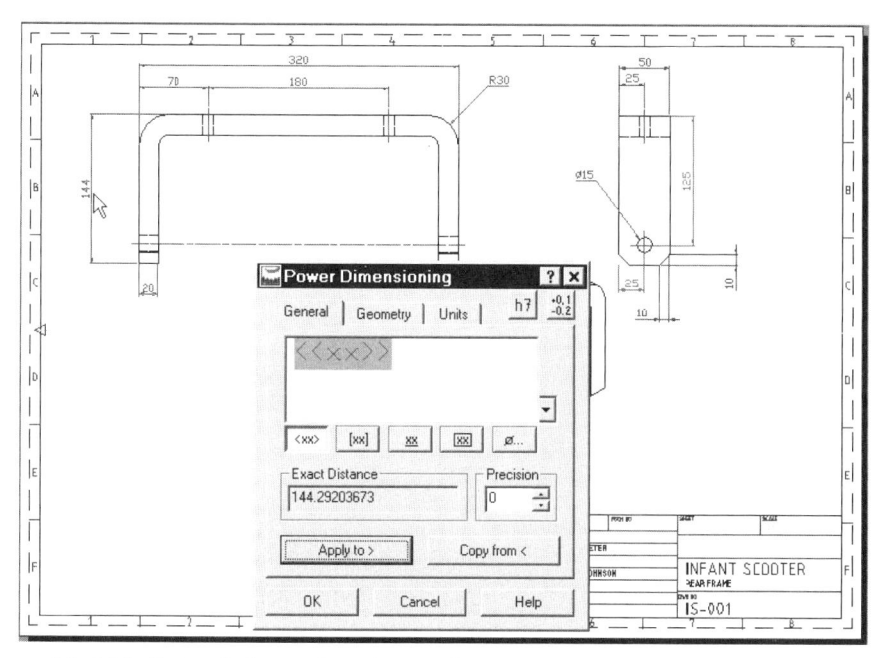

Figure 10–106 *Decimal places of a reference dimension changed*

TOLERANCE

There are two kinds of tolerances: dimensional tolerance and geometric tolerance.

Dimension Tolerance

A dimension tolerance specifies the allowable deviation from the nominal size specified by the dimension value. You use the Power Dimensioning dialog box to specify the deviation from the nominal size or specify the standard tolerance in terms of a standard deviation and a standard tolerance grade.

Now you will add dimension tolerance value to a parametric dimension.

25. Referring to Figure 10–107, select the dimension and double-click.
26. In the Power Dimensioning dialog box, select the Add Tolerance button.
27. In the Deviation area, specify 0.2 in the upper deviation box and specify 0.2 in the lower deviation box.

A tolerance of +0.2 and −0.2 is added to the dimension.

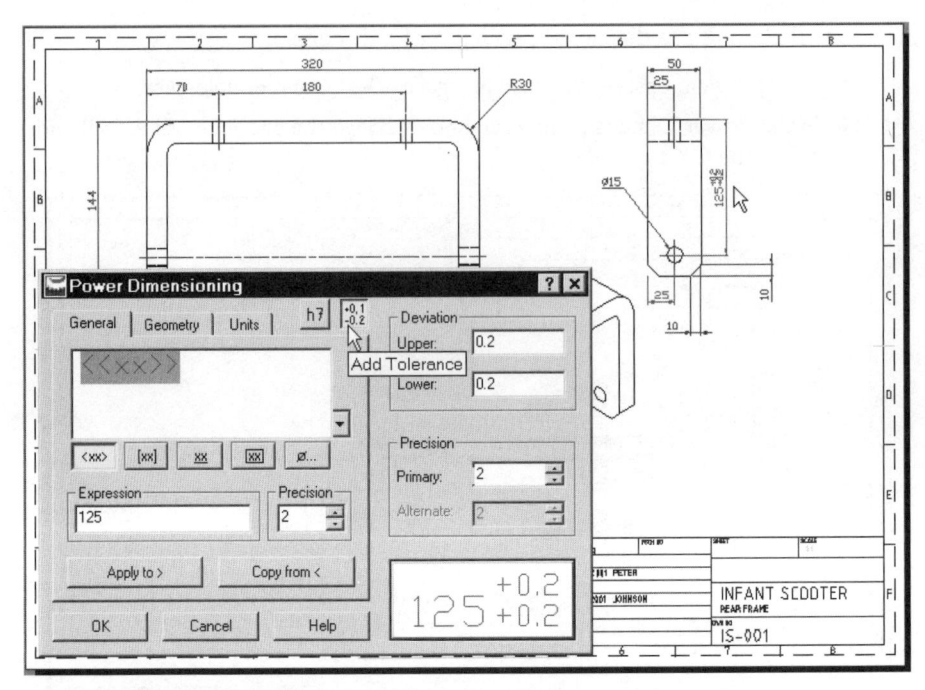

Figure 10–107 *Specifying tolerance value*

Now you will use the standard deviation and standard tolerance grade to indicate the dimension tolerance. A standard tolerance consists of a set of letters and a number. The letter(s) depicts the standard deviation from nominal size and the number depicts the standard tolerance grade. To have a better understanding of how the toleranced feature fits with its mating part, you can use the Fit dialog box.

28. Select the Add Fit button to specify standard tolerance (see Figure 10–108).
29. Select the button in the Fit area to display the Fit dialog box (see Figure 10–109).
30. Select the Mate button to expand the dialog box.

Engineering Drafting I 555

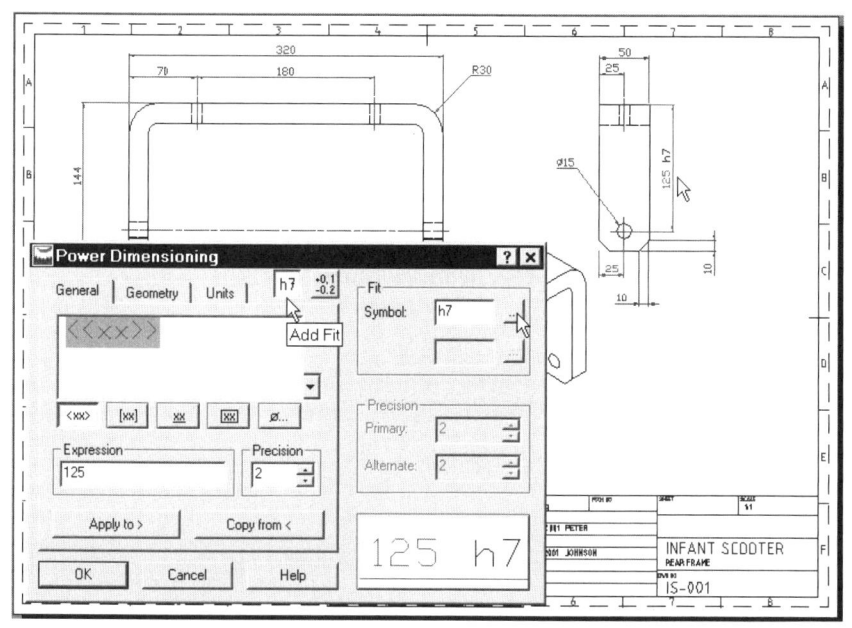

Figure 10–108 *Use standard fit*

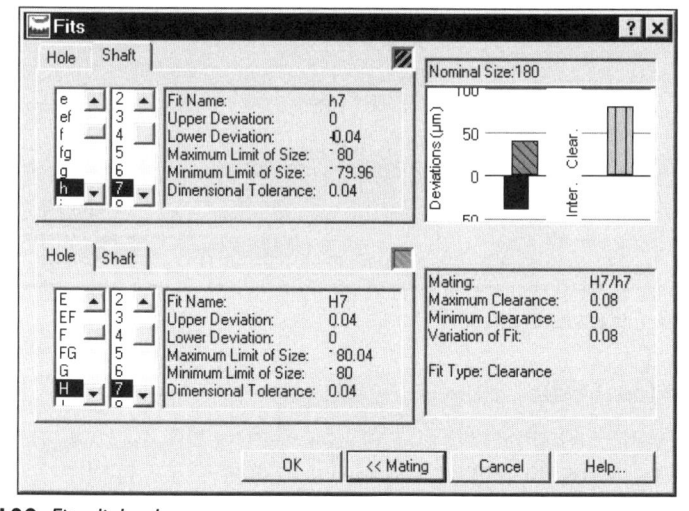

Figure 10–109 *Fits dialog box*

After the Fit dialog box is expanded, you can specify the mating feature's tolerance in the lower portion of the dialog box and discover how the select feature mates with its mating feature in the graphical chart.

31. Select the OK button.

A standard tolerance is specified.

Fits List

Now you will put existing fits of the drawing and their respective dimension values into a fits list and insert the list into your drawing.

    32. Select Annotate>Fits List

        Command: AMFITSLIST

    33. Press the ENTER key to construct a new fits list.

    34. Referring to Figure 10–110, select a location.

A fits list is constructed.

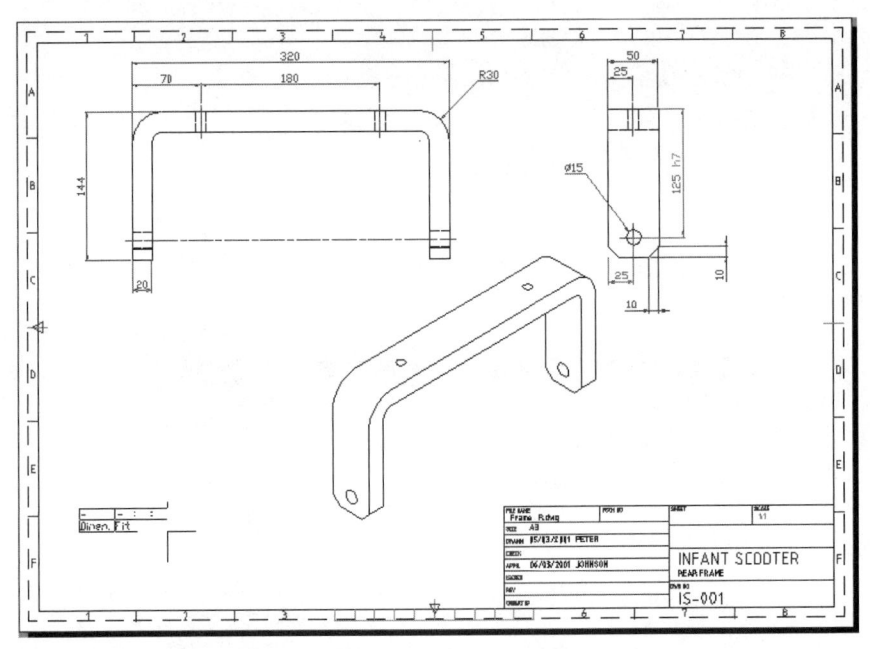

Figure 10–110 *Fits list constructed*

Toleranced Model Edit

We normally construct the solid part in accordance with the nominal sizes of the object. To cater for the deviation in size after tolerances are applied to the drawing, you can edit the solid part to the toleranced size (minimum, middle, or maximum size, or manually select a size).

    35. Select Annotate>Toleranced Model Edit

        Command: AMTOLCONDITION

The Transformation of Model dialog box displays (see Figure 10–111).

Engineering Drafting I 557

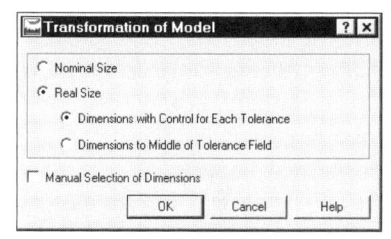

Figure 10–111 *Transformation of Model dialog box*

Using the Transformation of Model dialog box, set the solid part to the nominal size or the real size. If you select real size, you must decide to use the dimensions for each tolerance or the middle of the tolerance field.

36. Select Dimensions with Control for Each Tolerance.

The Manual Control of Dimension dialog box displays (see Figure 10–112).

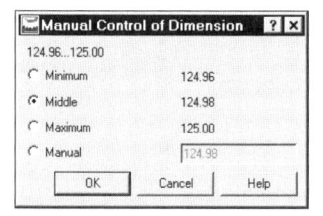

Figure 10–112 *Manual Control of Dimension dialog box*

37. Select Middle and the OK button.

The solid part is transformed to the middle size of the tolerance zone (see Figure 10–113).

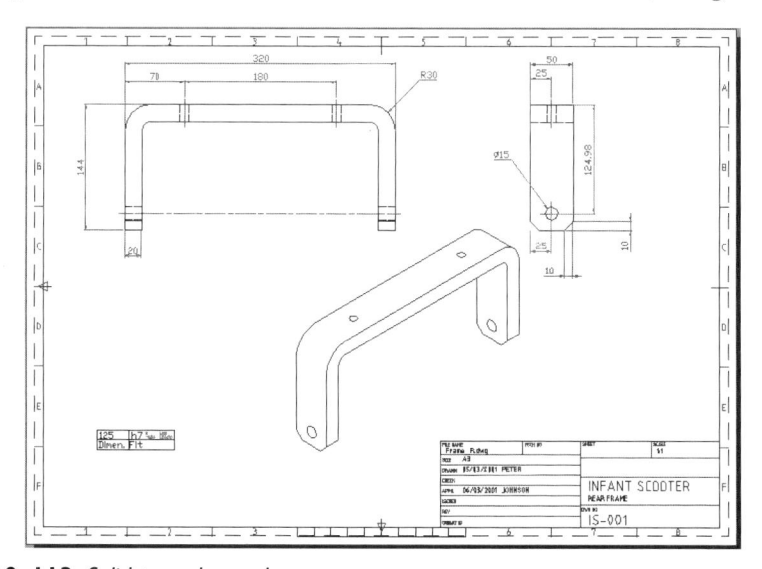

Figure 10–113 *Solid part changed*

Geometric Tolerances

A geometric tolerance indicates the allowable deviation in form and shape from the theoretical perfect shape. In a drawing, a geometric tolerance symbol is a feature control frame (in the form of a set of rectangular compartments) consisting of a symbol depicting the kind of geometric tolerance to be controlled, a tolerance value, and, where appropriate, a reference. To specify a reference, you use a datum identifier, datum target, or feature identifier.

Now you will construct a datum identifier to edge X (shown in Figure 10–114) and construct a feature control frame to edge Y (shown in Figure 10–115) to specify a perpendicularity tolerance to edge Y with reference to edge X.

38. Select Annotate>Symbols>Datum Identifier

    Command: AMDATUMID

39. Referring to Figure 10–114, select the front view for the datum identifier to attach.

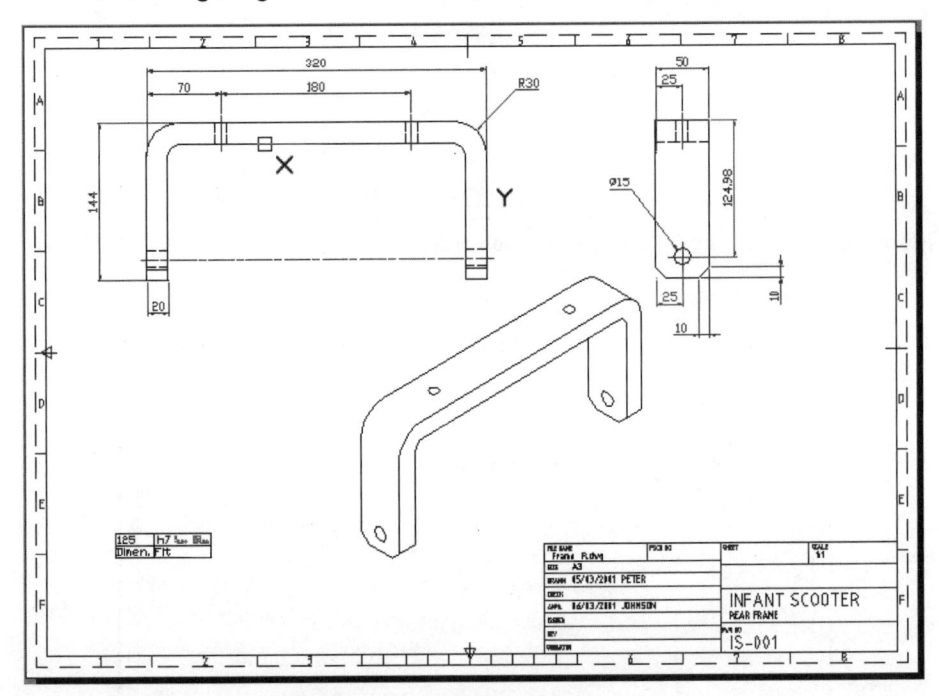

Figure 10–114 *An edge of the front view selected*

40. Press the SHIFT key, right-click, and select NEA.
41. Select edge X (shown in Figure 10–114) of the front view to specify the start point.
42. Select a point below the start point to specify the next point.

43. Press the ENTER key.

The Datum Identifier dialog box displays (see Figure 10–115).

44. Select the OK button.

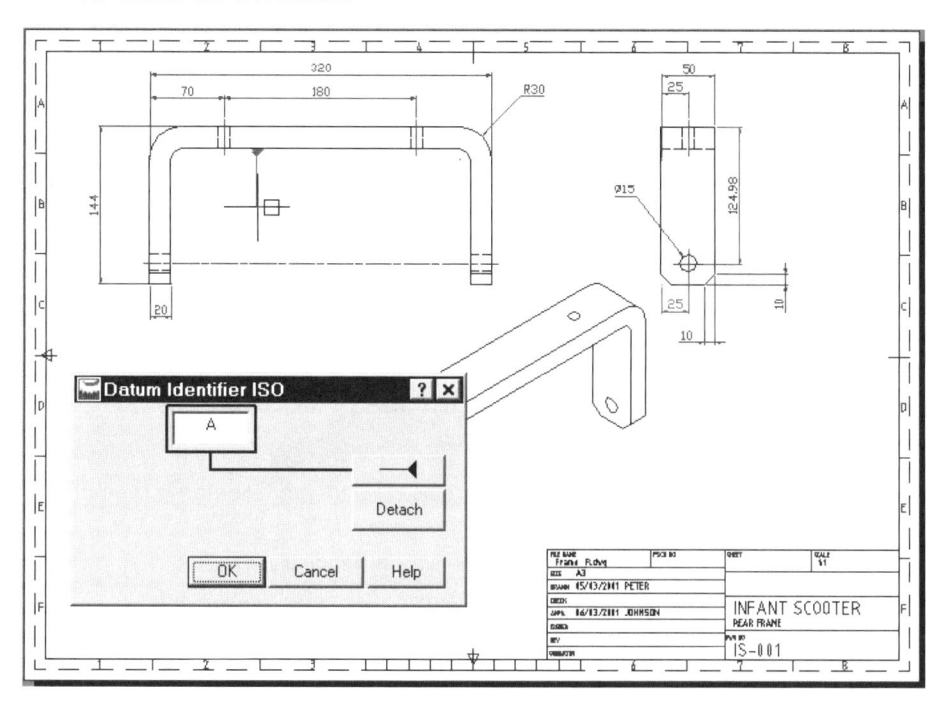

Figure 10–115 *Location of datum identifier selected*

A datum identifier is constructed. Now you will construct a feature control frame that references to the datum identifier.

45. Select Annotate>Symbols>Feature Control Frame

    Command: AMFCFRAME

46. Select the top view for the feature control frame to attach (see Figure 10–116).
47. Press the SHIFT key, right-click, and select NEA.
48. Select a point near the right vertical edge of the front view.
49. Select a point to the right of the first point.
50. Press the ENTER key.

The Feature Control Frame dialog box displays (see Figure 10–117).

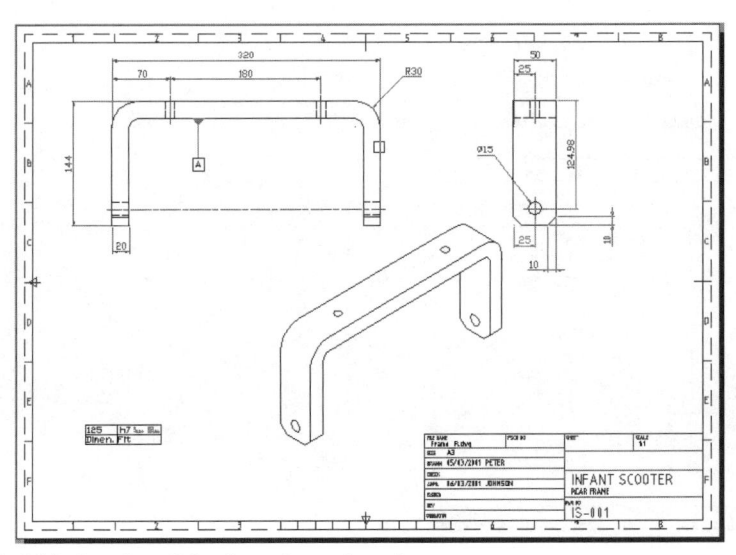

Figure 10-116 *An edge of the front view selected*

51. In the Feature Control Frame dialog box, select the Geometric Symbol Palette button.
52. In the Geometric Symbol Palette, select Perpendicularity.
53. Set Tolerance 1 to 0.2 and Datum 1 to A.
54. Select the OK button.

A feature control frame is constructed to control the perpendicularity of an edge with another edge.

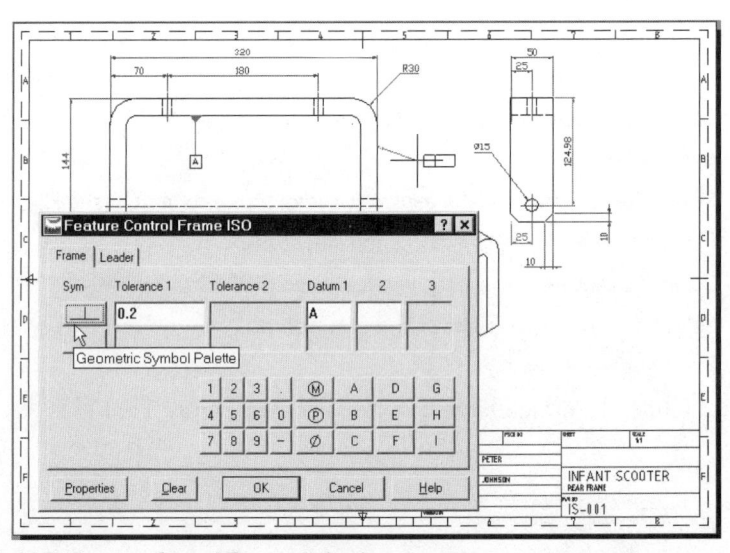

Figure 10-117 *Feature Control Frame dialog box*

Now you will establish a datum target and construct a feature control frame that references to the datum target.

55. Select Annotate>Symbols>Datum Target

    Command: AMDATUMTGT

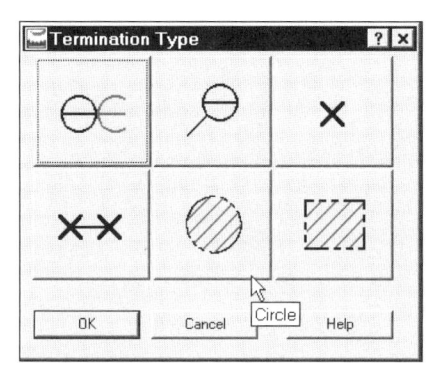

Figure 10–118 *Termination Type dialog box*

56. In the Termination Type dialog box, select the Circle button.
57. Select the front view for the datum target to attach.
58. Referring to Figure 10–119, select a location to specify the center point.
59. Select a point to specify the diameter (see Figure 10–120).
60. Select a point to indicate the next point and press the ENTER key.

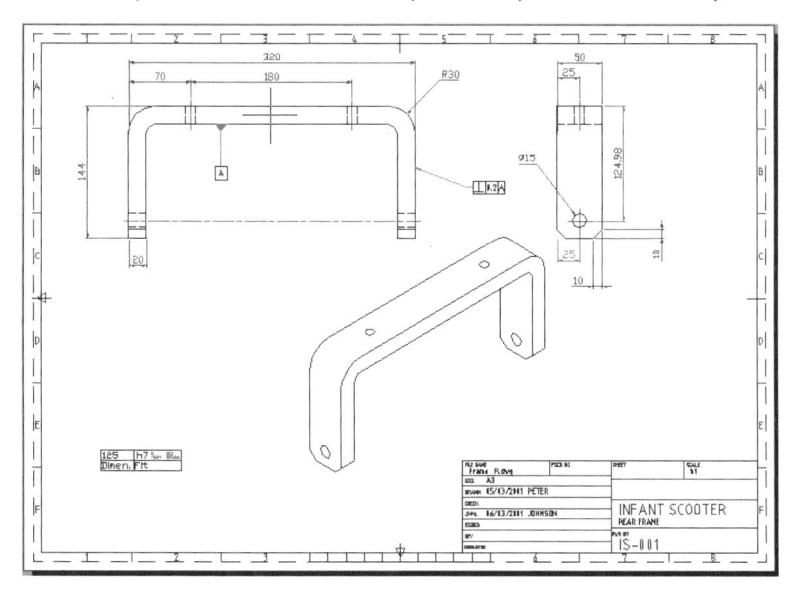

Figure 10–119 *Center selected*

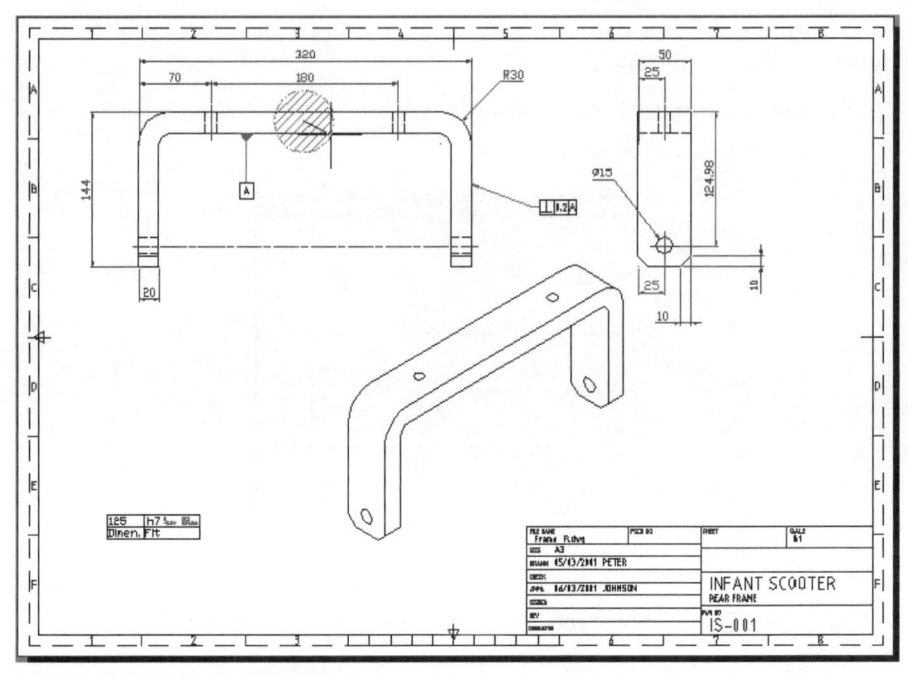

Figure 10-120 *Diameter specified*

61. In the Datum Target dialog box, specify a diameter of 5 mm and a datum name B, and select the OK button (see Figure 10-121).

A datum target is constructed (see Figure 10-122).

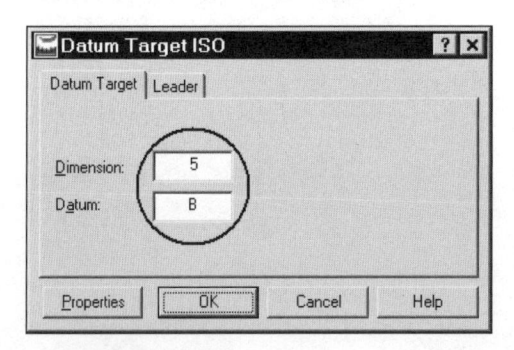

Figure 10-121 *Datum Target dialog box*

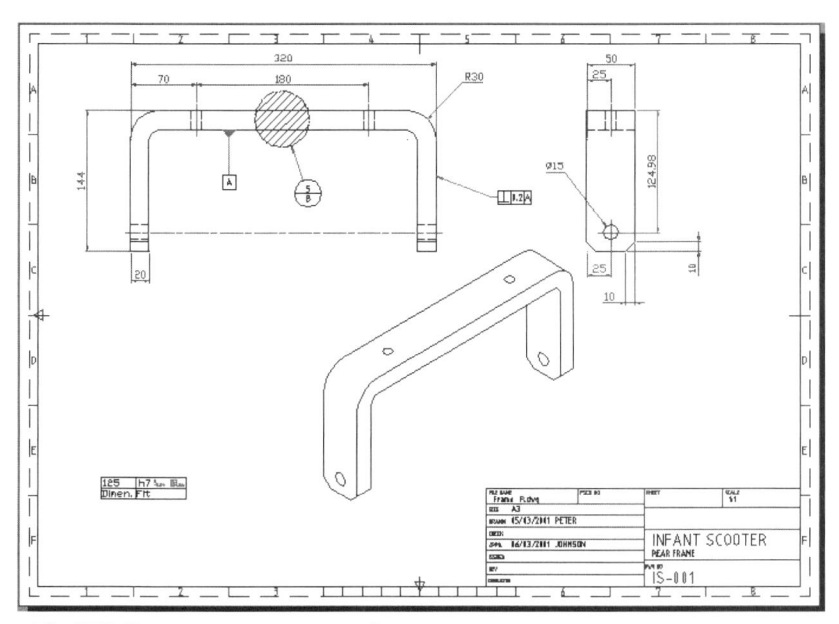

Figure 10–122 *Datum target constructed*

62. Repeat steps 55 through 61 to construct two more datum targets (see Figure 10–123).

Three datum targets are constructed. Together, they make up a datum plane B. Now you will construct a feature control frame with reference to the datum targets.

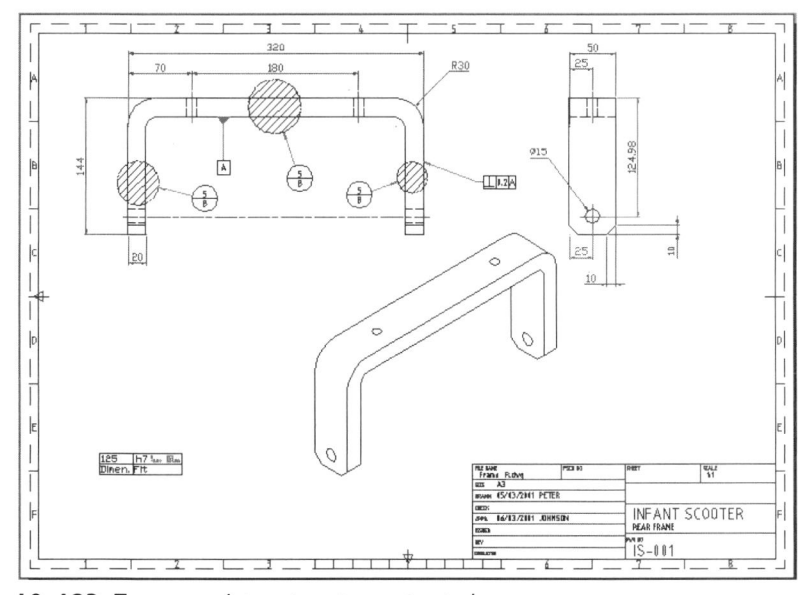

Figure 10–123 *Two more datum targets constructed*

63. Referring to Figures 10–124 and 10–125, construct a feature control frame.

Figure 10–124 *Feature control frame being constructed*

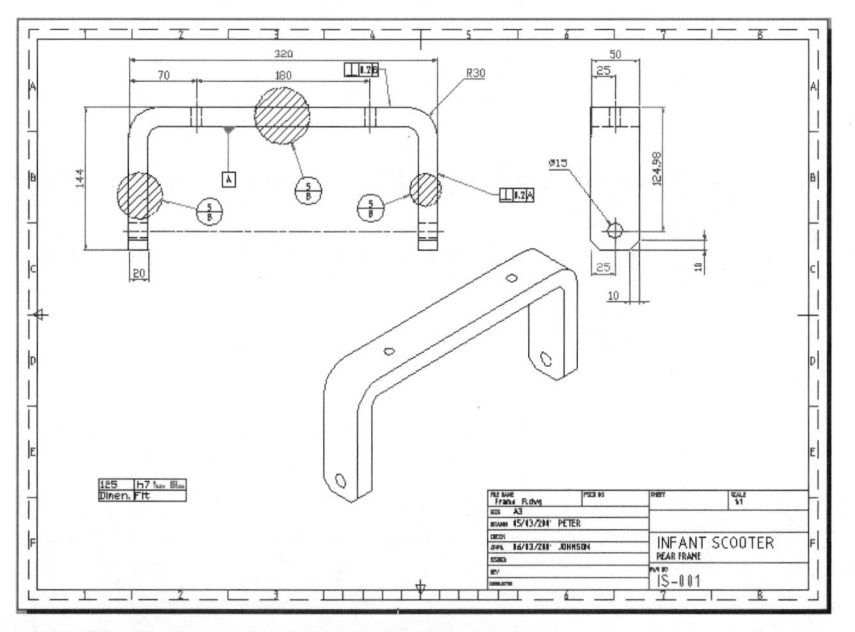

Figure 10–125 *Feature control frame constructed*

Now you will construct a feature identifier and a feature control frame.

64. Select Annotate>Symbols>Feature Identifier

    Command: AMFEATID

65. Referring to Figure 10–126, select a feature, select a start point and a second point, and press the ENTER key.

66. In the Feature Identifier dialog box, set feature name to C and select the OK button (see Figure 10–127).

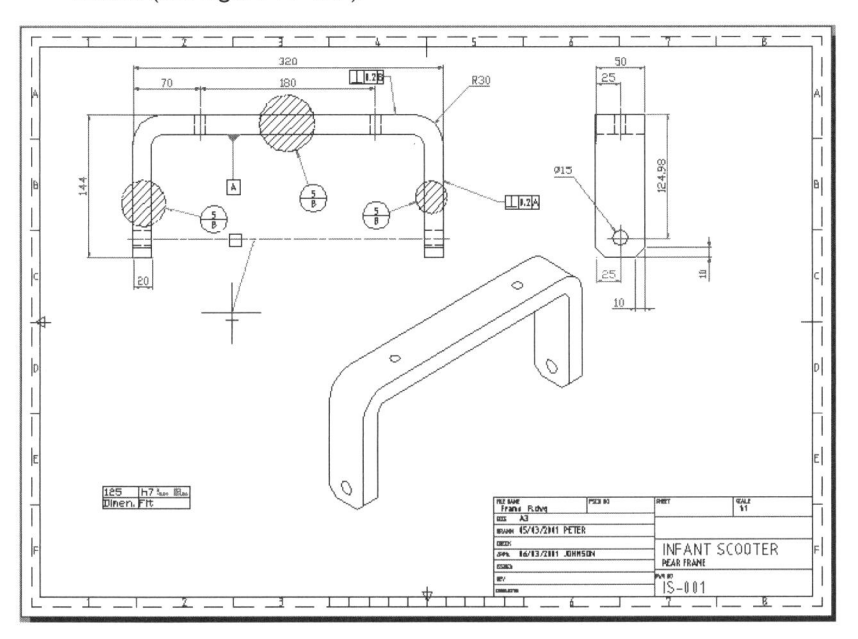

Figure 10–126 *Feature selected*

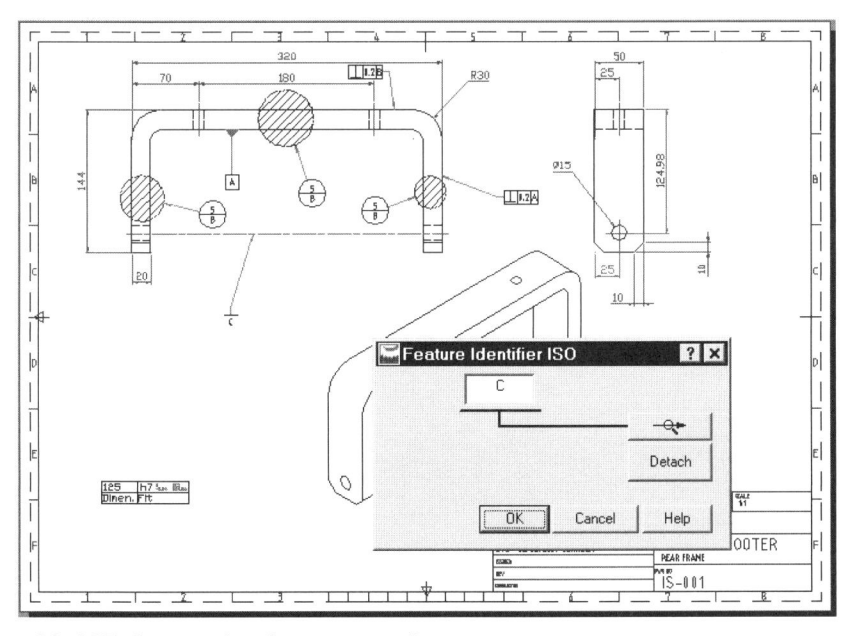

Figure 10–127 *Feature identifier constructed*

67. Construct a feature control frame in accordance with Figure 10–128.

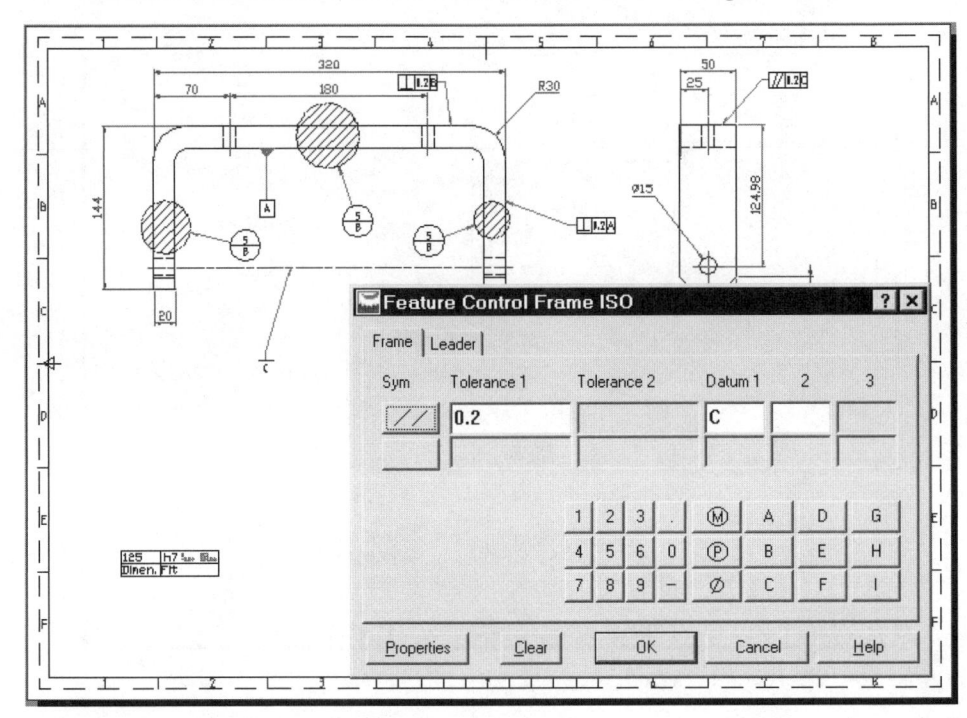

Figure 10–128 *Feature control frame constructed*

Now three feature control frames are constructed. They reference to the datum identifier, datum target, and feature identifier. Save your file.

SURFACE FINISH SYMBOL

To control the surface finish of components, you use surface finish symbols.

68. Select Annotate>Symbols>Surface Texture

    Command: AMSURFSYM

69. Referring to Figure 10–129, select the front view and select an edge as the start point.
70. Press the ENTER key.
71. In the Surface Texture dialog box, specify the roughness value and select the OK button (see Figure 10–130).

The surface texture symbol is constructed. Save your file.

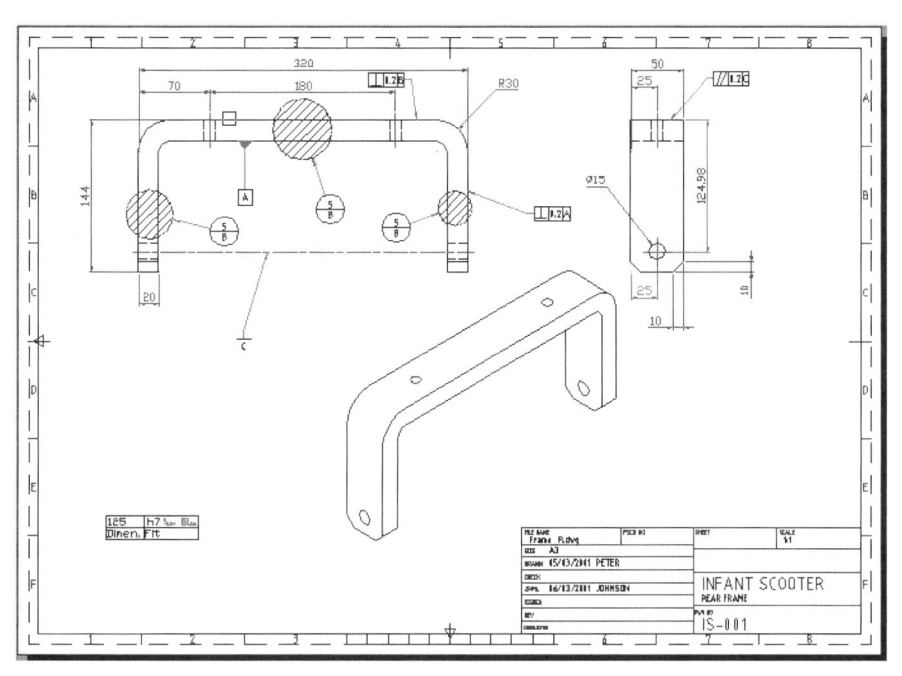

Figure 10–129 Location of surface texture symbol selected

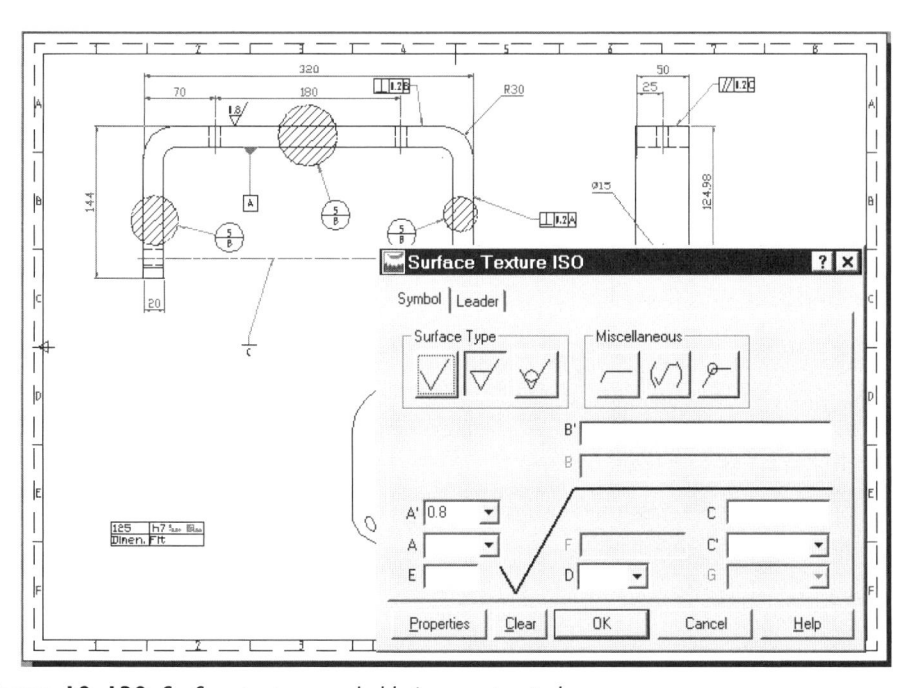

Figure 10–130 Surface texture symbol being constructed

EXPORTS

The drawing views you constructed in drawing mode are associative to the 3D object you constructed in model mode. Both the drawing views and the 3D objects reside on the same Mechanical Desktop file. If you wish to export a purely 2D drawing, you use the AMVIEWOUT command.

72. Select Drawing>Export View

    Command: AMVIEWOUT

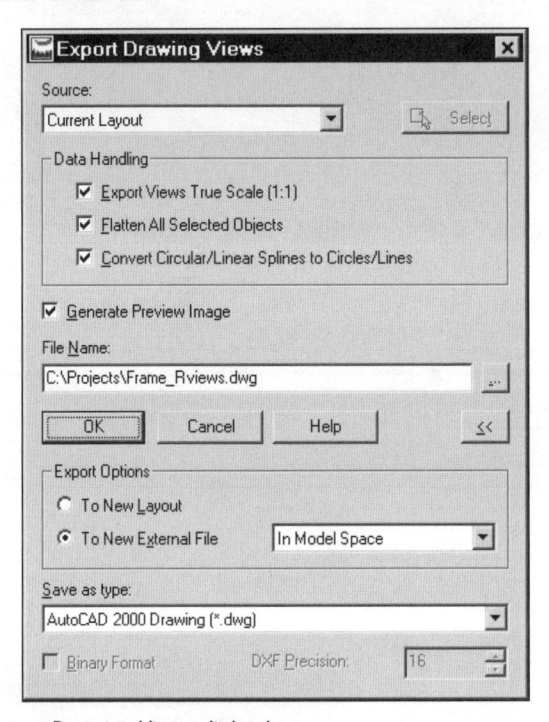

Figure 10–131 *Export Drawing Views dialog box*

Using the Export Drawing Views dialog box, you can export the current layout or selected views or entities of the drawing. Optionally, you can export the drawing views to their true scale, flatten all selected objects, and convert circular or linear splines to circles or lines.

73. Specify a file name and select the OK button.

A 2D drawing is exported. Now save and close your file.

> **SUMMARY**
>
> There are two working environments in a part file: model and drawing. In the model environment of a part file, you construct solid parts or surfaces. (Although you can use an assembly file, it is recommended that you construct solid parts and surfaces in part file.) After you have constructed 3D objects in the model mode of a file, construction of engineering drawing is semi-automated. You can switch to drawing mode, setup the page layout, insert a title block, set appropriate options, specify the type of drawing views, let the computer construct the drawing views, and insert annotations.
>
> There are several kinds of engineering drawing views. The view drawing view is the base view. From the base view, you construct orthographic, auxiliary, and isometric views. If you have a very long object to display, you can break the drawing view. To display internal details of an object, you can use section views. There are several kinds of section views. To construct an offset section view, you can construct a cut line. To construct a breakout section view, you can construct a break line.

PROJECTS

Now you will enhance your knowledge by working on the following projects.

INFANT SCOOTER

Complete the engineering drawing views of the component parts of the infant scooter that you constructed in this chapter and construct engineering drawing views (including orthographic, auxiliary, and isometric views) and section views (if necessary) for the remaining component parts of the infant scooter. Add dimensions, annotations, tolerances, and surface finish requirements.

TOY CAR

Construct a front view, side view, top view, and an isometric view for the parts of the toy car that you constructed in Chapter 5. Add dimensions, annotations, tolerances, and surface finish symbols to the drawing.

MODEL CAR BODIES

Construct a front view, a side view, a top view, and an isometric view for the model car bodies that you constructed in Chapter 8.

CAMERA CASING

Construct a front view, side view, top view, and an isometric view for the solid parts of the camera casing that you constructed in Chapter 9.

REVIEW QUESTIONS

1. There can be only one base view in a drawing. True or false?

2. The color and linetype of layers that Mechanical Desktop information is placed on cannot be changed. True or false?

3. The only way to create an isometric view is to first create a base view and then create an isometric view from it. True or false?

4. A tolerance condition can only be applied to the drawing views, not the part itself. True or false?

5. A Mechanical Desktop file can have a maximum of two layouts. True or false?

6. There can be only one base view per layout. True or false?

7. When plotting a drawing with Mechanical Desktop views, you can use any plot scale. True or false?

8. Every center line needs to be placed manually. True or false?

9. Changing reference dimensions cannot modify the part. True or false?

10. When will you use the Select option when creating drawing views?

11. Explain how a dimension can be moved from one view to another and also how to reattach a dimension.

12. Explain how to remove a line section from a dimension.

13. Explain what an annotation is in reference to Mechanical Desktop drawing views.

CHAPTER 11

Engineering Drafting II

OBJECTIVES

The aims of this chapter are to explain the ways to construct associative engineering drawing views and exploded views for an assembly, and to depict the way to construct welding symbols, balloons, and a parts list in an assembly drawing. After studying this chapter, you should be able to do the following:

- Construct associative engineering drawing views of an assembly
- Construct exploded views of an assembly
- Add welding symbols
- Insert balloons and parts lists in an assembly drawing

OVERVIEW

The process of constructing an engineering drawing for an assembly is similar to the construction of drawings for solid parts and 3D surfaces. You can use model mode to construct the assembly and use drawing mode to construct a drawing. Because an assembly has, in addition to model mode and drawing mode, a scene mode in which you can construct scenes of an assembly, you can select a scene to construct an exploded drawing view. Mainly, the annotations that you add to an assembly are a parts list delineating the components used to construct the assembly and a set of balloons referencing the components to the parts list. In addition, you can add a welding symbol to an assembly of welded structure.

ASSEMBLY DRAWING CONSTRUCTION

A Mechanical Desktop assembly file has three working environments: model mode, scene mode, and drawing mode. You can construct an assembly of components in model mode, construct assembly scenes in scene mode, and construct engineering drawings in drawing mode.

LAYOUT AND ENGINEERING STANDARD

Similar to making an engineering drawing for a solid part or a surface model, you perform several tasks before you construct engineering drawing views, including selecting a plotting device and a paper size, inserting a title block, and setting relevant standard and drawing options.

DRAWING VIEWS

To construct engineering drawing views of an assembly, you select components manually from the assembly or use the components arranged in the assembly scene. The type of drawing views that you construct for an assembly is the same as that for a solid part. You can construct a base view, multiple views, broken view, auxiliary view, detail view, and various kinds of section views. With regard to the section view, you can specify a unique hatch pattern for each component in the assembly drawing so that they can be distinguished from each other when they are sectioned.

Now you will construct the engineering drawing of an assembly.

1. Open the assembly file *FrontEnd.dwg* that you constructed in Chapter 6.

In the assembly file, you already constructed two scenes of the components. Figure 11–1 shows Scene2 of the assembly.

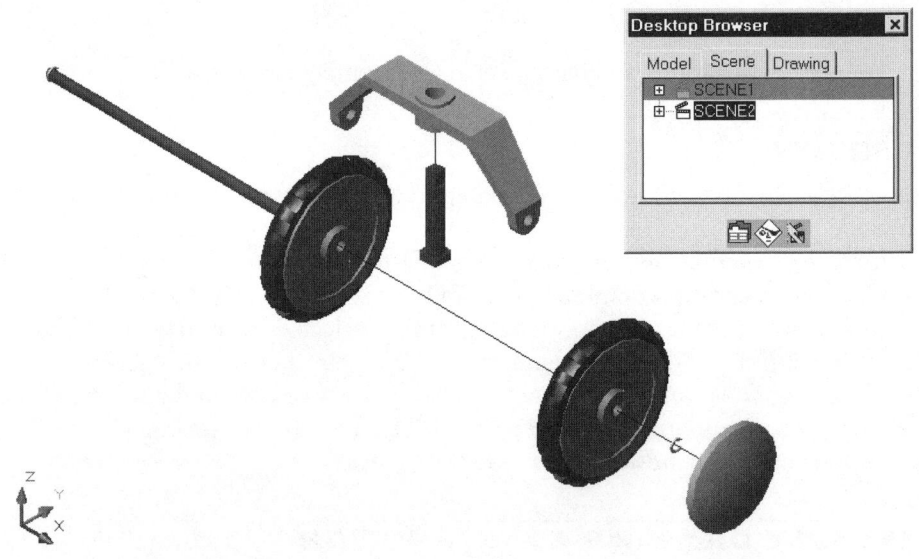

Figure 11–1 *Exploded scene of the assembly*

2. Select the Drawing tab to change to drawing mode.
3. Set up the page and insert a title block.
4. Select Layout1 from the Browser, right-click, and select New View (see Figure 11–2).

5. In the Create Drawing View dialog box (see Figure 11–3), select Base view and Scene2, set drawing scale to 0.2, and select the OK button.
6. In model mode, select a face and an edge to specify an orientation of the drawing view in accordance with Figure 11–4.

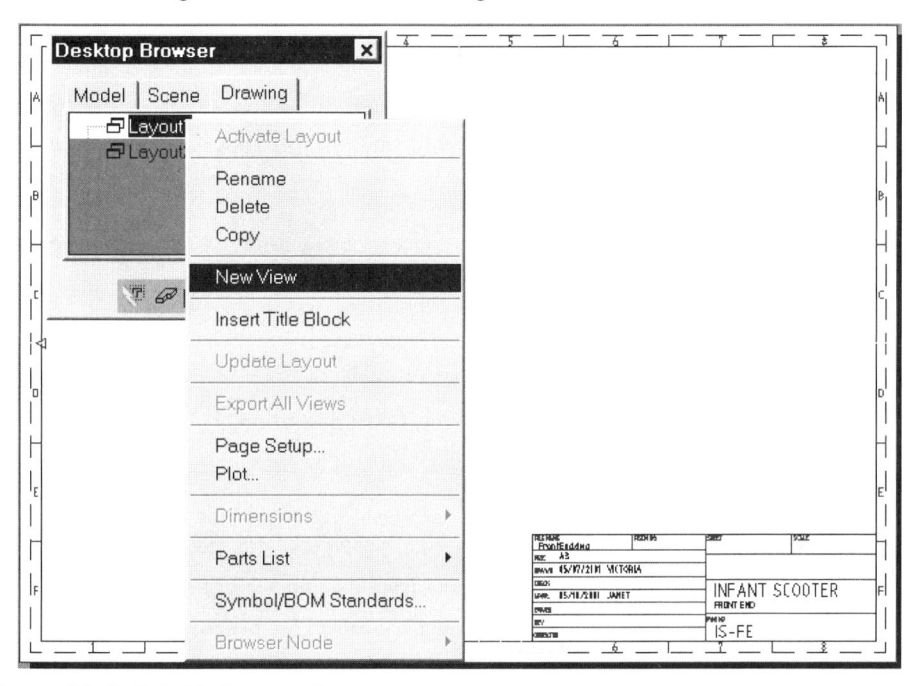

Figure 11–2 *Title block inserted*

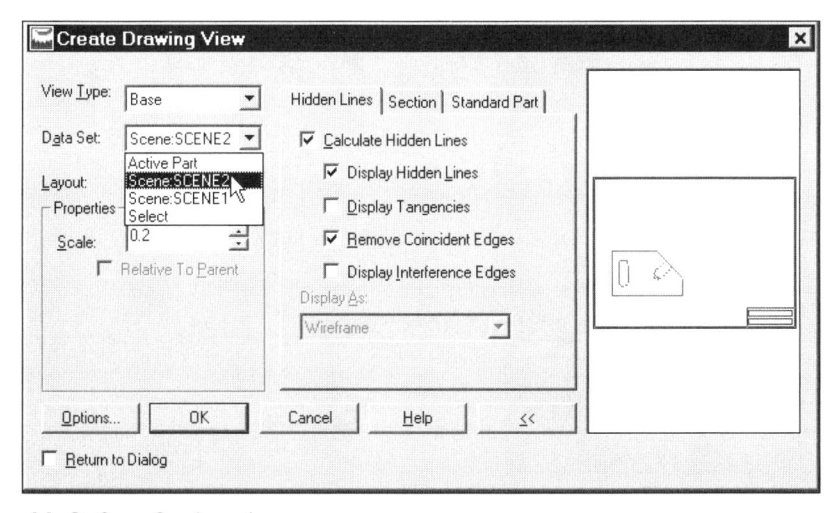

Figure 11–3 *Scene2 selected*

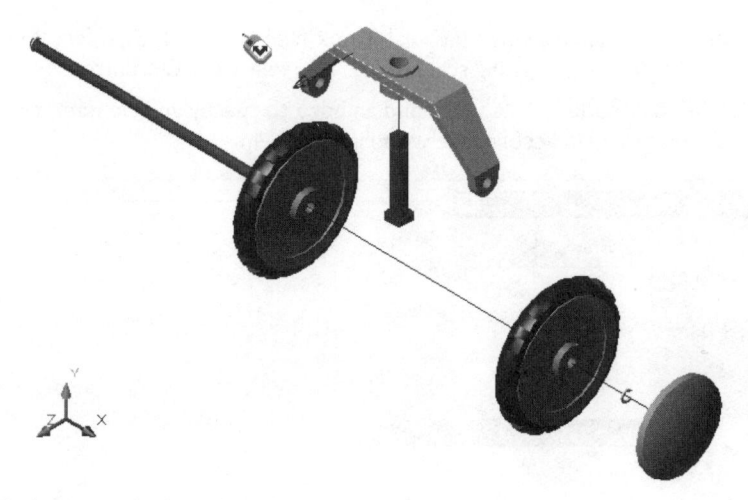

Figure 11–4 *Face and edge selected*

7. After returning to the drawing layout, select a location.

An engineering drawing view of the exploded scene is constructed (see Figure 11–5). Now you will construct an isometric view.

8. Select the base view from the Browser, right-click, and select New View.
9. In the Create View dialog box, select Iso view and select the OK button.
10. With reference to Figure 11–6, select a location for the isometric view.

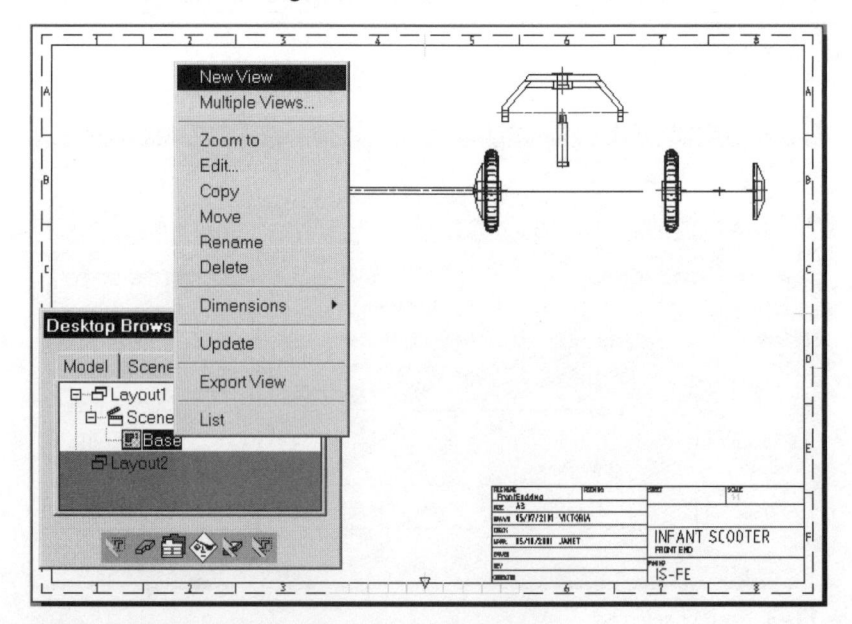

Figure 11–5 *Engineering drawing view of an exploded scene*

Engineering Drafting II 575

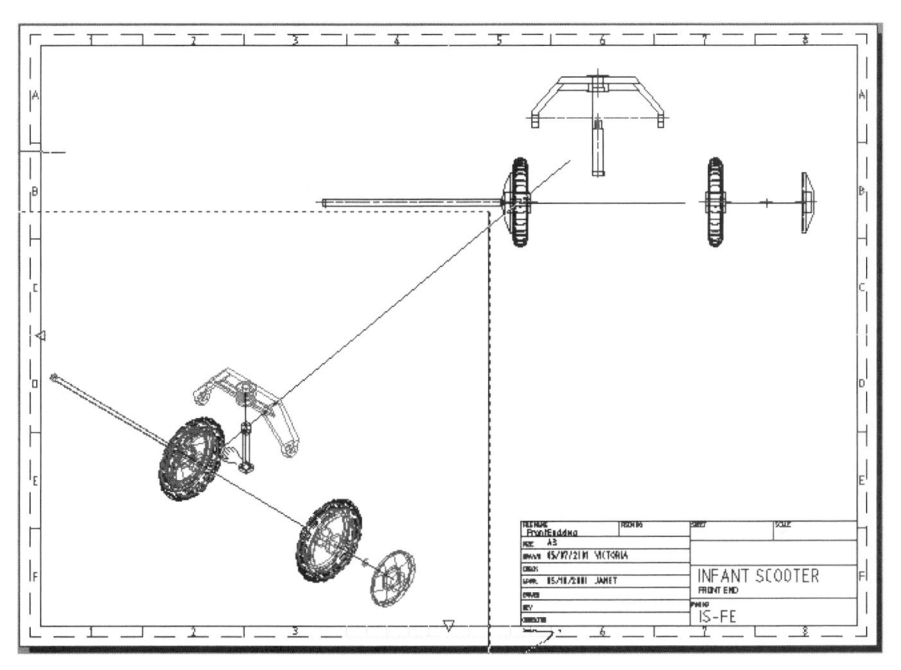

Figure 11-6 *Isometric view being constructed*

The base view, while required for making the isometric view, is not required once construction is complete. Therefore, you will erase it.

11. Select the base view from the Browser, right-click, and select Delete.
12. In the Delete Dependent Views dialog box, select the No button (see Figure 11-7).

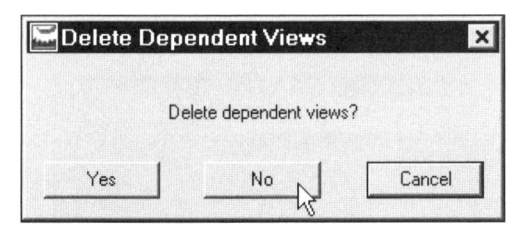

Figure 11-7 *Delete Dependent Views dialog box*

13. Referring to Figure 11-8, construct three drawing views of Scene1 of the assembly.

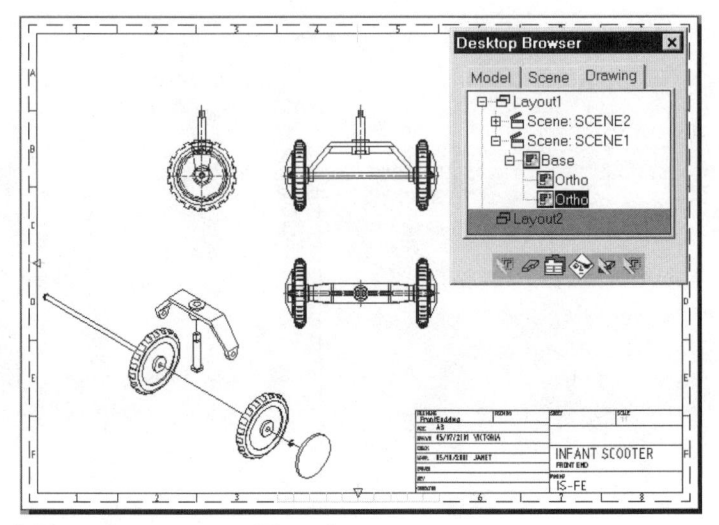

Figure 11-8 *Three drawing views of Scene1*

The drawing views of the assembly are complete. Save and close your file.

HATCH PATTERN OF COMPONENTS IN AN ASSEMBLY

While constructing a section view for an assembly of more than two component parts, you need to set a different hatch pattern for each of the solid parts. To set the hatch pattern, you use the AMPATTERNDEF command. This command is applicable to instances of external solid parts in the assembly drawing. The hatch pattern for the drawing of the original solid part is not affected.

1. Open the file *Frame.dwg* that you constructed in Chapter 5.
2. Select Assembly>Assembly>Hatch Patterns

   Command: AMPATTERNDEF

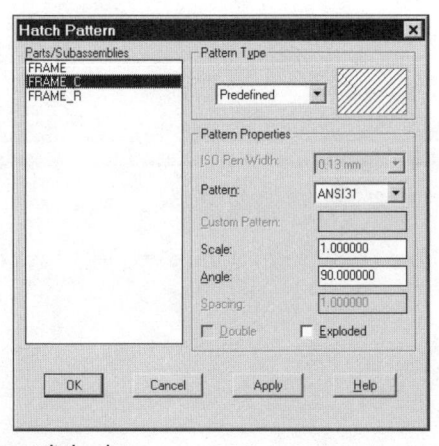

Figure 11-9 *Hatch Pattern dialog box*

3. In the Hatch Pattern dialog box (see Figure 11-9), select the parts Frame_C and Frame_R one by one, set the angle to 90 and 0 degrees respectively, and select the OK button.
4. Select the Scene tab from the Browser (see Figure 11-10).

In the Scene tab, you will find that there is no scene constructed.

Figure 11-10 *Scene tab*

5. Switch to drawing space, set up the layout, and insert a title block.
6. Select Layout1 from the Browser, right-click, and select New View.
7. In the Create Drawing View dialog box, select Base view and Scene, set the drawing scale to 0.2, and select the OK button.
8. Construct a base view in accordance with Figure 11-11.

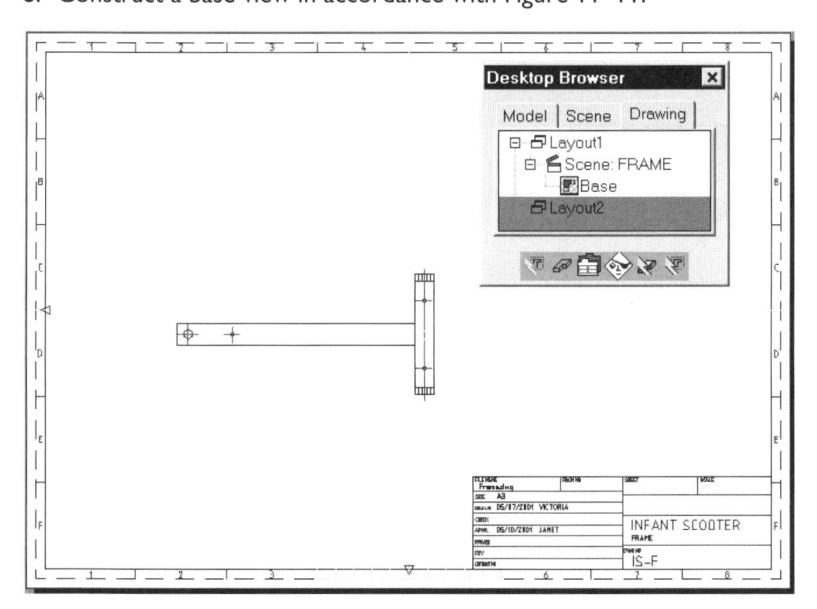

Figure 11-11 *Base view constructed*

After you select Scene as the data set in a drawing view, a scene with no explosion will be constructed automatically. If you select the Scene tab now, you will find a scene constructed.

Now you will construct a work plane and then construct a sectional view of the assembly.

9. Select the Model tab of the Browser to switch to model mode.
10. Select Frame_C from the Browser and double-click to activate part modeling mode (see Figure 11–12).

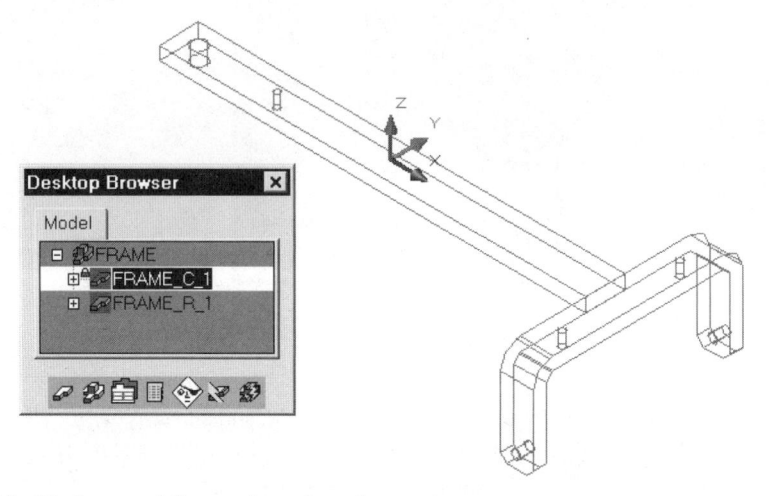

Figure 11–12 *Part modeling mode activated*

11. With reference to Figure 11–13, construct two work axes.
12. Construct a work plane (see Figure 11–14).

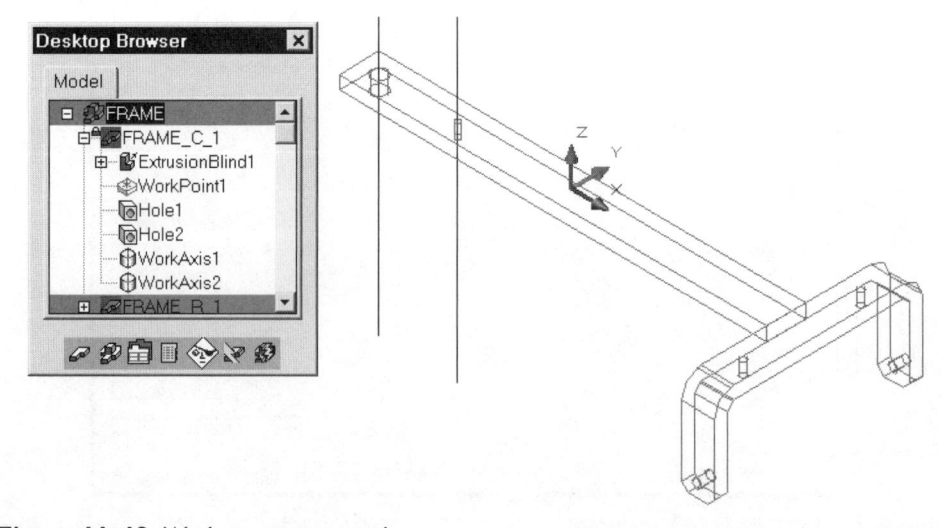

Figure 11–13 *Work axes constructed*

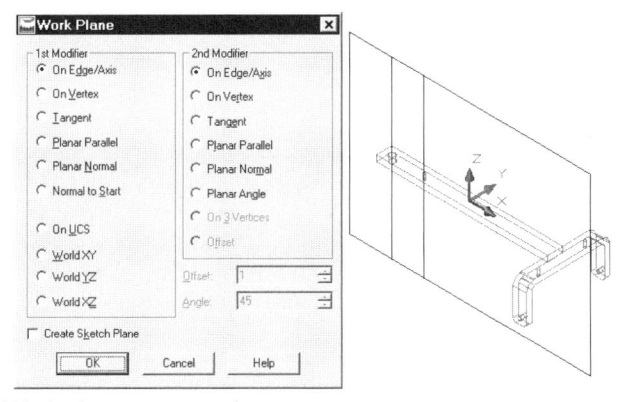

Figure 11–14 *Work plane constructed*

13. Select the assembly from the Browser and double-click to activate assembly mode.
14. Select the Drawing tab from the Browser to switch to drawing mode.
15. Select the base view from the Browser, right-click, and select New View.
16. In the Create Drawing View dialog box, select Ortho view and select the Section tab.
17. In the Section tab, select Full section, set label to A, and select the OK button.
18. Select a point above the base view to specify the location for the section view and press the ENTER key.
19. Press the ENTER key to use the Workplane option.
20. Referring to Figure 11–15, select the work plane.

A section view is constructed (see Figure 11–16).

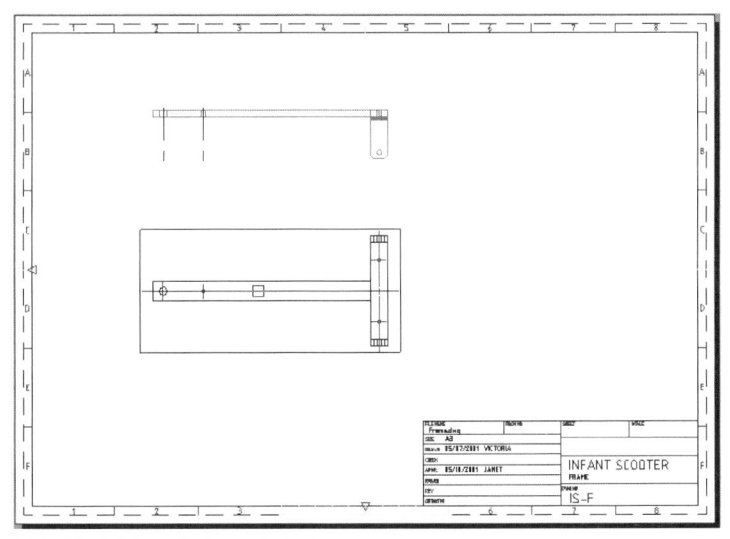

Figure 11–15 *Work plane being selected*

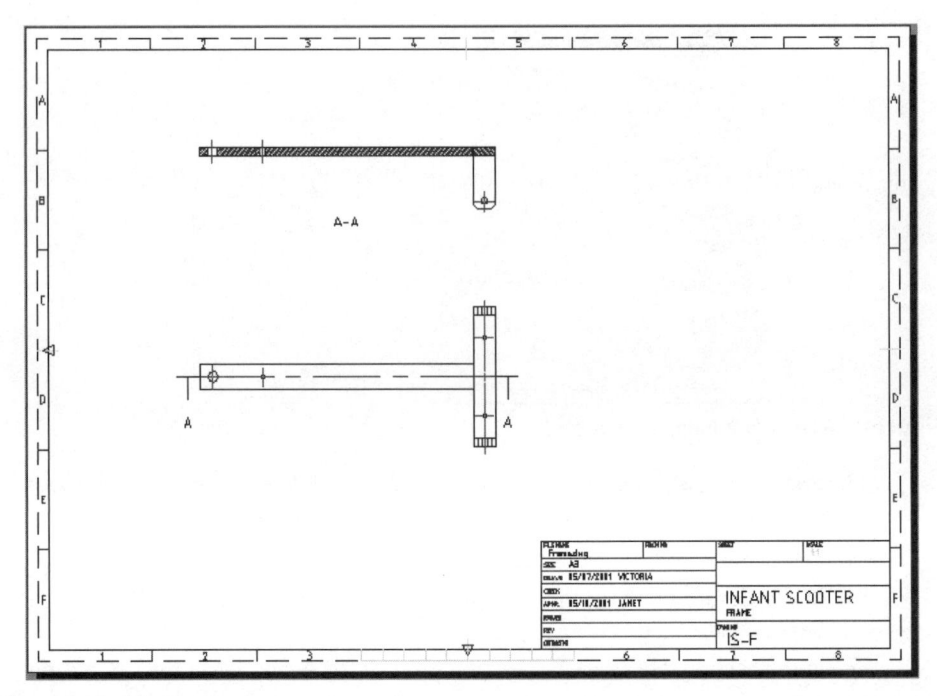

Figure 11–16 *Section view constructed*

The engineering drawing views are constructed. Save your file.

21. In the External File Save dialog box (see Figure 11–17), select the OK button to save the modified part file.

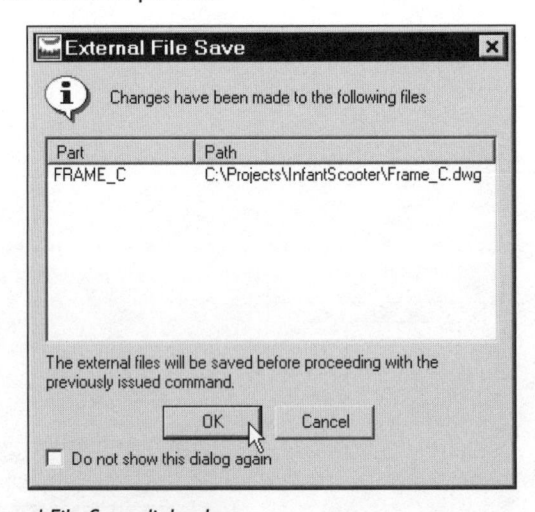

Figure 11–17 *External File Save dialog box*

ANNOTATION

To list the component parts used in an assembly, you add a bill of material. Construct a set of balloons as references to accompany the bill of material. To specify a welding requirement, you can add weld symbols. Apart from these annotations, you may add center lines, text, dimensions, tolerances, and surface finishes.

Now you will add a weld symbol to an assembly drawing.

22. Open the file *Frame.dwg*, if you have already closed it.
23. Select Annotate>Symbols>Welding

    Command: AMWELDSYM
24. Referring to Figure 11–18, select the base view and select the edge between the two component parts.
25. Select a point to indicate the end point of the leader line and press the ENTER key.
26. In the Weld Symbol dialog box, select the Arrow side data button (see Figure 11–19).
27. Select Square Butt Weld and select the OK button (see Figure 11–20).

A weld symbol is constructed. Save your file.

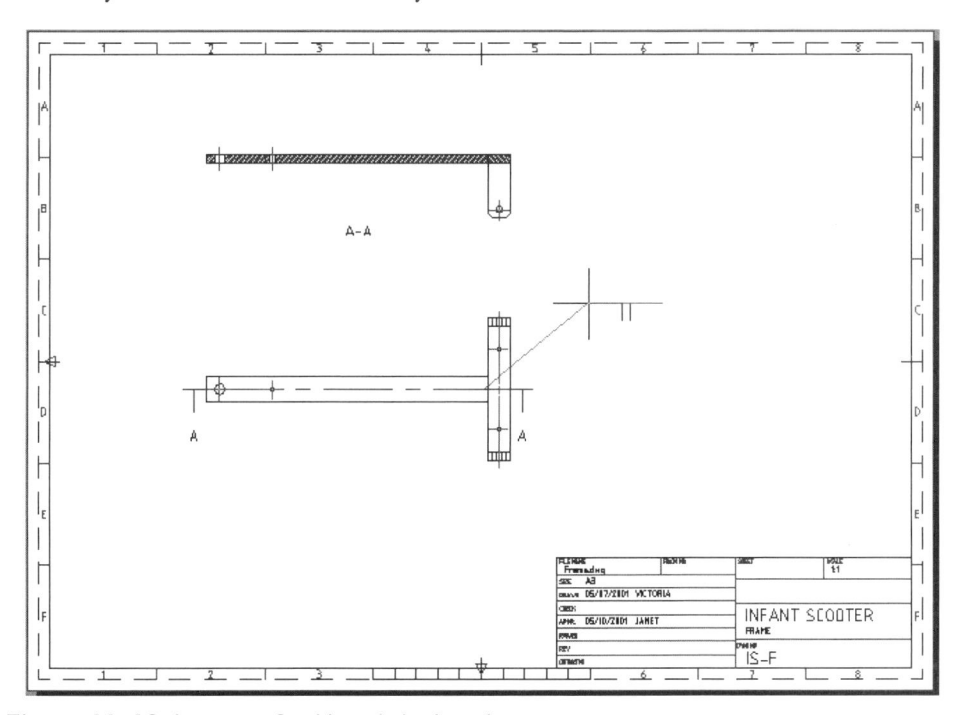

Figure 11–18 *Location of weld symbol selected*

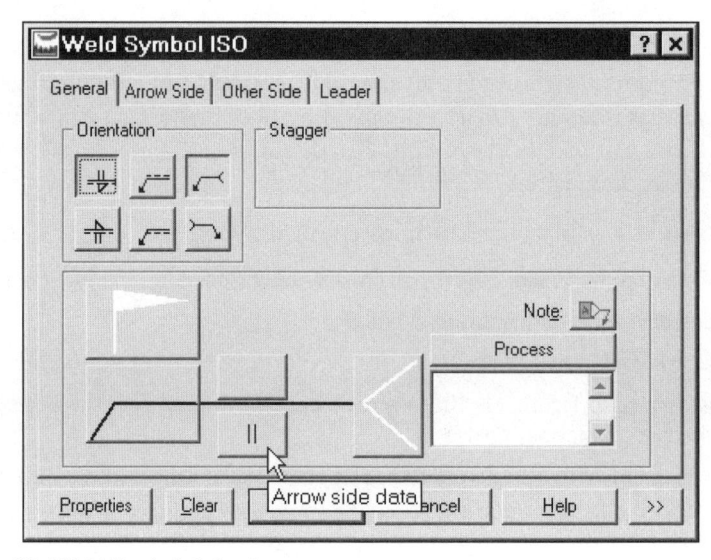

Figure 11-19 *Weld Symbol dialog box*

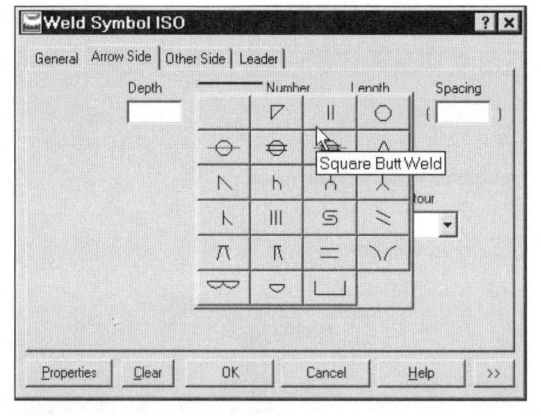

Figure 11-20 *Square Butt Weld joint selected*

BILL OF MATERIAL

A bill of material delineates details about the components used in the assembly. Now you will add a bill of material.

    28. Select Annotate>Parts Lists>BOM Database

        Command: AMBOM

    29. In the BOM dialog box (shown in Figure 11-21), select the first row under the Description column.

    30. Select the Create Formula button.

31. Select Name from the Variables pull-down list.
32. Select the OK button.

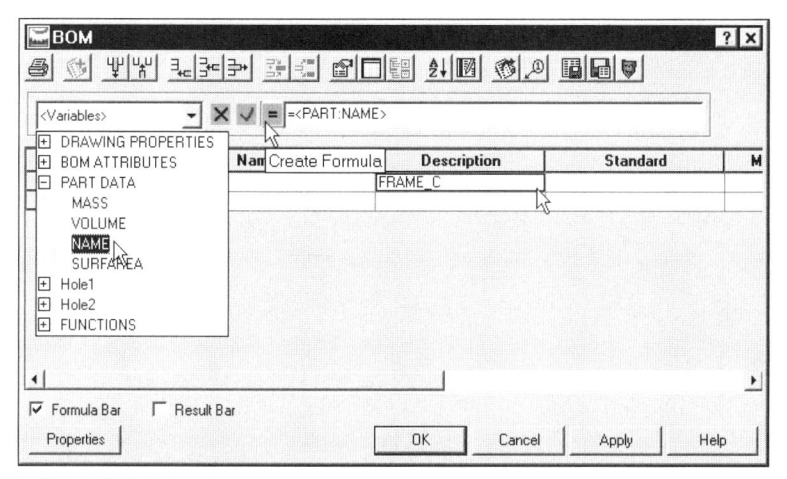

Figure 11-21 *BOM dialog box*

The BOM data is modified.

33. Select Annotate>Parts List>Part Reference

    Command: AMPARTREF

34. Select a part.
35. In the Part Ref Attributes ISO dialog box (shown in Figure 11-22), fill in the attributes and select the OK button.

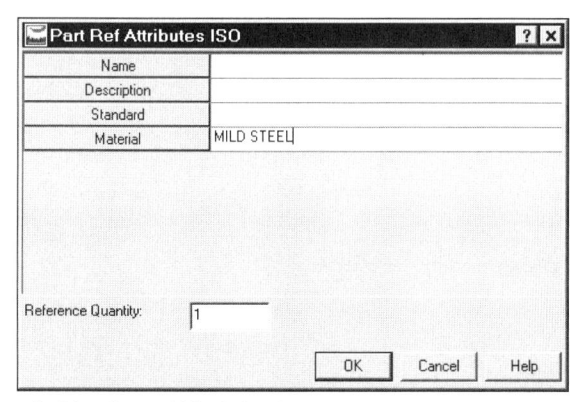

Figure 11-22 *Part Ref Attributes ISO dialog box*

Now you will insert a bill of materials.

36. Select Annotate>Parts List>Parts List

    Command: AMPARTLIST

37. In the Parts List ISO dialog box (shown in Figure 11-23), fill in the required information and select the OK button.

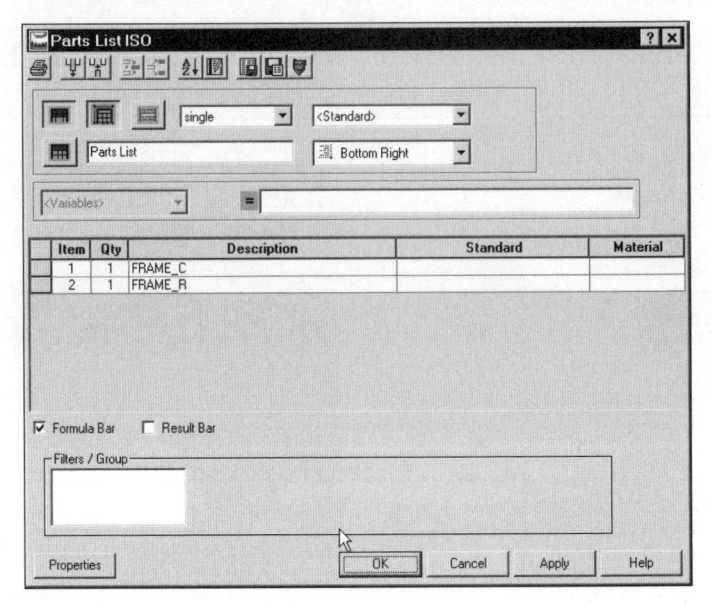

Figure 11-23 *Parts List ISO dialog box*

38. Referring to Figure 11-24, select a location for the parts list.

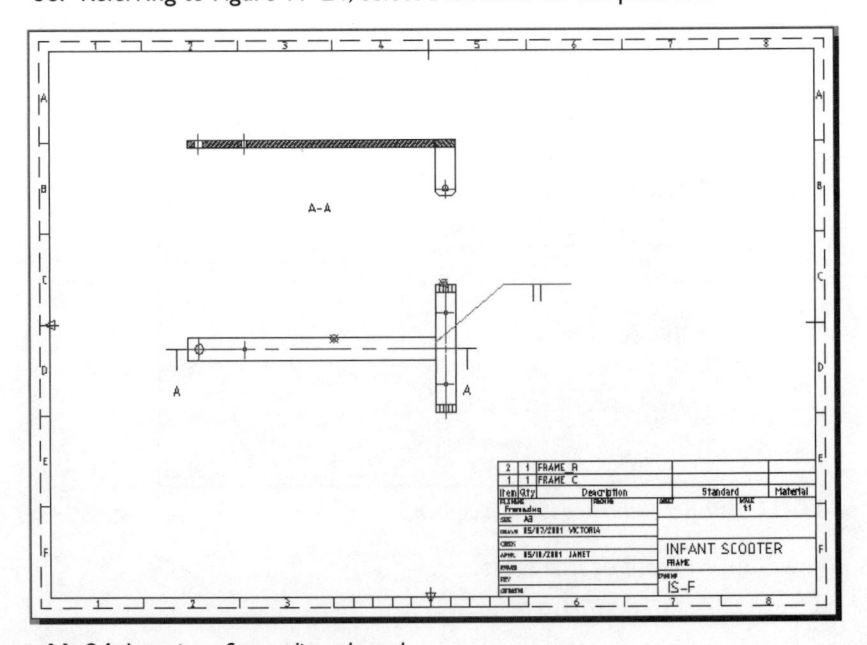

Figure 11-24 *Location of parts list selected*

BALLOON

Analagous to the bill of materials is a set of balloons referencing the component parts. Now you will construct a set of balloons.

    39. Select Annotate>Parts List>Balloons

        Command: AMBALLOON

    40. Referring to Figure 11–25, select a component, indicate a position for the end point of the leader, and press the ENTER key.

    41. Construct another balloon (see Figure 11–26).

The assembly drawing is complete. Save and close your file.

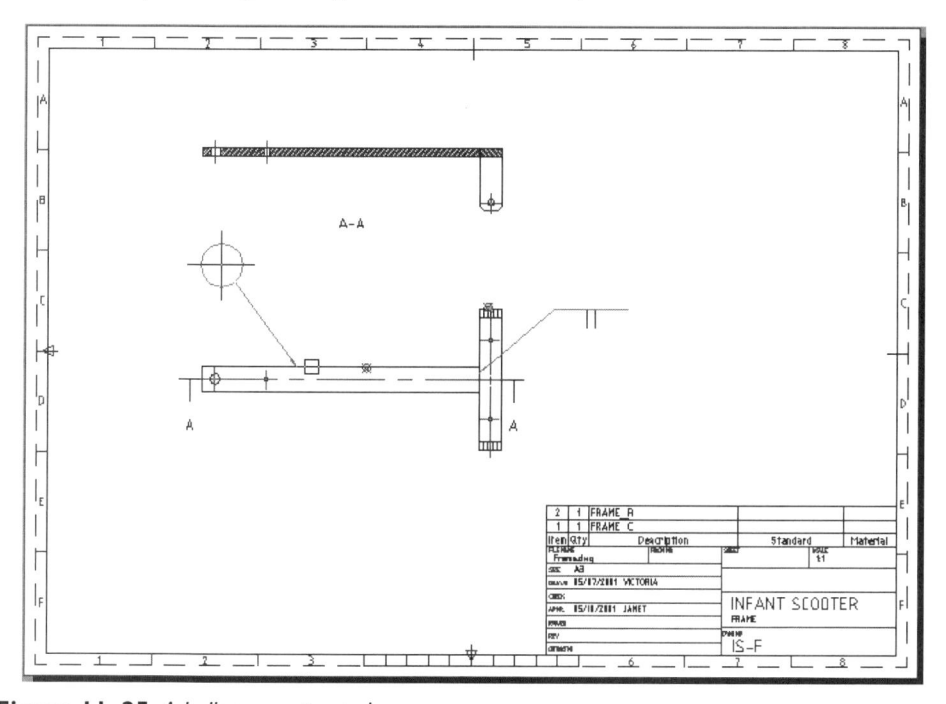

Figure 11–25 *A balloon constructed*

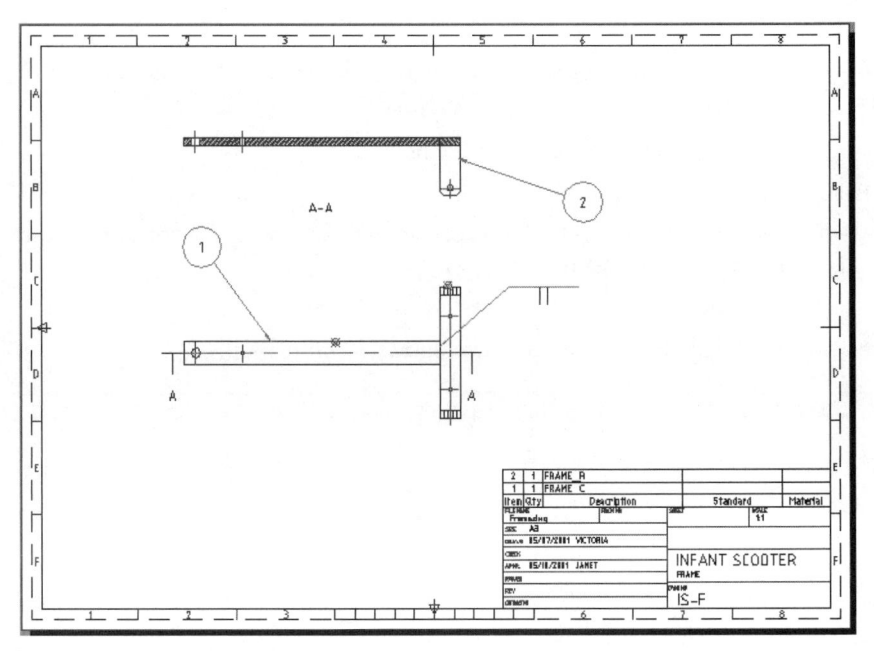

Figure 11-26 *Balloons constructed*

> **SUMMARY**
>
> Making the engineering drawing of an assembly is similar to making the engineering drawing for a solid part or a surface model. The differences are you select object(s) in the assembly or use the scene as a data set in making the drawing views, and you add a parts list and a set of balloons to the drawing in addition to other annotations such as centerlines, text, dimensions, tolerances, surface finishes, and weld symbols.

PROJECTS

Now you will enhance your knowledge by working on the following projects.

INFANT SCOOTER

Complete the assembly drawing *Frame_C.dwg* that you constructed in this chapter by adding a bill of materials and a set of balloons.

Construct the assembly drawings of all the other assemblies of the infant scooter. Include in your drawings a bill of materials and a set of balloons.

TOY CAR

Construct an assembly drawing of the toy car. Add a parts list and a set of balloons.

REVIEW QUESTIONS

1. The properties of the default parts list cannot be modified. True or false?

2. Balloons need to be placed before a parts list can be generated. True or false?

3. A bill of material entry is fully associative to the parts. True or false?

4. Exported AutoCAD 2D drawing views of a Mechanical Desktop drawing view are not associative to the Mechanical Desktop file. True or false?

5. Name three file formats that can be exported from a parts list.

INDEX

Note: Commands and system variables appear in SMALL CAPS.

?? dimension or constraint parameter, 35, 37-38
2D path, sweeping, 129-132
2D representation, 2, 9, 501. *see also* engineering drawings
2D wires, projecting, 305
3D edge path, sweeping, 134-137, 150, 189-190
3D helix path, sweeping, 132-134
3D manipulator, 213-215
3D models. *see also* solid models; surface models
    assembly of, 8 (*see also* assemblies)
    choosing, 7
    designing
        2D approach to, 2, 9, 22
        methods for, 7-8
        tools for, 1
    displaying, tools for, 62-64
    types of, 3-7
3D Orbit, 62-63
3D pipe path, sweeping, 137-144, 150-154, 195-198
3D polylines, constructing pipe path from, 137, 140-141
3D space
    scaling surfaces in, 296-297
    translating components in, 213-215
3D spline path, sweeping, 145-149
3D wires. *see* wires, 3D
3D work planes, basic, 120-121, 237. *see also* work planes

A

ACIS file format
    described, 466-467
    importing and converting, 469-474
Activate Scene, 262
active parts. *see also* solid parts
    combining parts with, 423-429
    design variables for, 441-442
    drawing views of, 498, 500, 502-503
adding
    attributes, 94
    geometric constraints, 37-38
    objects to sketches, 47-48, 75-76

    parametric dimensions, 39-41
    reference dimensions, 550-552
    search directories, 209-210
adjusting surface edges, 297, 337-338
aligned section views, 531-532
AM2DPATH, 129-132
AM2SF, 355, 398-399
AMACISOUT, 467
AMADDCON, 37-38
AMADJUSTSF, 337
AMANALYZE, 313-314
AMANGLE, 215, 249
AMASSMPROP, 240-241
AMAUDIT, 239
AMBALLOON, 585-586
AMBASICPLANES, 120-121, 237
ambient light, controlling, 63-64
AMBLEND, 336-338
AMBOM, 582-583
AMBREAK, 385-386
AMBREAKLINE, 534
AMCENLINE, 539-540
AMCHAMFER, 166-168
AMCHECKFIT, 396
AMCOPYIN, 233
AMCOPYOUT, 233-234, 430-431
AMCOPYSKETCH, 74-75
AMCORNER, 342
AMCUTLINE, 528-529
AMDATUMID, 558-559
AMDATUMTGT, 561-563
AMDELCON, 37-38
AMDIST, 242
AMDWGVIEW, 499
AMEDGE, 401
AMEDIT, 330
AMEDITAUG, 323
AMEDITSF, 329
American National Standards Institute, 467
AMEXTRUDE, 65

589

AMEXTRUDESF, 377-378, 391
AMFACESPLIT, 437-440
AMFCFRAME, 559-560
AMFEATID, 564-565
AMFILLET, 165-166
AMFILLET3D, 362-363
AMFILLETSF, 339-341, 391-392
AMFITSLIST, 556
AMFLOW, 332
AMFLUSH, 215, 248, 425, 427
AMHOLE, 158-162
AMHOLECHART, 545-547
AMINSERT, 215, 225
AMINTERFERE, 241
AMINTERSF, 397
AMJOIN3D, 363
AMJOINSF, 338
AMLENGTHEN, 389
AMLISTASSM, 238-239
AMLOFTU, 328, 367
AMMANIPULATE, 213-215
AMMATE, 215, 217-218, 246-247
AMMIRROR, 452
AMMODDIM, 41, 42
AMNOTE, 543-545
AMOPTIONS, 24, 32
AMPARDIM, 39-40, 122
AMPARTLIST, 583-584
AMPARTPROP, 91-93
AMPARTREF, 583
AMPATTERN, 170-177
AMPATTERNDEF, 576-577
AMPOWERDIM, 39
AMPOWEREDIT, 41
AMPRIMSF, 333
AMPROFILE, 35
AMPROJECT, 400
AMREFDIM, 550-552
AMREFINE3D, 323, 402, 407
AMREFINESF, 331
AMREFRESH, 239
AMREPLACE, 237-238
AMREPLAY, 91
AMREVOLVE, 66
AMREVOLVESF, 335-336
AMRULE, 412
AMRULEMODE system variable, 33
AMSCALE, 416-417
AMSHELL, 169-170
AMSHOWCON, 36

AMSKANGTOL system variable, 31, 32
AMSKMODE system variable, 33
AMSKPLN, 71
AMSOLCUT, 346
AMSPLINEDIT, 307-311
AMSPLITLINE, 434
AMSTITCH, 345
AMSTLOUT, 467
AMSURFCUT, 422
AMSURFSYM, 566-567
AMSWEEP, 131-132
AMSWEEPSF, 325, 350-351, 405-406
AMTEXTSK, 49-50
AMTHREAD, 163-164
AMTOLCONDITION, 556-557
AMTRAIL, 267-268
AMTUBE, 321-322
AMTWEAK, 265
AMVARS, 441
AMVIEWOUT, 568
AMVRMLOUT, 467
AMWELDSYM, 581-582
AMWHEREUSED, 238
AMWORKPLN, 111-112
AMWORKPT, 124-125
analyzing
    assemblies, 240-242
    surfaces, 313-314
Anchor fit or control point, 309
angle constraint, 215, 216, 249
angles, chamfer, specifying, 166, 168
angular tolerance, 32, 33
Annotate menu, 23
annotation
    assembly drawing, 571
        balloons for, 585-586
        bill of material for, 582-584
        weld symbols for, 581-582
    part drawing
        centerlines for, 539-540
        dimensions for, 548-553
        hole charts for, 545-547
        hole notes for, 543-545
        leaders for, 542-543
        surface finish symbols for, 566-567
        text for, 541-542
        tolerances for, 553-566
    tools for, 22, 23
Annotation toolbar, 23
appending. *see* adding

Apply Constraint Rules, 33
arcs
   constructing surfaces with, 361-362
   resolving sketches with, 29, 30
arrays, rectangular. *see* rectangular arrays
assemblies, 8. *see also* components
   analysis of, 240-242
   constraints for (*see* assembly constraints)
   constructing
      process for, 17-18, 211-219
      projects for, 245-257
   design approaches for
      bottom-up, 219-220
      choosing, 232
      hybrid, 228-231
      top-down, 220-228
   engineering drawings of
      annotating, 581-586
      constructing, 9, 571-580
      projects for, 586
   evaluating mass properties in, 92-93
   linking files to, 208-211
   manipulating definitions of, 232-239
   organizing parts in, 207-208
   scenes of (*see* scenes, assembly)
   setting options for, 242-243
   tools for, 17-18
Assembly Catalog, 209-210, 211, 253-254
assembly constraints, 208
   combining parts using, 423, 425, 482-483
   degrees of freedom and, 217-218
   editing or deleting, 218-219
   manipulating, project for, 245-249
   replaced components and, 232
   types of, 215-217
assembly files, 9
   bottom-up design and, 219
   browser for, 11, 43, 44
   externalizing and localizing parts in, 232-233
   hierarchy for, 208
   linking part files to, 208-211
   top-down design and, 220
   working environments for, 10, 571
Assembly Filter toggle, 44
Assembly menu, 11, 18
Assembly Modeling toolbar, 13
Assembly toolbar, 18
associative engineering drafting tool, 22-23
Assume Rough Sketch, 33
attributes, assigning, 94

auditing components, 239
augmented lines
   constructing, 302
   constructing surface from, 322-325
   editing, 313
   setting length of, 318
Auto Recognize, 469, 471-472, 474
AutoCAD 2002
   compatibility and interoperation with, 1, 9, 23
   solid editing commands of, 462-466
   tools of, accessing and using, 15, 17, 20, 21
   tools of, using, 12
AutoCAD native solids
   characteristics of, 15
   converting, 19, 23, 453-461
   cutting, 345-347, 348
   tools for, 17
Autodesk Streamline, 474
automatic update of scene, 272
auxiliary views, 513-515
axial patterns
   independent instances of, 174-176
   placing, 172-174
   suppressing, 176-177
axle
   sketch for, 54
   solid part for, 96

B

balloons, 585-586
base parts
   converting native solids to, 453-461
   converting parametric solids to, 91, 452-453
   stitching surfaces to, 347-348
base solid features, 66, 70
   constructing, 282
   editing, 462-466
   intersecting with, 78-81
   joining with, 71, 73-74
base surfaces, 292
   constructing fillet surface to, 340-341
   truncating, 294
base views, 499
   assembly drawing, 574-575
   constructing, 502-503
   moving, 506, 507
   multiviews with, 511-513
   section views for, 500
Basic 3D Work Planes, 120-121, 237
bend features, 61, 64, 68-70

bidirectional associativity, 491, 548, 549
bill of material, 582-584
Blend Tolerance for surfaces, 317
blended surfaces, 289-290, 336-338
blind holes, placing, 157-159, 161-162, 185, 188, 202-203
blocks, inserting and converting, 459-461
bolt
   adding threads to, 191-194
   modifying, 100-102
BOM dialog box, 582-583
Boolean operations
   combining features with, 7
   cut, 77-78
   intersect, 78-81
   join, 70-77
   projects using, 99-105
   split, 81-83
bottom-up design, 219-220
   combining parts using, 426
   hybrid design and, 228-231
boundaries. *see also* edges
   editing, 290-296
   trimmed, extracting, 305-306
boxes, constructing and converting, 453-461
break lines
   purpose of, 28
   resolving section to, 533-536
breaking surfaces, 294-295, 341-342, 385-386
breakout section views, 533-536
broken views, 499, 518-521
browsers, desktop. *see* desktop browsers

C

camera casing
   constructing, 475-488
   drawing views for, 569
camera lens housing, constructing, 488-489
car bodies
   drawing views for, 569
   surface models of
      constructing, 360-417
      types of, 278, 279, 280, 281
   toy, 220-231, 569
Catalog button, 44
centerlines
   annotating drawing views with, 539-540
   settings for, 498
central frame, scooter
   adding holes to, 179-180

assembling, 251
constructing work point on, 124-125
chamfer features
   example of, 16
   placing, 166-168, 183-185
checking
   interference, 207, 241
   normal surface direction, 352-353, 414-415
   solids, 463
   surface fit, 316, 395-396
circles, resolved sketches with, 29, 30
Clean, 463
clearance, analyzing, 241-242
clip
   sketch for, 52-53
   solid features for, 95-96
closing desktop browser, 44
collaboration, exporting data for, 474
color
   drawing, settings for, 497
   edge or face, 462
combining solid parts, 423-429, 482-483
commands
   AutoCAD, 12, 15
      solid editing, 462-466
      surface modeling, 21
   Mechanical Desktop, entering, 14
comments, design variable, 442, 443
components, 17. *see also* assemblies; solid parts
   adding features to, 178-205
   analyzing, 240-242
   annotating, 581-586
   assembling, 211-219
      design approaches for, 219-232
      project for, 250-257
   checking interference between, 207, 241
   constructing, 65-70, 95-106
   designing, 207
   exploded, adding trail lines to, 267-268, 275
   exploding, 259, 263-266, 273-275
   files for (*see* part files)
   hatch patterns for, 572, 576-580
   manipulating definitions of, 232-239
   organizing, 207-208
   placing instances of, 210-211
   setting options for, 33
   sketches for, 51-59
   suppressing section views of, 268, 270-271
compressed files, 33, 474
computer-aided design, 1-9, 280-281

concentric holes, placing, 161-162, 202-203
conical surfaces, 284
connected shapes, 284
constant radius fillet feature, 165-166
constant radius fillet surface, 341, 352
constraints. *see* assembly constraints; geometric constraints
construction geometry, 44, 45-47
consumed sketch, concept of, 48
context menus. *see* right-click menus
control points
    constructing spline from, 303
    converting fit points to, 310-311
    spline, editing, 308-309
    wire, refining, 311-312, 407
converting
    ACIS files, 469-474
    AutoCAD native solids, 453-461
    control or fit points, 310-311
    parametric parts to base, 91, 452-453
    quilts to surfaces, 355, 398-399
    sketches (*see* resolving sketches)
    solids to surfaces, 19-20, 23, 290, 301, 358
Copy In, 233
Copy Out, 233-234, 430-431
copying. *see also* mirroring
    components, 210-211, 247
    drawing views, 506-508
    edges, 462
        constructing pipe path from, 150-151
        sketched, 50
        surface, 305, 343-344, 401
    faces, 462
    polylines, 408
    scenes, 269
    sketches, 50, 74-75
    splines, 364, 365, 367
    surfaces, 329
corner fillet surfaces, 289, 341-343
counterbore holes, placing, 159-160, 188, 204
countersink holes, placing, 160-161
cross sections, sweep surface, options for, 405
cubical variable fillet features, 165
cubical variable fillet surfaces, 288, 339-340
cursor, blinking mouse, 71, 72
curvature, analyzing, 314
curves
    constructing solids from, 6
    constructing surfaces from, 4
Customize Toolbars, 13-14

cut lines
    constructing section views with, 528-529, 531
    controlling visibility of, 125
    purpose of, 28
cutting
    features from solids, 77-78, 99-100, 104-105
    section lines, 306
    solids with surfaces (*see* surface cut)
    sweep solids, 136, 137
cylinders
    constructing and converting, 453-461
    constructing work axis on, 122-123
cylindrical surfaces, 284

D

data sets, drawing view, 500
datum identifier, 558-559
datum target, 558, 561-563
DDIM, 38-39
DDPTYPE, 381
degrees of freedom
    constraints for, 215, 217-218
    displaying symbols for, 33, 212-213
deleting. *see also* erasing
    all trim, 291, 293-294, 305-306
    assembly constraints, 218-219
    drawing views, 508, 575
    faces, 462
    features, 90, 125
    geometric constraints, 37-38
    sketched objects, 47-48
    source objects, 303, 305
DELOBJ system variable, 303, 305, 408
dependent views, 506, 507, 575
derived surfaces, 288-290, 301-307, 358
design
    assembly, approaches to, 219-232
    computer-aided, 1-9, 280-281
Design menu, 17
design variables, maintaining
    files for, 441-447
    spreadsheets for, 447-451, 488-489
desktop browsers
    features of, 10-11, 43
    right-click menu for, 12
    synchronizing, 24
    types of, 43-44
    user interface for, 1
Desktop Visibility. *see* visibility
detail views, 515-518

deviation, surface, checking, 396
Dimension Style Manager, 38-39
dimension tolerance, 553-557
dimensions
  parametric
    adding to feature, 122
    adding to sketch, 30-31, 35
    construction geometry for, 45-47
    design variables for, 441-451
    displaying, 498
    engineering drawings with, 548-550
    manipulating, 38-42
    suppressed, setting color for, 33
    tolerance values for, 553
  reference, 548, 550-553
direct light, controlling, 63-64
direction
  auxiliary view, specifying, 513-514
  drawing view, setting, 502
  normal surface, 299, 318, 352-353, 414-415
  wire, changing, 312
display modes
  shaded, checking normal direction in, 352-353, 414-415
  types of, 62, 63
displaying
  constraint symbols, 36-37
  degrees-of-freedom symbols, 212-213
  dimensions, settings for, 39, 498
  surfaces, 315, 318
  toolbars, 12-14
distance
  chamfering, specifying, 166-168
  minimum component, checking, 241-242
  placing holes using, 160-161
  tweak, editing, 265-266
distance and angle chamfer, 168
distance tolerance, 31-32
DOF. *see* degrees of freedom
draft angle
  solid, adding, 436-437
  surface, analyzing, 314
Drafting Settings dialog box, 383
Drawing Layout toolbar, 13, 22
drawing limits, setting, 34
Drawing menu, 11, 22
drawing mode
  changing dimensions in, 548, 549
  engineering drawings and, 492, 571
  part browser tab for, 44

purpose of, 10
drawing views
  3D representation in, 2, 9, 548
  annotating (*see* annotation)
  assembly
    constructing, 572-580
    suppressing components in, 268, 270-271
  associating text and leader with, 541-542
  constructing
    auxiliary, 513-515
    basic types of, 502-506
    broken, 518-521
    detail, 515-518
    multiple, 509-513
    section, 521-538, 576-580
    settings for, 499-501
  displaying dimensions in, 498
  exporting, 568
  manipulating, 506-509
  tools for, 22-23
drilled holes, placing, 159, 178-180, 185, 187-188

E

edge nodes, displaying, 315
edge path, sweeping, 134-137, 150, 189-190
edges. *see also* boundaries
  adjusting, 297, 337-338
  constructing face draft from, 436
  constructing work plane through, 111-112
  copying
    for pipe path, 150-151
    for sketches, 50
    for surfaces, 305, 343-344, 401
  editing, commands for, 462
  mating, 246-247, 248
  orienting drawing views with, 502, 573, 574
  placing holes using, 159-160
  replacing, 297
editing. *see also* modifying
  3D wires, 307-313, 359
  assembly constraints, 218-219
  augmented lines, 313
  design variables, 447
  drawing views, 508-509
  features, 85-86, 462-466
  in-place, 234-237
  sketches, 83-84
  solid parts, 83-91, 100-102
  splines, 307-311, 326-327
  surfaces, 290-301, 329-331, 358-359

toleranced model, 556-557
tweak distance, 265-266
ellipse, constructing work axis on, 122-123
end points
   centerline, specifying, 540
   resolved sketches with, 29, 30
engineering drawings
   annotating, 539-567
   assembly
      annotating, 581-586
      constructing, 571-580
   constructing, process for, 492
   layouts for, 492-496, 572
   projects for, 569, 586
   relating 3D objects to, 2, 491
   setting options for, 496-498
   standards for, 496-499, 539, 572
   tools for, 22-23
   views in
      constructing, 502-506, 509-538
      exporting, 568
      manipulating, 506-509
      setting options for, 499-501
equal distance chamfer, 167
equation display for dimensions, 39
equations, design variable, 442, 443
erasing. *see also* deleting
   surfaces, 398
   wires, 402
Excel spreadsheets, defining design variables in, 447-451, 488-489
exploded views
   assembly scene, 259
      adding trail lines to, 267-268, 275
      constructing, 263-266, 273-275
      drawing views of, 572-576
   surface model, 5, 280, 361, 380
explosion factor, setting, 261, 263-264, 273
exporting
   design variables, 443
   engineering drawings, 568
   local definitions, 430
   solid parts, file formats for, 466-468, 474
external files, saving, 580
external parts, 219. *see also* solid parts
   auditing, 239
   copying in, 233
   hybrid design and, 228
   importing, 426-427
   in-place editing of, 234-237

   localizing, 232-233, 253, 427-428
   refreshing, 239
externalizing local definitions
   definition of, 220, 221
   process for, 225-228, 232-233
   project for, 255
extracting trim boundaries, 305-306
Extrude and Join, 73-74, 76-77
Extrude Faces, 462
extruded solid features
   constructing, 61, 64-66, 95-96, 99-105
   cutting, 77-78
   examples of, 6, 7, 15
   intersecting, 80-81
   joining, 70, 73-74, 76-77
   sketches for, 45-46
extruded surfaces
   constructing, 285, 377-378, 390-391, 392
   mirroring, 378

F

face draft feature, 436-437, 439-440
face split feature, 64
faces
   aligning, constraints for, 248, 425
   constructing sketch planes on, 71
   editing, commands for, 462
   mating vertices to, 247
   moving, 463
   orienting drawing views with, 502, 573, 574
   removing and reclaiming, 465
   splitting, 437-440
feature control frame
   components in, 558
   constructing, 559-560, 563-564, 566
feature identifier, 558, 564-565
feature recognition, 469-474
file formats, exportable, 466-468, 474
files. *see also* assembly files; part files
   basic types of, 9-10
   compressed, 33, 474
   design variable, 441-447
   hierarchies for, 43-44, 208
   locked, 235, 236
   saving, 33, 580
   template, 226-227
fillet solid features, 16, 164-166
fillet surfaces
   constructing, 288-289
   projects for, 339-343, 352, 391-393

filleting 3D wires, 304, 362-363
filter toggles, 44
finding components, 238
fit
    surface, checking, 316, 395-396
    tolerances for, 554-555
fit points
    constructing spline from, 303
    converting control points to, 310-311
    pipe path, 141-142
    spline, editing, 308-309
    spline path, 147-148
fits list, constructing, 556
fitting splines, 303, 317-318
flat-shaded modes, 63
flipping surface direction, 299, 353, 415
Float fit or control point, 309
flow lines
    breaking surfaces along, 385-386
    generating, 306, 332-333
flush constraint
    aligning faces with, 248
    combining solids with, 425, 427-428, 483
    function of, 215, 216
foam models, 281
folding line, bending solid around, 68-70
free-form objects
    engineering drawing of, 2
    examples of, 277
    wireframe model of, 3
free-form surfaces
    constructing, 285-288
        3D wires for, 301-307
        process for, 277, 279-280
        tools for, 18, 19, 358
    representations of, 278-279, 357
front end, scooter
    assembling, 250
    constructing scenes of, 273-274
frontal frame, scooter
    sketches for, 59
    solid features for, 102-105, 180-182
full-section views, 521-525

G

geometric constraints
    assembly constraints *vs.*, 215
    construction geometry for, 45-47
    determining need for, 35
    manipulating, 30, 36-38
    rules for, 29-30
geometric tolerance, 558-566
geometry, construction, 44, 45-47
global design variables, 441-442, 443
Gouraud shaded modes, 63, 352-353, 414-415
grip points
    editing surfaces with, 329-330, 331
    modifying, 298, 328-329

H

half-section views, 525-528
handle, constructing, 195-198
hatch patterns
    applying, 572, 576-580
    selecting, 523, 524
    settings for, 497
helical path
    constructing, 132-134
    thread feature with, 162, 193-194
helical spring, constructing, 322-325
Help, Feature Recognition, 469, 470
hidden lines, settings for, 500
hiding
    parametric dimensions, 550
    scene components, 262-263
    surface objects, 319, 334, 364-365, 368, 385
    unhiding and, 319, 338-339, 368
    work features, 116, 118, 125
hierarchy, file, 43-44, 208
hole charts, 545-547
hole features, 16
    annotating, 543-547
    design variables for, 444-445
    interactive recognition of, 473-474
    placing, 157-162, 178-185, 187-188, 202-204
hole notes, 543-545
hybrid design, 228-231, 426

I

IGES (Initial Graphics Exchange Specification) file
    format, 467
IGESOUT, 467-468
importing
    ACIS files, 469-470
    external parts, 426-427
Imprint, 462
in-place editing, 234-237
independent pattern instances, 174-176

independent view display, 516
infant scooter
    applying constraints to, 217-218
    assembling, 250-257
    Boolean operations for, 70-81
    drawing views for, 569
    linking part files for, 209-211
    placed features for, 178-205
    scenes for, 273-275
    sketched features for, 65-70, 95-106
    sketches for, 45-48, 51-59
insert constraint, 215, 217, 225
instances
    copying, 247
    making independent, 174-176
    placing, 210-211, 224-225
    suppressing, 176-177
Interactive Recognition, 469, 472-474
interference, checking, 207, 241
International Organization for Standardization, 468
Intersect and Trim
    described, 293
    projects using, 334, 367-369, 389-390, 397-398
intersecting
    solid features, 78-81
    solid parts, 428, 462
intersection, surface, constructing wire from, 304
isometric views
    constructing, 505-506, 574
    manipulating, 506-509
    multiviews with, 512-513

J

Join Gap Tolerance, 317
joining
    objects, resolving sketches and, 30
    solid features, 70-77, 100-102, 103
    solid parts, 426
    surfaces, 291, 295, 338
    wires, 302, 303, 363-364

L

Launch Toolbar, 13
layer names, 14
layouts, engineering drawing, 492-496, 572
leaders, 542-543
lengthening surfaces, 296, 389, 390
lights, controlling, 63-64
linear reference dimensions, 551
linear variable fillet feature, 165-166

linear variable fillet surface, 288, 339-340
lines, resolved sketches with, 29, 30
linking
    design variables to parts, 443, 445, 448-449
    parts to assemblies, 208-211
loading menus, 11
local parts, 220. *see also* solid parts
    automatic creation of, 237
    constructing, 221-224
    copying out, 233-234, 430-431
    externalizing, 220, 221, 225-228, 232-233, 255
    placing instances of, 224-225
    removing, 431
localizing external definitions, 232-233, 253, 427-428
locked files, 235, 236
locking scenes, 268, 270
loft solid features
    constructing, 154-157, 199-204
    examples of, 6, 15
    sketches for, 64, 109
loft U surfaces
    checking fit of, 395-396
    constructing, 286, 328-329, 364-368, 393-395
    editing, 329-331
    mirroring, 395
    offset surfaces from, 290
    trimming, 397-398
loft UV surfaces
    breaking, 385-386
    constructing, 287-288, 372-373, 375-376, 381-388

M

Make Base Part, 452-453
mass properties, analyzing
    assembly, 240-241
    solid part, 91-93
    surface, 315
mate constraint
    applying, 217-218, 246-247, 248
    combining solids using, 482
    function of, 215, 216
material properties, analyzing, 240-241
Mechanical Desktop 6, features of, 9-24
Mechanical Main toolbar, 12, 13
Mechanical Options dialog box. *see* options, setting
Mechanical View toolbar, 62
Menu Customization dialog box, 11, 12
MENULOAD, 11
menus, 11-12. *see also* right-click menus
    Annotate, 23

Assembly, 18
Design, 17
Drawing, 22
Feature Recognition, 469
Part, 16
Solids cascading, 17
Surface, 20
Surfaces cascading (AutoCAD), 21
mesh surfaces, 21, 278, 357
Microsoft Excel, defining design variables in, 447-451, 488-489
minimizing desktop browser, 44
MIRROR3D, 378, 404
mirroring
   points, 383, 386-387
   solid parts, 452
   surfaces, 378, 395
   wires, 404-405
model car bodies. *see* car bodies
model mode
   changing dimensions in, 548, 549
   engineering drawings and, 492, 571
   part browser tab for, 44
   purpose of, 10
Model Size button, 316-317
models. *see* 3D models; solid models; surface models
modifying. *see also* editing
   parametric dimensions, 41-42, 444-446, 450, 548-550
   pipe paths, 141-142
   reference dimensions, 553
   resolved sketches, 47-48
   spline paths, 147-148
   surface grip points, 298, 328-329
   sweep solids, 153
mouse, constructing model of, 348-355
moving
   drawing views, 506, 507
   faces, 462, 463
   splines, 364, 365
Multiline Text Editor, 541
multiple views, 499, 509-513

N

naming scenes, 269
native solids. *see* AutoCAD native solids
Node mode, Object Snap, 383
non-parametric sketches, resolving, 27, 28-30
non-parametric solids. *see* AutoCAD native solids
normal direction, surface
   checking, 352-353, 414-415
   determining, 299
   setting vector for, 318
Note Symbol dialog box, 544-545
Notepad, 443
numeric display for dimensions, 39
NURBS mathematics, 279, 317
NURBS surfaces. *see also* surface models; surfaces
   characteristics of, 279
   constructing, 283-290
      3D wires for, 301-313, 359
      fitting splines for, 317
      tools for, 18-20, 357-358
   editing, 290-301, 358-359
   solid operations using, 19, 23, 290, 301, 360
   types of, 277, 283
nut
   sketch for, 56-57
   solid features for, 99-100

O

object selection for mass property analysis, 92
Object Snap modes, using, 122, 383
offset-section views, 528-530
offsetting
   3D wires, 304
   faces, 462
   surfaces, 290, 355
   trail lines, 267
oil cooler surface model, 321-322
OPTIONS, 31-32
options, setting
   assembly, 242-243
   assembly scene, 260, 271-272
   engineering drawing, 496-498
   engineering standard, 498-499
   Feature Recognition, 469, 470-471
   parametric sketch, 31-33
   part browser button for, 44
   surface, 316-318
   system preference, 24
orbiting in 3D, 62-63
orientation, drawing view, 502, 573, 574
orthographic views
   auxiliary views with, 513-515
   constructing, 504-505
   moving, 506, 507
   multiviews with, 511-513
output spline, 401

P

page setup, 492-494
paper size, selecting, 492-493, 494-495
parameter display for dimensions, 39
parametric dimensions. *see* dimensions, parametric
parametric sketches. *see* sketches, parametric
parametric solids. *see* solid models; solid parts
Part Catalog, importing to, 426-427
part files
   assembly files and, 208
   bottom-up design and, 219
   browser for, 11, 43, 44
   externalizing, 220, 221, 225-228, 232-233, 255
   linking to, 208-211, 443, 445, 448-449
   localizing, 232-233, 253, 427-428
   purpose of, 9
   top-down design and, 220
   working environments for, 10, 492
Part Filter toggle, 44
Part menu, 11, 16
Part Modeling toolbar, 13, 16
parting lines, generating, 307
parts, solid. *see* solid parts
parts list, inserting, 583-584
patches, surface, refining, 298, 331
paths, sweeping
   constructing, 129-149
      comparative methods for, 150-154
      projects for, 189-190, 195-198
   purpose of, 28
pattern features. *see also* hatch patterns
   axial, 172-174
   independent instances of, 174-176
   polar, 171-172, 186-187
   rectangular, 170-171
   suppressing, 176-177
perpendicularity tolerance, 558, 560
persistent surface display, 315
Physical Materials List, 93
pickbox size, setting, 31-32
pipe path, sweeping, 137-144, 150-154, 195-198
placed solid features, 15-16. *see also specific type*
   combining, 423-429
   constructing, 157
      chamfer, 166-168
      face draft, 436-437, 439-440
      fillet, 164-166
      hole, 157-162
      pattern, 170-177
      shell, 168-170
      surface cut, 420-422, 476-482, 484
      thread, 162-164
   projects for, 178-204
   splitting, 429-435, 486-488
   types of, 109-110, 419
planar surfaces
   constructing, 284
   constructing face draft from, 436-437
   splitting parts along, 433-434
plotters, selecting, 492, 493
point style settings, 381
points, constructing surfaces from, 381-388, 393-394
polar patterns
   independent instances of, 174-176
   placing, 171-172, 186-187
   suppressing, 176-177
polygon mesh surfaces, 21, 278, 357
Polyline Fit, 317-318
polylines
   constructing pipe path from, 137, 140-141
   constructing surfaces from, 371-372
   converting splines to, 311
   copying, 408
   refining control points on, 407
Power Dimensioning dialog box
   modifying dimensions in, 444, 446, 450, 548-549, 553
   specifying tolerances in, 553-555
Power Manipulator, 213-215
Power Pack, 13
pre-constructed solid features. *see* placed solid features
Preferences, setting, 24
primitive surfaces
   characteristics of, 358
   constructing, 283-284, 333
   hiding, 334
   trimming, 333-334
profiles
   lofting along, 154, 156
   purpose of, 28
   resolving sketches to, 29, 35
   surface, 285, 296-299
   sweeping along, 129, 134, 144, 197
Project and Trim, 293, 408-412
projecting wires, 400
   constructing 3D wires by, 305
   trimming surfaces by, 293, 408-412
projects
   assembly, 245-257, 586
   assembly scenes, 273-275
   camera casing, 475-488, 569

camera lens housing, 488-489
constraint manipulation, 245-249
engineering drawing, 569, 586
helical spring, 322-325
infant scooter
    assembly, 250-257
    assembly drawings, 586
    assembly scenes, 273-275
    engineering drawings, 569
    sketches, 51-59
    solid features, 95-106, 178-205
model cars, 360-417, 569
mouse, 348-355
oil cooler, 321-322
remote control, 326-344, 345-348
sketching, 51-59
solid features, 95-106, 178-205
solid parts, 475-489
surface models, 321-355, 360-417
toy car, 220-231, 569, 586
pull-down menus. *see* menus

Q

QLEADER, 542-543
querying components, 238-239
quilts
    converting to surfaces, 355, 398-399
    cutting solids with, 345-347, 348
    stitching surfaces into
        process for, 295-296
        projects for, 345, 354, 367-369, 391-392
    thickening, 354-355
    trimming, 369, 397-398
    uses for, 277

R

radial reference dimensions, 551
radial section views, 536-538
radius, variable, specifying, 164-166
rail wires, 286-287
rear end, scooter, scenes for, 274
rear frame, scooter
    assembling, 251-252
    solid features for, 68-70, 183-184
reclaiming faces, 465
rectangular arrays
    block, converting, 460-461
    constructing surface from, 372
    subtracting from, 457-458
rectangular patterns
    independent instances of, 174-176

    placing, 170-171
    suppressing, 176-177
reference dimensions, 548, 550-553
refining
    lines, 323
    surfaces, 298, 331
    wires, 311-312, 402, 407
reflection line, analyzing, 314
refreshing components, 239
remote control
    solid models for, 345-348
    surface model for, 326-344
removing faces, 465
renaming scenes, 269
rendered images
    correcting surfaces in, 414-415
    of solid model, 6
    of surface models, 5, 281
reordering features, 86-89
replacing
    components, 232, 237-238
    surface edges, 297
Replay, 91
representation
    2D, 2, 9, 501
    3D, 3-8, 548
    ambiguous, 4
    surface, 21, 278-279, 357
    volume, 282-283
resolved sketches, modifying, 47-48
resolving sketches, 27
    objects used in, 28
    to profile, 29, 35
    rules for, 29-30
    to split line, 434
revolved solid features
    constructing, 61, 64, 66-67, 96-98, 201
    examples of, 6, 7, 15
revolved surfaces, 285, 335-336
rib features
    constructing, 126-129, 180-182
    sketches for, 64, 109
right-click menus
    desktop browser, 11, 12, 43
    Spline Edit, 309-310
Rotate Faces, 462
rule surfaces
    constructing, 286, 412-414
    mirroring, 378

S

SAT (Save As Text) files, 466-467
saving files, 33, 580
scale sweep surface, 405, 406
scaling
   drawing views, 502, 518
   solid parts, 451
   surfaces, 296-297, 416-417
   title blocks, 494, 495
scene mode, 10, 44, 571
scenes, assembly
   activating, 262
   adding trail lines to, 267-268
   constructing, 260-262
   controlling visibility in, 262-263
   dimensions for, 498
   drawing views for, 572-580
   exploded views of, 259, 263-266
   manipulating, 268-271
   projects for, 273-275
   setting options for, 271-272
Scenes toolbar, 13
screw cap
   sketch for, 54-55
   solid features for, 96-97
search directories, adding, 209-210
seat, scooter
   assembling, 253-255
   sketch for, 45-46
   solid features for, 65-66, 77-78, 187-191
section lines, cutting, 306
section views
   aligned, 531-532
   assembly
      constructing, 576-580
      suppressing components in, 268, 270-271
   breakout, 533-536
   dimensions for, 498
   full, 521-525
   half, 525-528
   hatch patterns for, 572, 576-580
   offset, 528-530
   radial, 536-538
   settings for, 500-501
Separate, 463
shaded display modes
   checking normal direction in, 352-353, 414-415
   types of, 62, 63
Shadow option, Face Draft, 436
shapes, surface, 283-284

shell features, 16
   combining, 423-426
   placing, 168-170, 187, 189
   splitting, 486-488
shelling solid parts, 463, 464-466
shrinkage, accounting for, 451
side views, 512-513
silhouettes, surface
   defining, 285
   editing, 296-299
sketch plane
   constructing work axis on, 121
   establishing, 34, 50
   intersect operation using, 79-80
   join operation using, 71-73
sketched solid features. *see also specific type*
   bend, 68-70
   combining, approach for, 423
   cutting, 77-78
   deleting, 90
   editing, 85-86
   extruded, 64-66
   intersecting, 78-81
   joining, 70-77
   loft, 154-157
   projects for, 187-191, 195-204
   reordering, 86-89
   revolved, 66-67
   rib, 126-129
   sketches for, 61
   splitting, 81-83, 437-440
   suppressing, 89-90
   sweep, 129-154
   tools for, 15-16
   types of, 64, 109
sketches, parametric
   appending objects to, 47-48, 75-76
   browsing for, 43-44
   constructing
      for intersect operation, 79-80
      for join operation, 71-73
      steps in, 34-42
      techniques for, 44-50
   constructing features from, 61, 64-70, 109
   constructing solids from, 15
   construction geometry for, 45-47
   copying, 74-75
   dimensions for, 30-31
   editing, 83-84
   geometric constraints for, 30

modifying, 47-48
multiple, 48-49
path, types of, 129
projects for, 51-59
resolving, 27, 28-30
setting options for, 31-33
text in, 49-50
sketches, rough, 28
constructing, 34
settings for, 33
sketching techniques, 44-50
Solid Edits fly-out toolbar, 462-463
solid features. *see* base solid features; placed solid features; sketched solid features
solid models
advantages of, 7
characteristics of, 3, 5-7
constructing
advanced techniques for, 419
approach for, 8
tools for, 15-17
solid parts. *see also* active parts; components
assigning attributes to, 94
combining, 423-429, 482-483
constructing, projects for, 95-106, 178-205, 475-489
constructing from surfaces
methods for, 19-20, 23, 282-283, 300-301, 360, 420
projects for, 345-355
converting native to parametric, 453-461
converting to surfaces, 19-20, 23, 290, 301, 358
cutting features from, 77-78, 99-100, 104-105
deleting features from, 90
design variables for, 441-451
drawing views for, 499-506
editing features of, 85-86, 462-466
editing sketches of, 83-84
evaluating mass properties of, 91-93
exporting, 466-468, 474
face draft feature for, 436-437
feature recognition for, 469-474
features for, types of, 109-110
intersecting features with, 78-81
joining features to, 70-77, 100-102
making base part from, 91, 452-453
manipulating definitions of, 232-239
mirroring, 452
relating 2D drawings to, 491, 548, 549
reordering features in, 86-89
replay feature for, 91
scaling, 451

sharing data on, 474
splitting, 81-83, 429-435, 486-488
splitting faces on, 437-440
suppressing features in, 89-90
surface cut for (*see* surface cut)
updating, 44, 84-85, 101-102, 153-154
Solids cascading menu, 17
Solids toolbar, 17
span, surface, modifying, 298-299, 330
Spatial Technology, Inc., 466
spherical surfaces, 284
spline path, sweeping, 145-149
splines
constructing surfaces from
extruded, 377-378
loft U, 326-328, 364-368, 394-395
loft UV, 369-376, 384-385, 387-388
revolved, 335-336
sweep, 350-351, 405-406
editing, 307-311, 401-404
fitting, 303, 317-318
joining wires into, 363
mirroring, 404-405
projecting, 399-400
surface cut with, 420, 421, 477-478, 481
tangent, constructing, 303
unsplining, 311
split lines
purpose of, 28
splitting faces along, 437-439
splitting parts along, 434-435
splitting
faces, 437-440
solid parts, 81-83, 429-435, 486-488
spreadsheets, defining design variables in, 447-451, 488-489
standard parts, setting drawing views for, 501
standard tolerance, specifying, 553, 554-555
standards, engineering, 496-499, 539, 572
static parts. *see* base parts
steering shaft
sketch for, 51-59
solid features for, 70-77, 185
steering wheel, constructing, 199-204
STEP (STandard for Exchange of Product Model Data) file format, 468
STEPOUT, 468
stitching surfaces
constructing solids by, 19, 282, 283, 300, 347-348
cutting *vs.*, 348

process for, 295-296
to quilts, 345, 354, 367-369, 391-392
STL (Stereolithography) file format, 467
stretch sweep surface, 405, 406
subassemblies
   concept of, 17
   linking files to, 208-211
   organizing, 207-208
Subtract, 458, 462
subviews, broken views with, 519-521
suppression
   engineering drawing, settings for, 497
   feature, 89-90
   pattern instance, 176-177
   sectional view, 268, 270-271
surface cut
   described, 19, 20, 282, 300
   process for, 420-422
   projects for, 345-347, 478-482, 484
   stitching *vs.*, 348
Surface Display, 315
surface finish symbols, 566-567
Surface menu, 11, 20
Surface Modeling toolbar, 13, 20, 283
surface models
   characteristics of, 3, 4-5
   constructing
      approach to, 8, 277, 279-280
      projects for, 321-355, 360-417
   multiple views of, 509-513
   NURBS (*see* NURBS surfaces)
   representations of, 21, 278-279, 357
   setting options for, 316-318
   significance of, 280-281
   tools for, 18-21, 279, 357-360
   uses for, 7
   utilities for, 313-316
   visibility control for, 319
   volume for, 282-283
surface texture symbols, 566-567
surfaces. *see also* NURBS surfaces
   computer-aided design of, 280-281
   constructing solids from, 19-20, 277, 282-283, 360
   converting quilts to, 355, 398-399
   converting solids to, 19-20, 23, 290, 301, 358
   cutting solids with (*see* surface cut)
   splitting parts along, 429-430
   types of, 277-278
Surfaces menu and toolbar (AutoCAD), 21

sweep solid features
   constructing
      2D path for, 129-132
      comparative methods for, 150-154
      edge path for, 134-137
      helical path for, 132-134
      pipe path for, 137-144
      projects for, 187, 189-190, 195-198
      spline path for, 145-149
   examples of, 6, 15
   sketches for, 64, 109
Sweep Surface dialog box, 405-406
sweep surfaces
   2-rail, 287
   cutting solids with, 421-422
   projects for, 322-325, 350-351, 373-374, 376-377, 405-406
   single-rail, 286-287
symbols
   degrees-of-freedom, 33, 212-213
   geometric constraint, 30, 33, 36-37
   geometric tolerance, 558-566
   surface finish, 566-567
   weld, 581-582
System Tolerance for surfaces, 317

T

tangent splines, 303
tangential work planes, 117-118
Taper Faces, 462
tapped holes, placing, 158-159, 185
target assembly, selecting, 261
template files, 226-227
temporary surface display, 315
Termination Type dialog box, 561
text, annotation, 541-542, 543
Text Sketch, 44, 49-50
text window, listing data in, 239
thickening quilts, 354-355
thickening surfaces
   analyzing mass properties by, 315
   constructing solids by, 19, 282, 300
thread features
   controlling visibility of, 125
   placing, 162-164, 191-194
   selecting, 158-159
through holes, placing, 159-161, 178-184, 187-188
tires
   mate constraint for, 217-218

sketch for, 46-48
solid features for, 66-67, 186-187
title blocks, inserting, 494-496, 510, 573
Toggles Keep Original button, 303, 408
Toleranced Model Edit, 556-557
tolerances
    angular, 32, 33
    dimension, 553-557
    distance, 31-32
    geometric, 558-566
    surface, 317
tool body
    combining active part with, 423-429
    definition of, 120
toolbars, 12-14
    Annotation, 23
    Assembly, 18
    Drawing Layout, 22
    Feature Recognition, 469
    Mechanical View, 62
    Part Modeling, 16
    Solid Edits fly-out (AutoCAD), 462-463
    Solids, 17
    Surface Modeling, 20, 283
    Surfaces (AutoCAD), 21
    synchronizing, 24
top-down design, 220-228
    combining parts using, 423-426
    hybrid design and, 228-231
    project for, 245-249, 476-482
toroidal surfaces, 284
toy car
    assembling, 220-231
    drawing views for, 569
trail lines, 266, 267-268, 275
Transformation of Model dialog box, 556-557
translation
    component, 211-215
    file, 468
trim boundaries, extracting, 305-306
trimmed surfaces
    constructing
        3D wires for, 301-307
        methods for, 291-293, 358
        projects for, 333-334, 367-369, 389-390, 397-398, 408-412
    planar, 284, 343-344
    truncating, 294
    untrimming, 291, 293-294, 305-306

trimming
    quilts, 369, 397-398
    wires, 402-403
truncating surfaces, 294
tubular surfaces, 284, 321-322
tweaking components apart, 264-266, 274, 275
two-distance chamfer, 167

U

U patches, refining, 298, 331
U wires
    constructing surfaces from, 285, 286-288
    setting display of, 318
undo
    adjusted surfaces, 337-338
    deleted features, 90
    dimension values, 40
unhiding surfaces, 319, 338-339, 368
Union, combining solids with, 454-455, 459, 462
unlocking scenes, 268
unsplining, 311
unsuppressing features, 89-90
untrimming surfaces, 291, 293-294, 305-306
updating
    assemblies, 44
    scenes, 260, 270, 272
    solid parts, 44, 84-85, 101-102, 153-154
uploading files, 474
user interface, features of, 1, 10-14

V

V patches, refining, 298, 331
V wires
    constructing surfaces from, 285, 286-288
    setting display of, 318
vectors
    augmented line, editing, 313
    normal surface, 299, 318, 414
vertices
    constructing work planes through, 112, 114
    mating, 246-247
views, drawing. *see* drawing views
virtual assemblies, 1, 207-208. *see also* assemblies
visibility. *see also* hiding
    assembly, 212
    assembly scene, 260, 261, 262-263
    degrees-of-freedom symbol, 212
    desktop, button for, 44

surface object, 319
work feature, 125
visualization, 3D, tools for, 62-64
volume, representing, 282-283
VRML (Virtual Reality Modeling Language) file format, 467

W

weld symbols, 581-582
wheel cap
   sketch for, 55-56
   solid features for, 97-98
wheels
   mate constraint for, 217-218
   sketch for, 57-58
   solid features for, 98, 178-179
Where Used, 238
wireframe display mode, 62, 63
wireframe models, 3-4, 7, 8
wires, 3D
   constructing, 301-307
   constructing surfaces from, 277, 279
      process for, 285-288
      projects for, 326-328, 360-417
      smooth defining wires for, 290-293
   editing, 307-313, 359
   setting display of, 318
   tools for, 18, 359
work axes
   constructing, 121-123, 186-187, 193-194
   controlling visibility of, 125
   for section views, 521, 525-526, 533, 578
work features
   constructing, 110-125, 178-205
   controlling visibility of, 125
   purpose of, 16

work planes
   constructing
      basic 3D, 120-121, 237
      methods for, 110-120
      projects for, 180-184, 193-204
   controlling visibility of, 125
   for section views, 521-528, 533, 536-538, 578-580
   splitting faces along, 439
   splitting parts along, 431-432, 486-488
work points
   constructing, 124-125, 179-180, 183-184, 187-188, 195-198
   controlling visibility of, 125
   placing holes on, 158-159, 179-180
   for surface cut, 420, 421, 476, 479, 480
working environments, 10
   for assemblies, 571
   changing dimensions in, 548, 549
   for engineering drawings, 492
   part browser tabs for, 44
   settings for, 24
   viewing, 62
WRL files, 467

X

X axis, orienting, 71-72
XGL (X Windows Graphics Library) file format, 474
XY/XZ/YZ planes, defining, 120-121

Y

Y axis, orienting, 71-72

Z

Z axis, orienting, 71-72
ZGL file format, 474
zooming, 34

LICENSE AGREEMENT FOR AUTODESK PRESS
A Thomson Learning Company

Educational Software/Data

You the customer, and Autodesk Press incur certain benefits, rights, and obligations to each other when you open this package and use the software/data it contains. BE SURE YOU READ THE LICENSE AGREEMENT CAREFULLY, SINCE BY USING THE SOFTWARE/DATA YOU INDICATE YOU HAVE READ, UNDERSTOOD, AND ACCEPTED THE TERMS OF THIS AGREEMENT.

Your rights:

1. You enjoy a non-exclusive license to use the enclosed software/data on a single microcomputer that is not part of a network or multi-machine system in consideration for payment of the required license fee, (which may be included in the purchase price of an accompanying print component), or receipt of this software/data, and your acceptance of the terms and conditions of this agreement.

2. You own the media on which the software/data is recorded, but you acknowledge that you do not own the software/data recorded on them. You also acknowledge that the software/data is furnished "as is," and contains copyrighted and/or proprietary and confidential information of Autodesk Press or its licensors.

3. If you do not accept the terms of this license agreement you may return the media within 30 days. However, you may not use the software during this period.

There are limitations on your rights:

1. You may not copy or print the software/data for any reason whatsoever, except to install it on a hard drive on a single microcomputer and to make one archival copy, unless copying or printing is expressly permitted in writing or statements recorded on the diskette(s).

2. You may not revise, translate, convert, disassemble or otherwise reverse engineer the software/data except that you may add to or rearrange any data recorded on the media as part of the normal use of the software/data.

3. You may not sell, license, lease, rent, loan, or otherwise distribute or network the software/data except that you may give the software/data to a student or and instructor for use at school or, temporarily at home.

Should you fail to abide by the Copyright Law of the United States as it applies to this software/data your license to use it will become invalid. You agree to erase or otherwise destroy the software/data immediately after receiving note of Autodesk Press' termination of this agreement for violation of its provisions.

Autodesk Press gives you a LIMITED WARRANTY covering the enclosed software/data. The LIMITED WARRANTY can be found in this product and/or the instructor's manual that accompanies it.

This license is the entire agreement between you and Autodesk Press interpreted and enforced under New York law.

Limited Warranty

Autodesk Press warrants to the original licensee/ purchaser of this copy of microcomputer software/ data and the media on which it is recorded that the media will be free from defects in material and workmanship for ninety (90) days from the date of original purchase. All implied warranties are limited in duration to this ninety (90) day period. THEREAFTER, ANY IMPLIED WARRANTIES, INCLUDING IMPLIED WARRANTIES OF MERCHANTABILITY AND FITNESS FOR A PARTICULAR PURPOSE ARE EXCLUDED. THIS WARRANTY IS IN LIEU OF ALL OTHER WARRANTIES, WHETHER ORAL OR WRITTEN, EXPRESSED OR IMPLIED.

If you believe the media is defective, please return it during the ninety day period to the address shown below. A defective diskette will be replaced without charge provided that it has not been subjected to misuse or damage.

This warranty does not extend to the software or information recorded on the media. The software and information are provided "AS IS." Any statements made about the utility of the software or information are not to be considered as express or implied warranties. Delmar will not be liable for incidental or consequential damages of any kind incurred by you, the consumer, or any other user.

Some states do not allow the exclusion or limitation of incidental or consequential damages, or limitations on the duration of implied warranties, so the above limitation or exclusion may not apply to you. This warranty gives you specific legal rights, and you may also have other rights which vary from state to state. Address all correspondence to:

AutodeskPress
3 Columbia Circle
P. O. Box 15015
Albany, NY 12212-5015